Volume III
Diseases of Fruit Crops

PLANT DISEASES OF INTERNATIONAL IMPORTANCE

J. KUMAR

Department of Plant Pathology, Hill Campus, G. B. Pant University of Agriculture and Technology, Ranichauri, India

H. S. CHAUBE

Department of Plant Pathology, G. B. Pant University of Agriculture and Technology, Pantnager, India

U. S. SINGH

Department of Plant Pathology, International Rice Research Institute, Los Banos, Philippines

A. N. MUKHOPADHYAY

Department of Plant Pathology, G. B. Pant University of Agriculture and Technology, Pantnagar, India

PRENTICE HALL, Englewood Cliffs, New Jersey 07632

Library of Congress Cataloging-in-Publication Data
(Revised for vol. 2–4)

Plant diseases of international importance.

    Vol. 2 edited by Hriday S. Chaube — [et al.];
Vol. 3. edited by Jatinder Kumar . . . [et. al.]; Vol. 4
edited by A. N. Mukhopadhyay . . . [et. al.].
    Includes bibliographical references and index.
    Contents: v. 1. Diseases of cereals and
pulses — v. 2. Diseases of vegetables, and oil seed
crops — v. 3. Diseases of fruit crops — v. 4. Diseases
of sugar, forest, and plantation crops.
    1. Plant diseases. I. Singh, Uma S. II. Chaubether,
Hriday S. III. Kumar, Jatinder. IV. Mukhopadhyay,
A. N. (Amar Nath), 1940–
SB731.P67 1992   632'.3   91-22598

ISBN 0–13–678582–4 (v. 1)
ISBN 0–13–678558–1 (v. 2)
ISBN 0–13–678566–2 (v. 3)
ISBN 0–13–678574–3 (v. 4)

Editorial production
  and interior design: *bookworks*
Acquisition editors: *Ken Tennity and Betty Sun*
Editorial assistant: *Maureen Diana*
Cover designer: *Joe D. Domenico*
Copy editor: *Patricia M. Daly*
Marketing Manager: *Alicia Aurichio*
Prepress buyer: *Mary Elizabeth McCartney*
Manufacturing buyer: *Dave Dickey*
Indexer: *WordFinders*

Printed in the United States of America

10 9 8 7 6 5 4 3 2 1

ISBN   0-13-678566-2

Prentice-Hall International (UK) Limited, *London*
Prentice-Hall of Australia Pty. Limited, *Sydney*
Prentice-Hall Canada Inc., *Toronto*
Prentice-Hall Hispanoamericana, S.A., *Mexico*
Prentice-Hall of India Private Limited, *New Delhi*
Prentice-Hall of Japan, Inc., *Tokyo*
Simon & Schuster Asia Pte. Ltd., *Singapore*
Editora Prentice-Hall do Brasil, Ltda., *Rio de Janeiro*

CONTENTS

Contents

dedicated to

GOBIND BALLABH PANT UNIVERSITY OF AGRICULTURE AND TECHNOLOGY, PANTNAGAR

—birthplace of green revolution in India
Norman E. Borlaug

AND

IT'S MAIN ARCHITECTS

Dr. Dhyan Pal Singh
Dr. N. K. Anant Rao
Dr. R. L. Paliwal
Dr. Y. L. Nene

PREFACE

Plant diseases are as old as agriculture and are important to humans because they cause damage to plants and plant products. If we go by numbers, plant diseases run into thousands. However, only a handful of them are of much economic and historic significance and consume a major proportion of the funds and scientific power invested on plant diseases. This is because of their widespread occurrence and devastating effects. During the past two decades, we have seen dramatic advances in the study of many internationally important diseases. Unfortunately, however, the information generated on these diseases is widely dispersed in diverse scientific journals, in several languages. This book is an attempt to bring this information together in four volumes, to summarize and evaluate recent developments, to integrate them with significant developments of the past, and to attempt some projections for the future.

Leading plant pathologists from over 20 countries with expertise in individual diseases have contributed chapters. All the volumes, therefore, contain highly authoritative and thought-provoking articles by experts in their respective fields. The number of topics presented and the in-depth coverage they receive are ample testimony that plant disease research has come of age. The text, besides serving as a reference source for research workers or scientists already in the field, might also interest those who work in allied areas. We have considered the growing numbers of postgraduate students who are interested in plant diseases of international importance.

In dealing with each disease, pertinent facts about the history and distribution (including a disease distribution map), symptoms, etiology, and disease cycle are given, and control measures used in different countries have been combined into programs that are as simple as possible. In presenting their own research findings and those of others, contributors were encouraged, when appropriate, to draw conclusions and propose hypotheses that might stimulate additional research or otherwise further our understanding of the diseases. The chapters are profusely illustrated, and pertinent literature is cited for readers who may need supplemental information on specific matters.

In preparing the four volumes, the editors and contributors drew heavily on numerous sources of information. Many of them are given as references or credited throughout the text, but space limitations preclude mentioning all. The editors and contributors wish to acknowledge their indebtedness to these sources. Numerous illustrations are reproduced from the literature and from several other sources. Wherever necessary, the source has been duly acknowledged.

The response of contributors to our request for articles was so overwhelming that we received more articles than expected. The editors thank all the eminent scientists for their valuable contributions in making this project a success.

Editorial responsibilities in the preparation of these volumes were shared as follows: A. N. Mukhopadhyay was responsible for preparation of the volume, *Diseases of Sugar, Forest, and Plantation Crops;* H. S. Chaube was responsible for *Diseases of Vegetables and Oil Seed Crops;* U. S. Singh was responsible for *Diseases of Cereals and Pulses;* and J. Kumar was responsible for *Diseases of Fruit Crops.* Finally, all the volumes were reviewed by each editor to bring uniformity to the text.

We are indebted to Kenneth J. Tennity, Senior Editor at Prentice Hall, and other skillful staff at Prentice Hall for their unexcelled cooperation in the creation of this four-volume set. Our sincere thanks to P. K. Mukherjee and T. K. Misra, who spent many hours reading the manuscripts, and to D. B. Parakh for performing many other essential functions. Finally, our special thanks and appreciations to our wives Neeta, Urmila, Namita, and Sumitra, who were highly cooperative and supportive of our efforts during the last five years.

Of course, it would be unrealistic to suppose that the text is free from errors. Notification of errors of omission or commission will be highly appreciated.

J. Kumar
H. S. Chaube
U. S. Singh
A. N. Mukhopadhyay

1

APPLE SCAB
(Venturia inaequalis (Cke.) Wint.)

G. K. GUPTA

Regional Horticultural Research Station, Mashobra, Shimla, India 171007

1-1 HISTORY AND DISTRIBUTION

The disease commonly known as "apple scab" has also been referred to as "scurf," "black spot," "Tasmanian black spot," and "mildew." "Black spot" is the name still used in Australia and New Zealand, while the term *scab*, referred to in horticultural literature between 1850 and 1860, has been in general use since then. Although an age-old disease, the scab fungus was first reported and named as *Spilocaea pomi* in Sweden in 1819 by Fries; subsequently, it was recorded in Europe (Germany) in 1833 by Schweinitz; in England in 1855 by Berkeley; and in Australia in 1862 as reviewed by Morris.[1] By 1971, the scab was known to occur throughout Europe, Australia, Cyprus, India, Malta, New Zealand, North and South America, Pakistan, and South Africa[2] (Fig. 1-1).

In India, although the commercial cultivation of apple was begun by European settlers and missionaries as early as 1870[3] and the rapid expansion of the industry came only during the last three to four decades, the scab was first recorded in the Kashmir valley (J&K state) in 1930.[4] After that its occurrence was deemed to be sporadic and not of much consequence, as it mostly affected the indigenous cultivar "Ambri." However, in 1973, the existence of the elite cultivars, such as Red Delicious, came into jeopardy by the onslaught of scab appearing suddenly in epidemic form in the valley.[5] During 1977, scab was also detected in Himachal Pradesh in an orchard and in a few localized pockets of Kullu, Mandi, and Chamba districts.[6] During the last two to three years, the disease has also been recorded in other apple-growing areas of the country,[7,8] and the scab fungus

1

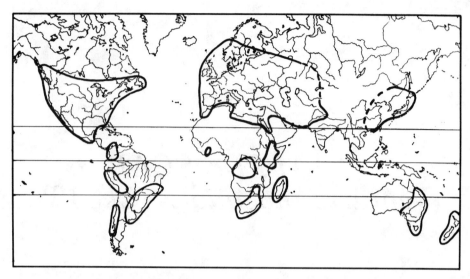

Figure 1-1   CMI geographical distribution map of apple scab.

seems to have been introduced in these areas through apple-planting material imported
from Kashmir where twig infection is not uncommon.[9]

1-2 ECONOMIC LOSS

Losses from scab over a period of years far exceed those from any other disease or insect
of apples. Losses result from (1) devaluation in fruit quality, (2) reduction in fruit production, (3) mid-season defoliation of trees, (4) weakening of the tree, (5) failure of fruit bud
formation, and (6) increased expenses due to spraying operations.

Wet weather conditions in spring and early summer may result in heavy scab development on leaves and pedicels, leading to drop of leaves and most of the fruit by mid-
summer or earlier. In some years, early scab infection on the fruit pedicels, particularly in
varieties such as Golden Delicious (which is not especially susceptible to scab), may
almost entirely prevent the setting of fruit. Early loss of leaves leads to the weakening of
the tree, which may fail to form fruit buds, resulting in crop failure in the following
season. Year-to-year loss of leaves in young, nonbearing trees results in stunted and
reduced growth.

The most serious aspect of the disease is a reduction in the quality of fruits. Often,
scabby fruits either fetch poor prices or are no longer marketable. Another burden scab
places on the orchardist is the increased cost of production due to expenditure incurred on
repeated sprays of fungicides at short intervals.

In India, the first epidemic of scab was observed in Kashmir valley in 1973, and this
ruined the apple crop worth Rupees 54 lakhs (U.S. $540,000) in a single season. Since
then, the disease has continued to cause enormous damage to the apple crop in the unsprayed orchard. In 1982 and 1983, this disease struck the apple crop in Himachal Pradesh, where it engulfed almost all the potential apple-growing areas (40,000 ha) of the
state and endangered the existence of this crop.[10] Although the scabbed fruit ranged from

10 to 50%, nearly 10% of the apple crop (30,000 of 300,000 metric tons of production) was rendered unfit for market consumption in 1983 and was destroyed on the spot. The state suffered a loss of Rupees 1.50 crores (U.S. $1.5 million).

1-3 HOST PLANTS AND SUSCEPTIBLE APPLE VARIETIES

The apple scab fungus is confined to apples (cultivated and crab) within the genus *Malus*, and it does not infect almond, apricot, peach, pear, plum, and other cultivated and forest plants.

Most of the apple varieties which are commonly grown in India and other countries are susceptible to scab. In India, varieties of the Delicious group, particularly Red Delicious, Starking (Royal) Delicious, Starkrimson Delicious, and Rich-a-Red Delicious, are highly susceptible. The reaction of most other varieties is similar, including Ambri, American Api-Rouge, Black Ben Davis (Kali Devi), Benoni (Hazratwali), Chamure (Chamura), Crimson, Gloster, Golden Delicious, Granny Smith, Indo Apple, Iwai Apple, Karkachu (a Ladakh variety), Kerry Pippin (Phokla), Khera Seb, King of Pippins (No. 13), Lal Ambri (Ambri × Red Delicious), Lal Cidar, Lord Lambourne, McIntosh, Megumi, Mollie's Delicious, Radhakrishan, Razakwari, Red Gold, Royal Mishri, Shan Shah, Sunheri (Ambri × Golden Delicious), Tydeman's Worcester, Winter Commercial, Winter Delicious, and Yellow Newton (Rus Pippin). Newly introduced spur-type varieties and some of the promising ones (including Hardeman, Red Chief, Red Spur Delicious, Skyline Spur Delicious, Stark Spur Golden, Stark Spur Delicious, Top Red, and Vance Delicious) are also susceptible to scab. The other susceptible varieties which are grown in Coonoor and Kodaikanal areas of Tamil Nadu are Carrington, Gravenstein, Grime's Golden, Irish Peach, Israel Type III, Kiddy's Orange Red, Parlin's Beauty, Rome Beauty, Signetillisch, and Winterstein.

Certain varieties, such as Cox's Orange Pippin (Kesri) and Jonathan, which are known or deemed to be highly susceptible to scab in many European countries and other continents, show either freedom or mild infection. Versifield, popularly known as Maharaji, used to be free from scab, but in recent years break of resistance has been indicated by a mild infection on fruits.[11]

1-4 SYMPTOMS

The most striking symptoms appear on leaves and fruits and rarely on one- to three-year-old shoots (twigs). Petioles, pedicels, and sepals of the flower also exhibit scab symptoms.

1-4-1 On Leaves

New and tender leaves in spring and summer are more susceptible to infection than the mature ones. Symptoms develop primarily on the undersurface of the leaf, the side exposed when the fruit buds first open. Once the entire leaf has unfolded, both sides may be infected. Light brown too live green powdery spots, which turn mousy black with age, appear on the leaf lamina (Fig. 1-2a). At first, the margins of lesion are feathery and

Figure 1-2a   Scab on leaves as velvety brown to olive spots.

Figure 1-2b   Scab spots on calyx end of apple fruit.

Figure 1-2c   Mousy black secondary scab lesions on the fruit surface of Delicious apple. (Courtesy: V. P. Goel, New Delhi)

Figure 1-2d   Misshapen knotty Delicious apple with small secondary lesions. (Courtesy: V. P. Goel, New Delhi)

Figure 1-2e   Scabbed area on fruit showing cracks of skin. (Courtesy: H. S. Adwinckle, U.S.A.)

indefinite, but later they exhibit distinct limits. On the lower surface of the leaf, the lesions are not very conspicuous because of the hairy character of this surface and because of the fungus growth merging more or less into healthy tissue without a definite margin. The lesions tend to extend along the vein and midrib. In the early stage, the reverse reflects the chlorotic effect on viewing the leaf against light. On the upper surface of the leaf, the lesions are more conspicuous and definite in outline. The apparent symptoms in susceptible varieties are similar in character. Lesions always turn dark with plenty of sporulation, while in resistant varieties such as Versified (Maharaji) and Cox's Orange Pippin (in which sporulation is scarce or even absent), lesions are light in color and chlorotic, sometimes with reddish-brown specks.[11]

As the infected leaf ages, the tissues in the neighborhood of the scab lesion thicken, resulting in an upward bulging of the affected area, with a corresponding cuplike depression on the lower surface. Severe spotting may result in curling, dwarfing, and distortion of leaves and may cause premature yellowing and defoliation. Such conditions may be further aggravated if damp conditions exist in the orchard and the scab fungus, instead of giving rise to definite lesions, covers the large portion of the leaf lamina with radiating black powdery growth, a condition known as *sheet scab*. Trees severely defoliated for two or three days are weakened and fail to produce fruit buds. The appearance of scab on petioles closely resembles those on leaf lamina.

1-4-2 On Fruits

Tips of sepals are likely to be infected first with the appearance of greyish lesions, which are not distinct due to hairy covering of the surface. These lesions serve as the inoculum source for the developing fruitlets. Early in the season, the scab spots often develop around the calyx end of the fruit (Fig. 1-2b), but later in the season they develop any-where on the fruit epicarp (Fig. 1-2c). Scab spots on fruit resemble leaf spots and are small, circular, and olive colored but with age become dark, brown, and corky. The color of the lesions varies from olive green to dark brown depending on the abundance of spores. The spots are partly covered by the cuticle of the fruit, giving a grey appearance to the young lesions. In older spots, cuticle eruption in rings is seen at the edge of the infected area, exposing a partially or completely black powdery mass in the center. In-fected fruit may have one or more small lesions or may have large, dark, warty lesions which distort the fruit's appearance. Severe early infection results in misshapen, knotty fruits (Fig. 1-2d). Deep cracks often develop in the corky, scabbed areas which serve as avenues for the entry of other organisms, causing rot of the fruit (Fig. 1-2e). Under damp conditions in the orchard, the appearance of pink fungus, *Trichothecium roseum*, on the scab lesions of the fruit is not uncommon, making fruit rot there. Appearance of shallow to deep cracks on the scabbed lesions is common in varieties like Golden Delicious which have thin skin. Long wet periods (due to monsoon rains) near fruit harvest often help in the development of numerous small, black erupting pustules toward the pedicel end of the mature apples. Such scabby spots are also favorable avenues for the entry of other fungi during transportation, resulting in the development of storage rots.

On some varieties, such as Versifield and Cox's Orange Pippin (which are rated resistant under Indian conditions), the scab spots are small, light brown to black, and appear mostly as healed patches on the skin of the fruit without affecting the shape and size of the fruit. The fruit pedicel, on being infected, exhibits similar symptoms as that of the petiole, and may show yellowing and dropping of fruit.

1-4-3 On Twigs

Scab also occurs on twigs, but in many regions or localities this is uncommon and of little importance. Infection of twigs is common in the United Kingdom, Ireland, Israel, Switz-erland, Denmark, and Italy but rare in the United States and India. In India, twig infection is restricted to the Kashmir Valley, where it is observed on certain seedling apple plants and on current-year growth of Red Delicious in a few orchards.[9,11] In the United King-dom, varieties susceptible to twig infection include Allington Pippin, Blenheim Orange, Cox's Orange Pippin, James Grieve, King of Pippins, Laxton's Superb, Lord Suffield, McIntosh, Warrer's King, Willington, and Yellow Ingestre. It is rare on Bramley's Seed-ling apple. Lesions develop in the form of small, raised, reddish-brown blisters com-monly on terminal growth but sometimes on one- to three-year-old twigs. In the spring, the blisters rupture the upper surface, with silvery-grey peeling and an olive green to black powdery mass underneath. This powdery mass consists of fungal spores. The blisters, when coalesced, give a scurfy appearance to the bark and wood. The small, individual lesions on the terminal growth of the twigs may easily be confused with enlarged lenticels or other malformations.

1-4-4 On Fruits in Storage

Late-season fruit infections, called pinpoint scab, can develop if long wetting periods occur in late summer near the fruit harvest. Resulting apple scab lesions, which are small, rough, black, and circular, may be apparent on the fruits at harvest or may go unnoticed until the fruits have been stored for some time.

The lesions may be brown or jet black depending on the variety of apple, and are often shiny because of the intact cuticle. They have definite borders and are circular and darker in color than lesions developing on fruit in the orchard. Apparently healthy fruits picked from infected trees exhibit scab symptoms even at low temperatures (1 °C) in cold storage and may not last long. The lesions on the fruit grow larger, and the areas surrounding these lesions show depression, leading to fruit shrinkage.[12] The affected fruits develop rot due to invasion of scab lesions by secondary rot pathogens, the most common being the pink mold rot fungus *Trichothecium roseum*. Such rotten fruits become a source of infection to the healthy fruits in storage.

1-5 CAUSAL ORGANISM

Apple scab is caused by the fungus *Venturia inaequalis* (Cke.) Winter (Ascomycotina, Dothidiales, Venturiaceae) with the imperfect stage as *Spilocaea pomi* Fr. For many years, the conidial stage of the fungus was considered to be *Fusicladium dendriticum* (Wallr.) Fuck. However, Hughes[13] corrected the nomenclature of pathogen and placed it under *Spilocaea pomi* Fr. Aderhold[14] in Germany and Clinton[15] in the United States demonstrated the connection between the conidial stage of *Spilocaea pomi* (reported as *Fusicladium dendriticum*) and the ascigerous stage, *Venturia inaequalis*, which was found developing on overwintering fallen leaves as saprophyte.

1-5-1 Morphology

Mycelium.   The scab fungus has septate and branched mycelium which is light in color (dull white) when very young but later turns mousy grey to smoky black in culture and olive to reddish brown in the host tissues. In culture, the growth is very slow and is slightly granular to smooth, radiating, and dendroid. The mycelium in the living tissue is located between the epidermal cells and the cuticle. In young foliage, the mycelium develops in branched, radiating ribbons of parallel strands, thus giving the characteristic dendroid appearance to a young lesion. With age, the mycelium becomes compact and several layers in thickness, giving rise to a cushioned stroma. This stromatic mass may be 10 or more cells in thickness at the center of the scab lesion, but near the edge where it forms a wedge between the cuticle and epidermal cells it is usually a single cell in thickness. The stroma is thicker on the fruit than on the foliage and provides more definite and uniform outline to the edge of the lesion.

Conidia and conidiophores.   The conidiophores arise from the ribbon-like strands or from the more compact stroma as simple, cylindrical structures, pale to mild olivaceous brown, variable in length, up to 90 μm long and 5 to 7.5 μm thick and sometimes swollen at the base (up to 10.5 μm). The conidia are produced singly at the tip of the conidiophores and then successively by proliferation through scars of the fallen conidia, resulting in characteristic and distinct annellations in conidiophores. They are obypyriform to obclavate, pale to mid olivaceous brown, smooth, 0 to 1 sepates, with a truncate base and somewhat pointed apex. They measure 13 to 36 μm $\times$ 6 to 12 μm in culture, and on host 12 to 30 μm $\times$ 6 to 10 μm with a truncate base 4 to 5 μm wide. Conidial pustules on the fruit show the dark olive conidia surrounded by a fringe of silvery white, torn cuticle.

Pseudothecia, asci, and ascospores.   Pseudothecia develop in dead leaves, maturing chiefly in early spring. They appear as small, dark brown or black pimples embedded in the leaf, opening by a short beak toward either side of the leaf which happens to be uppermost on the orchard floor. The pseudothecia develop separately or in groups and are spherical to subspherical, from 90 to 225 μm in diameter, with a short ostiolate beak around which are many single-celled appendices (setae). The pseudothecial wall, 10 to 18 μm thick, is composed of three to four layers of dark brown, thick-walled, 4 to 9 μm wide pseudoparenchymatous cells. The asci, which are thin walled, bitunicate, and with a short stipe, arise from a cushion of ascogenous cells at the base of a pseudothecium in varying numbers (usually 100 to 150, but maybe as many as 242) interspersed with pseudoparaphyses which disappear as pseudothecium matures (these are saccate, subhyaline, 45 to 78 $\times$ 5 to 13 μm, and contain eight ascospores). The ascospores are arranged in a single row in the upper portion and are bisereate in the lower portion of the ascus. When the mature ascus elongates in the presence of water, the spores assume a single-ranked position. The ascospores are oval to boat shaped, at first hyaline or pale green, later olive brown, but always translucent, unevenly bicelled, with the upper cell smaller than the lower in the ratio of 1:1.25 or 1:1.50 but slightly broader and forming a pointed apex. The ascospores measure 10 to 16.5 (rarely up to 19.5) $\times$ 4.5 to 6 (up to 8) μm.

1-5-2 Ontogeny of Pseudothecia

With the ensuing leaf fall when the leaf cells begin to die, the mycelium of the subcuticular stromatic tissue penetrates deep into the leaf tissue, growing throughout the mesophyll of dead leaf, and proceeds to form pseudothecia. A small coil in a hypha consisting of uninucleate cells initiates the formation of the stroma. As this develops, a coil of multinucleate cells representing the ascogonium differentiates inside the young stroma and gives rise to a well-defined trichogyne which protrudes from the stromatal wall. The antheridium arises from another hypha, becomes multinucleate, and a contact is soon established between the antheridium and the trichogyne. For plasmogamy to take place, the ascogonium and the antheridium must originate from individuals belonging to different mating types (+ and -), as *V. inaequalis* is heterothallic.[16,17] As a pore in the adjoining wall forms, the nuclei pass from the antheridium into the ascogonium through the trichogyne and pair with the ascogonial nuclei. The ascogenous hyphae carrying the nuclear pairs arise, and croziers form at their tips. A binucleate penultimate cell forms which is destined to become the ascus and is termed the ascus mother cell, and the two nuclei fuse promptly to form the diploid stage. In the meantime, the stroma continues to develop and forms the pseudothecium. The young ascus, with its diploid, zygote nucleus, begins to elongate, emerging in between the pseudo-paraphyses. Three successive divisions in the nuclei occur, the first of which is meiosis. The cytoplasm is delimited around each of the eight nuclei, and the same number of uninucleate haploid ascospores is formed. The nucleus in each spore divides, and the formation of a septum results in a spore with uninucleate cells of approximately equal size. The spores elongate rapidly, the cells growing at different rates, to form the characteristic mature ascospore with two cells of unequal size.

1-5-3 Variability

The fungus consists of many strains which differ in their morphologic, physiologic, cultural, and pathogenic reactions. In India, morphological and biological studies have revealed the presence of at least two strains of the fungus, the virulent one infecting Red Delicious, and the less virulent one infecting Ambri cultivar of apple in the Kashmir Valley.[17] There is another host specific strain on *Cotoneaster affinis* var. *bacillaris* (Wallich ex Lindl.) C. K. Schneid (Syn. *C. bacillaris* Wall.)[18,19] and on *Cotoneaster aitchinsonii* Schreider.[20] The fungus in its *Spilocaea* state has also been recorded on a crab apple, *Docynia indica*,[21] which seems to be another host-specific strain. Menon[22] has proposed formae speciales for some strains showing pathogenicity to different genera in cross-inoculation tests, viz. *Venturia inaequalis* f. sp. *mali* on apple, *V. inaequalis* f. sp. *aucupariae* on *Sorbus*, *V. inaequalis* f. sp. *cotoneasteris* on *Cotoneaster intergerrima*, and erroneously *V. inaequalis* f. sp. *crataegi* on *Crataegus oxycantha*. The latter fungus has a different conidial state of *Fusicladium* type. According to Sharma and Gupta,[19] the strain occurring on *Cotoneaster affinis* var. *bacillaris* (Riyush plant) is a different strain within the fungus, having been referred to as *Venturia inaequalis* (Cke.) Wint. f. sp. *cotoneasteris* Menon. Besides various strains, five physiologic races of the fungus have been determined by their differential pathogenicity to a range of *Malus* spp.[23] The Indian isolates of the scab fungus belong to Race 1, which is commonly encountered in the United States and various other countries.[24]

1-6 DISEASE CYCLE

There are two distinct cycles in the life history of the apple scab fungus, the primary cycle and the secondary cycle (Fig. 1-3). The primary cycle deals with the initiation of the disease and also includes the *Saprophytic*, or *Venturia*, stage or *perfect* stage (overwintering stage), which perpetuates on the infected fallen leaves in winter and spring. The secondary cycle deals with the *parasitic*, or *Spilocaea*, stage or *imperfect* stage (summer stage), which perpetuates on the fruits and leaves throughout the growing season.

1-6-1 Primary Cycle

Dormant period of the scab fungus.   The scab fungus remains alive through winter in the dead and fallen leaves of apples on the orchard floor, and enters the saprophytic phase soon after leaf fall in autumn.[25] Most of the pseudothecia develop on the leaves remaining in shady portions of the orchard floor. They appear to be black, pimple-

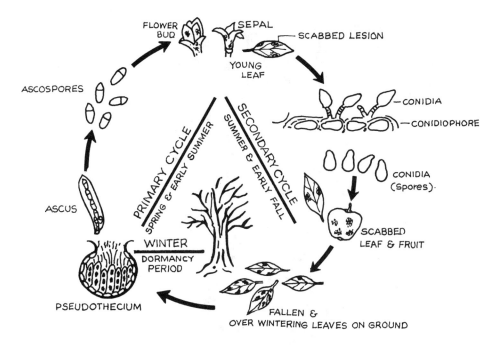

Figure 1-3   Disease cycle of apple scab.

like structures embedded in the leaf tissue and can be observed with the help of a hand lens or microscope. This stage is prevalent in India in March to April, while in other countries mid-April to May. Each pseudothecium contains 100 to 150 asci, each with eight ascospores (Figs. 1-4a and b), thereby yielding 800 to 1200 ascospores. On a single leaf, there may be production of nearly 2000 pseudothecia with a total load of 2 million ascospores. The ascospores produce the first, or primary, scab infections on the new growth in the spring and can be rightly termed spring spores.

Good snow coverage and alternate periods of wetness and dryness in winter and early spring favor the development of pseudothecia in the leaf tissue. Pseudothecia are formed within 30 days of leaf fall mainly on leaves which fall in September, October, and November, as the leaves of earlier dates do not survive and are decomposed quickly.[26] Pseudothecia are more readily formed in leaves which are moist enough to be fully pliable, as in leaves which are alternately wet and dry, than in those which are continuously wet. Drought conditions retard the development of pseudothecia and may also delay the ascospore maturity. In the presence of adequate moisture, pseudothecia may be initiated at temperatures below 0°C, but the pseudothecial density is maximum at temperatures between 4 and 8°C. There appears to be a transition from a lower to a higher range of thermal requirements as the pseudothecia develop. The thermal requirement of pseudothecia shifts during winter from 10°C for early diameter increase to 15 to 20°C for ascospore maturation.[26,27] Temperatures of 24°C and above are deterimental to normal ascospore development.

Initiation of the disease.   In most years, some ascospores are mature by the

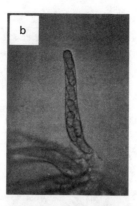

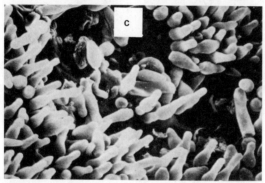

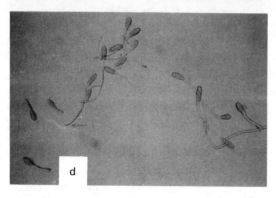

Figure 1-4a Cross-section of a pseudothecium of the apple scab fungus with mature ascospores. (Courtesy: A. L. Jones, U.S.A.)

Figure 1-4b Micrograph of a single ascus with eight ascospores.

Figure 1-4c Micrograph of a sporulating scab lesion on apple leaf producing flame-shaped conidia on short stalks or conidiophores. (Courtesy: A. L. Jones, U.S.A.)

Figure 1-4d Micrograph of germinating condia with few (arrow) having formed oval appressorium at the tip of long germ tube.

time the fruit buds start putting out green tissues with the break of dormancy in early spring.[28-30] They continue to mature throughout the spring and early summer. Once mature, they are forcibly ejected from the pseudothecia upon wetting of overwintering leaves on the orchard floor and may then be wind dispersed within the orchard. Pseudothecia, when wet with dew, can release some ascospores, but very few as compared to the number released by rain. As little as 0.13 mm (0.005 inch) of rain is sufficient to release some spores, but for larger releases at least 0.20 mm (0.008 inch) or more is necessary.[31] Wetting of pseudothecia just for 4 seconds by a thin film of water is enough for the release of some ascospores, while such a process may start after 45 to 120 seconds of wetting if the pseudothecia are covered with a thick film.[32] A water film greater than 40

μm is too deep, and the spores as released are more likely to be submerged than to become airborne. Thus, the best condition for the fastest discharge of ascospores is a thin film of water over the pseudothecia allowing the asci to extend through the ostiole and release the spores into the air. The successive emergence of asci and discharge of ascospores may continue as long as the water film remains. In general, few ascospores are discharged within the first hour of wetting, but concentrations rapidly rise to a maximum in the second and third hours. In prolonged wetting, about 75% of the discharge occurs within 3 to 6 hours, but spores may be released intermittently. Humid, comparatively warm conditions prior to wetting, with longer daylight hours during wetting, result in discharge of most or all of the mature spores; while at night, few or no ascospores are discharged irrespective of the degree of wetting.[32,33] Maturation and discharge of ascospores usually lasts 3 to 9 weeks. During the final stages of pseudothecial maturation, the discharge mechanism becomes erratic and inefficient, and the ascospores may be oozed rather than forcibly ejected, resulting in great numbers of spores being released into water rather than into the air. The final 5 to 25% of the ascal initials to mature may never discharge their spores into the air.[33] Periods of thorough yet not prolonged wetting followed by dry intervals during which the surface water evaporates (a condition often encountered during heavy showers) is the type of weather most likely to favor air dispersal of ascospores. During such conditions, air turbulence at ground level collects the newly released ascospores and distributes them in the orchard. These airborne spores rapidly decline in numbers with increasing distance from the source. The percentage of spores that travel 100 meters or more from the source is 0.005, 10.0, or over 10.0 under calm, dull and windy, or warm and windy conditions, respectively.[34] As such, it can be expected that the level of scab development within an orchard is the outcome almost entirely of the inoculum produced within the orchard provided there is no nearby abandoned orchard serving as a source of primary inoculum. Wherever wood infection is there, scab lesions on twigs start producing conidia, which may cause primary infection in the spring in addition to ascosporic infection. This is of significance in England, Israel, and other European countries[35,36] and in the Kashmir Valley of India.[9,17]

Germination of spores (ascospores and conidia) begins as soon as they land on susceptible green tissues (leaves, flower sepals, fruits, petiole, and pedicel), provided a water drop or film of moisture is present. The ascospores, as they alight on a surface, develop an adhesive material which fastens them firmly to substratum. Under the prevailing temperature conditions, the number of hours of wetting (presence of water drop) necessary for infection varies. The spore, on germination, produces a germ tube with disc-like appressorium at the tip, and a very slender mycelial tube (peg-like structure) emerging out of appressorium punctures the cuticle of the plant tissue to start primary infection. This forms a plate of cells from which mycelial ribbons develop in a radiating manner between the cuticle and the outer cell wall of the epidermal cells. The fungus largely remains in the subcuticular position, and by the time the lesion appears the epidermal cells below show a gradual depletion of their contents, and they eventually collapse and die. Soon the palisade, and later the mesophyll cells, exhibit the same reactions. The fungus seems to cause death of cells by secreting enzymes or toxic substances that alter the permeability of the cell membranes and break down macromolecular components of the cell into small molecules which then move along osmotic pressure gradients, reach the mycelium, and are absorbed by it. Once the mycelium is established and forms stromatic cushion subcuticularly, it starts producing conidiophore and conidia, whose outward growth results in bulging and rupturing of the cuticle, and giving rise to olive green, velvety scab lesions. Such lesions are not visible until several days after the spore first infects the leaf or fruit. Usually 8 to 17 days are required from inoculation to the appearance of the first scab lesion.

1-6-2 Secondary Cycle

Secondary infection and spread of the disease.   A large number of conidia which are produced on primary scab lesions (Fig. 1-4c) are washed by rain or sometimes even detached by wind[37-39] and are spread to other susceptible tissues (rapidly growing leaves and fruits) where, under appropriate environmental conditions, they cause secondary infection. Conidial germination (Fig. 1-4d) and infection occurs in about the same time of wetting as required for ascospores. The spread of disease through additional infection by conidia may continue at intervals throughout the growing season following a rain of sufficient duration; and in valley areas, even night-long dew may favor such buildup of infection.[40]

Conidia production begins before lesions are macroscopic. Humidity and temperature conditions which favor conidia production in scab lesions range from 60 to 100% RH and 4 to 28°C with optima being 90% and 16 to 20°C, respectively.[39,41] Wetting is not required for spore formation. As lesions mature, sporulation in the center declines and some necrosis often appears. Although apple cultivar influences lesion number and size on the infected portion,[30] lesion growth and spore production stops generally within 30 to 36 days, and lesions become necrotic or bronzed and reddish brown. Many secondary cycles of the conidial stage occur during the growing season, and each scab lesion produces thousands (50,000 to 100,000) of conidia.[39] Thus, the potential for disease incidence in a favorable environment is extremely high.

Secondary infection in fruits can occur in late summer (monsoon months) but may not show up until the fruits have been stored for one or two months.[12] There is no spread of scab from affected fruits to clean fruit in storage, and the scab developing during storage is the outcome of initial infection having taken place in the orchard while the fruit is on the tree.[42] In India, leaf fall usually takes place after 1 1/2 to 3 months of harvest, and in many orchard areas, dew keeps the leaves wet for sufficient length of time for initiation and buildup of secondary infection. Such late infection on leaves assumes importance in providing enough numbers of pseudothecia to start the new season even though a good spray program has been followed the previous year.

1-7 FAVORABLE WEATHER CONDITIONS

Scab is a wet-weather disease. For initiation of infection, the leaves must remain wet due to rain for a couple of hours (minimum 9 hours) at particular temperature conditions. Such correlations between wetting periods and temperature conditions are referred to as Mills periods or table (Table 1-1).[43] According to Mills,[44] one third less wetting time is needed for infection with conidia than with ascospores. Actually, about the same wetting period is required by both types of spores, and perhaps longer for conidia.[45] But Mac-

TABLE 1-1: APPROXIMATE HOURS OF WETTING AT INDICATED TEMPERATURES REQUIRED FOR LEAF SCAB INFECTION, AND DAYS REQUIRED FOR LESIONS TO APPEAR*

Average temperature (°C)	Minimum wetting hours[a] of leaves required for light infection		Days required for lesions (symptoms) to appear[b]
	From primary inoculum (ascospores)	From secondary inoculum (conidia)	
25.5	13	8.7	
25.0	11	7.3	
24.4	9½	6.3	
17.2–23.8	9	5.9	9
16.6	9	5.9	10
16.1	9	5.9	10
15.5	9½	6.3	11
15.0	10	6.6	12
14.4	10	6.6	12
13.8	10	6.6	13
13.3	11	7.3	13
12.7	11	7.3	14
12.2	11½	7.7	14
11.6	12	7.9	15
11.1	12	7.9	15
10.5	13	8.7	16
10.0	14	9.3	16
9.4	14½	9.7	17
8.8	15	9.9	17
8.2	17	11.3	
7.7	19	12.6	
7.2	20	13.3	
6.6	22	14.6	
6.1	25	16.5	
5.5	30	19.9	
0.55–5	More than 2 days	?	

[a] Leaves remain wet for varying lengths of time after the rain stops, depending on conditions. Wetting periods from intermittent showers should be added together. Average temperature for the period should be determined from hourly readings.

[b] Days required for lesions to appear once infection has been established. No further wetting is required. For this column, daily maximum and minimum temperatures are adequate for determining the average.

* From W.D. Mills, Cornell University; commonly referred to as Mills table or Mills infection periods.

Hardy[46] holds that infection occurs in 3 hours less the minimum hours for infection indicated by Mills curve. The shorter time for secondary infection is a factor of inoculum level, as accumulation of airborne ascospores on any given leaf is limited in contrast to conidia, which are present on established lesions and may accumulate on leaves and fruits in tremendous numbers (100,000 and more spores) during wet conditions.

Short wetting periods occurring at close intervals may constitute an infection period. However, the periods are not added together in calculating an infection period if the leaves are dry for about 8 hours between the wet periods, though earlier it was suggested to be a period of less than 4 hours or one half day of sunny weather.[43] Once the rain stops, it is best to measure the relative humidity, and if during dry periods the level of relative humidity does not fall below 90%, the entire period can be added toward the wet period.[47] Orchardists can record the beginning of rainfall and average temperatures and determine the length of time it takes for infection to occur (see Mills table).

Snow or rain during winter and early spring favors the pseudothecial development in the overwintering leaves, while frequent rains in spring and early summer lead to the maturity and discharge of ascospores, causing primary scab infection on emerging leaves and fruits. Such wet weather conditions also favor vigorous tree growth, which is more susceptible to scab, leads to quick wash-off of the fungicides, and often interferes with the fungicidal treatments, reducing the effectiveness of disease control procedures.

1-8 DISEASE PREDICTION

Three main criteria—quantity and relative maturation of pseudothecia as a source of primary inoculum, phenology of the tree, and the occurrence of infection periods (Mills criteria)—form the basis of apple scab prediction in the U.S.A., U.K., and several other countries.[10,23,35] Such prediction, which usually begins in early spring, determines the occurrence of infection prior to the appearance of scab symptoms and helps the growers to adopt suitable control measures.

1-8-1 Criteria of Prediction

Quantity of primary inoculum.   The quantity of primary inoculum is measured as (1) ascospore discharge (productivity) based on the number of mature spores/cm^2 overwintered leaf area, and (2) ascospore dose, which is the number of spores/volume air. In most countries, greater reliance is given to the first aspect (ascospore maturity and discharge) in the prediction of primary infection, whereas the second aspect is studied to develop a disease prediction system (after initiation of infection) and for determining the threshold level of inoculum.[23] Examination of 20 asci from 20–25 pseudothecia either picked up from overwintered apple leaves or observed in the transverse section of the leaf is commonly used to categorize ascospores/asci maturity level. Ascospore dose, in fact, measures the actual inoculum concentration in the orchard air at different stages of host phenology; this is dependent on (1) ascospore productivity, reflecting the effect of weather on perennation during the overwintering period; and (2) factors that influence spore release (i.e., air temperature, time of day, climatic date, and leaf wetting by rain/dew). Traps for measuring ascospore dose in the air are available, and among them Rotorod sampler is satisfactory for routine monitoring of ascospores.[23,48,49]

Phenological stages.   Ascospore maturity is accelerated by the temperature conditions, which also influence apple host development, and it coincides with the break of dormancy of apple trees. The colored spores increase week by week until bloom and to the early petal-fall stage of tree development, and then diminish.[29,30,50] Exceptions may occur when a few ascospores mature earlier than the bud burst due to unusually warm temperature conditions.[23,36] Since ascospores start maturing by bud burst, there is much reliance on beginning the spray program at the silver-tip to 1/4" gree-tip stage of the apple tree, and continuing fungicidal sprays at short intervals until the primary scab season is over. Tree phenological stages, however, have not been utilized in developing a predictive equation for improving chemical control strategy.

Infection periods.   Various weather monitoring equipment, such as hygrother-mograph deWit seven-day recording leaf wetness meter, Geneva leaf wetness detector,[51] leaf wetness recorder, or modified hygrothermograph,[52] are widely used for identifying infection periods. The occurrence of infection periods, the type and rate of fungicide, and the frequency of application are taken into account for keeping orchards almost free of scab infection.

1-8-2 Predictive Models and Predictive Procedures

Empirical models for predicting ascospore maturity from weather have been developed, and such approach is without the adoption of laborious methods of microscopic examina-tion. Using multiple regression analysis, the model of Massie and Szkolink[53] has shown that maturity depends to a larger extent on the number of accumulated degree days (base 32°F) from 50% leaf fall and to a much lesser extent on accumulated precipitation during the same period. Differing from this model, which was developed from the Geneva area of New York State, Sutton et al.[54] at North Carolina State University, Raleigh, have dem-onstrated the effect of precipitation equally important to that of temperature on pseudothe-cial ontogeny. As the Geneva model failed to predict satisfactorily the ascospore maturity in North Carolina and predicted mature ascospores much earlier than they occurred in nature, the results of their studies both under field and laboratory conditions have re-vealed that environmentally driven models should not be used in another climatic region without evaluation in that region. Similarly, Gadoury and MacHardy,[55] not being satisfied with the Geneva model in New Hampshire, have developed a linear statistical model based on accumulated degree days from the maturation of the first ascospore. Experience shows that these models work reasonably well but are specific to the location and culti-var.[54,56] Empirical models for predicting the amount of ascospore inoculum are still lack-ing, and other approaches which have been suggested include the development of mecha-nistic models of scab perennation,[33] which may eventually lead to simple summary models for practical use. Ultimately, a system must be developed by which one can decide on the spray strategy in the spring by looking into the previous year's infection level at leaf fall in the orchard.[23]

Among various predictive procedures,[25] automated data collection and prediction is the faster method of detecting infection periods. This overcomes the problem of time consumed in interpreting weather charts of environmental monitoring equipment for rightly calculating the infection periods. An electronic scab warning instrument (biomat SWG), which was constructed by Richter and Haussermann[57] in Germany, consists of a sensing unit for monitoring the ambient temperature, the duration of leaf wetness, and RH above 85%. Weather data are stored electronically, and the status of an apple scab infection period is provided through a display on the console. Based on the same princi-ple, a battery-powered microprocessor-based unit which uses mean air temperature, the hours of leaf wetness, rainfall, and the percent relative humidity as inputs to a resident predictive model has been developed.[58] This microcomputer (Apple Scab predictor) not only provides information on weather conditions but also gives information on how much time is left for after-infection spray and what type of fungicides will have to be used.

Although the microcomputer is a convenient and reliable tool to detect scab infection periods automatically for rationalizing the fungicide sprays and providing saving of an average of 3.3 sprays each season once the disease predictor's curative spray is substituted for a standard protective program,[59] its high cost and placement of number of such predictors in an orchard either having large area or different microclimate due to being planted on slopes of mountains are the main deterrents to its practical acceptability. Efforts are being made to incorporate the programs for other diseases (fire blight, cherry leaf spot, and black spot of grapes) and pests (codling moth) in the predictor to provide benefit of integrated control measures to the orchardists and to increase its acceptability.

1-8-3 Scab Warning Services

Apple scab warning services based on the three criteria of disease prediction are in vogue in many countries of the world.[35,60]

In Germany, a mean day temperature aggregate of 105°C (over 15 days) after March 1 favorable for pseudothecial ripening is also taken into account for issuing warnings. The warning system is one of local advice. In Holstein, for example, one wetness recorder is maintained for each 0.5 sq km for making use of Mills criteria.

In the Netherlands, warnings are issued by radio when (a) the ascospores are ready to mature, (b) an ascospore release is expected, (c) ascospore release has begun in the open, (d) infection periods have occurred, and (e) ascospore discharge has ceased.

In England, Smith periods[47] substituting leaf wetness duration of Mills criteria with the hours of 90% or more following rain in spring have been found equally satisfactory. This enables meterological data to be used for prediction instead of a surface wetness recorder. In general, both the criteria of Mills and Smith are taken into account for formulating scab warnings, and advice is issued locally by the regional advisory plant pathologist wherever needed through the radio, press, phone, TV (PRESTEL and CEEFAX—computerized programs managed by BBC), and by subscription (postcard, telegram, and TV service, etc.).

In the U.S.A., the regional extension plant pathologists, or country agents, give advice regarding the possibility of disease appearance. Arrangements exist for orchardists to send leaf samples in early spring to laboratories for examination, and to provide information regarding pseudothecial ripening and ascospore discharge, which are used for advisory radio, TV, and press reports and for communicating through a coded phone number.

In India, a scab warning service is in vogue in Himachal Pradesh. Although efforts have been made to carry on prediction in some of the fruit-growing areas by establishing apple scab monitoring laboratories equipped with modern predictive devices, such a system would be more accurate if an on-site approach, as is in operation in the U.S.A. and the U.K.,[36,61] is adopted and infection periods are recorded in individual orchards every spring.

1-9 DISEASE CONTROL

Since primary infection is brought about in most cases by ascospores, control strategies have been aimed toward (1) inhibition of pseudothecial production and ascospore release, and (2) protection against primary and secondary infections in the growing season. The use of resistant varieties of apple is becoming available[23,62] and will eliminate the need for adoption of such strategies and dependence, to some extent, on chemical sprays to manage scab.

1-9-1 Inhibition of the Primary Source of Inoculum

To eliminate or reduce the primary source of inoculum in spring, orchard sanitation and/or chemical spray can be used for quick decomposition of leaves harboring fungus. Recently, research efforts have been made to inhibit the perfect stage of the fungus by using biological agents.[63]

Sanitation and fall application of nitrogen.   Since the scab fungus produces pseudothecia on the fallen and overwintered leaves, it would appear that the disease can be managed by destroying such leaves. Unfortunately, complete elimination of dead leaves and the pseudothecia is not possible under orchard conditions with available methods. Just 1% of leaves can provide sufficient pseudothecia, releasing millions of spores for the buildup of severe scab under ideal weather conditions. However, the potential for severe scab can be reduced by making a spray of 5% urea or of 5 to 7% Ankur (liquid fertilizer with 32% N in ammonical nitrate form) in autumn shortly before general leaf fall, to hasten leaf decomposition in addition to promoting sanitary measures, such as ploughing or raking and destroying the fallen leaves. The autumn spray of urea has a detrimental effect on leaves, as the treated leaves turn dark brown and decompose by spring.[64-66] The mechanism of urea, though imperfectly understood, mainly promotes the decomposition process of leaves. Quick decomposition has been linked with a higher initial nitrogen content and a much increased microflora, particularly bacterial population per unit leaf area.[66] Since urea does not directly hit the fungus, there is no possibility of scab pathogen developing tolerance with its use. In fact, such preleaf fall urea sprays not only accelerate leaf decomposition but also supply nitrogen to the tree.[67]

Postharvest fungicide sprays.   In orchards which have a high nitrogen content, the application of urea may make plants more prone to cankers and other diseases. Such preleaf fall sprays can be substituted by two sprays of 0.05% Bavistin (carbendazim) or 0.10% Syllit (dodine), given just after harvest at one-month intervals for obtaining inhibition of pseudothecia and ascospore release.[65,68,69] The utility of such postharvest sprays has not been ascertained in other countries, since harvest and leaf fall in apple trees usually coincide and there is not much gap between the two stages, which is contrary to Indian conditions.

Preleaf fall fungicide spray.   Among fungicides, benzimidazoles (benomyl, carbendazim, and thiophanate-methyl) and dodine are very effective in destroying the overwintering stage of scab fungus and preventing ascospore release, but their use in preleaf fall spray has been limited to avoid rapid development of resistance in the scab fungus. This is particularly true with the use of benzimidazoles. Other fungicides which are effective as preleaf fall eradicants and can be used especially in areas where the scab fungus has developed resistance to benzimidazoles are Baycor (bitertanol) and Funginex or Saprol (triforine), the ergosterol-biosynthesis inhibitors.[23,69] However, since these compounds have limited site-inhibition characteristics and the potential for resistance in these new fungicides also exists, it is necessary to use these fungicides in only a few applications to enhance their minimum selective pressure on the pathogen.

Biological agents.   The microbial antagonism of *Athelia bombacina* Pers. (isolated from litter) and *Chaetomium globosum* Kze. (leaves and litter) to the apple scab pathogen has also been explored.[63] Among these, *A. bombacina* is effective in preventing the ascospore production mainly by softening the leaves and enhancing their decomposition. This antagonist tolerates 5% urea and thus could substitute as one component of an integrated program for apple scab control.

1-9-2 Protection Against Primary and Secondary Infections

Apple scab is thoroughly managed, even on the most susceptible apple cultivars, by timely sprays with proper fungicides. Every effort must, therefore, be made to manage the primary scab in early stages to prevent additional infections (secondary infections) during the remainder of the growing season.

Fungicide spray schedules which can be followed have either of three basic strategies—protective, postinfection, and eradication[10,70]—which differ in the site of chemical action and the time of fungicide application (see Tables 1-2 and 1-3).

Protective.   In a protective spray program, the fungicide is applied on apple

TABLE 1-2:   PROPERTIES OF FUNGICIDES FOR APPLE SCAB CONTROL

Fungicides	Retention	Redistribution	Postinfection activity (Kick-back) from 11.6°C to 17.2°C
Benomyl (Benlate) 50% WP	Poor–Fair	Poor–Fair	18–24 hrs.
Carbendazim (Bavistin, Agrozim, JKstein, Bengard, Derosal, B-sten) 50% WP	Poor–Fair	Poor–Fair	18–24 hrs.
Captafol (Difolatan, Foltaf) 80% WP	Excellent	Excellent	18–24 hrs.
Captan (Hexacap, Captaf, Deltan) 50% or 75% WP	Poor	Good	18–24 hrs.
Dithianon (Delan) 75% WP	Good	Good	36–48 hrs.
Dodine (Syllit, Cyprex, Venturo, Carpene) 65% WP	Good	Good	36–48 hrs.
Mancozeb (Dithane M-45, Manzeb) 75% WP	Good	Good	18–24 hrs.
Sulphur fungicide	Fair–Good	Good	None
Thiophanate-methyl (Topsin M) 70% WP	Poor–Fair	Poor–Fair	18–24 hrs.
Sterol-biosynthesis inhibitors (Baycor, Rubigan, Saprol, Systhane, Topas, etc.)	—	—	36–72 hrs.

TABLE 1-3:   FUNGICIDE SPRAY STRATEGIES FOR APPLE SCAB

Spray schedule	Point of chemical action	Time interval determination
Protective	Between the spore and the leaf or fruit surface	Follow spray schedule (Table 1–4) at approx. regular intervals depending on weather and tree stages
Postinfection (Curative)	On the already germinated spores	Determine infection period and select fungicides for spray as given in Table 1-2
Eradication	On established scab lesions	As soon as established scab lesions are discovered, repeat within 7 days either Dodine (100 g/100 liter water) or Carbendazim or thiophanate-methyl (50 g/100 liter water) or mixture of carbendazim + manocozeb (25 g + 250 g/100 liter water) for complete inactivation of lesions

trees before, during, or immediately after a wetting period (rain) from the time of bud break until all the ascospores are discharged from the pseudothecia. The strategy is to maintain a protective chemical barrier on all susceptible leaves and fruits by repeated sprays of fungicides timed according to tree growth stages. Repeated spraying is needed partly because of weathering of old deposits, but mainly because new tissues on rapidly expanding young leaves and fruits are being continuously exposed to spores. Such sprays prevent the spores from germination or penetration of leaf tissue. Although protective sprays (routine sprays) require considerable investment of fungicides, labor, and equipment, if they are followed properly, they effectively control the disease. The main disadvantage is that, at times, sprays are applied routinely even though they are not needed because weather is unfavorable for infection.

Postinfection (after infection and presymptom).   A postinfection spray strategy is to use the curative fungicides after an infection period instead of giving routine sprays. Curative fungicides kill and inhibit fungus in host tissues after infection has occurred. Such action (after-infection activity) for most fungicides is limited to a few hours or days after infection. Weather-monitoring equipment that can record temperature, RH, and leaf wetness, or electronic devices for automatic weather data collection and prediction (apple predictor) are needed to determine favorable conditions for infection. Once an infection period has been determined, a fungicide with sufficient postinfection activity (kick-back, back-action, or curative action) is applied if no other fungicide has been applied during the past 10 days (see Table 1-2).

Presymptom activity is an elaboration of after-infection activity.[71] Some fungicides applied two to three days before the appearance of symptoms or five to eight days after inoculation and infection as determined with the initiation of Mills infection period result in the development of atypical lesions, which may appear as chlorotic spots or flecks that produce few or no secondary conidia. This interval exceeds the after-infection limit of all fungicides, and such presymptom activity is common with sterol-biosynthesis inhibitors, benomyl, and dodine. This approach, though it may not completely prevent symptom expression (chlorotic lesions have viable scab fungus[72] and get invaded quickly by leaf-spotting fungi[73]), nonetheless offers an advantage in greatly reducing the secondary scab inoculum which, in turn, reduces the pressure on fungicidal sprays applied later.

The postinfection method of scab management is suited to only those orchard areas where (1) infection periods can be clearly defined and warnings quickly circulated; (2) infection periods are relatively few in number (e.g., two or three) and occur at a considerable interval during the period of ascospore discharge; (3) curative fungicides are readily available; and (4) the spray machinery is effective enough to cover the whole orchard within 24 to 48 hours (depending on the fungicides) from the beginning of an infection period. Such an approach entails risk, as at times it may not be possible to carry on sprays soon after the start of an infection period or two to three days prior to symptom appearance, particularly when weather conditions continue to be rainy. Thus, the best way is to use routine sprays as insurance against infection. Such a practice is commonly followed by almost all the fruit growers in the U.K., U.S.A. and many other countries.

Eradication (postsymptom).   Eradicative action may be necessary to manage the fungus if scab lesions appear, because the secondary spores (conidia) produced on these lesions can cause additional infections during summer. Benzimidazole fungicides (e.g., benomyl, carbendazim, and thiophanate-methyl) and dodine are used to inactivate established lesions, to suppress spore production, and to reduce the viability and germination of existing spores.[74,75] This way, the inoculum pressure is greatly reduced, which improves chances for better scab control with succeeding sprays. Some of the EBI fungicides also have postsymptom antisporulant activity[74] and can be utilized in such a spray strategy.

1-9-3 Suggested Spray Schedule

Timely chemical sprays.   Based on the field performance of various fungicides and their ability to stick to the leaves, to be redistributed by rain (redistribution), and to suppress sporulation on established lesions (postsymptom activity), protective or routine spray programs (see Table 1-4), as adopted in apple orchards in India, are formulated to include seven to eight sprays of nonsystemic and systemic fungicides in the growing season.

According to the schedule, the first spray of protectant fungicides must start at the green-tip (or even silver-tip) stage and be repeated at 10- to 14-day intervals until the fruitlet stage. Only three or four sprays are required to cover the stages of bud break to petal fall, as these stages end within 30 to 40 days. For satisfactory management of the disease, it is essential that the fungicides are applied at the proper time and interval. Once scab lesions appear, it becomes difficult and costly to avoid losses from scab. As such, every effort must be made to contain the disease early in the season when growth is rapid. Fungicides are sprayed at short intervals to provide a coat of chemical on the newly emerging leaves, blossoms, and green fruits and to prevent the spores from germinating or penetrating.

The single application technique (SAT), which involves application of captafol (Di-

TABLE 1-4: PROTECTIVE SPRAY SCHEDULE TO CONTROL APPLE SCAB

Spray No.	Tree Stage	Fungicide/100 litre water
I	Silver tip to green tip	Captafol (300 g) or mancozeb (400 g) or captan (300-400 g) or dodine (150 g) or dithianon (75–100 g)
II	Pink bud to early bloom (10 days after 1st spray)	If Captafol has not been used in 1st spray, then repeat mancozeb (300 g) or captan (200–300 g) or dithianon (50 g)
III	Petal fall (14 days after 2nd spray)	Carbendazim (50 g) or dodine (100 g) or thiophanate-methyl (50 g)
IV	Fruit set (10 days after 3rd spray)	Carbendazim or thiophanate-methyl (25 g) + mancozeb (250 g) or mancozeb (300 g) or captan (200–300 g) or dithianon (50 g) or dodine (75–100 g)
V	Fruit development (14 days after 4th spray)	Repeat captan or dithianon or mancozeb or dodine (75–100 g)
VI	Fruit development	Repeat captan or dithianon or or mancozeb
VII	Preharvest (15–20 days prior to harvest)	Captafol (150 g) or captan (200–300 g) or dithianon (50 g) or Mancozeb (250–300 g)
VIII	Preleaf fall (7–10 days prior to general leaf fall)	Urea (5 kg.) or Ankur (5–7 kg)

Captan 75% WP to be used in 1st spray at 300 g and in 2nd and onward at 200 g; captan 50% WP in 1st spray at 400 g and in 2nd & onward at 300 g.

Sticker-like Selwet-99, Agrowet (12.5 g), Teepol (25 ml), or Selwet-E/Tritone/Tohfa/Uphaar (50–75 ml) can be added in 100 litre fungicidal solution (except dodine) for enhanced solubility and retention.

folatan or Foltaf) at a higher rate in the green-tip stage, is used in commercial fruit production in the U.S.A. and other countries every year.[23,28,76,77] Such application does not need repetition every 10 to 14 days. Since a high rate of fungicide is deposited on the branches and redistributed by rain, the new growth is protected to a reasonable degree for about 20 to 30 days. In exceptionally wet or dry springs, complete dependence on this technique instead of resorting to a second spray at a 10-day interval may involve a slight risk, because the fungicide deposit may not persist for long, allowing the establishment of some primary scab lesions. In India, since most of the tree bud development stages end within 25 to 30 days, a concentration of 0.3 to 0.5% of formulated fungicide (80 W) is commonly used by apple growers in SAT, which is the rate of the product used in RSAT (Reduced Single Application Technique) in other countries.[28] In most of these countries, the bud development stages take longer and end within 35 to 50 days, and a comparatively higher concentration of the formulated fungicide, ranging from 0.6 to 1.5%, is used. The optimum time of spraying of captafol in SAT is after the appearance of green bud tissue (green-tip), but not after 1 cm green. Sprays applied at later stages (1/2" green; tight cluster; pink-bud, etc.) result in moderate damage to the cluster leaves and, sometimes, fruit russeting. The SAT usually does not lead to a reduction in cost for fungicide, but provides saving in costs of fuel, labor, and equipment. Other protectant fungicides (man-cozeb, dodine, dithianon, captan), if sprayed at the higher concentration in the first spray followed by a reduced rate in the second at a 10-day interval, satisfactorily prevent pri-mary infections.[68,77] This approach of RSAT is likely to be more useful to the growers in India than the SAT because of lower initial costs, high flexibility in spray programming, better coverage of rapidly growing tissues in an unusually wet spring, and effective con-trol of other diseases and pests (e.g., powdery mildew, thrips, etc.) by incorporating a suitable fungicide or insecticide in the spray mixture. Primary scab infections occur from the green-tip stage until a month after the petal-fall application. If infections are prevented during this period, special applications against scab are not needed for the rest of the season. On the other hand, if the leaves and fruits have become infected, a spray of carbendazim fungicide or thiophanate-methyl or its mixture with protectants or a spray of dodine should be carried out to inhibit established lesions and suppress spore production. This should be followed by a second spray of either of these fungicides given at a short interval of seven to ten days to inactivate most of the lesions, which turn to purplish brown in color.

Depending on the weather and disease status, subsequent sprays may be continued with protective fungicides at a 14-day interval. Under the dry weather conditions of summer months, the interval may be extended to 21 days, mainly in those orchards which have been planted on mountain slopes, but not in valley areas near to the river where dew is of sufficient length to keep leaves wet and result in continuous spread of the disease.[39] Such an extended interval may allow fresh scab lesions to appear and poor control of the disease, particularly in shady portions of the tree. The fungicide spray program as adopted for scab management also helps in management of other diseases (see Table 1-5), which include cankers, sootyblotch and flyspeck, fruit rots, and leaf spots. Some of the fungicides, particularly mancozeb, suppress the red spider mite as well.[77] The sprayed apples have better finish and more attractive color, and the leaves remain green for a longer time. From an economic point of view, such sprays do provide a high ratio (8–12:1) of benefit/cost and a good remuneration from the apple crop thereof.[78] Precautions which must be adhered to while using fungicides have been listed,[10] and these reveal the safety of mancozeb, carbendazim, thiophanate-methyl, and zineb fungicides for bloom spray in addition to safety for all cultivars of apple. Captan may cause spotting, yellow-ing, and dropping of Red Delicious leaves when used at full strength early in the season; while dithianon, dodine, and captafol may cause russet or scald of Golden Delicious apples and of light-colored cultivars if used beyond a certain tree stage. Captafol applied after the green-tip stage may cause fruit russet or injury to primary leaves on most of the cultivars, and use of this material, particularly in summer months and near harvest, may irritate the skin and eyes of the person handling it. Wettable sulfur fungicide can be mixed with mancozeb and used at green-tip and pink-bud stages of most of the cultivars, except Jonathan, for effective control of powdery mildew in addition to apple scab. However, in dry, hot summer it causes injury to leaves and fruits, and its use should be abandoned.

Orchard management practices.   In spite of an adequate and timely spray program, certain orchard management practices may create problems in satisfactory con-trol of the disease. Tree growth, including pruning and nutritional status, does affect the level of disease control, as a close correlation exists between the increase in the number of

TABLE 1-5: ACTIVITY SPECTRUM OF APPLE FUNGICIDES AND SAFE PREHARVEST INTERVAL

Fungicide	Control of					Safe interval of days from last application to harvest
	Scab	Powdery mildew	Black rot (Frog eye)	Sooty-blotch & Fly speck	Mite suppression[1]	
Benomyl	+++	++++	++	++++	++	0
Carbendazim	+++	++++	++	++++	+	0
Captafol	+++	0	−	++++	0	15
Captan	+++	0	+++	+++	0	NTL
Dithianon	+++	+	+++	+++	−	14
Dodine	++++	0	−	−	0	7
Mancozeb	+++	0	+++	+++	+++	15–25[2]
Sulfur	++	+++	++	++	0	7
Thiophanate-methyl	+++	++++	++	++++	+	0
Ziram	++	0	+	++	+	NTL
Zineb	+++	0	+++	+++	++	15

− = Unknown; 0 = None; + = Slight; ++ = Fair; +++ = Good; ++++ = Excellent; NTL = No time limitations, follow label recommendations

[1] Mite suppression only possible when used on a full season schedule.

[2] Depending on monsoon rains. In dry weather, interval will be 25 days.

Adapted in part from: Burr, T. J., 1980, Fruit diesase control recommendations for New York (The "Red Book").

leaves on terminal shoots and the increase in apple scab on those shoots. Young leaves are most susceptible to scab, and application of 1% urea in summer sprays further increases the susceptibility of plant tissues to scab pathogen and must be avoided if conditions favorable for scab exist and optimum application of nitrogen has been done through soil application in winter.[39] Unpruned trees or trees intermingling with each other usually have dense foliage, plenty of shade, and poor drying conditions, which interfere with proper coverage of the spray material and lead to inadequate control of scab.

New group of fungicides. New fungicides are now available for use on apple, and the most significant ones are the ergosterol-biosynthesis inhibitors (EBI), which have a curative action against scab.[71] They work very well in the control of powdery mildew, as well. The fungicides which have been tested extensively in field trials and are now in use in most of the apple-growing countries (including the U.S.A., the U.K., the Netherlands, Poland, and South Africa[23]) are bitertanol (Baycor), etaconazole (Sonax or Vangard) of triazole, triforine (Funginex, Saprol, or Denarin) of piperazine, and fenarimol (Rubigan) of pyrimidine compounds.

These fungicides have good protective activities, are highly effective at very low concentrations, and can stop infection late in the incubation period but prior to symptom appearance. Some of them may inactivate lesions if they should appear.[71-74,79,80] With the aid of predictive devices for the determination of scab infection periods, the outstanding postinfection control activity of these fungicides can be exploited fully. As the cost of these compounds comes down to the level of protective fungicides, their use is bound to increase, making monitored control programs economical and risk free.

1-9-4 Resistance to Scab Fungicides

Resistance in apple scab pathogen is known to occur with benzimidazole fungicides and dodine when one or the other closely related fungicides are used exclusively year after year, resulting in high and continuous selection pressure. Such resistance, which has occurred primarily in apple orchards where benzimidazoles (benomyl, carbendazim, and thiophanate-methyl) have been used intensively for three to four years, is now common in the U.S.A., Canada, Australia, and many other countries. There are instances of much rapid buildup of resistance in apple orchards of certain areas in the U.S.A., where benomyl has been used only in three or four prebloom sprays. Other fungicides unrelated to benomyl have been used for the remainder of the season, and despite that, benomyl-resistant strains have developed in about one third of the orchards within three to four years. In orchards sprayed with benomyl or other benzimidazole fungicides during the overwintering period, development of resistance may occur very rapidly from one growing season to the next by elimination of sensitive strains within the overwintering fungus population. Hence, every effort must be made to prolong the useful life of these chemicals, and their use must be judiciously restricted as to dose, frequency of application, and treated area. The use of the benzimidazoles should be restricted to critical periods in the development of the epidemic. These should be used in mixture with protectant fungicide or in alternating schedules with other fungicides.[81]

Dodine is another fungicide to which resistance in apple scab fungus is known to occur by its exclusive use in an apple scab control program for about 10 years or when it has been applied about 80 times in orchards before resistance is demonstrated. Until now, dodine resistance has not been reported in areas where the compound has had only limited use or has been used in mixed programs with other fungicides. Such is the case in the Hudson Valley of New York State, where dodine used for the early three or four sprayings and followed by other fungicides in combination with dithiocarbamates and captan in later sprayings for more than 20 years has not resulted in resistance to dodine in the scab fungus. This only supports the belief that the resistance has been avoided by alternating and mixing fungicides. If resistance is to develop to a point where apple scab no longer can be controlled satisfactorily with this fungicide, the fungus must be exposed to dodine as a principle fungicide for a considerable time.

Several sterol-inhibiting fungicides, now available for use on apple, have limited site-inhibition characteristics. They are inhibitors of C-14 demethylation in sterol biosynthesis.[82] Thus, potential for resistance in these fungicides exists. The detection of a less sensitive strain of *V. inaequalis* in an experimental orchard is a warning of potential problems in practice due to the intensive use of the C-14 demethylation group of sterol-inhibiting fungicides.[83] Thus, steps taken early in the use of new fungicides to avoid resistance would be appropriate, and as experience shows, the strategies (see Table 1-6)[84] that reduce selective pressure on the pathogen by the fungicide offer a hopeful approach.

1-9-5 Resistant Varieties

Recently, 30 resistant varieties of apple have been released (see Table 1-7). Most of the varieties have monogenic resistance derived from *Malus floribunda* 821 (Vf gene), while Nova Easygro has one or more scab-resistant genes (Vr) from the Russian seedling R 12740-7A. Freedom is one variety which has the genes for resistance to scab from both parents, the male parent carrying the "*Malus floribunda*" type of resistance and the female carrying the "Antonovka" type of resistance.[85] These varieties are highly resistant

TABLE 1-6: SCHEMATIC PRESENTATION OF STRATEGIES TO AVOID OR DELAY RESISTANCE

S = Specific-site agent with high risk for resistance
M = Multisite agent with little chance of resistance problem

	EXCLUSIVE	
S − − −S − − −S − − −S		Greatest chance for selection
	MIXTURES	
S+M − −S+M − − −S+M − − −S+M		Reduced selection if started soon M should be strong
S+S′ − −S+S′ − −S+S′ − −S+S′		S and S′ have different modes of action
		Start when no resistance to S or S′
	ROTATIONS	
M − − −S − − −M − − −S		S for critical use in schedule
M S M S		More effective if S has short residual
M S+M M M		Most effective in high-risk situations

(From Delp).

TABLE 1-7: SCAB-RESISTANT APPLE VARIETIES

Variety	Genome	Year of introduction	Country of origin
Prima	Vf	1970	U.S.A.
Generos	Polygenic	1972	Rumania
Priscilla	Vf	1972	U.S.A.
Macfree	Vf	1974	Canada
Priam	Vf	1974	France
Nova Easygro	Vr	1975	Canada
Sir Prize	Vf	1975	U.S.A.
Florine-Querina	Vf	1977	France
Gavin	Vf	1977	England
Liberty	Vf	1978	U.S.A.
Moira	Vf	1978	Canada
Novamac	Vf	1978	Canada
Jonafree	Vf	1979	U.S.A.
Trent	Vf	1979	Canada
Britegold	Vf	1980	Canada
Murray	Vm	1980	Canada
Mac Shay	Vf	1981	U.S.A.
Red Free	Vf	1981	U.S.A.
Freedom	Vf+Polygenic	1983	U.S.A.
Pionier	Vf	1983	Rumania
Rouville	Vm	1983	Canada
Richelieu	Vf	1983	Canada
Romus 1	Vf	1984	Rumania
Romus 2	Vf	1984	Rumania
Voinea	Vf	1985	Rumania
Judeline (R)	Vf for juice	1986	France
Judaine (R)	Vf for juice	1986	France
Dayton	Vf	1988	U.S.A.
Primicia	Vf	1988	Brazil
William's Pride	Vf	1988	U.S.A.

to scab but differ in their reaction to other diseases (see Table 1-8). Freedom is resistant to four canker diseases as well: European canker, anthracnose, collar rot, and papery bark or silver leaf. Four apple hybrids (Ambstarking, Ambrich, Ambroyal, Ambred) evolved at Mashobra, India[86,87] have high tolerance (field resistance) to scab. In comparison to susceptible Red Delicious, these exhibit few lesions, each with a scanty spore load only under severe epiphytotic conditions.[8] Emra (name given to Co-op-12) and Red Free are promising on the low- and mid-altitude hills of Himachal Pradesh, where cultivars of the Delicious group do not develop a good skin color.[88,89]

So far, none of the resistant varieties has found much favor in commercial plantations in the U.S.A. and other countries, mainly because of uncertainty about their market value, difficulty in replacing the existing plantation, and resistance confined only to scab (except Liberty and Redfree, which are also resistant to powdery mildew, cedar apple rust, and fire blight). Moreover, the fungicides are cheap and can be frequently sprayed with tractor-driven sprayers without much difficulty, covering a large area (100–180 acres) of the orchard within a day or two. But in India, adoption of a spraying program is

TABLE 1-8: RESISTANCE OF SOME OF THE SCAB-RESISTANT APPLE VARIETIES
TO OTHER DISEASES

		Reaction rating*			
Variety	Scab	Cedar apple rust	Fire blight	Powdery mildew	Susceptible to
Prima	1	4	2	2	Bitter pit
Priscilla	1	1	2	3	Quince rust
Macfree	1	3	2	—	
Nova Easygro	1	1	2	3	
Sir Prize	1	3	4	2	
Liberty	1	1	2	2	
Jonafree	1	—	3	3	
Redfree	1	1	2	2	
Freedom	1	2	2	3	

*1 = Field immune; 2 = Resistant; 3 = Susceptible; 4 = Very susceptible

difficult not only because of orchards having been raised on slopes and the rough terrain of the mountains, but mainly due to scarcity of water, high cost of fungicides, labor problems, and nonavailability of suitable spraying equipment. As such, the resistant varieties, if evolved to the liking of the populace, will have more acceptability, and these will eliminate or reduce the dependence on chemical sprays to manage apple scab.[10]

1-10 REFERENCES

1. Morris, H. E., A contribution to our knowledge of apple scab, *Mont. Agr. Exp. Sta. Bull.*, 96, 1914.

2. Atkinson, J. D., *Diseases of fruits in New Zealand*, A. R. Shearer Govt. Printer, Wellington, New Zealand, 1971.

3. Singh, R., *Fruits*, National Book Trust, New Delhi, India, 1969.

4. Nath, P., Studies in the disease of apples in Northern India. II. A short note on apple scab due to *Fusicladium dendriticum* Fcl., *J. Indian Bot. Soc.*, 14, 121, 1935.

5. Joshi, N. C., Malik, A. G., Kaul, M. L. and Anand, S. K., Some observations on the epidemic of scab disease of apple in Jammu and Kashmir, during 1973, *Indian Phytopath.*, 28, 288, 1975.

6. Gupta, G. K., Present status of apple scab (*Venturia inaequalis*) in Himachal Pradesh and strategy for its control, *Pesticides*, 12, 13, 1978.

7. Gupta, G. K., Apple scab-economics and technology, *Pesticides*, 33, 1989.

8. Gupta, G. K., Diseases of pome-fruit orchards in India and research objectives. In *Integrated control of pome fruit diseases* (C. Gessler, D. Y. Butt and B. Koller), IOBC/WPRS Bulletin 1989/XII 16, pp. 272–285, Phyto-medicine/Pathology, ETH, Zurich, Switzerland, 1989.

9. Gupta, G. K. and Lele, V. C., The scab fungus (*Venturia inaequalis* (Cke) Wint.) on apple twigs in Kashmir, *Curr. Sci.*, 45, 565, 1976.

10. Gupta, G. K., "Apple Scab (*Venturia inaequalis*)," E. Merck (India) Ltd., Bombay, 36, 1985.

11. Gupta, G. K. and Lele, V. C., Prevalence, distribution and intensity of apple scab in Kashmir Valley, *Indian J. Agric. Sci.*, 50, 45, 1980.

12. Gupta, G. K. and Verma, K. D., Studies on the keeping quality of apple fruits affected with scab (*Venturia inaequalis*) in later summer season. *Proc. Natl. Symp. Temperate Fruits, March 15–18, 1984, HPKVV, Solan*: 74 (Abstr.), 1984.

13. Hughes, S. J., Some foliicolous hyphomycetes, *Can. J. Bot.*, 31, 560, 1953.

14. Aderhold, R., Revisions der species *Venturia chlorospora inaequalis* etc., *Hedwigia*, 36, 67, 1897.

15. Clinton, G. P., Apple Scab. *III, Agr. Expt. Sta. Bull.* 67, 109, 1901.

16. Keitt, G. W. and Jones, L. K., Studies on the epidemiology and control of apple scab. *Wis. Agr. Exp. Sta. Res. Bull.*, 73, 104, 1926.

17. Gupta, G. K. and Lele, V. C., Morphology, physiology and epidemiology of the apple scab fungus, *Venturia inaequalis* (Cke) Wint. in Kashmir valley. *Indian J. Agric. Sci.*, 50, 51, 1980.

18. Gupta, G. K., Scab of *Cotoneaster bacillaris* in India caused by *Venturia inaequalis* (*Spilocaea* stage), *Plant Dis. Reptr*, 63, 156, 1979.

19. Sharma, K. K. and Gupta, G. K., Comparative morphology, physiology and biology of cotoneaster scab fungus and apple scab fungus (*Venturia inaequalis*), *Proc. IIIrd Int. Symp. Pl. Path. IARI, New Delhi*, 36–37, 1981.

20. Qasba, G. N., Dar, G. N. and Shah, A. M., Scab on *Cotoneaster aitchinsonii* in India caused by *Venturia inaequalis* (*Spilocaea pomi*), *Indian Phytopath.*, 35, 698, 1982.

21. Chohan, J. S., and Kaul, S., Occurrence of Indian crab-apple scab in the plains of the Punjab. *Indian Phytopath.*, 29, 440, 1930.

22. Menon, R., Studies on *Venturiaceae* on Rosaceous plants. *Phytopath. Z.*, 27, 177, 1956.

23. Gupta, G. K., Recent trends in forecasting and control of apple scab (*Venturia inaequalis*), *Pesticides*, 19, 19, 1985.

24. Singh, J. R. and Williams, E. W., Identification of three physiological races of *V. inaequalis*, *Phytopathology*, 46, 190, 1950.

25. Gupta, G. K., Effects of weather on the perennation and prediction of apple scab (*Venturia inaequalis* (Cke) Wint). In *Perspectives in mycology and plant pathology* (V. P. Agnihotri, A. K. Sarbhoy and D. Kumar), Malhotra Publishing House, New Delhi, 353–369, 1988.

26. Gadoury, D. M. and MacHardy, W. E., Effects of temperature on the development of pseudothecia of *Venturia inaequalis*, *Plant Dis.*, 66, 464, 1982.

27. James, J. R. and Sutton, T. B., Environmental factors influencing pseudothecial development and ascospore maturation of *Venturia inaequalis*, *Phytopathology*, 72, 1703, 1982.

28. Gilpatrick, J. D., Single application treatments for apple scab control. In *Proceedings apple and pear scab workshop* (A. L. Jones and J. D. Gilpatrick, eds.), *Spec. Rep. No. 28*, pp. 30–32, N.Y. Agric Exp. Stn. Geneva, USA, 1978.

29. Gupta, G.K., Maturation and discharge of ascospores of *Venturia inaequalis*, *Indian Phytopath.*, 34, 502, 1981.

30. Gupta, G. K., Studies on apple scab. Maturation and discharge of ascospores. *A rep. Reg. Fruit Res. Stn. Mashobra, Shimla, 1981–82*, 42, 1982.

31. Hirst, J. M. and Stedman, O. J., The epidemiology of apple scab (*Venturia inaequalis* (Cke.) Wint). II. Observations on the liberation of ascospores. *Ann. Appl. Biol.*, 50, 525, 1962.

32. Brook, P. J., Effects of light, temperature, and moisture on the release of ascospores by *Venturia inaequalis* (Cke.) Wint., *N.Z. J. Agric. Res.*, 12, 214, 1969.

33. MacHardy, W. E. and Jeger, M. J., Integrating control measures for the management of primary apple scab, *Venturia inaequalis* (Cke.) *Wint., Prot. Ecol.*, 5, 103, 1983.

34. Burchill, R. T., Air-dispersal of fungal spores with particular reference to apple scab (*Venturia inaequalis* (Cooke) Winter), *Colston Papers,* 18, 135, 1966.

35. Gupta, G. K., Epidemiology, forecasting and possible control of apple scab (*Venturia inaequalis* (Cke.) Wint.), *Pesticides*, 9, 31, 1975.

36. Jeger, M. J. and Butt, D. J., The effects of weather during perennation on epidemics of apple mildew and scab, *EPPO Bull.*, 13, 79, 1983.

37. Hirst, J. M. and Stedman, O. J., The epidemiology of apple scab (*Venturia inaequalis* (Cke.) Wint.). I. Frequency of airborne spores in orchards, *Ann. Appl. Biol.*, 49, 290, 1961.

38. Sutton, T. B., Jones, A. L. and Nelson, L. A., Factors affecting dispersal of conidia of the apple scab fungus, *Phytopathology*, 66, 1313, 1976.

39. Kumar, J. and Gupta, G. K., Influence of host response and climatic factors on the development of conidial stage of apple scab fungus (*Venturia inaequalis* (Cke.) Wint.), *Indian J. Mycol. Pl. Path.*, 15, 1985.

40. Gupta, G. K., Verma, K. D. and Pal, J., Some observations on the prevalence and severity of apple scab in Himachal Pradesh during the years 1977 to 1983, *Indian J. Mycol. Pl. Path., 14*:XIII–XIV (Abstr.), 1984.

41. Studt, H. G. and Weitzien, H. C., Der Enfluess der Umweltfatoren temperatur, relative Luftfeuchtigkeit und Licht auf die Konidienbildung bien Apfelschorf, *Venturia inaequalis* (Cooke) Winter, *Phytopath. Z.*, 84, 115, 1975.

42. Anderson, H. W., *Diseases of fruit crops,* McGraw-Hill, New York, 1956.

43. Mills, W. D. and La Plante, A. A., Control of diseases and insects in the orchard, *Bull. N.Y. Agric. Exp. Stn.,* 711, 1951.

44. Mills, W. D., Efficient use of sulfur dusts and sprays during rain to control apple scab, *Bull. N.Y. Agric. Exp. Stn.* 630, 1944.

45. Roosje, G. S., Research on apple and pear scab in the Netherlands from 1938 until 1961, *Neth. J. Pl. Path.*, 69, 132, 1963.

46. MacHardy, W. E., "Research needs for improving apple scab warning systems," 10 B C Bulletin, 18, 1989.

47. Preece, T. F. and Smith, L. P., Apple scab infection weather in England and Wales, 1956–1960, *Pl. Path.* 10, 43, 1961.

48. Sutton, T. B. and Jones, A. L., Evaluation of four spore traps for monitoring discharge of ascospores of *Venturia inaequalis, Phytopathology,* 66, 453, 1976.

49. Gupta, G. K. and Verma, K. D., Studies on apple scab. Measurement of ascospore dose in the orchard air, *A. Rep. Reg. Fruit Res. Stn. Mashobra, Shimla, 1983–84,* 37, 1984.

50. Gilpatrick, J. D. and Szkolnik, M., Maturation and discharge of ascospores of the apple scab fungus. In *Proceedings apple and pear scab workshop* (A. L. Jones and J. D. Gilpatrick, eds.), *Spec. Rep. No. 28*, pp. 1–6, N.Y. Agric. Exp. Stn., Geneva, USA, 1978.

51. Smith, C. A. and Gilpatrick, J. D., Geneva leaf-wetness detector, *Plant Dis.*, 64, 286, 1980.

52. Zuck, M. G. and MacHardy, W. E., Recent experiences in timing sprays for the control of apple scab: equipment and test result, *Plant Dis.*, 65, 995, 1981.

53. Massie, L. B. and Szkolnik, M., Production of ascospore maturity of *V. inaequalis* utilizing cumulative degree days, *Proc. Am. Phytopath. Soc.*, 1, 40, 1974.

54. Sutton, T. B., James, J. R. and Nardacci, J. F., Evaluation of a New York ascospore maturity model for *Venturia inaequalis* in North Carolina, *Phytopathology*, 71, 1030, 1981.

55. Gadoury, D. M. and MacHardy, W. E., Preparation and interpretation of squash mounts of pseudothecia of *Venturia inaequalis, Phytopathology,* 72, 92, 1982.

56. Gadoury, D. M. and Machardy, W. E., A model to estimate maturity of ascospores of *Venturia inaequalis, Phytopathology*, 72, 901, 1982.

57. Richter, Von. J. and Haussermann, R., Ein Elektronische Schorfwarngerat. *Umweitschutz*, 48, 107, 1975.

58. Jones, A. L., Lillevik, S. L., Fisher, P. D. and Stebbins, T. C., A micro computer-based instrument to predict primary apple scab infection periods, *Plant Dis.*, 64, 69, 1980.

59. Jones, A. L. Fisher, P. D., Seem, R. C., Kroon, J. C. and Van De Motter, P. J., Development and commercialization of an in-field microcomputer delivery system for weather-driven predictive models, *Plant Dis.*, 68, 458, 1984.

60. Post, J. J., Allison, C. C., Burchhardt, H. and Preece, T. F., The influence of weather conditions on the occurrence of apple scab, *Tech. Notes World Meterol. Organ. Geneva No. 55*, 41, 1963.

61. Jones, A. L., Disease prediction: Current status and future directions. In *Challenging problems in plant health* (T. Kommedahl and P. H. Williams, eds.) pp. 362–367, The American Phytopathological Soc., St. Paul, MN, 1983.

62. Williams, E. B., Current status of apple scab resistant varieties. In *Proceedings apple and pear scab workshop* (A. L. Jones and J. D. Gilpatrick, eds.), *Spec. Rep. No. 28,* pp. 18–19, N.Y. Agric. Exp. Stn., Geneva, USA, 1978.

63. Heye, C. C. and Andrews, J. H., Antagonism of *Athelia bombacina* and *Chaetomium globosum* to the apple scab fungus, *Venturia inaequalis, Phytopathology*, 73, 650, 1983.

64. Burchill, R. T., Hutton, K. E., Crosse, J. E. and Garrett, C. M. E., Inhibition of the perfect stage of *Venturia inaequalis* (Cke.) Wint. by urea, *Nature*, 205, 520, 1965.

65. Gupta, G. K., Role of on-season, post harvest and pre-leaf fall sprays in the control of apple scab (*Veturia inaequalis*), *Indian J. Mycol. Pl. Path*, 9, 141, 1979.

66. Crosse, J. E., Garrett, C. M. E. and Burchill, R. T., Changes in the microbial population of apple leaves associated with the inhibition of the perfect stage of *Venturia inaequalis* after urea treatment, *Ann. appl. Biol.*, 61, 203, 1968.

67. Oland, K., Response of cropping apple trees to post harvest urea sprays, *Nature*, 198, 1281, 1963.

68. Gupta, G. K., Strategy, efficacy and profitability of chemical spray programme in the control of apple scab (*Venturia inaequalis*), *Indian Phytopath.*, 37, 385, 1984.

69. Gupta, G. K., Investigations on the effect of urea and fungicides in suppressing the ascigerous stage of apple scab pathogen, *Int. J. Trop. Pl. Diseases,* 5, 93, 1987.

70. Biggs, A. R., Apple scab. In "Compendium of apple and pear diseases," The American Phytopathological Soc., St. Paul, MN, 1990.

71. Szkolnik, M., Physical modes of action of sterol-inhibiting fungicides against apple diseases, *Plant Dis.*, 65, 981, 1981.

72. Kelley, R. D. and Hones, A. L., Evaluation of two triazole fungicides for postinfection control of apple scab, *Phytopathology* 71, 737, 1981.

73. Rosenberger, D. A. and Meyer, F. W., Evaluation of Baycor and Vangard on reduced spray schedules, 1981, *Fungicide and Nematicide Tests*, 37, 13, 1982.

74. Gupta, G. K. and Kumar, J., Studies on the curative (after infection) and eradicant (post-symptom) activity of fungicides against apple leaf infection by *Venturia inaequalis, Pesticides*, 19, 18, 1985.

75. Gupta, G. K. and Verma, K. D., Post-symptom antisporulant activity of fungicides on apple scab foliage lesions, *Indian J. Agric. Sci.*, 55, 381, 1985.

76. Marz, H. R., Difolatan—a broad spectrum and persistent fungicide, *Proc. IInd. Int. Symp., IPS., IARI,* New Delhi, 16–17, 1971.

77. Gupta, G. K., Behaviour of fungicides and various spray schedules in the control of apple scab (*Venturia inaequalis*), *Int. J. Trop. Pl. Diseases,* 1, 181, 1983.

78. Gupta, G. K., Economics of apple scab control programme in Himachal Pradesh, *Pesticides*, 18, 31, 1984.

79. Brandes, W. and Paul, V., Studies on the effect of Baycor on apple scab pathogenesis, *Pflanzenschhutz Nachrichten*, 34, 48, 1981.

80. Schwabe, W. F. S., Jones, A. L. and Jonker, J. P., Greenhouse evaluation of the curative and protective action of sterol-inhibiting fungicides against apple scab, *Phytopathology*, 74, 249, 1984.

81. Gilpatrick, J. D., Case study 2: *Venturia* of pome fruits and *Monilinia* of stone fruits. In *Fungicide resistance in crop protection* (J. Dekker and S. G. Georgopoulos, eds.), pp. 195–206, Centre for Agricultural Publishing and Documentation, Wageningen, 1982.

82. Kellor, W. and Scheinpflug, Fungal resistance to sterol biosynthesis inhibitors: A new challenge, *Plant Disease*, 71, 1066, 1987.

83. Stanis, V. F. and Jones, A. L., Reduced sensitivity to sterol-inhibiting fungicides in field isolates of *Venturia inaequalis, Phytopathology*, 75, 1098, 1985.

84. Delp, C. J., Strategies to avoid or delay resistance. Lecture given at International Course of Fungicide Resistance in Crop Plants for South-East Asia, Univ. Pertanian, Malaysia, Serdang at 17–24, 1984.

85. Lamb, R. C., Aldwinckle, H. S. and Terry, D. E., 'Freedom,' a new disease-resistant apple, *New York's Food & Life Sciences Bull.*, 102, 1983.

86. Chadha, T. R. and Sharma, Y. D., Breeding of apple varieties in Himachal Pradesh, *Indian J. Hort.*, 35, 178, 1975.

87. Chadha, T. R. and Sharma, Y. D., New cultivars of apple. In "Some research contributions on apple" (H. S. Verma, R. L. Sharma and G. K. Gupta), Farmers communication No. 101, pp. 6–7, Reg. Fruit Res. Stn., Mashobra, Simla, 1980.

88. Sharma, R. L., Kumar, J. and Ram, V., Performance of some scab resistant cultivars of apple in Kullu valley of Himachal Pradesh, *J. Tree Sci.*, 7, 45, 1988.

89. Gupta, G. K., Major temperate fruit diseases in India and their management strategies, *Pl. Dis. Res.*, 5 (*Special*), 136, 1990.

2

FIRE BLIGHT
OF APPLE AND PEAR

Sherman V. Thomson

Department of Biology
Utah State University
Logan, Utah

2-1 HISTORICAL BACKGROUND

Erwinia amylovora (Burr.) Winslow et al., the cause of fire blight, apparently evolved on indigenous American plants such as hawthorn (Crategus), mountain ash (Sorbus), service berry (Amelanchier), and native crab apples (Malus). Damage to these native plants was probably minimal and usually only caused the loss of a few flower clusters or in some years loss of some of the branches. The bacteria and hosts had evolved to a stage where they coexisted and, indeed, the disease may have been beneficial to the pomaceous hosts by accomplishing some random pruning and thinning.

When the colonists arrived in North America in the early 1600s, they brought with them seeds and plants of pear, apple, and quince from England and France.[1] These introduced hosts apparently grew without problems for about 100 years. At least there was no mention of a problem until William Denning wrote a letter in 1793 describing a disorder of apples, pears, and quince in the Hudson River Valley that appeared to be fire blight.[2] Apparently, fire blight was still not a major problem because 25 years passed before a report by Coxe was presented in his book "Cultivation of Fruit Trees" in 1817.[3] From this time forward, the reports of fire blight increased sharply and the devastation caused by the disease was mentioned frequently.

The reason for a lapse of about 100 years before fire blight became a problem in the eastern United States is a bit of a puzzle, but one must consider how fruit was grown by the early colonists. Trees were planted in meadows and pastures and rarely fertilized, pruned, or cultivated. Fruit was grown mostly for cider or perry, and the appearance or size of the fruit was inconsequential. When grown under these conditions, even the most susceptible varieties were often not seriously damaged. Fire blight becomes a problem when trees are pruned, fertilized, and stimulated to grow rapidly.

The localization of the fire blight bacterium in a limited geographical area in the eastern United States is also unique. Fire blight was not reported on the west coast of the United States until 1887, almost 90 years after the reports on the eastern United States, despite the presence of pears in California for at least 100 years. George Vancouver, the English navigator, found apples and pears growing at Spanish missions in California where he visited in 1792.[1] These fruits arrived in California from Europe by way of Franciscan fathers from Mexico. The absence of *E. amylovora* in the indigenous California flora is unusual but is thought to be due to the geographical and botanical barriers of the Great Plains and the Rocky Mountains.[4]

Pear and apple plantings became more intensively cultivated as communities grew and the demand for fruit exceeded that which was produced in "wild" plantings. Orchards were planted and trees were grown in ways to make them more susceptible to fire blight. By about 1900, fire blight was so severe in the eastern United States that the planting of the European pear was discontinued as a commercial orchard tree. Apples are generally more resistant to fire blight, and therefore the fruit industry in the northeastern United States expanded with extensive apple plantings. Even today, with the knowledge we have about control of fire blight, there is only small-scale production of lower-quality, blight-resistant pears in the eastern United States.[5]

Before the discovery that bacteria cause disease, fire blight was thought to be the result of insects (1794–1847), lightning and electricity (1794–1879), cold winters causing frozen sap (1844–1845), various fungi (1863–1875), and rays of the sun that were concentrated by vapors in the atmosphere.[4] These conditions surely existed in other parts of the world where pears and apples were grown, but the local colonists must have thought that the environment in North America was radically different than Europe.

Fire blight was the first disease of plants demonstrated to be caused by bacteria. Thomas J. Burrill and J. C. Arthur demonstrated independently in the late 1880s that bacteria infect plants, similar to the discovery made by Robert Koch in 1876 that a bacterium was responsible for anthrax in sheep.[1]

Fire blight has been instrumental in training plant pathologists and establishing new concepts in plant pathology. There have been over 2400 papers published on the disease, with 20 Ph.D. dissertations and master's theses.[4] *Erwinia amylovora* was the first bacterium demonstrated to cause a disease in plants and was also the first disease where insects were shown to be vectors.[1]

2-2 DISTRIBUTION

2-2-1 Confirmed Reports

Fire blight continues to spread throughout the world and is considered a major concern where pome fruits are grown. Almost each year a new country or area reports the first occurrence of fire blight. It is now found in 19 countries on four continents; all of North

America, much of Central America, Europe, the Near East, and New Zealand (Fig. 2-1).

It took fire blight 135 years to move from the eastern United States to the west coast of North America, generally following the settlement of the country. The first valid report of fire blight outside of North America was from New Zealand in 1919.[6,7] Although the disease was found on several hosts, it was localized in a small area on the North Island near Auckland. It took only 10 years for the disease to make its way to most other fruit-growing areas in New Zealand, including the South Island. It is speculated that the organism was introduced to New Zealand on planting stock.[4]

Australia was quick to establish a quarantine in 1924 to prevent the importation of susceptible hosts or fruit from any country where fire blight was reported. Even with the close proximity and extensive trade between New Zealand and Australia, fire blight has not yet been reported in Australia.

A major jump in the distribution of fire blight occurred when the disease was found in England in 1958.[8] This exposed the entire European continent, which has ideal fire blight weather and a uniform distribution of highly susceptible hosts. Fire blight was subsequently reported in the following countries: Poland and the Netherlands, 1966; Denmark, 1968; West Germany, 1971, France and Belgium, 1972; Sweden, Norway, and Ireland, 1986. Movement into the Middle East began in 1964 in Egypt, followed by Cyprus, 1984; Israel, 1985; and Greece, 1986.[4,9,10]

The means of long-range dissemination of fire blight to England and its spread throughout Europe remains obscure. It is most likely that fire blight was introduced into England by contaminated fruit boxes from North America or New Zealand, since it is known that fruit boxes from these countries were used in the Kent orchards prior to the initial fire blight outbreaks in 1956–1957.[11] Movement by infected propagating wood is a good possibility and has been proposed as the source in several outbreaks.[12–14]

Migratory starlings or other birds have been strongly implicated, without experi-

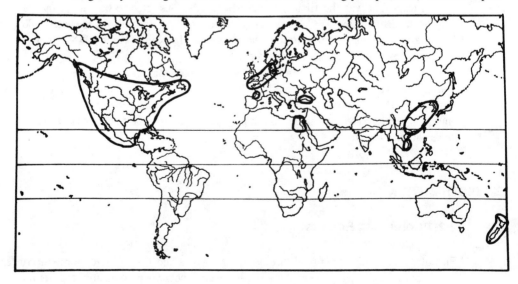

Figure 2-1 Geographical distribution of fire blight of apple and pear.

mental proof, in the long-distance spread of fire blight across the English channel to the coastal regions of northwestern Europe. Bech-Anderson[15] provided strong circumstantial evidence by showing that viable *E. amylovora* were present in starling excrement and on their feet eight days after the birds were artificially infested with the organism. Starlings normally complete their migration from southern England to continental Europe in two to three days and utilize hawthorns as a primary shelter and feeding site in the migration.

2-2-2 Unconfirmed Reports

There are numerous reports of fire blight from other countries, but often the outbreaks have not been confirmed or the disease was erroneously diagnosed. Most often, fire blight is confused with blossom blast caused by *Pseudomonas syringae* pv. *syringae*, another common and widely distributed disease of pears and apples. This is apparently the case in Chile,[16] Japan, South Africa, U.S.S.R., Italy, and possibly Turkey.[4]

It appears that fire blight will eventually move into all European countries. There are large plantings of highly susceptible hosts, and the environment appears to be adequate for the disease to develop. Strict quarantines may keep the pathogen restricted for a limited time, but even quarantines will not last forever. For example, Norway and Sweden enacted strict quarantines in the 1960s, but the disease still made its way into the countries. Eradication programs have been underway in these newly infested countries, but only time will tell how successful they are. There have been no cases where fire blight has been eradicated from a country.

Areas not contiguous to current fire blight outbreaks may remain free of the disease for longer periods. For example, South Africa, Australia, and Japan are more likely to exclude fire blight because of the relative isolation from other land masses where the disease is present.

2-3 ECONOMIC IMPORTANCE

The damage caused by fire blight is very erratic, and in many cases an actual loss cannot be assessed. The current season's production can be eliminated with the blighting of virtually every blossom as well as the death of spurs and subsequent loss of future production. Tree structure can be permanently affected due to the blighting of central leaders or major scaffold branches. Twig blighting eliminates potential fruiting spurs, and often the infection advances into major scaffolds or the trunk. Flowers that develop on the trunk or major scaffolds are serious infection sites because the disease frequently progresses into the structural branches of the tree. Fruit infections occur anytime up to harvest, although the fruit usually becomes more resistant as it matures.

Fire blight in the nursery can be particularly devastating because of the proximity and succulence of the trees. Infected trees must be destroyed, and adjacent trees in the block should not be sold until they are known to be free of blight.

Fire blight occasionally kills mature trees, especially when the infection enters the lower trunk, collar, or roots.

The mere report of fire blight in a country can be an extremely important economic factor. Many countries will not accept fruit or nursery stock from countries where fire blight is known to occur. Therefore, it is becoming increasingly important for fire-blight–free countries to retain their status because the number of blight-free countries is dwindling each year.

Fire blight was an important economic factor in the development of horticulture in the United States. Early production of pears was extensive in the East, but frequent fire blight outbreaks caused the elimination of orchards and the planting of orchards in new areas where fire blight was easier to control. Eventually almost 90% of the Bartlett pear production ended up on the West Coast, where weather conditions are less conducive for fire blight.[5]

Fire blight was instrumental in eliminating some early horticultural ventures where stockholders were left holding the core. In 1873, the Old Dominion Fruit Growing Company attracted investors to finance the planting of 20,000 pear trees. This company paid 20 and 60% on stock in 1880 and 1881, respectively. However, fire blight soon eliminated the pear orchards and at the same time eliminated the hopes of investors.[4]

California growers thought that it was too dry for fire blight to develop in the warm central California valleys, and extensive pear plantings were made in the 1890s. However, soon after their establishment, fire blight became as severe as any previous outbreaks in the Northeast. Between 1901 and 1904, 95% of the pear trees were killed by fire blight in the four counties (Fresno, Kern, Kings, Tulare) of the southern San Joaquin valley alone.[1] These areas were never replanted to pears, although in the last few years there have been some major plantings of apples and asian pears. A historical analysis of weather and fire blight outbreaks indicates that the southern and central valleys of California are high fire blight risk areas, and plantings of susceptible pome fruits should be avoided.[17]

Even in the east and midwest United States where growers relied on plantings of blight-resistant varieties, there have been some devastating epidemics of fire blight. Pickett[18] estimated a loss of $1,500,000 in the apple and pear crop in Illinois in 1914. Estimates for losses in the entire United States are difficult to determine, but in 1936 it was suggested that 14% of the pear production or $4 million was lost due to fire blight.[19] This did not take into account the losses to other hosts, such as apple or ornamentals.

The European experience with fire blight has been worthy of attention. England experienced a serious epidemic on pears and ornamentals in 1966 when nearly 12,000 infected pear and apple trees were identified.[20] Surveys by the Ministry of Agriculture, Fisheries and Food from 1958 to 1969 indicated losses of 20,000 pear trees, 20,000 hawthorn, 15,000 cotoneaster, and 2,000 pyracantha to fire blight.[21]

Some of the serious losses have not been solely due to fire blight infection but occurred as a result of the intentional destruction of hosts to preserve quality reputations. Meijneke[22] indicated that more than 2 million cotoneaster, 13,000 pyracantha, 8,700 stranvaesia, and 4,500 mountain ash were destroyed in 1975 due to a fire blight outbreak in Netherland nurseries. The loss to the nursery industries in Europe because of fire blight has undoubtedly amounted to millions of dollars.

Some losses are tangible, but there are many other costs of fire blight that are often overlooked. The effort of many researchers, agriculture quarantine inspectors, and others amounts to millions of dollars every year. There has probably been more money spent on fire blight research control, quarantine, and eradication than on any other pome fruit disease.

2-4 SYMPTOMS

The name *fire blight* was apparently coined by William Coxe in 1817.[3] His early depiction is still very descriptive; leaves suddenly turn brown, "as if they had passed through a hot flame and causing a morbid matter to exude from the pores of the bark." Probably the most obvious symptom of fire blight is the rapid wilt and browning or blackening of infected blossoms or leaves. Infections occur on flowers, twigs, fruit, limbs, and trunks. The host range includes susceptible plants in at least 174 species and about 40 genera, all in the family Rosaceae.[4]

The most susceptible hosts and those most important economically include species of *Pyrus*, *Malus*, *Cotoneaster*, *Crataegus*, *Cydonia*, *Pyracantha*, and *Sorbus*. Fire blight has also been occasionally reported on some *Prunus*, *Fragaria*, and *Rubus* species.[23,24] In some cases inoculations of plants, both in and out of the Rosaceae family, have been made with the development of symptoms. Interpretation of such studies should be made with caution because inoculations under optimal conditions with high levels of inoculum may yield spurious results either because of a hypersensitive response or false symptoms.

Flowers are usually the first part of the plant to become infected. Individual flowers or the entire cluster becomes water soaked and turns slightly darker green. Usually white to clear-colored droplets of ooze form on the fruit, flower, or the pedicel. This is probably the most diagnostic symptom, since other pome fruit pathogens do not produce ooze. The ooze may turn amber to black and dry, leaving a slightly shiny area. The infection progresses into the spur and causes wilting of leaves and possibly canker formation on the stem (Fig. 2-2). Infected tissues usually turn dark brown to black in pears and brown in apples. Leaves remain attached on infected twigs even into the winter, which aids in eradication when pruning.

Current year's growth and especially water sprouts are also attacked. Symptoms of water soaking, ooze, and wilting are similar to flower infections. The terminal portion of the infected shoot wilts or recurves, forming a shepherd's crook. The infection progresses very rapidly in succulent growth and may advance 10 to 30 cm in one day. The rapid advance of the infection and blackening of tissues account for the reports that the disease seemed to occur overnight.

Under optimal environmental conditions and in susceptible hosts, the blossom or shoot infections may progress into limbs or trunks, killing large portions of the tree. Tissues distal to the infections die soon after the branch is girdled. These tissues do not turn as dark, nor is ooze produced on them. In highly susceptible pear varieties, infections may progress into the trunk, resulting in death of the tree. The cankers formed in large branches are sunken and usually darker than the surrounding bark. Cracks often form along the margins of cankers as they age. Cankers without cracks on the periphery are termed indefinite and are more likely to have viable bacteria surviving on the margins.[25] The interior bark adjacent to cankers is often marked with reddish-brown streaks that extend several centimeters into the healthy wood.

Figure 2-2   Flower clusters and leaves of pear infected with fire blight quickly wilt and turn dark brown as if scorched by fire.

Fire blight can also infect the roots or base of the tree, resulting in rapid death of the entire tree.[26]

Fruit may become infected through the pedicel or directly through lenticels. Older fruit is generally more resistant, but even harvested fruit can become infected. An extremely sticky and stringy ooze is often produced, which may drip and produce long, thin strands. Upon drying, these brittle strands splinter into small fragments that may become windborne and serve as inoculum.[27] Moisture loss from infected fruit is fairly rapid, resulting in fruit mummies which hang on the tree through winter. Pear mummies are dark brown to black, whereas apple mummies turn reddish brown.

Symptoms on pyracantha, hawthorn, and cotoneaster are usually less severe, and often only the flower clusters and adjacent spurs are killed. Leaves turn reddish brown and remain attached for several months. Shoot tips are occasionally infected, often with the characteristic shepherd's crook. Ooze may form on infected tissues, but less frequently than on pear.

2-5 THE ORGANISM

Erwinia amylovora is a gram negative short rod 1.0 to 3.0 μm long by 0.5 to 1.2 μm in width. Cells are typically motile with peritrichous flagella. It grows best at 21 to 28°C

(70–83 °F) with a minimum growth temperature of 3 to 12 °C (38–54 °F) and maximum temperature for growth of 35 to 37 °C (95–99 °F). The generation time under optimal conditions is 70 to 90 minutes. Nicotinic acid is required for growth. The G + C content of the DNA is 53.6 to 54.1 moles per cent.[28]

Erwinia amylovora is a homogeneous bacterial species which can be easily and quickly identified with a few simple tests. Growth on selective and differential media is a fairly reliable preliminary screen. The complex Miller and Schroth selective medium (with sorbitol as the carbon source, MSS) eliminates a large number of common bacterial species but allows growth of *E. amylovora*.[29,30] Colonies of *E. amylovora* on the medium have smooth surfaces with uniform margins, dark orange centers, and translucent margins.

The differential high sucrose medium of Crosse and Goodman[31] is generally not sufficiently selective to eliminate contaminants, but *E. amylovora* colonies, when observed under oblique light, have a surface with characteristic craters. Ishimaru and Klos[32] modified and simplified the MSS medium to improve detection of *E. amylovora* in the presence of high populations of *E. herbicola*. Colonies of *E. amylovora* (72 hours old), when viewed from the underside of the petri dish, have blue striations radiating from the centers. Craters are also produced within 24 to 48 hours on this medium. Colonies of *E. amylovora* on nutrient agar or Kings medium B[33] are smooth, creamy white, and rounded with a glistening shine.

Hypersensitivity on tobacco is a most useful supplementary test to check tentative colonies from selective or differential media.[34] The hypersensitivity test is best performed with actively growing young tobacco (cv "Burley") plants. The plants should be grown in bright light and moved to the lab just before use or tested in the greenhouse. Inject approximately 10^9 colony-forming units (CFU)/ml tap water of freshly cultured bacteria into the intercellular space with a 25-gauge needle and syringe. In most cases, it is easier to remove the needle and place the syringe perpendicular against the leaf surface. Other cultivars of tobacco and even other species of plants will give positive hypersensitive reactions. *Erwinia amylovora* elicits a positive reaction within 8 to 12 hours, whereas saprophytes may only produce chlorosis after 48 hours or more.

Slide agglutination with specific antisera can be used to verify isolates quickly.[30] Pathogenicity tests with immature pear fruits or tender shoot tips of pear, apple, or other hosts provide definite results and quick confirmation when the plant material is available.

2-5-1 Serology

Serological studies of *E. amylovora* have revealed a great deal of homology. Elrod[35] could not detect any significant antigenic differences and concluded that *E. amylovora* was an exceedingly homogeneous species. Laroche and Verhoyen[36] also found serological homogeneity. Samson[37] compared 25 isolates of *E. amylovora* and found a common antigen in the isolates, but other surface antigens were also present, which allowed the separation into five serotypes. There are some antigens which are common with other Erwinia species, but *E. amylovora* is unique and can easily be distinguished serologically.

Immunofluorescence using polyclonal and monoclonal antibodies has been used to detect *E. amylovora* populations in infected fruit and leaf tissue and screen apparently healthy tissues for the presence of low populations of *E. amylovora*.[38-41]

2-5-2 Avirulence

Avirulent *E. amylovora* have been reported since the 1930s,[42-45] but whether all the isolates mentioned were indeed *E. amylovora* is uncertain. Some of the isolates may have actually been *E. herbicola* or other related species. There are isolates of *E. herbicola* which are also white rather than yellow and have serological cross reactions with *E. amylovora*.[46]

There appears to be a relationship between the extracellular polysaccharide (EPS) capsule of *E. amylovora* and virulence. Isolates without the EPS are often avirulent[47-49] and also have a rough colony rather than the smooth colonies of the virulent strains.[42,43,50]

Bennett[51] showed that EPS capsulation was one factor necessary for virulence, but not the only one since he found capsulated avirulent forms. When an avirulent capsulated isolate was coinoculated with an avirulent non-capsulated strain, a typical virulent response resulted. Billing[52] suggested that a cell leakage inducing agent and EPS were both necessary for virulence.

2-5-3 Toxins

Toxin(s) have been suggested as the physiological basis for symptom induction by *E. amylovora*. Pierstroff[53] demonstrated that a dilution of the bacterial ooze from infected, immature pear fruit caused wilting of apple shoots. Goodman et al.[54] reported the purification of amylovorin, a host-specific phytotoxic polysaccharide, from ooze. Beer et al.[55] subsequently showed that amylovorin lacks the host specificity requisite to explain the typical fire blight host reaction. It is clear, however, that *E. amylovora* produces an EPS that is involved in pathogenicity. Bennett and Billing[47] provided evidence that the EPS produced by the bacterium is one virulence factor necessary for symptom expression. There is also a correlation between the levels of EPS production and virulence when comparing virulent and avirulent strains of the pathogen.[50]

Buchanan and Starr[56] described the isolation and partial purification of a necrotoxin which produced necrosis of pear cell suspension cultures unlike the wilting produced by EPS. Bauer and Beer[57] provided evidence that the necrotoxin may have been an artifact of the preparation technique and that the real cause of necrosis of the cells was the inorganic salts in the tissue culture medium.

2-5-4 Molecular Genetics

The development of new techniques in molecular biology has made it possible to address the mechanisms responsible for pathogenicity in *E. amylovora*. The transfer of chromosomal genes determining virulence in *E. amylovora* was accomplished in 1973 by using conjugation between recipient, avirulent auxotrophs and prototrophic, virulent donor strains.[58,59] These studies are precursors to the mapping of gene order as well as analysis of pathogenicity. Current research involves the use of transposon mutagenesis with Tn5 to characterize and clone the genes responsible for virulence and the induction of hypersensitivity.[60-62] These studies may reveal the mechanisms of virulence of *E. amylovora* and provide clues for controlling the disease.

2-5-5 Bacteriophages

Bacteriophages specific to *E. amylovora* seem to be relatively common on diseased parts of pear and apple trees. Ritchie and Klos[63] found up to 10^6 plaque-forming units per terminal. Erskine[64] described a bacteriophage isolated from soil beneath diseased pear trees lysogenized a yellow saprophytic bacterium *E. herbicola* but was lytic towards *E. amylovora*. Thus, the yellow bacterium which is commonly associated with *E. amylovora* on fire blight hosts possibly serves as a reservoir of phage which may lyse *E. amylovora* and influence the development of epidemics.

The high specificity of bacteriophages toward *E. amylovora* make them useful tools for identification.[65] However, there are some cross-reactions, and therefore caution should be used when relying on phage typing alone for identification of *E. amylovora* strains.

2-5-6 *Erwinia amylovora* and *Erwinia herbicola*

A yellow mucoid bacterium is often associated with the fire blight organism in infected and healthy tissues and has been found in cultures of *E. amylovora* assumed to be pure.[66-70] The organism is thought to be *E. herbicola*, a closely related Erwinia species.[67,70]

This relationship seems to be more than casual in the epidemiology of *E. amylovora* and may be important in natural control of fire blight. Shaw[71] was one of the first to show that the yellow organism had an inhibitory effect on infections by *E. amylovora*. Several studies have shown that *E. herbicola* may be involved in natural biocontrol of fire blight.[67,70]

2-6 DISEASE CYCLE

The primary overwintering sites of the fire blight organism are the cankers originating from the previous seasons' infections (Fig. 2-3). There is also some speculation that resident organisms in healthy tissues are sources of new infections.

Holdover cankers are well documented as a source of overwintering inoculum.[25,53,72-75] These cankers often exude bacterial ooze in the spring as temperatures warm. This ooze is usually produced from indeterminant cankers without a definite margin[25] and is composed of viable bacteria in a matrix of polysaccharide.[76] It is attractive to many insects such as Pegomyia and Syrphid flies, which become contaminated with the bacteria during feeding. These same insects are also attracted to flowers as a source of pollen or nectar, and they inadvertently deposit the bacteria on flower surfaces.[29]

The bacterium enters tissues through natural openings such as nectaries, hydathodes, and lenticels[74,77,78] or through wounds caused by insects.[79,80] The nectarial surfaces of the hypanthium were more frequently infected in pear flowers; whereas in apple flowers the bacteria were found to penetrate through stigmas, anthers, and receptacle walls.[74] Meteorological events such as hail or driving rain create minute injuries, allowing entry of the pathogen.[81,82]

Miller and Schroth[29] demonstrated the presence of epiphytic bacteria on apparently healthy pear flowers, fruits, and leaves as well as on the surface of inactive cankers and

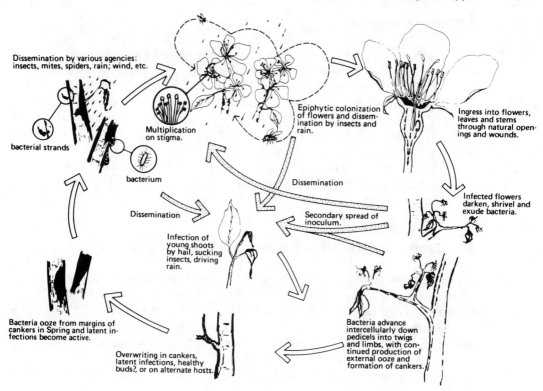

Figure 2-3   The fire blight disease cycle caused by *Erwinia amylovora*.

on insects collected from orchards. Thomson et al.[30] showed that the occurrence of fire blight outbreaks could be directly related to the presence of epiphytic populations of bacteria on pear flowers. In many cases, the populations exceeded 10^5 to 10^6 CFU per healthy flower as much as two weeks prior to evidence of disease.

Epiphytic populations of bacteria were isolated almost exclusively from the stigmatic surfaces of the pistils of flowers of pear, apple, and ornamental hosts.[83] Simulated rain in the field promoted movement of the epiphytic bacteria on pear flowers from the stigmatic surface (Fig. 2-4) to other flower parts resulting in infections. Inoculations with *E. amylovora* populations of $<10^3$ CFU placed directly on the hypanthium of pear flowers declined, especially when the relative humidity was less than 30%; whereas the same populations placed on the stigmas under dry conditions resulted in multiplication to 10^5 to 10^6 CFU per flower without causing infection. Rundle and Beer[84] also found that the multiplication of *E. amylovora* and *E. herbicola* occurred primarily on the stigma rather than on the stamens, calyx, or styles. In their studies, the flowers did become infected and the infection appeared to originate from the stigma and through the style, as indicated in previous microscopic studies made by Rosen in 1930.[77,85] Infections probably occurred more readily in their studies because the level of inoculum was higher at 10^7 to 10^8 CFU/ml and the relative humidity was kept near 75 to 80%.

Hattingh et al.[86] showed with scanning electron microscopy that *E. herbicola* and *E.*

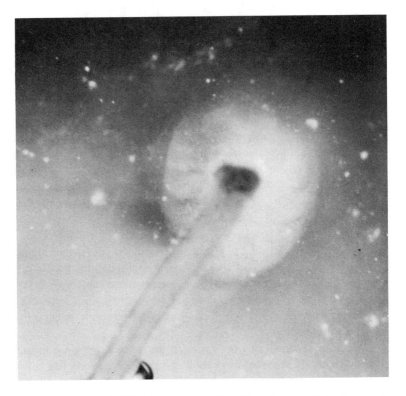

Figure 2-4   "Swarm" of *Erwinia amylovora* released from the stigmatic tip of pear pistil one minute after being placed in water.

amylovora multiplied predominantly on the stigmatic surfaces when apple blossoms were sprayed with suspensions of bacteria.

Lelliot[87] demonstrated that under cool conditions pear flowers were inhibitory to low populations of *E. amylovora* inoculum. Norelli and Beer[88] also found that infections were infrequent when the concentrations of the inoculum on pear and apple flowers were less than 10^3 to 10^4 CFU per flower, especially when the relative humidity was less than 60%.

Fire blight outbreaks almost always occur with a nearly simultaneous expression of symptoms on flowers or shoot tips.[29] Random distribution by insects would not result in a sudden outbreak if infection occurred as a result of the mere deposition of the bacteria on the flower. Rain dispersal would result in localized outbreaks near oozing holdover cankers or the distribution of infections in an inverted, cone-shaped area.[25]

These results explain why epidemics occur so rapidly following a rain storm. Insects or rain are responsible for inoculating flowers with low populations of bacteria, which at the time may not initiate an infection. The stigma provides the proper environment and nutrients for the populations to increase to levels of 10^5 to 10^7, and when rain or heavy dew occurs these high populations are transferred en masse to the hypanthium where infections occur almost simultaneously in the orchard.[83] It may also be possible for infections to occur without rain if the bacteria on the stigma reach a high enough population.

After entry into the flower, the pathogen moves intercellularly through the parenchyma of the flower pedicel[26,89] and into the spur with subsequent invasion of the leaves and possibly the branch. Highly susceptible varieties or succulent growth allow the pathogen to progress into large branches. Infected leaves turn black or brown rapidly, and cankers are formed in twigs and branches.

There are numerous reports where *E. amylovora* was isolated from healthy tissues. Baldwin and Goodman[66] isolated *E. amylovora* from apparently healthy apple buds in Missouri. A study in Ontario, Canada detected *E. amylovora* in healthy apple and pear buds using isolation on selective media.[90] The importance of these bacteria in fire blight outbreaks is uncertain since there are no studies showing that new infections originate from resident bacteria in buds or in other tissues.

Inoculum for secondary infection originates from droplets of ooze produced on infected flowers or terminals. This ooze, like that on overwintering cankers, is attractive to insects, which quickly spread the inoculum through the orchard. Rain or heavy dew randomly disseminates the bacteria to potential infection sites on flowers or terminals. Strong winds associated with thunderstorms and hail storms are ideal for initiating epidemic blight because these conditions generate the wounds necessary for infection as well as transfer the bacteria.[25,53,91,92]

The role of the honey bee in disseminating fire blight was a hotly contested area of research in the 1930s. That work showed rather conclusively that bees are instrumental in the secondary dissemination of the fire blight bacterium.[93-97]

The ability of the fire blight organism to survive in honey or beehives is somewhat ambivalent. Some studies showed that the organism could survive for long periods and even overwinter in the hive.[94,98] In contrast, Pierstorff and Lamb[97] could not detect the organism in combs, frames, or in honey taken from beehives 24 hours after inoculation.

Insects with piercing or sucking mouthparts have often been implicated in fire blight outbreaks, but researchers are not in agreement regarding their importance. As early as 1913 to 1916, there were reports about aphids and other sucking insects being instrumental in spreading blight.[79,80] Plurad et al.[99] fed *E. amylovora* to apple aphids and demonstrated that the fire blight bacterium could be detected in the aphids within five minutes after feeding and survived in the insect for at least 72 hours. However, they often found that contaminated aphids did not inject enough bacteria into the host to cause disease.

The pear psylla and tarnished plant bug have been shown to initiate fire blight infections during normal feeding activity on flower clusters and leaf axils.[100] Fruit growers in Colorado are advised to control populations of tarnished plant bugs when mowing the ground cover in orchards or cutting adjacent alfalfa fields.[101]

However, studies in the midwest United States in 1928 indicated that aphids were rarely involved in either primary or secondary spread of the organism.[25,102] The reasons for these contrasting views may reside in technique but may also be due to subtle differences in insects or habits in various geographical areas. There is no question that in some situations, sucking and piercing insects serve as major agents in dissemination of fire blight.

Migratory birds have been proposed as agents capable of long-distance dissemination. The movement of fire blight from England to the coastal regions of northwest Europe may have been due to migrating starlings. Bech-Anderson[15] showed that *E. amylovora* could survive in starling excrement and on their feet for up to eight days. The migration across the English Channel is estimated to take only two to three days. It probably is no coincidence that their primary food source on these migrations was the fruit of hawthorns.

The fire blight bacterium is also disseminated in budding or grafting material,[13] fruit boxes,[103] pruning equipment,[104] and possibly inside healthy fruit.[105–107]

The numerous ways that the fire blight bacterium can be disseminated make it very difficult for blight-free countries to remain free of the disease. Every precaution should be taken to keep the disease out, but countries should also be ready to combat the disease if it enters.

The bacteria survive the winter in cankers which develop on all hosts. Large cankers are easy to locate and eradicate. However, there are many small cankers which are formed on twigs that go unnoticed. The inconspicuous cankers may be responsible for major sources of inoculum in the spring.[108] Reducing the number of overwintering cankers is one of the major control strategies that every grower should utilize. However, even when growers are extremely thorough in pruning, fire blight can still develop under conducive environments.

2-7 INOCULATION

Potted pear, apple, hawthorn, cotoneaster, and other hosts can be used for confirmation of pathogenicity of suspect *E. amylovora* isolates. Inoculation is performed by wounding the plant near the growing tip or on young leaves and placing or injecting the bacteria into the plant. Typical fire blight symptoms develop within three to seven days. The drawback of this method is the requirement for extensive greenhouse space and inherent variability of seedlings or even vegatatively propagated material.

Etiolated pear or apple seedlings provide a year-round source of susceptible plant material but may not be useful for some tests because of the wide variability resulting from seedlings.

Green pear fruit is very susceptible to *E. amylovora* and can be used to verify cultures in question. Inoculated fruit begins to "sweat" or ooze clear to turbid droplets of sticky material within two to five days. The green fruit test can be used to discriminate between virulent and avirulent *E. amylovora* isolates and has also been used as a bioassay for screening candidates for biological control.[109] Fruit can be picked when green and stored at 0°C (32°F) for several months.

Pathogenicity tests on flowers can be performed in the orchard, but the length of time available for such studies is short and without environmental control. Fresh bouquets of flowers can be cut and taken to laboratory growth chambers with environmental control, but the time limitations are still present. Some of these limitations can be overcome by cutting branches with flower buds in late winter and storing them in plastic bags at 0°C (32°F). Flowers are available within one to two weeks by placing branches in water at room temperature. Inoculated flowers develop typical fire blight symptoms within four to seven days. Branches produce flowers with reduced vigor after four to five months of storage and are not reliable.

Pear cell-suspension cultures (PCSC) can be maintained in the lab for a continuous supply of susceptible tissue.[56] They can be obtained from callus cultures from pith or fruit of susceptible hosts. Bauer and Beer[57] suggest caution in the use of PCSC because of the large variations in growth habit and susceptibility. They did note, however, that the PCSC system was able to distinguish between virulent and avirulent *E. amylovora*.

Norelli et al.[110] developed a system using meristem tissue cultures of apple for evaluating differential virulence of *E. amylovora*. This technique produces plant material with uniform susceptibility and can be available year round. The amount of space required is small, and a greenhouse is not necessary.

2-8 EPIDEMIOLOGY

The influence of environment is undoubtedly the most important factor in fire blight epidemics but is also the aspect of the disease we are least able to predict or control. These deficiencies in knowledge seem more acute when fire blight devastates an assumed resistant cultivar or attacks trees in a geographical area not normally plagued with the disease. Despite the numerous studies on the epidemiology of fire blight and improved bactericides, the disease is still not always satisfactorily controlled.

2-8-1 Temperature

The optimum temperature for the growth of *E. amylovora* in vitro is 21 to 28°C, although some growth occurs even at 3 to 5°C. Field studies are in agreement with laboratory investigations that blight develops most rapidly at temperatures between 18 and 30°C.[25,73]

Mills[111] was able to forecast fire blight epidemics in western New York during bloom or shortly after bloom by using a correlation between degree days above 18°C coincident with rain or high humidity. Luepschen et al.[112] confirmed the temperature relationship but found that high relative humidity was just as important as precipitation as a factor favoring fire blight development.

Powell[113] concluded that prebloom freezes influenced the inoculum level and subsequent disease development in Illinois apple orchards. This relationship was based on Reinhardt's report that freezing temperatures reduce the viability of *E. amylovora*.[114] Powell suggested that at least 16.5 degree days above 18°C (30 degree days above 65°F) between the latest prebloom freeze and early bloom was required for development of sufficient inoculum for blossom blight. He did not actually measure inoculum levels, and therefore he had no knowledge of the actual effect of these temperatures on inoculum levels. Studies in California have shown no effect of a single prebloom freeze on epiphytic populations of *E. amylovora* in blossoms.[30]

2-8-2 Rain

Rain is known to transfer bacteria from cankers to other tree parts. For example, Miller[25] showed that when the primary source of inoculum was in the upper portion of a tree, secondary infections developed in a cone-shaped pattern caused by the downward and outward spread of the bacteria.

Bauske[115] found that rain produced aerosols of microdroplets of water that carried inoculum of *E. amylovora*. Thus, the wind often associated with rain storms aids in the dispersal of *E. amylovora* to other potential infection sites.

Rain is often associated with fire blight outbreaks but is not necessary. High relative humidity or heavy dew are sufficient to promote infection.[30,83,112]

Wounds and natural openings are necessary for infection by the bacteria. Hail storms produce all of the necessary ingredients for an epidemic. Wind, moisture, and tissue damage occur during hail storms, which often results in infection of succulent shoot tips and normally resistant leaves.

2-8-3 Environment and Disease

Many of the early prediction methods attempted to correlate the effect of temperatures on the outbreak of fire blight without consideration of the effect on the inoculum level. The development of selective media to monitor population of *E. amylovora* on host tissues allowed for a systematic method to evaluate the effect of environment on inoculum levels.[29,30]

Beer and Opgenorth[116] found that bacteria were present on canker surfaces in New York before the cankers appeared active and before bacteria could be detected in blossoms. The population of *E. amylovora* on holdover cankers was positively correlated with warm (17°C), moist conditions.

Thomson et al.[30] showed that the development of epiphytic populations of bacteria in pear and flowers was related to environmental conditions. *E. amylovora* was detected in healthy pear flowers only after the daily mean temperature exceeded a line drawn from 16.7°C (62°F) on March 1 to 14.4°C (58°F) on May 1. This epiphytic population was present in healthy flowers 7 to 20 days before fire blight symptoms were noted. In orchards where the epiphytic population of bacteria in blossoms was low or absent, disease incidence was insignificant.

Zoller and Sisevich[117] found that 10% of the pear blossoms sampled contained epiphytic *E. amylovora* after an accumulation of 200°C degree hours were reached, whereas 40% of the blossom samples were positive when over 336°C degree hours were accumulated (one centigrade degree hour equals one degree above 18.3 for one hour).

The monitoring techniques used in California have not proved to be as effective in detecting epiphytic bacteria in flowers before infection in the northeast United States and Europe. In Michigan and the Netherlands, epiphytic *E. amylovora* was detected on flowers or leaves but only after infections were evident in the orchards.[118,119] Monitoring of flowers in New York revealed that epiphytic *E. amylovora* was not detected in blossoms in sufficient time to prevent infection by the immediate application of bactericides.[116]

Billing[120] determined the growth rates of *E. amylovora* in vitro and translated the generation times to potential doublings (PD) over one day. This PD value is an estimate of the number of times the bacterial population would be expected to divide each day if temperature were the only factor limiting multiplication. At temperatures below 18°C, the PD values are not sufficient to indicate a significant fire blight risk. However, between 18°C and 30°C, the PD value increases rapidly as does the risk for fire blight. The PD values are used to estimate the duration of an incubation period and subsequent development of fire blight. Billing found that rain was very important in her model and that potential infection days occur whenever daily rain exceeds 2.5 mm. Rain is not a prerequisite for infection during the bloom period, and any day with a PD $\geq$ 9 is also considered a potential infection day.

Schouten[121] modified the PD values of Billing to estimate more accurately the potential for growth of *E. amylovora*. The Billing model has the utility to aid significantly in the understanding and predictability of fire blight epidemics.

2-9 MANAGEMENT

Fire blight is an erratic but potentially devastating disease that has prompted pear and apple producers in the United States to spare little in their efforts to control the disease. The early control procedures centered around sanitation and cultural techniques, but with the advent of chemical controls, growers relied heavily on sprays during the bloom period to protect their orchards. Despite these efforts, fire blight is still a threatening disease that cannot be controlled by a few simple measures. Satisfactory control can only be obtained by utilizing an integrated management program that includes many cultural and chemical methods.

2-9-1 Quarantines

Large geographic barriers are effective in preventing the rapid dispersal of fire blight. However, once the disease is present in an area, it appears to become permanently established. Eradication attempts of new outbreaks have been made in newly infested countries, but efforts thus far have been unsuccessful, and eventually the disease has become well established. The most recent eradication attempt is underway in Norway and Sweden, where new fire blight outbreaks have been quickly destroyed in hopes that the organism did not become permanently established.

The eradication of fire blight, once it has become widespread, is probably futile. There are so many susceptible fruit and ornamental hosts that the cost of removing potential hosts may exceed the cost of losses caused by fire blight.[11] It may be better to channel labor and other resources into good management practices.

Quarantines have apparently been successful in keeping fire blight out of Australia, South Africa, Asia, and perhaps South America. In countries with geographic isolation, it seems possible to exclude fire blight by eliminating the importation of hosts. The restriction of pome fruits from countries with fire blight is also common. However, it seems very remote that contaminated fruit could be responsible for establishing new outbreaks.

The presence of *E. amylovora* on or in healthy fruit has not been shown to be a source of inoculum in fruit orchards.[105,107,122]

Australia is considering a relaxation of the quarantine and may allow importation of apple fruit from New Zealand orchards which have been inspected and found free of fire blight during the current growing season.[123]

In countries where fire blight is not yet established, plant protection personnel should become familiar with symptoms and diagnostic procedures for fire blight. In countries close to fire blight outbreaks, cultural techniques and varieties should be modified over time so when fire blight does gain entrance into the countries, the overall losses will not be too great. Growers should be educated to recognize the disease and to alert plant protection organizations promptly so the outbreak can be contained.

2-9-2 Resistance

The use of resistant cultivars of pear, apple, and ornamentals is unquestionably the most efficient and economical method of controlling fire blight. Indeed, most growers do not practice any fire blight control measures on Red Delicious sports because of the inherent resistance. Fire blight control on susceptible varieties is not complete even at its best, therefore; any relief through the use of resistance would be very beneficial.

Fortunately, some of the most important apple cultivars have considerable resistance to fire blight. Red Delicious and its red sports, the most important variety in the United States, are usually quite resistant to fire blight. Others with resistance include Northwest Greening, Cox's Orange Pippen, Stayman, Winesap, Britemac, Carroll, Hawaii, Primegold, Pricilla, Quinte, Splendor, and Viking.[4,124]

Growers soon learn which varieties are susceptible, but university tests have confirmed that Jonathan, Rhode Island Greening, Rome Beauty, York Imperial, Barry, Burgundy, Idared, Julyred, Lodi, Milton, Monroe, Niagra, Summerred, Vista Bella, and Webster are highly susceptible.[4] Fire blight will not be a problem each year on susceptible varieties, but disease can be serious when weather conditions are conducive for the pathogen.

Pear cultivars are much more susceptible to fire blight than apples. Because of the greater susceptibility of pears, they cannot be grown in some environments even with the use of modern control techniques and chemicals. Most of the commonly grown pears have been derived from *Pyrus communis* L. or the European pear. Although these varieties differ considerably in their resistance, none of the high-quality European pears are known to be sufficiently resistant to thrive in regions where fire blight is serious. Cultivars are categorized as very susceptible, susceptible, resistant, and highly resistant.

Culitvars such as Conference, DeVoe, Forelle, and Laxton's Superb are very susceptible; whereas Bartlett, Beurre Bosc, Beurre d'Anjou, and Clapp Favorite are slightly more resistant but still considered susceptible.[4,125]

Some cultivars show variable resistance where some trees are highly resistant, but others are moderately resistant or even very susceptible. Cultivars in this group include Comice, Dawn, Duchess, and Keiffer.[126]

Highly resistant cultivars include Orient, Richard Peters, Waite, and several new pear releases such as Maxine, Moonglow, and perhaps Mac and Spartlet. Magness was released from the USDA breeding program as resistant but subsequently was found to be susceptible to trunk blight or a direct infection of the mature trunk through growth cracks. Magness as well as Giant Seckel and 12 other clonal selections of Pyrus species have resistance to blossom and shoot blight but suffer from infections of the mature trunk.[127,128]

Van der Zwet[4] listed 390 cultivars of apple and the literature sources indicating their resistance. His summary indicated that of the 193 varieties available prior to 1920, about 28% were classified as resistant, 27% moderately resistant, and 28% were susceptible. Nine percent were considered variably resistant.

In contrast, only 11% of the 287 cultivars of pears available before 1920 were considered resistant, 40% moderately resistant, and 40% susceptible. Nine percent were variable in resistance.[4]

2-9-3 Cultural Management Practices

Fire blight will be less of a problem and fruit of better quality when orchards are planted on well-drained sites with light soils and with a slightly alkaline soil (pH of 6.0 to 6.5). Studies have shown that trees grown on heavy or poorly drained soils sustain more blight, and the damage from blight is more extensive than on trees grown on favorable sites.[129] This is partly due to the slower release and delayed availability of nitrate nitrogen on poorly drained sites. Nitrate nitrogen is the form used by the plant, whereas the ammonium ion must be converted to the nitrate form before being used. Under cool soil temperatures, the microorganisms that cause nitrification (conversion of ammonium ion to nitrate) are not active, and therefore the trees grow slowly. As soil temperatures increase, more nitrate becomes available and trees respond with vigorous and susceptible growth at a time when ambient air temperatures are more conducive to fire blight activity.[125]

Irrigation procedures may also influence the incidence of fire blight. Sprinkler irrigation which wets the foliage or blossoms spreads the organism and provides the moisture necessary for initiating infections.[130] Sprinklers with low angles should be used to avoid wetting the foliage. Overhead irrigation should not be used where blight is a problem. Excessive irrigation may increase the susceptibility of trees by delaying growth and keeping trees in an active growth stage longer. Active growth late in the season is more likely to become infected because temperatures are warmer and more favorable for the pathogen. Infections in trees with active growth are more likely to develop lesions which extend farther and result in indeterminant cankers that are more likely to overwinter successfully. Alternatively, insufficient irrigation will cause more losses in the quality of the crop than the potential losses caused by fire blight.

Tree nutrition has been studied extensively, and a definite relationship has been shown between the susceptibility of trees and nitrogen, potassium, phosphorus, and some of the micronutrients.[131,132] Fertilization with nitrogen or cultivation (which made more nitrogen available to trees) resulted in more fire blight infections and greater lesion length on apple.[133,134] Most growers are aware of the effect of nitrogen on increased succulence and subsequent increases in fire blight susceptibility, but some growers withhold nitrogen to the extent that tree growth and yields are seriously affected.

Nutrient application should be based on soil and leaf analyses and on levels of growth observed in previous years. Excess nitrogen or organic sources of nitrogen, such as animal manures, should not be applied to orchards where fire blight is likely to be a problem. This is especially true for orchards on heavy, poorly drained soils. When nitrogen must be applied on such sites, the ammonium form should not be used since it becomes available later in the growing season.[125]

Vigorous vegetative growth is very susceptible to blight infections. Therefore, avoid any pruning practice which stimulates rapid growth. It is best to use light annual pruning and avoid major cuts which stimulate sucker growth.

Fruiting spurs and suckers on the trunk or major scaffold branches should be pruned out to avoid infections which might kill the tree or cause serious damage to the tree form or structure.

Summer pruning or the removal of blight infections during the growing season may not be advisable because of the increased potential for new infections. The open wounds in succulent tissue are ideal sites for entry of the pathogen. Pruning at this time also stimulates sucker growth, which is very susceptible. Summer pruning should only be done when conditions are dry and then only if fire blight is not significant in the area. Summer removal of blight infections should generally only be done when there is a possibility of the infection progressing into the trunk or scaffold branches. In most cases, it is better to leave the blight in the trees and prune it out during the dormant season.

Pruning new blight infections in geographic areas where blight has been recently introduced is advised. Pruning should be accomplished immediately to remove the source of inoculum. In young orchards or highly susceptible blocks of trees or where blight is new, the grower should examine the orchard twice per week and remove any infections.[125] Secondary flowers or rattail bloom should be regularly removed in high-risk orchards to prevent late-season infection.[135]

The removal of overwintering cankers during dormant pruning operations is one of the cultural practices that has been shown to reduce subsequent blight in an orchard.[25,125,135] Most cankers can be detected easily and removed during the normal dormant pruning practices. However, it is difficult to eradicate all cankers in an orchard. Cankers with indeterminate margins and those on small twigs frequently go undetected. These cankers provide sufficient inoculum in an orchard for a serious epidemic to develop.[108]

The fire blight pathogen is present in tissues in advance of externally visible symptoms.[14] Therefore, cuts should be made at least 15 to 30 cm (6 to 12 in.) ahead of the visible lesion. Failure to do so may result in the extension of the infection or contamination of the pruning tools.

Sterilizing tools between cuts is especially important when pruning is being done in the summer or when blight is still active in the orchard. Denatured alcohol or sodium hypochlorite has been used successfully to disinfest tools.[104,136,137] However, if cuts are made properly, sterilizing tools is not necessary during the dormant season pruning.

2-9-4 Vector Control

Insects act as very efficient vectors of *E. amylovora* from ooze on overwintering cankers to flowers as well as transfer of the pathogen from flower to flower. In some cases, the same species of insects may be responsible for both the transfer from canker to flower or flower to flower.

Many insects are attracted to the bacterial ooze which occurs in the spring on overwintering cankers. The ooze is made primarily of a high molecular wight polysaccharide composed of galactose, uronide, and small quantities of glucose and possibly mannose.[54,76] Syrphid and Minettia species, dipteran flies, ants, etc. are attracted to this material as a source of nutrition and inadvertently become contaminated with the bacteria which are present in high numbers in the ooze.[29] The contaminated insects are also attracted to flowers as a source of pollen and nectar. Once the bacteria are present on flowers then many other pollinating insects, including honeybees, act as vectors. The vector relationship is nonspecific, and any insect which visits the flower is capable of disseminating the organism.[135] In the process of seeking pollen or nectar, the insects casually come in contact with flower parts and especially the stigmatic surfaces of the pistils where *E. amylovora* readily multiplies.[83]

The activity of these insect vectors is so closely intertwined with the essential pollination activity that early-season vector control is not recommended in fruit crops. It may be possible to reduce the risk of fire blight in ornamentals by controlling the vectors with insecticides, but this method is not currently recommended.

Aphids, leaf hoppers, psylla, tarnished plant bugs, and other sucking insects are considered important as the vectors responsible for shoot blight.[92,99,138,139] Sucking insects may inoculate plants in their feeding activities or create wounds which may serve as infection courts.[101,139] The control of sucking insects in the postbloom period is recommended partly to reduce the actual damage caused by the insects and partly to reduce the potential for initiating fire blight infections of vegetative shoots. This approach seems to be particularly important if an adjacent field of alfalfa or another crop is harvested, creating a mass migration of insects.

2-9-5 Chemical Management Practices

Orchards with a previous history of fire blight and highly susceptible cultivars should be sprayed with a Bordeaux mixture plus oil at silver- to green-tip stages.[125,140] This concentrated copper application apparently reduces the production of inoculum from overwintering cankers. Other copper compounds are also effective.[141]

The protection of flowers by bactericide sprays in most years is usually an effective method of controlling fire blight even in areas where conditions and inoculum are conducive for disease. However, the level of control obtained by the currently available bactericides is less than ideal. None of the registered products in the United States have eradicative or significant systemic properties. Good blossom control will greatly aid in reducing the incidence of twig blight later in the season.

Bordeaux mixture was one of the first compounds recommended for use as a blossom blight control spray in 1929.[142,143] Results were satisfactory in most cases, but when conditions were highly conducive for fire blight, Bordeaux mixture was not very effective. However, it is still recommended and in many areas it will provide adequate control when disease pressure is not too high or cultivars are not too susceptible. Other copper compounds have also been shown to be effective for controlling blight, but unfortunately all copper compounds cause russeting of many fruit varieties, especially pears.

The discovery of antibiotics for blossom sprays was a significant improvement in the control of blossom blight and without the problems of phytotoxicity associated with copper compounds. Research in the early 1950s demonstrated excellent control with streptomycin sprays when applied at the proper time and repeated every three to five days.[144,145] Streptomycin is only locally systemic and effective on open blossoms for two to five days. Therefore, applications must be repeated every three to five days to protect newly opened blossoms and provide a new protective residue.

The discovery of streptomycin resistant strains of *E. amylovora* in California orchards in 1972, where streptomycin was used exclusively and repeatedly for blight control, was not surprising.[146] The diversity of the streptomycin-resistant strains with respect to virulence, generation times, and colony morphologies indicates that they arose by mutation from a heterogeneous assortment of strains rather than from one resistant strain which spread from an epicenter.[147] Streptomycin-resistant strains were also found in Washington and Oregon a few years later.[148]

Oxytetracycline was included in the early formulations of streptomycin, but because the combination product did not improve control over streptomycin alone, it was eliminated. The combination of two antibiotics as a blossom spray in effective concentrations would probably reduce the possibility of resistance developing to either antibiotic. However, the early combination product contained 15% streptomycin sulfate and only 1.5% oxytetracycline, not enough to provide control if used alone. Therefore, it is likely that resistance to streptomycin would still have developed despite the use of the original combination product. It is interesting to note that streptomycin-resistant strains have been isolated from sources where streptomycin has not been used[147,149] but not found in some areas where streptomycin has been used routinely.[118,150]

Oxytetracycline is being used alone in some orchards where streptomycin-resistant strains are present. It seems likely that oxytetracycline resistance could become a problem in those orchards if it is used repeatedly and exclusively. The best procedure would be to utilize different bactericides throughout the year to reduce the chances for the selection of resistant strains.

Antibiotics cannot be used in most European countries because of the potential for the development of antibiotic resistance in medically important bacteria.[125,151]

There have been no new bactericides registered in the United States for fire blight control since streptomycin in the 1950s and oxytetracycline in the late 1970s. Only one new bactericide has been shown to be effective since the registration of streptomycin and oxytetracycline. MBR-10995, a synthetic antibiotic, was effective in several field tests in the United States and Europe.[152-154] It has been given temporary registration in France under the name Flumequine.[151]

2-9-6 Predicting Bloom Sprays

The devastation caused by fire blight arouses sufficient concern for most growers and requires that they apply preventive bactericide sprays on a regular basis without knowledge of the potential for disease or the inoculum level. Without the methodology to predict accurately disease occurrence, many California growers previously applied streptomycin or copper compounds to trees, beginning at 5% bloom with subsequent applications every three to five days during the bloom period. The bloom period in California (including the secondary or rattail bloom) may last up to three months and require as many as 18 applications to keep blossoms protected.[155] Routine spraying during bloom provides good control but is not always necessary and certainly not economical. Moreover, a calendar-based program loads the environment with undesirable bactericide residues and increases the propensity for the selection of resistant bacteria. Researchers have attempted to forecast when weather conditions are suitable for disease and more recently to determine when sufficient inoculum is present to justify bactericide applications.

Mills[111] devised one of the first forecasting systems for western New York. Growers were advised to delay application of streptomycin until the maximum temperature exceeded 18°C, precipitation was forecast, and the relative humidity exceeded 70%. Luepschen[112] confirmed and refined the system and found that high relative humidity for three days could replace the precipitation requirement. This system seemed to work quite well in the area where it was developed, but when tested in other areas, it was not accurate enough.

Powell[113] developed forecasting methods based on observation of the weather conditions and blossom blight incidence in Illinois apple orchards. A significant component of this method was Reinhardt's[114] report that freezing reduced the viability of E. amylovora. Powell felt that prebloom freezing conditions reduced the population of the pathogen and that an accumulation of at least 16.5 degree days above 18°C (30 degree days above 65°F) were necessary for the growth of sufficient inoculum for blossom blight to occur. Unfortunately, at the time there was no way to measure the effects of pre-bloom freezes on the inoculum.

Epidemiological work in the 1970s revealed that preventive applications of bactericides could be applied on the basis of weather and a knowledge of bacterial populations.[30,155] The presence of epiphytic populations of E. amylovora on apparently healthy apple and pear blossoms[29,30] was used as an empirical tool to evaluate the effects of the environment on inoculum production. Thomson et al.[30,156] demonstrated that E. amylovora was present in healthy blossoms as much as 14 days before any symptoms were noted.

Based on the monitoring of epiphytic populations in California, a prediction model was devised to indicate when epiphytic bacteria were likely to be present in healthy flowers. Applications of bactericides timed according to the initial presence of epiphytic bacteria provided control equal to or better than the normal program where sprays were initiated at first bloom.[156] The model is very simple and relies solely on the daily mean temperature. Applications are recommended when the daily mean temperature exceeds the temperatures on a straight line drawn from 16.5°C starting at full pink in apples or full white in pears and dropping 0.5°C every 15 days to a minimum mean temperature of 14.4°C. This model has also been used successfully to make appropriately timed bactericide applications in Washington[157] and Utah.[158] In many years this technique has reduced the number of applications by at least 50%. It should be kept in mind that this model predicts the occurrence of epiphytic populations of E. amylovora in flowers, and not disease.

Zoller and Sisevich,[117] also in California, determined that 10% of the pear blossom samples contained E. amylovora when 200°C degree hours (350°F degree hours) were reached, whereas 40% of blossom samples were positive when over 336°C degree hours were accumulated (one celsius degree hour equals one degree C above 18.3°C for one hour or 100°F degree hours equals 56°C degree hours). Three days of cool weather below 18.9°C (66°F) negates any accumulated degree hours. Applications of bactericides are based on the number of accumulated degree hours. Nothing is recommended until the number of degree hours ranges from 0.5 to 84°C (1–50°F degree hours), at which time growers are advised to spray if rain is forecast within 24 hours. The need for bactericides is more urgent when degree hour accumulations exceed 84°C (150°F) and are then recommended every three to four days. Alternate day treatments are recommended if the degree hours surpass 280°C (500°F).

Billing[159,160] developed a system in Europe to predict when outbreaks of fire blight would likely be found in the field. The method is based on experimental studies of the growth rate of *E. amylovora* in vitro and historical fire blight outbreaks in southeast England. This model determines the potential doubling (PD) of bacteria from daily maximum and minimum temperatures. The PD values are used to estimate the duration of an incubation period which starts after an infection period. An infection period is initiated when there is 2.5 mm or more of rain; except up to June 5 in the spring, when PD days of nine or more are also considered infection days. The incubation period is completed and infections should be apparent in the field when the PD value accumulated from an infection day exceeds a certain threshold determined by temperatures and rainfall.[159–161] There has been no published information on the use of this model for timing bactericide applications to prevent disease. Schouten[121,162] found that the PD values published by Billing[160,161] were underestimated by 0 to 80%. This provides an overestimation of the duration of incubation periods and may lead to warnings that are too late.

The original intent of the Billing model was not for timing of spray applications but to ". . . judge periods of high risk for each susceptible host so that a policy of seeking and destroying infections at an early stage could be implemented."[160,161] Revisions of the model, to be published soon, will take into account Schouten's correction of PD and provide methods for timing spray applications.[163] Field evaluations of this method to prevent blight by timing bactericide applications still need to be accomplished.

A warning system being developed by Schwager et al.[164] evaluates the relative risk of fire blight by considering cultural conditions in specific orchards and correlates the orchard risk factors with weather factors. This integration should provide a more accurate estimation of potential fire blight for individual orchards.

2-9-7 Biological Control

The use of bacteriophages for control of fire blight has been suggested, but field studies thus far have not been accomplished. Erskine[64] proposed that lysogenized yellow bacteria may harbor a phage that is pathogenic to *E. amylovora* and under certain conditions may influence the development of epidemics. Ritchie and Klos[63] examined aerial parts of apple tissues and found *E. amylovora* bacteriophages on every infected sample but could not detect phages on healthy tissues. The role the bacteriophages play in the development of fire blight epidemics is unknown, but their use as biological control agents is not as promising as antagonistic bacteria because of the selectivity of the phages and the propensity for bacteria to become resistant.[165]

The possibility of biological control of fire blight with antagonistic bacteria has been proposed for many years. Parker[92] conducted field tests from 1928 to 1931, in which he introduced antagonistic bacteria into blossoms prior to or simultaneously with *E. amylovora*. He was unable to obtain consistent control of fire blight but did show that there was an inhibitory effect on disease incidence and severity when the antagonists were placed in the flowers before *E. amylovora*. Unfortunately, the practical use of biological control is not much better today than it was in the 1930s. Field tests with *E. herbicola*[70,166,167] and *E. carotovora* subsp. *betavasculorum*[155] are often encouraging enough to continue research but have only achieved limited success at controlling fire blight.

A yellow saprophytic bacterium (presumed to be *E. herbicola*) is often associated with host tissue or in cankers and has been studied extensively as a potential biological control candidate.[67-70,168-170] Recent studies by Beer et al.[166] on the mode of antagonism and field tests of *E. herbicola* isolates are encouraging. In some cases, the level of control with antagonistic bacteria is as good as with streptomycin sprays. Additional studies on timing of applications of the antagonists and the influence of environment will aid in the future use of biological control. The use of molecular genetic techniques may provide information about the mechanisms of antagonism or identify the gene products of *E. herbicola* responsible for inhibition of *E. amylovora*. Natural biological control may be operating in New Zealand apple and pear orchards. The disease created some significant losses in the first years after introduction into the country, but after the initial epidemics there was a notable decline in the disease incidence and subsequently it has only been a minor problem.[171]

The incidence of fire blight in New Zealand is much less than predicted by the fire blight prediction models despite the plantings of susceptible apple and pear cultivars.[172] The consistently higher populations of saprophytic bacteria on the flowers may be inhibitory to *E. amylovora*.

2-10 CONCLUSION

Fire blight is still a major threat to pome fruit producers and the ornamental industry worldwide, despite the large accumulation of knowledge that is available regarding the epidemiology and control of the disease. There are some significant gaps in our understanding of the disease, and it continues to cause economic losses. Fire blight has become an important international disease and poses a constant threat to countries currently fighting the disease as well as those countries attempting to exclude it.

The previous history of fire blight expansion indicates that the disease will continue to spread to other countries of the world. Countries without fire blight should become prepared and start immediately to modify cultural techniques and varieties. Personnel should be trained in the recognition of the symptoms of fire blight and also how to culture and identify the pathogen.

The long-term solution to fire blight is the development of resistant varieties. Unfortunately, the host range is so large that only through coordinated international efforts and a commitment by agricultural support groups will accomplishments be made. Breeding programs are currently in progress for pear and apple, and several resistant cultivars are now available. Fire blight has been more of a problem in Europe on ornamentals, and several European countries have initiated breeding programs to develop fire-blight–resistant selections.

The prospects of obtaining fire blight resistance from naturally resistant cultivars and incorporating the beneficial characteristics into commercial fruit or ornamental cultivars is much more likely with the new genetic engineering techniques that are now available. Interspecific crosses, embryo rescue, gene transfer, and other techniques provide opportunities for the rapid development of new resistant varieties.

Understanding how *E. amylovora* causes the disease will also aid in the development of resistant varieties. For example, it may be possible to identify toxins or other specific proteins that are responsible for symptom expression. These compounds could then be used to screen progeny of selected crosses or perhaps even to select resistant single cells from cell culture suspensions and produce entire resistant plants.

The recent information generated about epidemiology of fire blight has greatly improved control of the disease. The ability to predict epidemics and instigate timely control measures has provided better control of the disease and also reduced the use of bactericides. Additional knowledge and a refinement of predictive methodologies should allow significant progress in control. Development of a model which takes into account the risks of different geographical, cultural, and environmental parameters is needed.

The sensitivity of diagnostic tools, such as selective and differential media, has dramatically improved our understanding of the epidemiology of *E. amylovora*. New methodologies, such as ELISA and monoclonal antibodies, have improved the detection of low populations of *E. amylovora*. DNA probes may improve detection even more. The importance of these low populations in apparently healthy tissues needs to be resolved. Thus far, no one has demonstrated that bacteria in healthy buds or other tissues are responsible for outbreaks of fire blight.

The future of fire blight control with bactericides is bleak. The current attitude regarding the use of pesticides in the production of agriculture commodities and the stringent regulatory climate curtail much optimism about the registration or even development of new bactericides. The currently registered bactericides are often inadequate to provide satisfactory control. Streptomycin resistance of *E. amylovora* is quite common in West Coast orchards, and oxytetracycline is often used as a substitute. It is likely that resistant strains of *E. amylovora* will be selected with the continued exclusive use of oxytetracycline. The copper compounds still provide acceptable control in most cases, but the propensity for them to be phytotoxic limits their use on sensitive fruits. It is also possible that resistance to copper bactericides could arise.[173]

Most of the European countries ban the use of antibiotics for control of bacterial diseases because of the concern for the selection of antibiotic resistance in medically important bacteria. Some of the R-factors are promiscuous and could be readily transmitted to potentially important human pathogens.

Most chemical companies do not even screen bactericides for efficacy against *E. amylovora* because they do not feel there is an economic advantage. There has been one new bactericide available for testing in the last 20 years and it will not likely be registered in the United States. It is improbable that many new bactericides will become available unless there is an unforeseen breakthrough or change in policy.

The discovery of biological control organisms and methodologies is an area with a bright future. The research performed thus far has shown that fire blight can be controlled with selected bacteria, but results are inconsistent. Further research into the mechanisms of action should reveal parameters that are important in the successful use of biological control agents. It is impossible that the active products of the antagonists can be identified and used as control measures without understanding or dealing with the complexity of the living organism.

It is often apparent that the spray coverage and distribution of bactericides in the orchard is often wasteful and perhaps ineffective. Research into new spray technologies and application techniques could result in equipment that will deliver the bactericide to the site where the bacteria are located or where infection takes place.

The spread of fire blight to additional countries may have some side benefits, because more scientists and resources will be available to solve some of the enigmas of this disease.

2-11 REFERENCES

1. Baker, K. F., Fire blight of pome fruits: The genesis of the concept that bacteria can be pathogenic to plants, *Hilgardia*, 40, 603, 1971.

2. Denning, W., On the decay of apple trees, *Trans. N.Y. Soc. Prom. Agr., Arts, and Mfrs.*, 2, 219, 1794.

3. Coxe, W., *A view of the cultivation of fruit trees, and the management of orchards and cider*, Pears, M. Carey and Son, Philadelphia, 174, 1817.

4. van der Zwet, T. and Keil, H. L., Fire blight: A bacterial disease of rosaceous plants, *U.S. Dept. Agri. Handbook*, 510, 200 pp., 1979.

5. Campbell, R. N., Fire blight, *Natural History*, 88, 62, 1979.

6. Cockayne, A. H., Fire blight, a serious disease of fruit trees, *New Zeal. J. Agr.*, 20, 156, 1920.

7. Campbell, J. A., The orchard; the outbreak of fire blight, *New Zeal. J. Agr.*, 20, 181, 1920.

8. Crosse, J. E., Bennett, M. and Garrett, C. M. E., Fire blight of pear in England, *Nature*, 182, 1530, 1958.

9. Shabi, E. and Zutra, D., Outbreaks of fire blight in Israel in 1985 and 1986, *Acta Hort.*, 217, 23, 1987.

10. Paulin, J. P., Fire blight spread in Europe, *Proc. XXI Intl. Hort. Cong.*, 1, 138, 1982.

11. van der Zwet, T., Recent spread and present distribution of fire blight in the world, *Plant Dis. Rep.*, 52, 698, 1968.

12. Burkowicz, A., The appearance and current situation of fire blight in Poland, *EPPO Newsletter*, 72 E, 3, 1972.

13. van der Zwet, T., Occurrence of fire blight in commerical pear seedling rootstocks following budding with symptomless scionwood, *Phytopathology*, 73, 969 (Abstr.), 1983.

14. Aldwinckle, H. S. and Preczewski, J. L., Reaction of terminal shoots of apple cultivars to invasion by *Erwinia amylovora*, *Phytopathology*, 66, 1439, 1976.

15. Bech-Andersen, J., Dissemination of the bacterial disease "fireblight" in Europe, *Grana*, 14, 46, 1974.

16. Latorre, B. A., Tizon bacterial de la flor del Peral, *Revista Fruiticola*, 3, 49, 1982.

17. Reil, W. O., Moller, W. J. and Thomson, S. V., An historical analysis of fire blight epidemics in the central valley of California, *Plant Dis. Rep.*, 63, 545, 1979.

18. Pickett, B. S., The blight of apples, pears, and quinces, *Ill. Agr. Expt. Sta. Cir.* 172, 10 pp., 1914.

19. Hsiong, S. L., Breeding for resistance to fire blight in pears, N.Y. (Cornell) *Agr. Exp. Sta. Quart. Bull.*, 4, 2, 1938.

20. Great Britain Ministry of Agriculture, Fisheries, and Food, 1966, The worst year yet for fire blight, says the Ministry, *Com. Grower*, Nov., 25, 921, 1966.

21. Great Britian Ministry of Agriculture, Fisheries, and Food, Fire blight of apple and pear, *Adv. Leaflet*, 571, 11 pp., 1969.

22. Meijneke, C., Fire blight: An isolated outbreak in the Netherlands, *EPPO Publ. Ser. A. (45-E)*, 17, 1967.

23. Ries, S. M. and Otterbacher, A. G., Occurrence of fire blight on thornless blackberry in Illinois, *Plant Dis. Rep.*, 61, 232, 1977.

24. Munn, M. T., Pathogenicity of *Bacillus amylovorus* (Burr.) Trev. for blossoms of the strawberry (Fragaria sp.), *Phytopathology*, 8, 33, 1918.

25. Miller, P. W., Studies of fire blight of apple in Wisconsin, *J. Agr. Res.*, 39, 579, 1929.

26. Nixon, E. L., Fire blight, *Pa. Agr. Expt. Sta. Bull.*, 203, 22 pp., 1926.

27. Keil, H. L. and van der Zwet, T., Aerial strands of *Erwinia amylovora*: Structure and enhanced production by pesticide oil, *Phytopathology*, 62, 355, 1972.

28. Buchanan, R. E. and Gibbons, N. E., Bergey's manual of determinative bacteriology, 8th ed., Williams & Wilkins, Baltimore, 1974, 333.

29. Miller, T. D. and Schroth, M. N., Monitoring the epiphytic population of *Erwinia amylovora* on pear with a selective medium, *Phytopathology*, 62, 1175, 1972.

30. Thomson, S. V., Schroth, M. N., Moller, W. J. and Reil, W. O., Occurrence of fire blight of pears in relation to weather and epiphytic populations of *Erwinia amylovora*, *Phytopathology*, 65, 353, 1975.

31. Crosse, J. E. and Goodman, R. N., A selective medium for a definite colony characteristic of *Erwinia amylovora*, *Phytopathology*, 63, 1425, 1973.

32. Ishimaru, C. and Klos, E. J., New medium for detecting *Erwinia amylovora* and its use in epidemiological studies, *Phytopathology*, 74, 1342, 1984.

33. King, E. O., Ward, M. K. and Raney, D. E., Two simple media for the demonstration of pyocyanin and fluorescin, *J. Lab. Clin. Med.*, 44, 301, 1954.

34. Klement, Z. and Goodman, R. N., Hypersensitive reaction induced in apple shoots by an avirulent form of *Erwinia amylovora*, *Acta Phytopath. Acad. Sci. Hungaricae*, 1, 177, 1966.

35. Elrod, R. P., Serological studies of the Erwineae. I. *Erwinia amylovora*, *Bot. Gaz.*, 103, 123, 1941.

36. Laroche, M. and Verhoyen, M., Identification serologique d'*Erwinia amylovora* par immunodiffusion, *Med. Fac. Landbouww. Riiksuniv. Gent*, 47, 1083, 1982.

37. Samson, R., Heterogeneity of heat-stable somatic antigens in *Erwinia amylovora*, *Ann. Phytopathol.*, 4, 157, 1972.

38. Calzolari, A., Peddes, P., Mazzucchi, U., Mori, P. and Garzena, C., Occurrence of *Erwinia amylovora* in buds of asymptomatic apple plants in commerce, *Phytopath. Z.*, 103, 156, 1982.

39. Calzolari, A., Mazzucchi, U. and Gasperini, C., Cross-reactions between *Erwinia amylovora* and other bacteria in immunofluorescence staining using different antisera, *Phytopath. Medit.*, 21, 110, 1982.

40. Lin, C. P., Chen, T. A., Wells, J. M. and van der Zwet, T., Identification and detection of *Erwinia amylovora* with monoclonal antibodies, *Phytopathology*, 77, 376, 1987.

41. Lin, C. P., Chen, T. A., Wells, J. M. and van der Zwet, T., Direct counts of *Erwinia amylovora* by immunofluorescent staining with monoclonal antibodies, *Acta Hort.*, 217, 81, 1987.

42. Ark, P. A., Dissociation in *Erwinia amylovora* (Burrill) Comm. S.A.B., *Science*, 80, 20, 1934.

43. Ark, P. A., Variability in the fire-blight organism, *Erwinia amylovora*, *Phytopathology*, 27, 1, 1937.

44. Hildebrand, E. M., Strains of the fire-blight organism, *Phytopathology*, 30, 9, 1940.

45. Hildebrand, E. M., Relative stability of fire blight bacteria, *Phytopathology*, 44, 192, 1954.

46. Gibbons, L. N., Variation in the occurrence of extracellular diffusible antigens in temperature induced varients of *Erwinia herbicola* strain Y-46 and observaton of their relationships with *Erwinia amylovora*, *Can. J. Microbiol.*, 20, 643, 1974.

47. Bennett, R. A. and Billing, E., Capsulation and virulence in *Erwinia amylovora*, *Ann. Appl. Biol.*, 89, 41, 1978.

48. Bennett, R. A., Quantitative studies on the virulence of *Erwinia amylovora*, *Proc. IV Intl. Conf. Plant Path. Bact.*, 527 (Abstr.), 1978.

49. Bennett, R. A., Characteristics of *Erwinia amylovora* in relation to virulence, *Proc. IV Intl. Conf. Plant Path. Bact.*, 479, 1978.

50. Ayers, A. R., Ayers, S. B. and Goodman, R. N., Extracellular polysaccharide of *Erwinia amylovora*: A correlation with virulence, *App. Environ. Microbiol.*, 38, 659, 1979.

51. Bennett, R. A., Evidence for two virulence determinants in the fire blight pathogen *Erwinia amylovora*, *J. Gen. Microbiol.*, 116, 351, 1980.

52. Billing, E., Studies on avirulent strains of *Erwinia amylovora*, *Acta Hort.*, 151, 249, 1984.

53. Pierstorff, A. L., Studies on the fire-blight organism, *Bacillus amylovorus*, *New York (Cornell) Agr. Exp. Sta. Memoir*, 136, 53 pp., 1931.

54. Goodman, R. N., Huang, J. S. and Huang, P. Y., Host-specific phytotoxic polysaccharide from apple tissue infected by *Erwinia amylovora*, *Science*, 183, 1081, 1974.

55. Beer, S. V., Sjulin, T. M. and Aldwinckle, H. S., Amylovorin-induced shoot wilting: Lack of correlation with susceptibility to *Erwinia amylovora*, *Phytopathology*, 73, 1328, 1983.

56. Buchanan, G. E. and Starr, M. P., Phytotoxic material from associations between *Erwinia amylovora* and pear tissue culture: Possible role in necrotic symptomatology of fireblight disease, *Current Microbiol.*, 4, 63, 1980.

57. Bauer, D. W. and Beer, S. V., Evidence that a putative necrotoxin of *Erwinia amylovora* is an artifact caused by the activity of inorganic salts, *Physiological Plant Pathol.*, 27, 289, 1985.

58. Pugashetti, B. K. and Starr, M. P., Conjugational transfer of genes determining plant virulence in *Erwinia amylovora*, *J. Bact.*, 122, 485, 1976.

59. Chatterjee, A. K. and Starr, M. P., Gene transmission among strains of *Erwinia amylovora*, *J. Bacteriol.*, 116, 1100, 1973.

60. Beer, S. V., Bauer, D. W. and Steinberger, E. M., Studies on the mechanism of pathogenesis of *Erwinia amylovora*, *Acta Hort.*, 151, 233, 1984.

61. Steinberger, E. M. and Beer, S. V., Transposon mutagenesis of *Erwinia amylovora* with TN5, *Phytopathology*, 73, 1347, 1983.

62. Bauer, D. W. and Beer, S. V, Cloning of *Erwinia amylovora* DNA involved in pathogenicity and induction of the hypersensitive reaction, *Acta Hort.*, 217, 169, 1987.

63. Ritchie, D. F. and Klos, E. J., Isolation of *Erwinia amylovora* bacteriophage from aerial parts of apple trees, *Phytopathology*, 67, 101, 1977.

64. Erskine, J., Characteristics of *Erwinia amylovora* bacteriophage and its possible role in the epidemiology of fire blight, *Can. J. Microbiol.*, 19, 837, 1973.

65. Billing, E., An association between capsulation and phage sensitivity in *Erwinia amylovora*, *Nature* [London], 186, 819, 1960.

66. Baldwin, C. H., Jr. and Goodman, R. N., Prevalence of *Erwinia amylovora* in apple buds as detected by phage typing, *Phytopathology*, 53, 1299, 1963.

67. Erskine, J. M. and Lopatecki, L. E., In vitro and in vivo interactions between *Erwinia amylovora* and related saprophytic bacteria, *Can. J. Microbiol.*, 21, 35, 1975.

68. Farabee, G. J. and Lockwood, J. L., Inhibition of *Erwinia amylovora* by bacterium sp. isolated from fire blight cankers, *Phytopathology*, 48, 209, 1958.

69. Goodman, R. N., In vitro and in vivo interactions between components of mixed bacterial culture isolated from apple buds, *Phytopathology*, 55, 217, 1965.

70. Riggle, J. H. and Klos, E. J., Relationship of *Erwinia herbicola* to *Erwinia amylovora*, *Can. J. Bot.*, 50, 1077, 1972.

71. Shaw, L., Studies on fire blight: I. The identity of a yellow schizomycete found associated with *Bacillus amylovorus*, M.S. Thesis, Univ. Ark., 1929.

72. Beer, S. V. and Norelli, J. L., Fire blight epidemiology: Factors affecting release of *Erwinia amylovora* by cankers, *Phytopathology*, 67, 1119, 1977.

73. Brooks, A. N., Studies of the epidemiology and control of fire blight of apple, *Phytopathology*, 16, 665, 1926.

74. Rosen, H. R., The life history of the fire blight pathogen, *Bacillus amylovorus*, as related to the means of overwintering and dissemination, *Ark. Agr. Exp. Sta. Bull.*, 244, 96 pp., 1929.

75. Rosen, H. R., Further studies on the overwintering and dissemination of the fire-blight pathogen, *Ark. Agr. Exp. Sta. Bull.*, 283, 102 pp., 1933.

76. Eden-Green, S. J. and Knee, M., Bacterial polysaccharide and sorbitol in fire blight exudate, *J. Gen. Microbiol.*, 81, 509, 1974.

77. Rosen, H. R., Mode of penetration and of progressive invasion of fire blight bacteria into apple and pear blossoms, *Ark. Agr. Exp. Sta. Bull.*, 331, 68 pp., 1936.

78. Hildebrand, E. M. and MacDaniels, L. H., Modes of entry of *Erwinia amylovora* into the flowers of the principal pome fruits, *Phytopathology*, 25, 20 (Abstr.), 1935.

79. Stewart, V. B., The importance of the tarnished plant bug in the dissemination of fire blight in nursery stock, *Phytopathology*, 3, 273, 1913.

80. Stewart, V. B. and Leonard, M. D., Further studies in the role of insects in the dissemination of fire blight bacteria, *Phytopathology*, 6, 152, 1916.

81. Gossard, H. A. and Walton, R. C., Fire-blight infection—rain proves an important carrier of disease in open orchards, *Ohio Agr. Exp. Sta. Mon. Bull.*, 2, 357, 1917.

82. Gossard, H. A. and Walton, R. C., Dissemination of fire blight, *Ohio Agr. Exp. Sta. Bull.*, 357, 126 pp., 1922.

83. Thomson, S. V., The role of the stigma in fire blight infections, *Phytopathology*, 76, 476, 1986.

84. Rundle, J. R. and Beer, S. V., Population dynamics of *Erwinia amylovora* and a biological control agent, *Erwinia herbicola*, on apple blossom parts, *Acta Hort.*, 217, 221, 1987.

85. Rosen, H. R., The mode of penetration of pear and apple blossoms by the fire-blight pathogen, *Science*, 81, 26, 1935.

86. Hattingh, M. J., Beer, S. V. and Lawson, E. W., Scanning electron microscopy of apple blossoms colonized by *Erwinia amylovora* and *E. herbicola*, *Phytopathology*, 76, 900, 1986.

87. Lelliott, R. A., Effect of temperature on populations of *Erwinia amylovora* in host blossoms, *Proc. IV Intl. Conf. Pl. Path. Bact.*, 527 (Abstr.), 1978.

88. Norelli, J. L. and Beer, S. V., Factors affecting the development of fire blight blossom infections, *Acta Hort.*, 151, 37, 1984.

89. Bachmann, F. M., The migration of *Bacillus amylovorus* in the host tissues, *Phytopathology*, 3, 13, 1913.

90. Dueck, J. and Morand, J. B., Seasonal changes in the epiphytic population of *Erwinia amylovora* on apple and pear, *Can. J. Plant Sci.*, 55, 1007, 1975.

91. Bauske, R. J., Dissemination of waterborne *Erwinia amylovora* by wind in nursery plantings, *Proc. Am. Soc. Hort. Sci.*, 91, 795, 1967.

92. Parker, K. G., Fire blight: Overwintering, dissemination, and control of the pathogen, *N. Y. (Cornell) Agr. Exp. Sta. Memoir*, 193, 42 pp., 1936.

93. Gossard, H. A. and Walton, R. C., *Bacillus amylovorus* in honey and in honeydew, *Phytopathology*, 6, 113 (Abstr.), 1916.

94. Hildebrand, E. M. and Phillips, E. F., The honeybee and the beehive in relation to fire blight, *J. Agr. Res.*, 52, 789, 1936.

95. Keitt, G. W. and Ivanoff, S. S., Transmission of fire blight by bees and its relation to nectar concentration of apple and pear blossoms, *J. Agr. Res.*, 62, 745, 1941.

96. Martin, J. P., The honey bee and pear blight, *Univ. Calif. J. Agr.*, 7, 4, 1921.

97. Pierstorff, A. L. and Lamb, H. N., The honeybee in relation to the overwintering and primary spread of fire blight organism, *Phytopathology*, 24, 1347, 1934.

98. Rosen, H. R., Relation of fire blight to honey bees, *Proc. Mo. State Hort. Soc.*, 67, 1932.

99. Plurad, S. B., Goodwin, R. N. and Enns, W. R., Persistence of *Erwinia amylovora* in the apple aphid (*Aphis pomi* de Geer), a probable vector, *Nature*, 205, 206, 1965.

100. Jones, A. L., Possible relation of insects to fire blight infection in pear orchards of New York, *Phytopathology*, 55, 1063 (Abstr.), 1965.

101. Stahl, F. J. and Luepschen, N. S., Transmission of *Erwinia amylovora* to pear by *Lygus* spp., *Plant Dis. Rep.*, 61, 936, 1977.

102. Tullis, E. C., Studies on the overwintering and modes of infection of the fire blight organism, *Mich. Agr. Exp. Sta. Tech. Bull.*, 97, 32 pp., 1929.

103. Lelliott, R. A., Fireblight in England: Its nature and its attempted eradication, *EPPO Publ. Serv. A.*, (45-E), 10, 1968.

104. Keil, H. L. and van der Zwet, T., Sodium hypochlorite as a disinfectant of pruning tools for fire blight control, *Plant Dis. Rep.*, 51, 753, 1967.

105. van der Zwet, T. and van Buskirk, P. D., Detection of endophytic and epiphytic *Erwinia amylovora* in various pear and apple tissues, *Acta Hort.*, 151, 69, 1984.

106. Dueck, J., Survival of *Erwinia amylovora* in assocation with mature apple fruit, *Can. J. Plant Sci.*, 54, 349, 1974.

107. van der Zwet, T., Thomson, S. V., Covey, R.P. and Bonn, W.G., Endophytic *Erwinia amylovora* not recovered from core tissues of apples from apparently healthy trees, *Phytopathology*, 76, 1140 (Abstr.), 1986.

108. Ritchie, D. F. and Klos, E. J., Overwinter survival of *Erwinia amylovora* in apple and pear cankers, *Proc. Am. Phytopathol. Soc.*, 2, 67, 1975.

109. Beer, S. V. and Rundle, J. R., Suppression of *Erwinia amylovora* by *Erwinia herbicola* in immature pear fruits, *Phytopathology*, 73, 1346 (Abstr.), 1983.

110. Norelli, J. L., Aldwinckle, H. S. and Beer, S. V., An assay for the virulence of *Erwinia amylovora* using Malus (apple) tissue culture, *Phytopathology*, 76, 656 (Abstr.), 1986.

111. Mills, W. D., Fire blight development on apple in western New York, *Plant Dis. Rep.*, 39, 206, 1955.

112. Luepschen, N. S., Parker, K. G. and Mills, W. D., Five-year study of fire blight blossom infection and its control in New York, *N. Y. (Cornell) Agr. Expt. Sta. Bull.*, 963, 19 pp., 1961.

113. Powell, D., Prebloom freezing as a factor in the occurrence of the blossom blight phase of fire blight of apple, *Trans. Ill. State Hort. Soc.*, 97, 144, 1963.

114. Reinhardt, J. F., The effect of sub-freezing temperatures on viability and pathogenicity of the fire blight pathogen, *Erwinia amylovora* (Burrill) Winslow et al., *Diss. Abstracts*, 18, 358, 1958.

115. Bauske, R. J., Wind dissemination of waterborne *Erwinia amylovora* from Pyrus to Pyracantha and Cotoneaster, *Phytopathology*, 61, 741, 1971.

116. Beer, S. V. and Opgenorth, D. C., *Erwinia amylovora* on fire blight cankers surfaces and blossoms in relation to disease occurrence, *Phytopathology*, 66, 317, 1976.

117. Zoller, B. G. and Sisevich, J., Blossom populations of *Erwinia amylovora* in pear orchards vs. accumulated degree hours over 18.3 C (65 F), 1972–1976, *Phytopathology*, 69, 1050, 1979.

118. Sutton, T. B. and Jones, A. L., Monitoring *Erwinia amylovora* populations on apple in relation to disease incidence, *Phytopathology*, 65, 1009, 1975.

119. Miller, H. J. and van Diepen, H., Monitoring of epiphytic populations of *Erwinia amylovora* in the Netherlands, *Acta Hort.*, 86, 57, 1978.

120. Billing, E., The effect of temperature on the growth of the fireblight pathogen, *Erwinia amylovora* , *J. Appl. Bact.*, 37, 643, 1974.

121. Schouten, H. J., A revision of Billing's potential doublings table for fire blight prediction, *Neth. J. Pl. Path.*, 93, 55, 1987.

122. Lin, C. P., Chen, T. A., Wells, J. M. and van der Zwet, T., In vitro and in situ detection of *Erwinia amylovora* with monoclonal antibodies, *Acta Hort.*, 217, 77, 1987.

123. Hale, C. N., personal communication, 1988.

124. Aldwinckle, H. S. and Beer, S. V., Recent progress in breeding for fire blight resistance in apples and pears in North Amercia, *EPPO Bull.*, 9, 27, 1979.

125. Aldwinckle, H. S. and Beer, S. V., Fire blight and its control, *Hort. Reviews*, 1, 423, 1978.

126. van der Zwet, T., Oitto, W. A. and Blake, R. C., Fire blight resistance in pear cultivars, *HortScience*, 9, 340, 1974.

127. van der Zwet, T. and Keil, H. L., Incidence of fire blight in trunks of 'Magness' pear trees, *Plant Dis. Rep.*, 56, 844, 1972.

128. van der Zwet, T. and Keil, H. L., Relative susceptibility of succulent and woody tissue of Magness pear to infection by *Erwinia amylovora*, *Phytopathology*, 60, 593, 1970.

129. Fisher, E. G., Parker, K. G., Luepschen, N.S. and Kwong, S.S., The influence of phosphorus, potassium, mulch, and soil drainage on fruit size, yield, and firmness of the Bartlett pear and on development of the fire blight disease, *Proc. Am. Soc. Hort. Sci.*, 73, 78, 1959.

130. Spotts, R. A., Stang, E. J. and Ferree, D. C., Effect of overtree misting for bloom delay on incidence of fire blight, *Plant Dis. Rep.*, 60, 329, 1976.

131. Shaw, L., Studies on resistance of apple and other rosaceous plants to fire blight, *J. Agr. Res.*, 49, 283, 1934.

132. Lewis, L. and Kenworthy, A., Nutritional balance as related to leaf compostion and fire blight susceptibility in the Bartlett pear, *Proc. Am. Soc. Hort. Sci.*, 81, 108, 1962.

133. Hildebrand, E. M. and Heinicke, A. J., Incidence of fire blight in young apple trees in relation to orchard practices, *Cornell Univ. Agr. Exp. Sta. Memoir*, 203, 36 pp., 1937.

134. Parker, C. G., Luepschen, N. S. and Fisher, E. G., Tree nutrition and fire blight development, *Phytopathology*, 51, 557, 1961.

135. Schroth, M. N., Moller, W. J., Thomson, S. V. and Hildebrand, D. C., Epidemiology and control of fire blight, *Annu. Rev. Phytopathol.*, 12, 389, 1974.

136. Kleinhempel, H., Nachtigall, M., Ficke, W. and Ehrig, F., Disinfection of pruning shears for the prevention of the fire blight transmission, *Acta Hort.*, 217, 211, 1987.

137. Beer, S. V. and Rundle, J. R., Disinfectants for treating pruning shears used for fire blight infections, *Acta Hort.*, 217, 243 (Abstr.), 1987.

138. Hildebrand, E. M., Fire blight and its control, *Cornell University Ext. Bull.*, 405, 1939.

139. Thomson, S. V. and Purcell, A. H., Survival and transmission of *Erwinia amylovora* by the leafhopper *Graphocephala atropunctata, Phytopathology*, 69, 921 (Abstr.), 1979.

140. van der Zwet, T., Review of fire blight control measures in the United States, *Trans. Ill. State Hort. Soc.*, 101, 63, 1967.

141. Powell, D. and Reinhardt, J. F., The effect of copper sulfate as a dormant spray for fire blight control, *Trans. Ill. State Hort. Soc.,* 88, 161, 1954.

142. McCown, M., Bordeaux spray in the control of fire blight of apple, *Phytopathology*, 19, 285, 1929.

143. McCown, M., Weak Bordeaux spray in the control of fire blight of apple, *Phytopathology*, 23, 729, 1933.

144. Ark, P. A., Use of streptomycin dust to control fire blight, *Plant Dis. Rep.*, 37, 404, 1953.

145. Goodman, R. N., Fireblight control with sprays of agri-mycin, a streptomycin-terramycin combination, *Plant Dis. Rep.,* 38, 874, 1954.

146. Moller, W., Beutel, J. A., Reil, W. O. and Perry, F. J., Fireblight—streptomycin-resistant control studies, 1972, *Calif. Agr.*, 27, 4, 1973.

147. Schroth, M. N., Thomson, S. V. and Moller, W. J., Streptomycin resistance in *Erwinia amylovora*, *Phytopathology*, 69, 565, 1978.

148. Coyier, D. L. and Covey, R. P., Tolerance of *Erwinia amylovora* to streptomycin sulfate in Oregon and Washington, *Plant Dis. Rep.*, 59, 849, 1975.

149. Moller, W. J., Schroth, M. N. and Thomson, S. V., The scenario of fire blight and streptomycin resistance, *Plant Dis.*, 65, 563, 1981.

150. Beer, S. V. and Norelli, J. L., Steptomycin-resistant *Erwinia amylovora* not found in western New York pear and apple orchrds, *Plant Dis. Rep.*, 60, 624, 1976.

151. Manceau, C., Paulin, J. P. and Gardan, L., The use of antibiotics to control fire blight in France. Environmental hazards and established legislation, *Acta Hort.*, 217, 195, 1987.

152. Luepschen, N. S. and Harder, H. H., Pear fire blight control studies in 1973, *Colo. Agr. Expt. Sta. Prog. Rept.*, 73, 1974.

153. Reil, W. O., Thomson, S. V., Schroth, M. N., Griggs, W. H. and Moller, W. J., Pear fire blight control tests, 1973, *Calif. Agr.*, 28, 4, 1974.

154. Deckers, T. and Porreye, W., Chemical control of *Erwinia amylovora* Burrill Winslow et. al. in pear orchards, *Acta Hort.*, 151, 215, 1984.

155. Thomson, S. V., Schroth, M. N., Moller, W. J. and Reil, W. O., Efficacy of bactericides and saprophytic bacteria in reducing colonization and infection of pear flowers by *Erwinia amylovora, Phytopathology*, 66, 1457, 1976.

156. Thomson, S. V., Schroth, M. N., Moller, W. J. and Reil, W. O., A forecasting model for fire blight of pear, *Plant Dis.*, 66, 576, 1982.

157. Covey, R. P., Feasibility of using mean orchard temperature for timing pear fire blight spray in Washington, *Phytopathology*, 71, 104 (Abstr.), 1981.

158. Thomson, S. V., unpublished data, 1988.

159. Billing, E., Weather and fire blight in England, *Ann. Appl. Biol*, 82, 259, 1976.

160. Billing, E., Fire blight in Kent, England in relation to weather (1955–1976), *Ann. Appl. Biol.*, 95, 341, 1980.

161. Billing, E., Fire blight (*Erwinia amylovora*) and weather: A comparison of warning systems, *Ann. Appl. biol.*, 95, 365, 1980.

162. Schouten, H. J., Confidence intervals for the estimation of the incubation period of fire blight following Billing's prediction system 1, *Neth. J. Pl. Path.*, 93, 49, 1987.

163. Billing, E., personal communication, 1988.

164. Schwager, S. J., Beer, S. V., Norelli, J. L., Aldwinckle, H. S. and Burr, T. J., Multiple regression analysis of factors associated with the occurrence of blossom infection in the northeastern United States, *Acta Hort.*, 217, 99 (Abstr.), 1987.

165. Vidaver, A. K., Prospects for control of phytopathogenic bacteria by bacteriophages and bacteriocins, *Annu. Rev. Phytopathol.*, 14, 451, 1976.

166. Beer, S. V., Rundle, J. R. and Norelli, J. L., Recent progress in the development of biological control for fire blight—a review, *Acta Hort.*, 151, 195, 1984.

167. Beer, S. V., Rundle, J. R. and Norelli, J. L., Orchard evaluation of five strains of *Erwinia herbicola* for control of blossom infection, *Acta Hort.*, 217, 219 (Abstr.), 1987.

168. Chatterjee, A. K., Gibbins, L. N. and Carpenter, J. A., Some observations on the physiology of *Erwinia herbicola* and its possible implication as a factor antagonistic to *Erwinia amylovora* in the "fire-blight" syndrome, *Can. J. Microbiol.*, 15, 640, 1969.

169. Goodman, R. N., Protection of apple stem tissue against *Erwinia amylovora* infection by avirulent strains and three other bacterial species, *Phytopathology*, 57, 22, 1967.

170. Wrather, J. A., Kuc, J. and Williams, E. B., Protection of apple and pear fruit tissue against fire blight with nonpathogenic bacteria, *Phytopathology*, 63, 1075, 1973.

171. Phillips, R., Fire blight in New Zealand, *EPPO Ser. A (45-E)*, 20, 1967.

172. Thomson, S. V. and Hale, C. N., A comparison of fire blight incidence and environment between New Zealand and western United States, *Acta Hort.*, 217, 93, 1987.

173. Marco, G. M. and Stall, R. E., Control of bacterial spot of pepper initiated by strains of *Xanthomonas campestris* pv. *vesicatoria* that differ in sensitivity to copper, *Plant Dis.*, 67, 779, 1983.

3

POWDERY MILDEW OF APPLE

K. S. YODER

*Virginia Polytechnic Institute
and State University, Winchester, Virginia*

3-1 INTRODUCTION

Apple powdery mildew caused by *Podosphaera leucotricha* (Ell. and Ev.) Salm. is present in all the apple-growing regions of the world.[1] Once considered primarily a disease of nursery stock and of relatively minor importance in most commercial apple-growing regions,[2,3] it now requires many applications of fungicides for control in some countries. The major factor in its rise to prominence in the apple disease spectrum of the mid-Atlantic region of the U.S. was the replacement of sulfur fungicides with organic fungicides (during the late 1940s and early 1950s) for control of scab (*Venturia inaequalis* (Cke.) Wint.), rusts (*Gymnosporangium spp.*), and other apple diseases.[4]

In some areas, powdery mildew is the major apple disease to be controlled, while in other areas, such as the mid-Atlantic region, it is one of 10 or more fungal diseases which are potential threats to apple production.[4] The relative prominence of these diseases varies with annual precipitation and temperature patterns throughout the region. The severity of powdery mildew and the need for control measures are related to susceptibility to powdery mildew and intended market for the apple cultivar grown.

Reviews of early literature on apple powdery mildew are presented by Fisher[3] and Woodward.[5]

3-2 DISTRIBUTION AND ECONOMIC IMPORTANCE

Apple powdery mildew occurs on all the continents where apples are grown: Africa (Angola, Ethiopia, Kenya, Libya, Morocco, South Africa, Tanzania, Zimbabwe); Asia (Afghanistan, Bhutan, China, India, Iran, Iraq, Israel, Japan, Korea, Lebanon, Nepal, Pakistan, Saudi Arabia, Syria, Taiwan, Turkey, the Soviet Union); Australasia and Oceania (Australia, New Zealand); Europe (Austria, Belgium, Britain and N. Ireland, Bulgaria, Cyprus, Czechoslovakia, Denmark, Finland, France, Germany, Greece, Hungary, Irish Republic, Italy, Netherlands, Norway, Poland, Portugal, Romania, Sweden, Switzerland, the Soviet Union, Yugoslavia); North and South America (Argentina, Belize, Brazil, Canada, Chile, Columbia, Costa Rica, Ecuador, El Salvador, Guatemala, Honduras, Mexico, Nicaragua, Panama, Peru, the United States)[6,7] (Fig. 3-1).

Economic damage from powdery mildew in bearing orchards results from reductions in tree vigor and blossom bud production, aborted blossoms, and fruit russetting (which results in a loss of grade and value). For example, U.S. Department of Agriculture standards downgrade fresh market fruit from extra fancy to No. 1 grade if the area affected by net-like russetting is 25% or more, or if the solid aggregate area is 15% or more.[8] The drop in grade from extra fancy to No. 1 reduces the value of downgraded fruit by approximately 50%. Cumulative effects of mildew infection on yield have been observed by Hickey[4] and van der Scheer.[9] Infection can reduce trunk growth, fruit size, crop weight, and value.[10] Severe infection can reduce the amount of bloom and almost eliminate the crop the following season.[11] In nurseries and young plantings, mildew stunts tree growth and causes poorly formed, misshapen trees.

Mildew infection reduces photosynthesis, transpiration, and carbohydrate content of the host,[12] which affects tree vigor and blossom production. Other physiological responses of the host include an increase in phenolic content,[13] accumulation of sulfur at the site of young mildew colonies, and reduced calcium transport to infected leaves.[14]

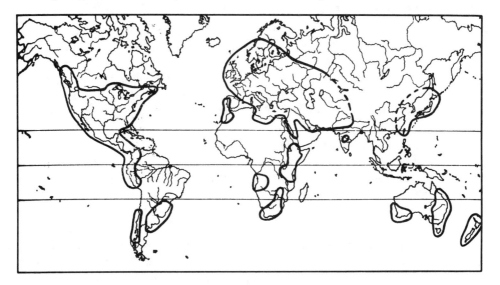

Figure 3-1   Geographical distribution of apple powdery mildew.

3-3 SYMPTOMS

Powdery mildew attacks foliage, blossoms, fruit, and growing shoots (Figs. 3-2 through 3-6). Infection appears on foliage as whitish lesions, typically on the lower leaf surface (Figs. 3-2 through 3-4), although the upper surface may also be infected. Lesions on the lower surface may be accompanied by mild chlorosis in the corresponding position on the upper surface.

Lesions expand into whitish or grayish felt-like patches often covering much of the leaf. Infection of young leaves causes curling, crinkling, reduction in leaf width, and longitudinal folding (Figs. 3-2 and 3-3). Severely infected leaves become brittle and sometimes fall from the trees in mid-season.

Petals of infected flowers are pale yellow or green and covered with mycelium. The flowers are shrivelled, may fail to set fruit, and are more susceptible to spring frosts. Severe blossom infection often also involves the flower cluster leaves (Fig. 3-2).

The effects of powdery mildew on fruit are related to the severity of the infection and developmental fruit stage at which infection occurs. Severely infected blossoms emerging from infected buds may give rise to small, severely russetted fruit. Less severe blossom infection occurring later results in milder, net-like russetting (Fig. 3-5)

Shoot infection occurs on succulent current-season growth (Figs. 3-2 and 3-3). Infected shoots are stunted, have shortened internodes, and are covered with grayish patches of mycelium, the color of which may persist until the following season (Fig. 3-6). Terminal and lateral buds are infected during the season they are produced. Infected buds and twigs are more susceptible to winter injury, resulting in poor tree form, particularly of young, non-bearing trees.

3-4 CAUSAL ORGANISM

Apple powdery mildew is caused by the ascomycetous fungus, *Podosphaera leucotricha* (Ell. and Ev.) Salm. The current nomenclature has been retained since 1900.[15] Earlier synonyms were *Podosphaera kunzei* (Bessey), *Sphaerotheca leucotricha* Ell. and Ev.,[16] *Sphaerotheca castagnei* Lev. f. *mali* Sorauer, *Sphaerotheca mali* Burr., *Albigo leucotricha* (Ell. and Ev.) Kuntze, *Oidium farinosum* Cooke, and *Oidium mespili* Cooke.[3,7]

P. leucotricha conidia are produced in long chains, measure 22–30 × 15–20 μm, and contain distinct fibrosin bodies. Cleistothecia bear both apically and basally inserted appendages. The apical appendages are three to seven times as long as the diameter of the ascocarp and are usually straight and undivided. The basal appendages are rudimentary, rarely well developed, short, and more or less tortuous. The single ascus in the cleistothecium is 55–70 μm × 44–55 μm and contains eight ovate to elliptic ascospores measuring 22–26 × 12–15 μm.[1,7] *P. leucotricha* is heterothallic.[17]

Peach (*Prunus persica*),[7] quince (*Cydonia vulgaris*),[7] *Photinia* spp. and pear (*Pyrus communis* L.)[18] are also hosts for *P. leucotricha*. Other powdery mildews reported on apple are *Erysiphe heraclei* (DC.) St.-Am. and *Podosphaera oxyacanthae* (DC.) de Bary.[1]

Figure 3-2 Primary and secondary mildew infection of Jonathan apple. Aborted blossom cluster at right.

Figure 3-3 Shoot with primary powdery mildew infection (right) compared with healthy shoot (left).

Figure 3-4 Secondary powdery mildew lesion on apple leaf.

3-5 DISEASE CYCLE

The apple powdery mildew fungus overwinters as mycelium in dormant terminal and lateral shoot buds and in blossom buds produced and infected the previous growing season (Fig. 3-7).[5]

Conidia are produced and released from emerging leaves from infected buds.[1,5] Conidia germinate in high relative humidity at 10 to 25 °C (optimum 19 to 22 °C).[2,19] Little germination occurs in free moisture.[5,19] The early-season epiphytotic is regulated more closely by temperature than by humidity. Abundant sporulation from overwintering shoots and secondary lesions on young foliage leads to a rapid buildup of inoculum.

Figure 3-5 Jonathan fruit with russeting caused by powdery mildew infection.

Figure 3-6 Healthy terminal shoot (left) compared with overwintering mildew-infected shoots. [From the files of A. B. Groves]

Secondary infection cycles may be repeated until susceptible tissue is no longer available. Because leaves are most susceptible soon after emergence, infection of new leaves may occur as long as shoot growth continues. Fruit infection occurs near the time of blossoming. Infection of overwintering buds occurs soon after bud initiation.[5]

Cleistothecia are produced on heavily infected shoots and leaves in mid-summer, but they are not regarded as an important inoculum source because the ascospores they contain do not germinate readily.[2,3,20,21]

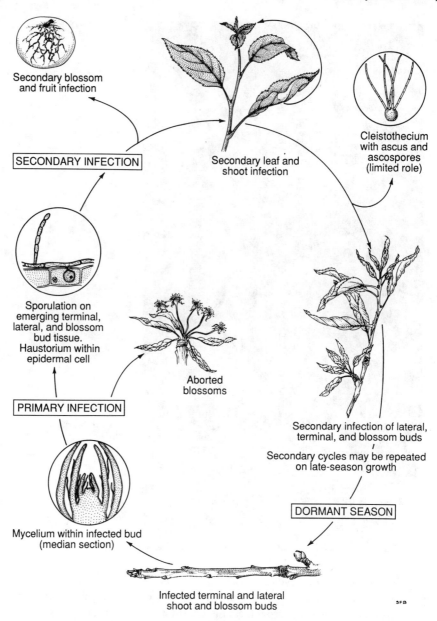

Secondary blossom
and fruit infection

SECONDARY INFECTION

Secondary leaf and
shoot infection

Cleistothecium
with ascus and
ascospores
(limited role)

Sporulation on
emerging terminal,
lateral, and blossom
bud tissue.
Haustorium within
epidermal cell

Aborted
blossoms

PRIMARY INFECTION

Secondary infection of lateral,
terminal, and blossom buds

Secondary cycles may be repeated
on late-season growth

DORMANT SEASON

Mycelium within infected bud
(median section)

Infected terminal and lateral
shoot and blossom buds

Figure 3-7   Disease cycle of apple powdery mildew (*Podosphaera leucotricha*) (Ell. and Ev.) Salm. [Drawn by S. F. Brown]

3-6 INOCULATION AND DISEASE RATING

P. leucotricha is an obligate parasite capable of producing infection in a typical green-house environment; therefore, certain special techniques have been developed for research purposes.

3-6-1 Greenhouse and Laboratory

A supply of uninfected host tissue is a prerequisite to nearly any research endeavor. This may be obtained by growing susceptible plants in an isolated, mildew-free area,[22,23] or under a fine water mist provided by misting nozzles.[24] A single-plant isolation chamber [25] is useful for maintaining healthy plants or individual mildew cultures. Techniques depending on chemical control to maintain a supply of uninfected foliage require adequate controls to assure noninterference by the protection method and to eliminate error from pretreatment infection.[26] Actively growing plants,[26] excised susceptible leaves,[23] or leaf discs[27] provide susceptible foliage.

Plants may be inoculated simply by blowing spores over them from infected leaves; however, results from this method may be too variable for quantitive research. Other methods of inoculation include the use of an air jet and fans to discharge and distribute conidia from infected plants randomly placed among replicated plants,[28] a settling tower,[26] brushing conidia,[27] and spraying with a conidial suspension.[23] A comparison of techniques favored inoculation of leaf discs with a mite brush technique rather than brushing or spraying a conidial suspension.[22]

A technique using parenchymatous apple leaf cells in a synthetic gel medium in petri dish culture has been developed.[29] This medium supports conidial germination and mycelial growth.

Disease may be assessed in greenhouse studies by counting lesions or estimating percent leaf area infected. Disease severity rating scales may be employed based on lesion number, area infected, or overall severity on shoots infected under continuous exposure to inoculum.

3-6-2 Field Studies

Techniques for establishing field plots primarily for the evaluation of fungicides for powdery mildew control have been described.[30–33] The general principles of field plot technique, including randomization and replication, apply especially to field research involving apple powdery mildew.

For field research purposes, a susceptible cultivar appropriate to the region should be selected. Jonathan is a cultivar which is regarded as very susceptible in many parts of the world. Susceptibility of a cultivar to other diseases may also be considered. It is desirable to have uniform or standardized inoculum levels in each test tree to reduce variability of the results. Disease pressure and chemical control are related to primary inoculum levels.[34] Because the fungus overwinters in buds, a year or two may be required to allow inoculum levels to increase and equalize in the research area. In areas where mildew is endemic, but inoculum levels are low, a minimal mildewcide for one year may permit sufficient inoculum buildup under favorable weather patterns.[32] Inoculum may be introduced into the proposed research area on infected potted trees grown in the greenhouse or on excised infected shoots placed in the trees in a container of water to prevent rapid wilting.[31]

It may be useful to record mildew incidence and severity more than once during the growing season. To establish disease progress curves, three or more counts may be required between the first and the last appearance of symptoms. The single most valuable count to assess seasonal control by mildewcides on leaves is in mid-summer after the growth of terminal shoots has ceased.[33] Secondary mildew may be assessed by randomly selecting shoots and recording incidence and percent area infected (severity) on each leaf on the shoot, or on the entire shoot as a whole. A rating system such as that developed by Horsfall and Barratt[35] facilitates mildew severity assessment. Periodically tagging leaves on monitoring shoots as they develop during the season may provide insight into critical periods for control or effectiveness of mildewcides applied to the same tree at different times during the season.

Other diseases and insects may interfere with secondary mildew development or severity assessment. In our experience, it is more likely that severe mildew infection will interfere with development of other foliar disease (e.g., cedar-apple rust, *Gymnosporangium juniperi-virginianae* Schw.).[36] In the mid-Atlantic region of the U.S., the leaf-curling response to the feeding of the potato leafhopper (*Empoasca fabae* Harris) is sometimes confused with mildew leaf curling. In making mildew ratings where leafhopper injury is present, the leaf-curling system is excluded from the severity rating unless a distinct mildew lesion is detected in the affected area. Heavy pubescence on the leaves of cultivars such as Granny Smith can also affect mildew severity assessment.

Treatment effect on potential carryover of primary mildew may be estimated at pink stage the following season by counting the percent of shoots that appear whitened or silvery due to the surface mycelial growth.[31] A more accurate assessment of treatment effect on carryover may be obtained by recording primary infection on tissue emerging from terminal, lateral, or fruit spur buds. The percentage of dead or infected buds can be determined by excising a random sample of whitened shoots, forcing the buds to open in a greenhouse or laboratory in early spring, and assessing infection on the emerging tissue. Infection of dead buds that remain unopened cannot be ascertained, so they must be recorded separately from emerging healthy and infected buds.[33]

Mildew incidence and severity are evaluated on fruit late in the season or at harvest. Evaluation of mildew russet on fruit is quite subjective, particularly on fruit treated with compounds which may cause phytotoxic russetting. Russet symptoms on fruit may be caused by mildew infection, chemical injury, a combination of both, or possibly a change in the appearance of mildew russet by an effective treatment. The investigator may choose to report an overall russet rating which is a composite of all types of russet, or to attempt to separate mildew russet from chemical russetting and to report them as separate readings. Some discussion about the russet based on its appearance and collected data may be helpful to the reader.[37]

3-7 EPIDEMIOLOGY

3-7-1 Effect of Cold on Overwintering

P. leucotricha overwinters as mycelium in infected buds,[5] and its survival is greatly affected by low temperatures.[38–41] Survival of healthy buds at −26°C was similar to infected bud survival at −22°C and, over a 16-year period, terminal mildew infection averaged 27% following winters warmer than −22°C but only 4% following winters with a minimum of −24°C or colder.[39] Lowest survival temperatures ranged from 2.5 to 9.8°C colder for healthy buds than diseased buds during individual winter months.[40]

Although the deleterious effect of extremely cold temperatures on overwintering of apple powdery mildew has long been recognized, the basis for the reduction in disease incidence has remained somewhat elusive. Jeger and Butt[41] provide tentative evidence that temperatures of near $-12\,^{\circ}C$ can kill mycelium in buds and allow them to produce healthy leaves.

3-7-2 Spore Production and Dispersal

There is a potential for production of conidia on tissues emerging from any infected fruit, bud lateral, (axillary) bud, or terminal (extension) shoot bud. Fruit buds are the earliest to emerge and may therefore provide the earliest source of inoculum. Tissues arising from fruit buds (the blossoms and flower cluster leaves) provide a relatively limited surface area for spore production. Terminal shoot buds may become active slightly later than fruit buds but constitute an important inoculum source because of their larger infected area and the continued growth and spore production on new growth well into the season.[42] Lateral buds typically emerge still later and usually do not grow as long but may become dominant if the terminal bud is killed. Infected buds of all types usually emerge slightly later than their healthy counterparts, assuring the presence of susceptible tissue when the first conidia are produced.

Initial spore dispersal has been reported at pink bud[43,44] and usually before or during bloom.[20] In Virginia, conidia have been observed on unfolding lateral bud leaves by tight cluster stage.[45]

A diurnal periodicity of spore release has been recognized for *P. leucotricha*[43,46–48] with a peak of spore collection at mid-day or early afternoon. The concentrations of conidia trapped were positively correlated with wind velocity, temperature and solar radiation and negatively correlated with relative humidity and leaf wetness.[48]

Greenhouse studies by Butt and Jeger[49] found the maximum number of attached conidia per conidiophore 7 to 12 days after the first sign of a colony. Numbers of conidia/conidiophore and conidiophores/mm² of colony on younger leaves and upper leaf surfaces were higher; however, total number of conidia/mm² colony, which also included detached conidia, was greater on lower leaf surface colonies.

3-7-3 Factors Affecting Spore Germination

Some early workers erroneously assumed that spore germination and development of apple powdery mildew in semi-arid regions were dependent on dews.[3] Woodward[5] found only 2 to 3 percent germination of conidia in water, but recognized that conidia would germinate and infect a dry healthy leaf in a moist petri dish.

Berwith[19] reported that high relative humidity (90%) was required for germination of conidia on dry glass slides but that germination in hanging water drops never exceeded 1 percent. Molnar[50] showed that although high relative humidity (96%) is required for germination of conidia on dry glass slides, germination can occur as low as 70% RH on apple leaves and apple and rose petals.

The optimum temperature range for germination of conidia is 20–22 C.[5,19,50] The minimum and maximum temperatures for germination are 4 and 30 C.

3-7-4 The Infection Process

Upon germination the conidium produces a germ tube which penetrates the cuticle.[5] Following pentration of the cuticle the hypha becomes thin and peg-like, penetrates the epidermal cell and forms a haustorium. Woodward observed a slight swelling of the cuticle, which suggests enzymatic action of the fungus, and found no evidence of an appressorium, which would indicated mechanical penetration. Fungal hyphae conformed closely to the surface of the leaf but were not firmly attached until haustoria had been formed. The infection process does not occur when the leaf surface is covered with a water film,

3-7-5 Time of Bud Infection

Lateral buds may be infected soon after they are formed on young shoots in the spring, a process described by Woodward.[5] He observed hyphae from mycelium on the young shoot invading lateral buds and entering a developing fruit bud from the mycelium on the preceding fruit bud. In both cases, the buds were not yet fully protected by scales. He also recognized occasional secondary infection of healthy fruit spurs and shoots by conidia from other sources but did not observe the important secondary infection cycles later reported by Burchill.[42]

Most fruit buds remain briefly susceptible to secondary (and primary) infection.[42] Individual lateral and terminal buds remain susceptible for approximately one month after initiation. However, the lateral buds collectively provide exposed susceptible tissue over a longer period of time because the most recently formed buds remain susceptible until after the terminal bud sets.

In some areas, terminal shoot growth may stop prematurely due to a dry period in early summer and then resume after a favorable rainfall, thus lengthening the period of terminal bud susceptibility.[4] In one study, 67% of all primary infection sites developed from buds that had set more than once during the growing season.[51] Mildew susceptibility of young shoots following a checking of growth due to insect attack has also been noted.[5]

3-7-6 Role of the Cleistothecium

The role of the cleistothecium in the disease cycle of apple powdery mildew has not been adequately demonstrated. Ample evidence has been provided that the fungus commonly overwinters as mycelium in buds, thereby diminishing the importance of the cleistothecium as a means of survival of the fungus. Numerous attempts to germinate ascospores or to inoculate foliage with them have been unsuccessful,[3,19–21] leading to the conclusion that ascospores are not significant in the spread of the disease.

Tsuyama et al.[52] observed spontaneous rupturing of *P. leucotricha* cleistothecia and forcible discharge of asci and ascospores starting in October, about three months after initial formation of cleistothecia in Japan. In periodic checks of discharged ascospore viability, they observed some germination in water in January and February during a period of study from July to February one year and in December, February, and March, but not in January, a second year. They concluded that the ascospores need at least four, and perhaps as long as seven, months to gain germinability from the initial time of cleistothecial formation. The maximum germination reported, 12.5%, was achieved in a study of the temperature effect on germination at 25 °C.

3-7-7 Mathematical Models

Several mathematical models have been presented on various aspects of the epidemiology of apple powdery mildew. These include incidence and severity relationships of secondary infection,[53] the relationship between disease progress and cumulative numbers of trapped conidia,[54] effects of leaf age and fungicidal control,[55] and the functional relationship between primary inoculum and control by the fungicide bitertanol.[34]

3-8 MANAGEMENT

Powdery mildew management strategies are related to cultivar susceptibility,[4,56] the intended market for the cultivar, and the presence or absence of other diseases to be controlled. For susceptible cultivars, the general management strategy is based on reduction of primary inoculum and protection from secondary inoculum. A major component of the management program involves the timely protection of foliage, fruit, and buds by fungicide sprays applied throughout the period of susceptibility.

Because powdery mildew is already present in most of the major apple-production regions of the world, there is little effort to exclude it from present production areas by regulatory means.

Practical integration of biological control into the overall management scheme has not been realized, although at least two possibilities exist. *Cicinnobolus cesatii*, a hyperparasite of *P. leucotricha,* was observed overwintering within mildewed apple buds and *P. leucotricha* cleistothecia.[58] Wet weather was found to favor the development of *C. cesatii,* and its artificial inoculation onto *P. leucotricha* conidiophores showed potential for mildew suppression. Effects of some fungicides on the development of these two fungi have been investigated.[59] Although Woodward[5] observed cecidomyid larvae feeding on *P. leucotricha* conidia and hyphae, he did not mention an effect on mildew incidence.

3-8-1 Disease Resistance

Differences in cultivar susceptibility to apple powdery mildew are widely recognized. In a survey of North American fruit pathologists,[56] cultivars which were ranked moderately to highly susceptible, for which the regular spray schedule was not always adequate, included Jonathan, Cortland, Idared, and Rome Beauty. The cultivars Red Delicious, Golden Delicious, and Winesap were ranked as slightly susceptible, meaning that the regular schedule can sometimes be reduced for powdery mildew control on these cultivars.

The Jonathan cultivar can be considered a standard of susceptibility throughout the world because it is widely planted and its high susceptibility has been noted in numerous locations.[56,60–63] Cox's Orange Pippin, Granny Smith, Gravenstein, and Cortland are also recognized as highly susceptible. By contrast, Red Delicious may serve as a standard of moderate susceptibility or resistance in different locations.[56,60,62–64]

Although differential susceptibility is well known, many highly susceptible culti-vars continue to be planted, even in regions where mildew is a serious problem, because they are valued for their pomological qualities or for pollination purposes.[4,62] This practice increases the disease pressure and the need for control measures on adjacent trees of less susceptible cultivars. In such cases, supplemental sprays may be required on the highly susceptible cultivar to avoid creating a mildew problem on the less susceptible cultivar.

Several cultivars have been shown to have heritable mildew resistance potentially suitable for incorporation in breeding programs. These include Delicious or Delicious progeny,[65–67] Winesap,[65] Lord Lambourne,[68] and Egri Red.[69] Sources of high-level resist-ance or immunity have been located in *Malus robusta* Rehd. and *M. zumi* Rehd.,[70] crabap-ple cultivars David, White Angel, *M. x robusta* (Robusta 5), *M. x robusta* Korea,[71] *M. zumi calocarpa* Rehd., *M. sargenti* Rehd 843 × self, and *M. baccata jackii* Rehd.[66]

Initial disease resistance emphasis was placed on apple scab in United States apple-breeding programs, but mildew resistance is now an objective of breeding programs in the U.S. and other countries.[57] Cultivars with commercial quality and resistance or immu-nity to scab, cedar apple rust, and resistance to powdery mildew are now becoming available.[72]

3-8-2 Cultural Practices

Pruning of dormant shoots infected with mildew the previous season has been recom-mended as a means of reducing primary inoculum.[3,42,61,73,74] Removal of these infected shoots reduced secondary infection by half on mildewcide-treated trees.[74] Pruning of dor-mant tips of all shoots longer than 15 cm reduced the early summer mildew level to only 10 to 20% of that in trees not pruned in this manner.

In situations of heavy infestation, pruning to remove overwintering inoculum has several negative aspects. These include poor economic returns for the amount of labor required, difficulty in locating all the overwintering inoculum sources, and interference of this pruning with preferred tree management practices, tree structure, and crop pro-duction.[3,42] A compromise recommendation involves removing infected terminal shoots from moderately infected small trees during the winter pruning operation, recognizing that thorough removal of infected buds may not be economically feasible on larger, heav-ily infected trees.[75] Although pruning infected dormant tissue is an effective means of reducing inoculum, it should be considered a supplement to, but not a replacement for, routine fungicide applications.

An influence of orchard ground cover management on incidence of apple powdery mildew has been demonstrated.[76] Where clean cultivation of soil was practiced, mildew incidence and nitrogen and potassium levels were significantly higher than where a green cover crop was planted.

3-8-3 Chemical Control

A major effort in powdery mildew management on susceptible apple cultivars is in the timely application of fungicides to protect fruit and foliage from secondary inoculum. Registration and usage of apple fungicides vary from country to country and are fre-quently changing; therefore, this discussion is not intended to be a complete listing of available or registered compounds but a representation of the current situation.

Fungicides currently registered for mildew control in the U.S. include sulfur, dinocap, oxythioquinox, benomyl, thiophanate-methyl, fenarimol, triadimefon, and triforine. Bupirimate, nitrothal-isopropyl, and pyrazophos are examples of compounds available for mildew control elsewhere but not registered in the U.S.[77] The sterol-biosynthesis inhibiting (SBI) fungicides are generally highly effective for control of apple powdery mildew,[78] scab, and rusts.[79] Advances in the development of new fungicides for use on apples in the U.S. are primarily in the SBI area of chemistry, as indicated by 1988 experimental use permits for flusilazol, myclobutanil and triflumazole.

The aforementioned fungicides can provide effective mildew protection when applied at frequent intervals, and the application schedule is generally similar regardless of the fungicide being used. The grower's choice of mildewcide may be based on other characteristics, such as the efficacy of the fungicide against other diseases, registration restrictions, compatibility with other pesticides, phytotoxicity factors, or price, rather than just its effectiveness against mildew.

Development of fungicide resistance by other apple pathogens, as in the case of the benzimidazole fungicides, affects the desirability of the fungicide for the broad-spectrum disease control program, if not specifically for powdery mildew control. The possible development of resistance to the benzimidazole fungicides by *P. leucotricha*[80] and population shifts in sensitivity to SBI fungicides reported for other powdery mildews[81] are ongoing concerns for fungicide control of apple powdery mildew.

In a study of postsymptom activity in the greenhouse, triadimefon reduced the number of normal conidia produced 10 days after treatment, and triadimefon, etaconazole, and sulfur gave reductions 20 days after treatment.[82] Vapor activity by etaconazole has been demonstrated in the greenhouse, but the significance of this effect on mildew control in the orchard is not known.

The most important period for applications of mildewcides on highly susceptible cultivars is from the tight cluster of prepink stages until terminal shoot growth ceases in mid-summer.[83] This may require 8,[4] 10 to 12,[84] or as many as 18 sprays.[85] Although not generally recommended, applications at the 1-cm green-tip stage have sometimes been beneficial.[4] Blossoms must be protected as early as the pink stage to prevent fruit infection.[86] In some years, mildewcides must again be applied in mid- to late season to protect a flush of new growth with highly susceptible tissue such as that produced with favorable rainfall following an early season drought period,[4,87] or following insect injury to shoots.[5] Efforts to reduce primary inoculum with dormant applications of surfactant-type materials to eradicate overwintering mildew have been successful experimentally, but this technique may also have phytotoxic limitations.[88]

Potential phytotoxicity of mildewcides used during the growing season is also an important consideration because the critical period to prevent mildew infection is also a critical time for potential inhibition of pollen germination[89] and spray russeting.[90]

With triadimefon, applications at tight cluster and pink stages could be eliminated without a significant reduction in mildew control on Rome Beauty foliage.[91] The most critical period for control with triadimefon was from bloom through the second cover spray. The earlier sprays during this period more effectively controlled mildew on vegetative shoots, and the later ones were more effective for the later-developing bourse shoots.[91]

Terminal buds were protected by fungicides applied at 10-day intervals during the period after bloom until terminal growth stopped.[92] Although terminal buds were not infected until later in the season, applications soon after bloom more effectively reduced terminal bud infection because they reduced the amount of inoculum available for infecting terminal buds later. The apices of lateral shoots were protected by sprays applied in June. Fruit buds were not protected by applications made after early June.

The length of the interval between applications is an important factor in mildew control, with control often being more enhanced by shortening the interval between sprays rather than increasing the fungicide rate. This phenomenon has been demonstrated with dinocap[93] and with dinocap-benomyl mixtures.[4] No benefit was achieved by doubling the rate of dinocap-benomyl mixtures applied at 14-day spray intervals; however, a reduction of mildew incidence on leaves and fruit was obtained by reducing both the amount of fungicides applied and the length of the application interval. Although application costs may become prohibitive at the shorter (three- to four-day) intervals, control is improved in severe disease situations.

Butt et al.[94] describe a "double-hit" application strategy in which bupirimate and triadimefon were each applied on consecutive days at two-week intervals. Treatments applied in this manner more effectively reduced sporulation than when the same total amount of fungicide was applied at standard intervals of one or two weeks.

To offset the cost of the overall spray program, which includes mildewcides, application times are coordinated for broad-spectrum disease and insect control. In the mid-Atlantic region of the U.S., the timing of mildewcide applications in a typically protective spray schedule for broad-spectrum disease and insect control is seven-day intervals from the green-tip to petal-fall stages (excluding insecticides during bloom) and 14-day intervals during the cover spray period, if both sides of the tree are sprayed on the same dates. If the materials are applied from only one side of the tree on each application date (half-spray), as in the alternate middle system of spraying,[95] intervals between half-sprays are reduced to five days before petal fall and 10 days for cover sprays. The benefits of the alternate middle system of application have been demonstrated by Hickey[4] for control of mildew and other diseases. Where tree size and sprayer capability are matched, this system permits better utilization of short residual pesticides, since applications from alternate sides are made at shorter intervals than sprays applied from both sides on the same date.

Although the application of pesticides in a lower volume of water per treated area is generally more economical, the use of low water volumes (e.g., 47 l/ha on trees requiring 3740 l/ha for complete wetting) for the application of mildewcides may be questionable.[95] High-volume sprays with thorough wetting are recommended,[96] and reduced effectiveness with low liquid quantities, as in helicopter spraying (50 l/ha), has been noted.[97] However, differences in effectiveness related to liquid application volumes ranging from 280 to 935 l/ha are not always evident[98-100] and may be related to differences in disease pressure or other factors.

Other application factors investigated for their possible effect on mildew control include spray velocity,[99] electrostatic charging of droplets,[101,102] and droplet size.[103]

3-8-4 Integrated Management

Successful management of apple powdery mildew involves the integration of many factors. These include disease forecasting based on inoculum assessment and weather conditions, potential crop loss (especially as related to intended market of fruit), varietal composition and susceptibility and age of the trees in the orchard,[104] control of other diseases by mildew fungicides, and timing of broad-spectrum fungicide and insecticide applications to optimize the effectiveness of both.

Butt[105] devised an orchard-scouting method and proposed tentative guidelines for supervised control decisions based on summer mildew ratings. The described scouting method suggests the number of trees to sample per given area. Sampling starts in late May and is repeated at two- to three-week intervals. In a mixed orchard block, the most susceptible cultivar is sampled. Trees to be sampled are marked and mapped to facilitate resampling on each sampling date. Four shoots are rated on each sampled tree. Only the five leaves immediately below a reference leaf near the shoot tip are assessed at each sampling. These leaves are selected because they provide the most recent evidence of mildew activity and will become a major secondary inoculum source. The proposed action thresholds and suggested actions based on this sampling technique are as follows: light disease rating (less than 8% of leaves infected)—opportunity for economy by reducing the fungicide rate or extending the spray interval, particularly during periods of slow shoot growth or unfavorable weather; moderate rating (8 to 20% incidence)—maintain program but increase control if weather becomes highly favorable or shoot growth rapid; potentially severe (more than 20% incidence)—improve control immediately if shoots are growing.

The successful integration of mildew assessment, fungicide selection, and application strategy in a supervised mildew control program provided savings of 9.5 applications and a 30% reduction in fungicide cost compared to a routine program over the three-year study period.[106]

It is important to prevent mildew from becoming established in young trees one to three years after planting. Early exclusion of the disease is particularly critical for highly susceptible cultivars.[104]

Cultivar differences have been detected in the efficiencies of fungicide control.[85] Although unsprayed Golden Delicious and Cox's Orange Pippin are generally considered similar in susceptibility, mildew was more effectively controlled by sprays on Golden Delicious. The cultivar Discovery is considered more resistant than Golden Delicious when unsprayed, but mildew severity was similar when both were sprayed.

Although varietal differences and similarities in mildew susceptibility are recognized, these factors may not be included in decisions regarding the fungicide schedule[85] or in varietal composition of the planting.[4] Orchards could be designed with greater attention to the fungicide requirements of the cultivars included and, once planted, could be maintained accordingly. Improvements in this area would reduce the need for separate applications to portions of a planting. The main difficulty is that some of the mildew-susceptible cultivars, such as Jonathan, have been planted as pollenizers, and their proximity to less susceptible cultivars as a pollen source also assures their proximity as a mildew inoculum source. The introduction and grower acceptance of disease-resistant crabapple pollenizers[107] may help to alleviate this problem.

Another hindrance to the successful integration of management practices for apple powdery mildew control in Virginia involves grower education about the differences between powdery mildew and wet-weather diseases. The grower tends to fear the wet-weather diseases such as scab, rusts, and fruit rots and is less aware that powdery mildew increases rapidly during rain-free periods. In areas where mildew is common, many days are favorable for infection.[108] Many growers are concerned about spray coverage and timing during wet weather but consider periods of dry weather as an opportunity to reduce spray costs as long as insect pests are under control. Growers are more aware of diminishing pesticide residues through rainfall wash-off than they are of the effects of photodegradation, microbial decomposition, and dilution by increased leaf area through growth. Mildew is less likely to affect the fruit directly than are scab and the summer rot diseases and thus receives less attention.

Additional research on the value of certain mildew control practices on individual cultivars would facilitate the grower-education process. Any increase in control practices represents an additional cost the grower is inclined to avoid. Economic data on the value of primary inoculum reduction by pruning or the value of additional early-season or late-season applications are needed in determining the number of applications and the type of fungicide needed for control.

Jeger and Butt[85] point out that although fungicides differ in their field efficacy and epidemiological mode of action (e.g., protectant, antisporulant, curative activity), such attributes are rarely exploited. Ideal characteristics of a mildewcide in the mid-Atlantic region of the U.S. would include (1) curative, antisporulant activity; (2) systemic translocation to unprotected shoot growth during dry weather; (3) residual activity and redistribution during wet weather; and (4) broad-spectrum activity to assure economic attractiveness.[4] Although this list may be too idealistic, several of the SBI fungicides possess several of these attributes. Because some of these compounds have excellent activity against apple powdery mildew, scab, and rust diseases, their use for general early-season disease control in this region will likely improve mildew control coincidentally.[75]

3-9 CONCLUSIONS

Powdery mildew is present in all major apple-producing regions of the world. It is unique among fungal apple diseases because infection occurs during dry weather, necessitating different management strategies than for wet-weather diseases such as scab. Where mildew is common, weather conditions are often favorable for infection while shoot growth is in progress, ensuring at least some infection of susceptible buds and potential overwintering of the disease. Cold winter temperatures are a major factor limiting the overwintering stage, primary inoculum and the subsequent epidemic the following year.

Mildew control strategy is related to varietal susceptibility and is based on reduction of primary inoculum and protection of fruit, foliage, and buds from secondary inoculum. General recommendations call for protection of susceptible tissue with applications of mildewcides from tight cluster stage until terminal shoot growth ceases. The registration and availability of mildewcides vary from country to country. Many of the sterol-biosynthesis inhibiting fungicides are active on mildew and several other important early-season diseases.

Several elusive aspects about apple powdery mildew have implications for control and warrant further research effort.

The role of cleistothecia and ascospores.   The relatively common production of cleistothecia but the rare germination of ascospores remains an intriguing biological curiosity. Several investigators reported little or no germination, and Tsuyama[52] achieved the maximum germination of only 12.5%. In areas where mildew is a common problem, perrenation can be adequately explained on the basis of infected overwintering buds. Although low ascospore germination would indicate low effective inoculum levels and reduced probability of a subsequent epidemic, the cleistothecia could serve as a means of long-term survival of extremely adverse environmental conditions. The possibility of genetic recombination within the cleistothecium, in spite of ascospore germination of only 5 to 10%, also holds implications for development of resistance to fungicides and the durability of host resistance. The trend toward development of more highly selective fungicides and the inclusion of mildew resistance in apple-breeding programs justifies additional investigation into the role of the cleistothecium.

Fungicide resistance.   Development of resistance to preferred mildewcides would greatly affect apple powdery mildew management decisions. Resistance to benomyl has been suspected[80] but not confirmed. Resistance to SBI fungicides has been confirmed in other powdery mildews.[81] Thus, the possibility of resistance to these or other mildewcides by *P. leucotricha* deserves ongoing vigilance.

Mildew-resistant cultivars.   The development and adoption of resistant cultivars may provide the best long-range solution to management of apple powdery mildew. Sources of resistance have been identified, and the selection of mildew-resistant progeny has become a part of apple-breeding programs. However, development of mildew-resistant apple cultivars with commercial pomological qualities may require 20 or more years,[109] and their commercial and public acceptance still longer.

Management programs.   While advances in management have been made since the flare-up of mildew following the switch from sulfur to organic fungicides such as captan in the 1950s, there remains potential for improved control in the areas of orchard planting design, grower education, supervised management, and continued fungicide development. Management can be facilitated by planting cultivars with matched susceptibilities or resistance to mildew and other diseases. In areas where wet-weather diseases are also common, growers must be reminded about subtle but significant losses to mildew through improper application timing and fungicide selection. Development of materials capable of dormant season eradication of *P.leucotricha* and with reduced phytotoxic potential would provide a valuable management tool. New fungicides should be surveyed for possible effects on hyperparasitic activities of *C. cesatii* or other organisms. Tactical application of fungicides at crucial times must be based on research involving individual cultivar susceptibility, growth stage, and environmental factors.

3-10 REFERENCES

1. Burchill, R. T., Powdery mildews of tree crops. In *The Powdery Mildews*. D. M. Spencer, ed., 565, Academic Press, New York, 1978.

2. Anderson, H. W., *Diseases of fruit crops*, McGraw-Hill, New York, 501, 1956.

3. Fisher, D. F., Apple powdery mildew and its control in the arid regions of the Pacific Northwest, *U.S. Dept. Agric. Bull.*, 712, 28, 1918.

4. Yoder, K.S. and Hickey, K. D., Control of apple powdery mildew in the Mid-Atlantic Region, *Plant Disease*, 67, 245, 1983.

5. Woodward, R. C., Studies on *Podosphaera leucotricha* (Ell. and Ev.) Salm., *Trans. Br. Mycol. Soc.*, 12, 173, 1927.

6. Commonwealth Mycological Institute, Distribution Maps of Plant Diseases, Map No. 118, 1987.

7. Kapoor, J. N., *Podosphaera leucotricha*, Commonwealth Mycological Institute Descriptions of Pathogenic Fungi and Bacteria, 158, 1967.

8. U.S. Department of Agriculture, *Apples, Shipping Point Inspection Instructions*, 153, 1978.

9. van der Scheer, H. A., Th., Threshold of economic injury for apple powdery mildew and scab. In *Integrated control of insect pests in the Netherlands*, 49, 1980.

10. Butt, D. J., Robinson, J. D., Souter, R.D., and Swait, A. A. J., Apple mildew crop loss study, *Rep. for 1982, East Malling Research Station, Maidstone, England.* 81, 1983.

11. Yoder, K. S., Byers, R. E., Cochran, A. E. II and Warren, J. R., Comparison of morning and evening concentrate applications of sterol-inhibiting fungicides on Jonathan apple, 1983–84, Fungicide and Nematicide Tests 40, *Am. Phytopathol. Soc.*, 30, 1985.

12. Ellis, M. A., Ferree, D. C. and Spring, D. E., Photosynthesis, transpiration, and carbohydrate content of apple leaves infected by *Podosphaera leucotricha*, *Phytopathology*, 71, 392, 1981.

13. Barnes, E. H. and Williams, E. B., A biochemical response of apple tissues to fungus infection, *Phytopathology*, 50, 844, 1960.

14. Wieneke, J., Covey, R. P., Jr. and Benson, N., Influence of powdery mildew infection on S and Ca accumulations in leaves of apple seedlings, *Phytopathology*, 61, 1100, 1971.

15. Salmon, E. S., A monograph on the Erysiphaceae, *Mem. Torrey Bat. Club*, 9, 1900.

16. Ellis, J. B. and Everhart, B. M., New species of fungi from various localities, *Jour. Mycol.*, 4(6), 58, 1888.

17. Coyier, D. L., Heterothallism in the apple powdery mildew fungus *Podosphaera leucotricha*, *Phytopathology*, 64, 246, 1973.

18. Fisher, D. F., An outbreak of powdery mildew *(Podosphaera leucotricha)* on pears, *Phytopathology*, 12, 103, 1921.

19. Berwith, C. E., Apple powdery mildew, *Phytopathology*, 26, 1071, 1936.

20. Cimanowski, J., Epidemiologia maczniaka jabloniowego *Podosphaera leucotricha* (Ell. et Ev.) Salm. w. Polsce, I. Zimowanie grzyba *Podosphaera leucotricha* (Ell. et Ev.) Salm., *Acta Agrobotanica*, 22, 253, 1969.

21. Molnar, J., Cleistothecia of the fungus *Podosphaera leucotricha* (Ell et Ev.) Salm. under the conditions of Czechoslovakia, *Ceska Mykologia*, 25, 211, 1971.

22. Goszczynski, W. and Cimanowski, J., Evaluation of inoculation techniques of apple leaves with conidia of the fungus *Podosphaera leucotricha* (Ell. et Ev.) Salm. under laboratory conditions, *Fruit Sci. Rep.*, 12, 103, 1985.

23. Yoder, K. S., Methods for monitoring tolerance to Benomyl in *Venturia inaequalis, Monilinia spp.*, and selected powdery mildew fungi. In *Methods for evaluating plant fungicides, nematicides, and bactericides* (E. I. Zehr, Ed.), American Phytopathological Society, 1978.

24. Torgeson, D. C. and Lindberg, C. G., A greenhouse method for evaluation of chemicals to control apple powdery mildew, *Contributions from Boyce Thompson Institute*, 21, 33, 1961.

25. Coyier, D. L., A portable, single-plant isolation chamber, *Can. J. Plant Sci.*, 53, 915, 1973.

26. Kirby, A. H. M. and Frick, E. L., Greenhouse evaluation of chemicals for control of powdery mildews I. A method suitable for apple and barley, *Ann. Appl. Biol.*, 51, 51, 1963.

27. Maszkiewicz, J., Blaszczak, W. and Millikan, D. F., Investigation on the apple proliferation disease I. Increased susceptibility of affected leaf tissue to *Podosphaera leucotricha, Phytoprotection* 60, 47, 1979.

28. Dooley, H. L., Greenhouse method of screening protective fungicides for apple powdery mildew. In *Methods for evaluating plant fungicides, nematicides, and bactericides* (E. I. Zehr, Ed.), American Phytopathological Society, 1978.

29. Chaumont, D. J., Taris, B. and Harada, H., Essais de cultures associees entre des parasites obligatoire et des tissues ou des cellules dissociees (de leur hote) cultives in vitro, Essais de cultures associees de *Podosphaera leucotricha* (Ell. et Ev.) Salm, parasite obligatoire du Pommier, et de cellules dissociees du parenchyme foliare de *Malus pumila Mill.*, *Comptes Rendus*, 278, 465, 1974.

30. Burchill, R. T. and Williamson, C. J., Comparison of some new fungicides for the control of scab and powdery mildew of apple, *Plant Pathology,* 20, 173, 1971.

31. Hickey, K. D., Method for field evaluation of fungicides for apple powdery mildew control. In *Methods for evaluating plant fungicides, nematicides, and bactericides* (E. I. Zehr, G. W. Bird, K. D. Fisher, K. D. Hickey, F. H. Lewis, R. F. Line, and S. F. Rickard, eds.), 141, American Phytopathological Society, St. Paul, MN, 1978.

32. Luepschen, N. S., Testing chemicals in the field for apple powdery mildew control in Colorado. In *Methods for evaluating plant fungicides, nematicides, and bactericides* (E. I. Zehr, G. W. Bird, K. D. Fisher, K. D. Hickey, F. H. Lewis, R. F. Line, and S. F. Rickard, eds.), 141, American Phytopathological Society, St. Paul, MN, 1978.

33. Hickey, K. D., Yoder, K. S. and Zehr, E. I., Methods for field evaluation of fungicides for control of foliar and fruit diseases of apple. In *Methods for evaluating pesticides for control of plant pathogens* (K. D. Hickey, ed.), American Phytopathological Society, St. Paul, MN, 1986.

34. Lalancette, N. and Hickey, K. D., An apple powdery mildew model based on plant growth, primary inoculum, and fungicide concentration, *Phytopathology,* 76, 1176, 1986.

35. Horsfall, J. G. and Barratt, R. W. An improved grading system for measuring plant disease (Abstr.), *Phytopathology,* 35, 655, 1945.

36. Yoder, K. S., unpublished data, 1983.

37. Yoder, K. S., Cochran, A. E. II, Warren, J. R., Schmidt, C. M. and Vann, M. A., Control of powdery mildew on Jonathan apple by SI fungicides, 1987, *Fungicide and Nematicide Tests,* 43, The American Phytopathological Society, 46, 1988.

38. Cimanowski, J., Epidemiologia maczniak jabloniowego *Podosphaera leuoctricha* (Ell. et Ev.) Salm. w Polsce, II. Rozwoj grzyba *Podosphaera leucotricha* (Ell. et Ev.) Salm. w okresie wegetacji, *Acta Agrobotanica* 22, 278, 1969.

39. Spotts, R. A., Covey, R. P. and Chen, P. M., Effect of low temperature on survival of apple buds infected with the powdery mildew fungus, *HortScience,* 16, 781, 1981.

40. Spotts, R. A. and Chen, P. M., Cold hardiness and temperature response of healthy and mildew-infected terminal buds of apple during dormancy, *Phytopathology*, 74, 542, 1984.

41. Jeger, M. J. and Butt, D. J., The effects of weather during perennation on epidemics of apple mildew and scab, *EPPO Bull.,* 13, 79, 1983.

42. Burchill, R. T., The role of secondary infections in the spread of apple mildew *(Podosphaera leucotricha* (Ell. and Ev.) Salm.), *J. Hortic. Sci.,* 35, 66, 1960.

43. Burchill, R. T., Seasonal fluctuations in spore concentrations of *Podosphaera leucotricha* (Ell. and Ev.) Salm. in relation to the incidence of leaf infections, *Ann. Biol.,* 55, 409, 1965.

44. Molnar, J., The occurrence, concentration of conidia, and course of secondary infection with *Podosphaera leucotricha* (Ell. et Ev.) Salm., Mucnatky Jablenovej. *Ochrana Rostlin,* 43, 207, 1970 (edited by Carole Steere, 1974).

45. Yoder, K. S., unpublished data, 1979.

46. Childs, James F. L., Diurnal cycle of spore maturation in certain powdery mildews, *Phytopathology,* 30, 65, 1940.

47. Pady, S. M., Spore release in powdery mildew, *Phytopathology,* 62, 1099, 1972.

48. Sutton, T. B. and Jones, A. L., Analysis of factors affecting dispersal of *Podosphaera leucotricha* conidia, *Phytopathology,* 69, 380, 1979.

49. Butt, D. J. and Jeger, M. J., Components of spore production in apple powdery mildew *(Podosphaera leucotricha), Plant Pathology*, 35, 491, 1986.

50. Molnar, J., The effect of environmental factors on the germination of conidia of *Podosphaera leucotricha* (Ell. et Ev.) Salm., *Ochrana Rostlin,* 44, 7, 1971 (edited by Carole Steere, 1974).

51. Berkett, L. P. and Hickey, K. D., Location of primary infection sites of apple powdery mildew and its significance in management strategies, *Phytopathology,* 72, 705, 1982.

52. Tsuyama, H., Nagai, M. and Aizawa, T., Germination of ascospore of apple powdery mildew, *Journal of the Faculty of Agriculture, Iwate University,* 8, 235, 1967.

53. Seem, R. C. and Gilpatrick, J. D., Incidence and severity relationships of secondary infections of powdery mildew on apple, *Phytopathology,* 70, 851, 1980.

54. Jeger, M. J., Relating disease progress to cumulative numbers of trapped spores: Apple powdery mildew and scab epidemics in sprayed and unsprayed orchard plots, *Plant Pathology,* 33, 517, 1984.

55. Lalancette, N., Jr. and Hickey, K. D., Apple powdery mildew disease progress on sections of shoot growth: An analysis of leaf maturation and fungicide effects, *Phytopathology,* 75, 130, 1984.

56. Aldwinckle, H., Field susceptibility of 51 apple cultivars to apple scab and apple powdery mildew, *Plant Dis. Rep.,* 58, 625, 1974.

57. Sutton, T. B. and Jones, A. L., Apple Disease Management. In *CRC handbook of pest management in agriculture,* Vol 3 (D. Pimentel, ed.), CRC Press, Boca Raton, Fl, P9-J, 1981.

58. Kobakhidze, D. M., O roli *Cicinnobolus cesatii* de Bary podavlenii razvitiya *Podosphaera leucotricha* (Ell. et Ev.) Salm., *Bot. Zh. SSSR,* 50, 1307, 1965.

59. Kobakhidze, D. M., Deistvie razlichnykw fungitsidor na muchnistuyu rosu yabloni, *Tr. Inst., Zashch. Rast. Tiflis,* 20, 24, 1964.

60. Docea, E., Mihaescu, G. and Ignat, V., Behavior of some apple varieties when attacked by *Podosphaera leucotricha,* Lucrari Stiintifice. Seria B. *Hoticultura,* 23, 61, 1980.

61. Penrose, L. J., Controlling apple mildew, *The Agricultural Gazette of New South Wales,* 159, 1973.

62. Sprague, R., Apple powdery mildew spray trials, *Plant Dis. Rep.,* 37, 601, 1953.

63. Dar, G. N. and Kaul, R. N., Powdery mildew disease of apple and its control in Kashmir, *Pestology,* 5, 9, 1981.

64. Wilcox, W. F., Riedl, H. and Stiles, W. C., 1986 *Cornell Recommendations for Commercial Tree-Fruit Production, Apple Production Systems,* Cornell University, Ithaca, NY, 1986.

65. Mowry, J. B., Inheritance of susceptibility of apple to *Podosphaera leucotricha, Phytopathology,* 55, 76, 1965.

66. Dayton, D. F., Genetic immunity to apple mildew incited by *Podosphaera leucotricha, Hort Science,* 12, 225, 1977.

67. Korban, S. S. and Dayton, D. F., Evaluation of *Malus* germplasm for resistance to powdery mildew, *HortScience,* 18, 219, 1983.

68. Visser, T., Verhaegh, J. J. and de Vries, D. P., Resistance to scab *(Venturia inaequalis)* and mildew *(Podosphaera leucotricha)* and fruiting properties of the offspring of the apple cultivar Antonovka, *Rev. Plant Pathology,* 54 (Abstr.), 190, 1975.

69. Kovacs, S., Breeding of new disease resistant apple varieties, *Fruit Varieties Journal,* 39, 26, 1989.

70. Knight, R. L. and Alston, F. H., Sources of field immunity to mildew *(Podosphaera leucotricha)* in apple, *Can. J. Genet. Cytol.,* 10, 294, 1968.

71. Gallott, J. C., Lamb, R. C. and Aldwinckle, H. S., Resistance to powdery mildew from some small-fruited *Malus* cultivars, *HortScience,* 20, 1085, 1985.

72. Brown, S., Lamb, R. C. and Way, R. D., *Pome fruit, new and noteworthy fruits,* New York State Fruit Testing Cooperative Assoc., Inc., Geneva, NY, 3–10, 1987–88.

73. Baker, J. V., Winter and spring pruning against apple mildew, *Proc. British Insecticide Fungicide Conf.,* 1, 179, 1961.

74. Ketskhoveli, E. B., Effectiveness of curative pruning against powdery mildew of apple *Podosphaera leucotricha* (Ell. and Ev.) Salm., *Rev. Plant Pathology,* 55, 525, 1976.

75. Yoder, K. S., Control of apple powdery mildew with sterol-inhibiting fungicides, *Proc. Michigan State Horticultural Society,* 117, 61, 1987.

76. Schmidle, A., Dickler, E., Seemuller, E., Krozal, H. and Kunze, L., Effect of fertilization and soil management on the occurrence of diseases and pests in apple orchards, *Zeitschrift fur Pflanzenkrankheiten und Pflanzenschutz,* 82, 522, 1975.

77. Thomson, W. T., *Agriculture chemicals;* Book IV, Fungicides, 1987–88 rev., 196, Thomson Publications, Fresno, 1987.

78. Hickey, K. D. and Yoder, K. S., Field performance of sterol-inhibiting fungicides against apple powdery mildew in the mid-Atlantic apple growing region, *Plant Disease,* 65, 1002, 1981.

79. Yoder, K. S. and Hickey, K. D., Sterol-inhibiting fungicides for control of certain diseases of apple in the Cumberland-Shenandoah region, *Plant Disease,* 65, 998, 1981.

80. Luepschen, N. S. and Smiley, E. T., Apple mildew studies 1978: Possible tolerance to Benlate and effectiveness of tank-mix and alternating-schedule programs. In *Progress report,* Colorado State University Exp. Station 24, 1978.

81. Koller, W. and Scheinpflug, H., Fungal resistance to sterol-biosynthesis inhibitors, a new challenge, *Plant Disease,* 71, 1066, 1987.

82. Cimanowski, J. and Szkolnik, M., Postsymptom activity of ergosterol inhibitors against apple powdery mildew, *Plant Disease,* 69, 562, 1985.

83. Jones, A. L. and Sutton, T. B., *Diseases of tree fruits,* North Central Regional Extension Publ. No. 45, Michigan State Cooperative Extension Service, East Lansing, 59, 1984.

84. Motte, G., Zimmerman, U., Jahn, M., and Burth, V., Der Einflub von Behandlungen gegen Apfelschorf (*Venturia inaequalis*) and Apfelmehltau (*Podosphaera leucotricha*) aud das Auftreten parasitarer Lagerfaulen, *Nachrichtenblatt fur den Pflanzenschutz in der DDR,* 37, 239, 1983.

85. Jeger, M. J. and Butt, D. J., Management of orchard diseases in the United Kingdom, *Plant Prot. Bull.* 32, 61, 1984.

86. Daines, R., Weber, D. J., Bunderson, E. D. and Roper, T., Effect of early sprays on control of powdery mildew fruit russet on apples, *Plant Disease,* 68, 326, 1984.

87. Berkett, L. P., Apple powdery mildew: Relation of time of application to efficacy of a sterol-inhibiting fungicide, the effect of late season mildew on apical bud survival, and an evaluation of a management tactic against late season secondary mildew, Ph.D. thesis, Pennsylvania State Univ., State College, 1985.

88. Burchill, R. T., Frick, E. L., Cook, M. E. and Swait, A. A. J., Fungitoxic and phytotoxic effect of some surface-active agents applied for the control of apple powdery mildew, *Ann. Appl. Biol.,* 91, 41, 1979.

89. Butt. D. J., Swait, A. A. J. and Souter, R. D., Apple powdery mildew, *Podosphaera leucotricha,* blossom phytotoxicity trial, *Rep. for 1978, East Malling Research Station,* Maidstone, England, 88, 1979.

90. Butt, D. J., Kirby, A. H. M. and Williamson, C. J., Fungitoxic and phytotoxic effects of fungicides controlling powdery mildew on apple, *Ann. Appl. Biol.,* 75, 217, 1973.

91. Berkett, L. P.,Hickey, K. D. and Cole, H., Jr., Relation of application timing to efficacy of triadimefon in controlling apple powdery mildew, *Plant Disease,* 72, 310, 1988.

92. Butt, D. J., The timing of sprays for the protection of terminal buds on apple shoots from powdery mildew, *Ann. Appl. Biol.,* 72, 239, 1972.

93. Groves, A. B., Wampler, E. L. and Lyon, C. B., The development of an efficient schedule for the use of Karathane in the control of apple powdery mildew, *Plant Dis. Rep.,* 42, 252, 1958.

94. Butt, D. J., Swait, A. A., J., Jeger, M. J., Wood, S. J. and Robinson, J. D., Apple powdery mildew *Podosphaera leucotricha,* control, *Rep. for 1982, East Malling Research Station,* Maidstone, England, 81, 1983.

95. Lewis, F. H. and Hickey, K. D., Fungicide usage on deciduous fruit trees, *Ann. Rev. Phytopathol.,* 10, 399, 1972.

96. Victoria Department of Agriculture, Fruit Crop Branch, Powdery mildew of apples, *Agnote,* 1979.

97. Schellenberg, G. H., Motte, G., Burth, U. and Kohler, S., Erfahrungen und Ergebnisse zum Einsatz des Hubschraubers bei der Bekampfung pilzlicher Erkrankungen in der Apfelproduktion, *Nachrichtenblatt fur den Pflanzenschutz in der DDR,* 32, 153, 1978.

98. Hickey, K. D., Apple fungicide test, 1969, Report No. 25, *Fungicide and nematicide tests,* 26, The American Phytopathological Society, 17, 1970.

99. Hickey, K. D., Apple fungicide test, 1970, Report No. 25, *Fungicide and Nematicide Tests,* 27, The American Phytopathological Society, 19, 1971.

100. Hickey, K.D., Apple fungicide test, 1971, Report No. 26, *Fungicide and Nematicide Tests,* 28, The American Phytopathological Society, 19, 1972.

101. Swait, A. A. J., Butt, D. J. and Wood, S. J., Spray application studies, *Rep. for 1982, East Malling Research Station,* Maidstone, England, 81, 1983.

102. Garretson, M., Hickey, K. D. and May, J., Evaluation of an electrostatic low volume sprayer for scab and mildew control, *Fungicide and Nematicide Tests,* 37, The American Phytopathological Society, 6, 1981.

103. Allen, J. G., Butt, D. J., Dicker, G. H. L. and Hunter, L. D., A comparison of the efficiency of sprays of two drop-size ranges in an apple orchard, *Pesticide Science,* 9, 545, 1978.

104. Zimmerman, V. and Motte, G., Befallsverlauf beim Apfelmehltau *(Podosphaera leucotricha* Ell. et Ev. Salm.) im Havellandischen Obstbaugebiet unter besonderer Beruck sichtigung von Apfeljunganlagen, *Nachrichtenblatt fur den Pflanzen schutz in der DDR,* 34, 223, 1980.

105. Butt, D. J., An improved apple mildew assessment method for use in supervised control, *Rep. for 1978, East Malling Research Station,* Maidstone, England, 211, 1979.

106. Butt, D. J., Souter, R. D., Jeger, M. J., Swait, A. A. J., Harris, M. A. and Robinson, J. D., Management of apple orchard diseases, *Rep. for 1982, East Malling Research Station,* Maidstone, England, 82, 1983.

107. Crassweller, R. M., Ferree, D. C. and Nichols, L. P., Flowering crab apples as potential pollenizers for commercial apple cultivars, *J. Amer. Soc. Hort. Sci.,* 105(3), 475, 1980.

108. Roosje, G. S., Besemer, A. F. H., Meijncke, C. A. R. and Post, J. J., *Observations and research on apple powdery mildew in the Netherlands, 1953 to 1963,* Agricultural Publications and Documentation Center, 79, 1965.

109. Brown, A. G., Apples. In *Advances in fruit breeding* (J. Janick and J. N. Moore, eds.), 623, Purdue University Press, West Lafayette, IN, 1975.

4

DOWNY MILDEW OF GRAPES

R. W. EMMETT

Sunraysia Horticultural Centre, Department of Agriculture
P.O. Box 905, Mildura, Victoria, Australia, 3502

T. J. WICKS

Northfield Research Laboratories, Department of Agriculture
G.P.O. Box 1671, Adelaide, South Australia, Australia, 5001

P. A. MAGAREY

Loxton Research Centre, Department of Agriculture
P.O. Box 411, Loxton, South Australia, Australia, 5333

4-1 INTRODUCTION

Downy mildew is a most destructive fungal disease of grapevines and occurs in most grape-growing areas of the world. Because of its economic importance, it is one of the most widely studied diseases.

Downy mildew is indigenous to eastern North America,[1,2] where it spread from wild grapes to cultivated vineyards. The disease was not reported in Europe until 1878, when it was apparently introduced into France on a grape cultivar imported from the U.S.A. for use as a rootstock resistant to grape phylloxera (*Daktulosphaira vitifolii*). This aphid pest was also introduced from North America prior to 1863. In western Europe, downy mildew spread rapidly with substantial economic and social impact during the 1880s and was reported in Britain by 1894.[3] It was recorded in Brazil in 1893, South Africa in 1907, and Australasia in 1916. Introduction into these countries was probably by way of infected nursery stock.[1]

4-2 DISTRIBUTION AND ECONOMIC IMPORTANCE

Downy mildew has been recorded in 91 countries,[4] from temperate zones to the tropics (Fig. 4-1). It occurs in most grape-growing areas of the world including eastern and western Europe, Asia, Africa, Australasia, and the Americas. Weltzien[5] reviewed factors associated with occurrence of the disease and its economic importance. Temperature,

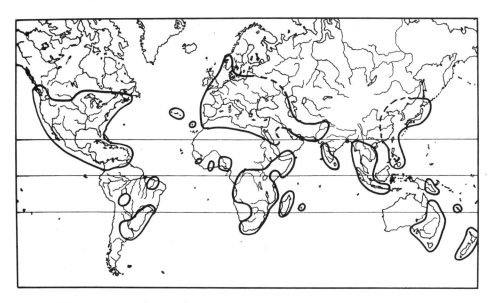

Figure 4-1   Geographical distribution of downy mildew of grape.

wetness duration, and relative humidity at night in relation to foliage and crop development are of particular significance as these influence infection, sporulation, and sporangial survival.

Localities with frequent summer rain and high humidity at night with temperatures
above 13°C are favorable for disease development (e.g., parts of France, Germany,
South Africa, and Australia). In areas with a relatively dry growing season, the incidence
of epidemics is infrequent (e.g., some parts of Australia and the Mediterranean). In
Mendoza, Argentina, the climate is very dry and downy mildew is of little importance. In
Afghanistan, California, Chile, Egypt, and Western Australia, the absence of suitable
conditions, usually through the lack of spring and summer rainfall, prevents disease.[5-7]

Most economic loss is associated with cluster destruction and loss of vine foliage
(or photosynthetic area).[8] When conditions favor downy mildew on flowers or young
berries, crop losses from 50%[9] to 100% occur. Damage is greatest following early infection. In epidemic situations, vines may be defoliated. This prevents the maturation of fruit
and canes and exposes fruit to sunburn. In the next season, vine vigor and crop potential
may be reduced because downy mildew infection and subsequent defoliation depletes
carbohydrate reserves. As a result, canes are weakened, become less winter hardy, and
die back more readily than those on healthy vines. In the following growing season, bud
burst may be retarded and crop potential reduced due to lower numbers of viable buds.[10]

4-3 SYMPTOMS

Leaves, tendrils, shoots, inflorescences, and clusters can be affected by downy mildew.
Leaves are most susceptible to infection during active growth.[11] Symptoms on leaves vary
according to leaf age and prevailing weather (Figs. 4-2a through 4-2c). On young leaves,

first symptoms of disease appear as oil spots—circular yellow translucent lesions, often surrounded by an oily brown halo (Fig. 4-2a). This halo fades as the oil spots expand. Later, oil spots may coalesce to cover the entire leaf surface. Under favorable conditions, masses of white downy sporangiophores and sporangia develop from stomata on the undersides (Fig. 4-2b) of oil spots. Given suitable weather and soft young growth, sporangiophores may develop up to one day before oil spots appear. During hot, dry weather or after sporulation, oil spots turn brown or reddish brown, dry out, and die. However, given suitable conditions, further sporulation may occur around the outer growing edge of necrotic tissues. On aged leaves, symptoms appear as a mosaic of small,

Figure 4-2a   Downy mildew oil spots on the upper surface of leaves.

Figure 4-2b   Masses of downy mildew spores on the underside of oil spots.

Figure 4-2c Bunch rot caused by downy mildew early in the season.

angular, yellow to reddish-brown lesions limited by veinlets. Severely affected leaves fall prematurely. When young shoots and tendrils are infected, they turn brown and become stunted, distorted, and necrotic. Shoots are rarely infected and then only when up to 10 to 15 cm long. Nodes are more susceptible than internodes.[11]

Flower clusters are highly susceptible. Infected inflorescences first turn oily yellow-brown and may develop white down (sporangia) during periods of high humidity at night. When young peduncles are infected, entire clusters may die. Occasionally, parts of clusters wither and die when only pedicels are diseased (Fig. 4-2c). Individual flowers are also susceptible to infection.

Young berries are most susceptible to infection from setting until 5 to 6 mm in diameter. Infected berries stop growing, harden, and become dull bluish-green before turning brown, withering, and falling from the clusters. Older berries become resistant to infection probably because stomata on berries are no longer functional,[11] but they may be killed when pedicels and cluster stems (which remain susceptible) are diseased. Sporangia are not produced on mature berries but may develop on diseased pedicels or cluster stems.

A simple diagnostic method to confirm the suspected presence of downy mildew is to lightly moisten leaves or clusters with water, enclose them in a polyethylene bag (to maintain high relative humidity), and incubate them overnight (in darkness) at 20 to 25°C. White downy growth (sporangia) will be produced on active infected tissue.

4-4 CAUSAL ORGANISM

Grape downy mildew is caused by the obligate fungus *Plasmopara viticola* (Berk. & Curt.).[12] Historically, *P. viticola* has also been classified taxonomically as *Botrytis cana* Llk. (Schweinitz),[13] according to Wilson,[14] *Botrytis viticola* Berk. & Curt.,[15] and *Peronospora viticola* De Bary.[1,9,11,16,17] Apart from original descriptions, the morphology of *P. viticola* has been summarized by various authors.[9,11,17,18]

Asexual sporangia are hyaline, ellipsoid, and papillate ($17-25 \times 10-16 \mu$m). They are produced at the tips of dichotomously branched sporangiophores (mostly 140–250 μm long) that usually arise from mycelial cushions in the substomatal cavity. Germinating sporangia produce 1 to 6 zoospores ($6-8 \times 4-5 \mu$m) with two lateral flagellae.

Zoospores encyst and produce a flexuous nonseptate germ tube 50 to 80 μm long. A substomatal vesicle, hyphae, and haustoria are produced in infected tissue. For illustrations, see Galbiati.[19] Tubular coenocytic mycelia vary from 1 to 60 μm in diameter as hyphae take up, and conform to, the shape of intercellular spaces in infected tissues. Haustoria are globular or pear shaped and 4 to 10 μm in diameter.

Diploid oospores produced in host tissues are uninucleate, pigmented, and enclosed by two membranes and covered by the wrinkled wall of the oogonium. They are 28 to 40 μm in diameter. Antheridia are rarely seen.[9] Oospores germinate to produce a slender germ tube of variable length, 2 to 3 μm in diameter, terminating in a piriform sporangium $35-40 \times 25 \mu$m, larger than secondary sporangia produced on host tissues. Sporangia from oospores germinate to produce 8 to 20 zoospores.[11] Little is known about the diversity of strains of *P. viticola*, although ecotypes may exist.[20,21] Strains of varying virulence to *Vitis* and related genera have been recorded.[22,23]

Apart from virulence to cultivated grapes (mainly *Vitis vinfera* L. and hybrids with other *Vitis* species), *P. viticola* has low virulence on numerous *Vitis* species native to North America such as *V. berlandieri* Planch., *V. aestivalis* Michx., *V. cordifolia* Michx., *V. riparia* Michx., and *V. labrusca* L.[1] *P. viticola* has also been observed on wild hosts such as *Ampelopsis veitchii* Hort., *Cissus (Vitis) caesia* Afzel., *Parthenocissus tricuspidata* (L.) Planch. (English ivy), *P. quinquefolia* (L.) Planch. (Virginia creeper), *Vitis coignetiae* Pulliat ex Planch., and *Vitis californica* Benth.[24] In Australia, the fungus has also been observed on *Cissus hypoglauca* A. Gray and on *Ampelocissus acetosa* (S. Mull.) Planch.[25] *C. hypoglauca* is infected by *P. viticola* from *V. vinifera*, but it is uncertain whether the reverse applies.[25] Attempts to inoculate *P. quinquefolia* with Australian isolates of *P. viticola* have not been successful in preliminary tests.[26]

In California, Santelli[22] found the physiologic form of *P. viticola* isolated from *V. californica* also infected *V. girdiana*, *V. arizonica* Engel, and *V. treleasei* but did not reproduce on *V. vinifera* cultivars. However, *P. viticola* from *V. vinifera* infected all *Vitis* species tested. Strains of *P. viticola* with different sensitivity to phenylamide (acylanilide) and phosphonate fungicides have also been reported in Europe.[27,28]

4-5 DISEASE CYCLE

In addition to early studies of the biology of grape downy mildew in France[29–32] and America,[33–35] a large number of epidemiological studies have contributed to our present understanding of the disease life cycle. The following is a synopsis, which is presented diagrammatically in Fig. 4-3.

P. viticola overwinters predominantly as thick-walled oospores within leaf debris. However, where the winter period is relatively indistinct, the mycelium of *P. viticola* may survive between bud scales or in diseased leaves that remain on the plant until bud burst.[9,36,37] There is a possibility that *P. viticola* overwinters in buds even in areas with distinct winters.[38] *P. viticola* also overwinters in the cortical tissue and cane buds of the wild species, *Vitis californica* Benth. in California,[39] but this is not a source of downy mildew on cultivated grapes. In other regions, the role of wild hosts in relation to the perpetuation of downy mildew on cultivated grapes is uncertain but appears to be of little importance.

Oospores form in infected leaves during late summer and autumn and survive in leaf debris for at least two years in or on the soil surface. When mature, they germinate to produce a sporangium that releases zoospores. Inoculum for primary infection arises from either airborne sporangia or splash-dispersed zoospores.[40] When the undersurface of leaves is wet, motile zoospores are attracted to stomata, encyst, and germinate to produce a germ tube that penetrates the stomatal aperture. The chemical and physical attraction of zoospores to stomata has been investigated by Royle and Thomas.[41]

After an incubation period, sporangiophores and sporangia are produced through stomatal openings on warm, humid nights. Secondary infection follows wind or water-splash dispersal of these sporangia.

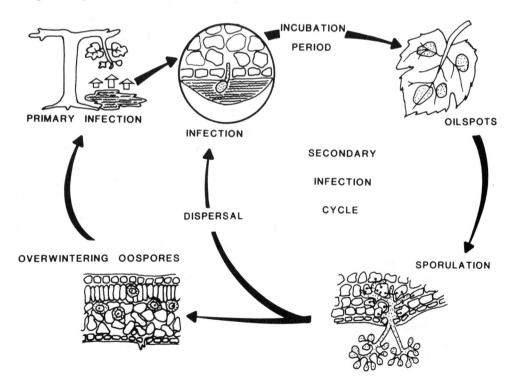

Figure 4-3 Disease cycle of grapevine downy mildew.

4-6 EPIDEMIOLOGY

4-6-1 Oospores

Formation.   Oospores are the main means of survival of *P. viticola* from season to season and are a source of primary infection. They are mostly formed in leaf tissues toward the end of the growing season when conditions are less suitable for the asexual reproduction of the fungus. However, oospore formation may occur at any time during the growing season.[42] Studies of oospore formation have been reviewed by Ronzon,[43] but conditions for formation are not completely understood. Some studies indicate that temperatures of 14 to 18°C favor development,[44] but others suggest that temperature may not be as important as autumn rainfall.[45,46] Generally, formation takes place in aging leaves with tapestry-like symptoms resulting from late-season infections, although it may occur in drought-affected leaves earlier in the season.[42] More than 200 oospores/mm² have been observed.[47-49] On potted plants in the glasshouse, oospores are produced throughout the year in older leaves where lesions remain small. They also occur more abundantly on *Vitis* species and on vine cultivars that inhibit asexual sporulation than on hosts supporting normal sporulation.[50] In the field, oospores are mostly formed in autumn. If formation occurs in summer, it is usually associated with low humidities that inhibit asexual sporulation.

In laboratory tests, Galbiati and Longhin[51] found that oospores formed most abundantly when inoculated grapevine leaves were kept at 16 to 18°C at 75 to 85% relative humidity (RH) with a photoperiod of 14 hours. However, Ronzon-Tran Manh Sung and Clerjeau[42] found that oogenesis in leaves was stimulated by thermal shock from 20 to 10°C.

Maturation and Survival.   Some authors believe that after formation, oospores enter a temperature-dependent dormant period. Maturation that follows is influenced primarily by winter rainfall and soil temperature.[43,52-56] It is uncertain whether maturation can proceed before dormancy is completed.[43] Under controlled conditions, oospores mature at alternating weekly temperatures of 10 and 5°C.[42]

Under field conditions, most oospores germinate over a period of two to three months in late winter or spring. The calendar months when germination occurs vary from region to region.[48] Generally, oospores germinate more abundantly and in shorter time early in the season when they have been subjected to frequent rains and mild winter and spring temperatures. Oospores not exposed to winter rain are unable to germinate. It has also been reported that low temperatures reduce the dormant period.[48] In the laboratory, Galbiati and Longhin[51] recorded optimum oospore viability after inoculated leavés containing oospores soaked in water were stored at 4°C for 20 to 30 days.

The longevity of oospores in soil is normally limited to the spring and early summer following their formation. Survival until spring of the second year after they have been protected from drying is considered exceptional by some authors.[50,56,57] In Europe, however, some oospores appear to persist in soils for at least three seasons. When conditions for germination are fulfilled, not all oospores germinate. Some remain viable and germinate in subsequent seasons.[47] In Australia, the sudden reappearance of downy mildew after more than two to three years of absence in vineyards is consistent with European experience.

Temperature has little effect on survival.[11,43] Gaumann[58] reported that oospores survived -26°C for five days. Survival appears to be favored by the presence of moisture but reduced by drought[11] or when leaves are deeply buried.[57]

Germination.   Oospores rapidly germinate in water to produce a short germ tube terminating in a single sporangium from which zoospores are released. Minimum temperatures for germination range from 7 to 13°C.[52] Most authors agree that germination commences when temperatures reach 8°C, although its progress is slow between 8 and 10°C, and vineyard soils must be saturated for long periods.[52] Muller and Sleumer[50] found the temperature range for germination was 11 to 32°C. Reported optimum temperatures range from 20 to 22°C[43,48,51] to 25°C.[50] At 21°C, the germination process takes 0.5 to 2 hours compared to 2 to 6 hours at 10°C.[18] High light intensities may retard or prevent germination.[40]

4-6-2 Primary Infection

Primary infection occurs when water-borne sporangia and/or zoospores produced by germinating oospores are splashed or blown onto wet vine foliage. Zoospore release from sporangia commences within 30 minutes at 20°C.[59] In the film of water on the surface of vine tissues, zoospores (propelled by their flagellae) swim in a whorling motion toward stomata. Zoospore swarming may continue for at least 3 to 5 hours.[40] Arens[60] positively correlated zoospore attraction with width of stomatal aperture. Royle and Thomas[41] considered the response independent of photosynthesis and predominantly associated with the oxygen regime at stomata, although other chemotactic and physical stimuli also appear to be involved.

Within 20 to 30 minutes, zoospores become motionless at stomata and encyst by absorbing their flagella and enclosing in a membrane. After approximately 12 minutes (under favorable conditions), they produce a germ tube that grows through the stomatal aperture into the substomatal cavity.[11,40] A single stomate may be penetrated by up to 17 germ tubes.[36] Host penetration is almost exclusively via stomata, and although other ways such as via flower stigmas, wounds, and phylloxera galls have been reported, these are considered exceptional.[11] Presence or absence of light has little effect on stomatal penetration and no significant effect on infection.[41] The threshold number of zoospores required for infection of young vine tissue is lower than for older tissue.[18]

In the substomatal cavity, a vesicle is formed, and this develops a short hypha at its distal end. Subsequently, a haustorium is initiated where this hypha contacts a host cell.[59] On susceptible hosts (e.g., *V. vinifera*) at 23°C, haustorium formation begins within 3.5 hours of sporangium deposition on the undersurface of leaves. Further hyphal development does not occur for the next 12 to 15 hours, although the size of the primary haustorium increases. Within 24 hours of sporangium deposition, hyphae extend beyond the first haustorium and begin forming intercellular hyphae and additional haustoria.[59] After 36 hours, grapevine cells are noticeably plasmolysed,[40] but as there is no obvious reaction of most host cells to the presence of haustoria, toxins do not appear to be produced.[9] Hypertrophy of cortical cells may cause a slight swelling of infected stems.[9] On resistant hosts (e.g., *V. viparia*), the substomatal vesicle and primary haustorium are formed as in susceptible hosts, but subsequently fungal development is restricted (i.e., from approximately 18 hours after sporangium deposition) and browning (hypersensitive) reactions of host cells appear after 24 to 48 hours.[59] Development on resistant hosts does not usually proceed sufficiently to allow sporangium production unless conditions for infection are extremely favorable (e.g., continuous high RH).[59]

Some biochemical changes in grapevine tissues associated with infection include increases in total phenolic content and organic acid content[61,62] and decreases in sugar content.[63] Several stress metabolites are also produced as components of the defense mechanism in grapevines in response to infection,[64] and these will be examined further in relation to resistance.

The infection process is dependent on wetness duration and temperature. Generally about four hours of surface wetness are considered necessary for the infection process in epidemiological models.[65]

Conditions required for primary infection (i.e., oospore germination followed by sporangium or zoospore dispersal, zoospore migration to stomata, encystment, germination, and host penetration) are important since their fulfilment signals the commencement of the disease cycle each season. They are therefore an important consideration in disease warning services used as a basis for planning treatment programs. Generally, the requirements for primary infection can be summarized as follows:

1. Viable and mature oospores

2. Grapevine foliage or shoot growth

3. Sufficient rainfall to saturate soils and/or create pooling of water in vineyards (usually rainfalls of 10 mm or more) that wets soil for 24 hours or more

4. Sufficient temperature during rainfall to permit oospore germination and subsequent dissemination of sporangia or zoospores (usually 10°C or more)

5. Subsequent leaf wetness for two to six hours to allow host penetration.[12,58,65-67]

Most authors agree that rainfall is a principal factor associated with primary infection. Other sources of moisture such as irrigation, dew, mist, or high humidity usually do not provide adequate soil saturation, splash, and/or surface wetness of sufficient duration to allow primary infection unless associated with rain.[40,50,66,68,69]

4-6-3 Incubation

The incubation period is the time between infection and the first appearance of disease symptoms or sporulation. During the incubation period, *P. viticola* invades the intercellular spaces of the host and establishes haustoria in host cells. The extent of growth within host tissues before tissue discoloration and the appearance of symptoms is dependent on tissue age. Gregory[18] found that mycelial growth was more abundant in young tissues.

Apart from tissue age, growth of the mycelium within leaves is also limited by temperature. Muller and Sleumer[50] found that growth ceased at 30°C although the mycelium remained viable for 12 days at 42 to 43°C.

The duration of incubation has been studied extensively because knowledge of this aspect is important when planning treatments for disease control. Depending on the author and conditions of observation, incubation periods from 4 to 21 days (mostly 7 to 10 days) have been reported.[11,18,21,22,59,70–74] The duration of incubation is affected by air temperature and humidity.[11] As temperature increases from 5 to 26°C, the duration of incubation decreases.

The relationship between temperature and duration of incubation has been described by Muller and Sleumer,[50] Magarey et al.,[73,75] and others. Optimum temperatures are 20 to 26°C.[75] Duration of incubation is increased when air humidity is reduced. For example, at 13°C, Muller and Sleumer[50] found that the incubation period was 6 days at 100% RH and 11 to 12 days at 80 to 90% RH. At 28°C, it was 5 to 6 days at 100% and 8 days at 70% RH.

As the infected tissues increase in age, the duration of incubation is also extended.[21] Longer incubation periods have been reported for resistant hosts than for susceptible ones.[18]

To help with the timing of fungicide treatments, incubation calendars have been prepared for various regions.[69,73,76]

4-6-4 Sporangia

Formation (Sporulation).   After incubation and the discoloration of tissues, sporangia and sporangiophores may develop from stomatal openings. When temperatures are optimum for sporulation and RH is near saturation, sporangiophores may develop prior to tissue discoloration.[11,59] In the latter situation, a hypersensitive (or browning) response of small groups of cells that have been in contact with the fungus the longest appears in the center of lesions as sporulation progresses.[59]

The sporulation process is influenced by RH, temperature, darkness, and the condition of infected tissues.

Sporulation requires continuous RH ranging from 95 to 100%[58,77] or saturation vapor pressure deficit (SVPD) ≤ 1.3 mb.[75] It also requires darkness,[22,79] a minimum of four hours darkness being required for sporangiophore formation.[71,80] Moist, dark conditions following a period of light favor maximum downy mildew sporulation. Under continuous light or darkness and high humidity, there is little or no sporulation.[81] Near ultraviolet light (310 to 400 μm) or green light (500 to 560 μm) at intensities more than 3 or 3.5 Wm^{-2} (radiant flux density), respectively, also inhibit sporulation.[79]

Sporulation occurs at temperatures ranging from 9 to 34°C,[18,21] though Zachos[56] and Hill[82] showed that sporulation was not initiated below 13°C. A subsequent decrease in temperature does not stop sporulation while low SVDP continues.[82] Optimum temperatures have been cited as 18 to 24°C,[50] and 18 to 28°C.[20] Blaeser[20] found that maximum sporangial production occurred between 23 and 28°C. At temperatures ranging from 13 to 27°C, sporangia form within 24 hours,[11,80,83] and often within 4 to 5 hours.[75] Temperature also influences the duration of the sporulation period.[82] With other downy mildews, sporulation periods are longer at low temperatures than at high temperatures.[48]

Since high RH (i.e., low SVPD) often precedes and follows rain, this has a significant influence on sporulation.[84] Darkness before dew formation is also expected to favor sporulation. Four or more hours of low SVPD prior to midnight favor production of sporangia before sunrise.[75,80]

Given favorable conditions, sporulation can occur for at least several successive nights on the same lesion.[50] Hill[82] reports repeated sporulation over a two-month period. There are few other reports on how long lesions continue to sporulate, although Marais and Knox-Davies[85] did not observe sporulation seven to nine weeks after lesion appearance. Sporulation is also favored by high water content of infected tissues.[86]

Dissemination.   Sporangia dehisce from sporangiophores at a cross-wall of callus that dissolves in water. This ensures that aerial liberation of sporangia takes place only in moist air.[11] The separation of sporangia occurs when sporangiophores twist in response to changes in SVPD, a common sporangial release mechanism for downy mildews.[87] Decreases in RH that stimulate sporangium release are frequently associated with increasing temperature and air movement in the early morning following the wet period that promoted sporulation. Sporangia are dispersed by wind or in wind-blown or splashed water droplets during rains, overvine irrigations, or dews.[71]

Levels of *P. viticola* sporangia in vineyard air ranging from 800 to $2400/m^3$ have been recorded during studies in Switzerland,[88] the maximum occurring during late afternoon (1800 to 2000 hours). Sporangia may be dispersed at least 40 m[89,144] to 100 m.[90]

Survival.   Sporangium viability decreases with increasing temperature and decreasing RH,[20,77,80] and attached sporangia survive longer than detached ones. Gregory[18] found that sporangium viability was retained in moist air for more than 14 days at 10°C but survival was reduced to 7 to 10 days at 20°C, although Blaeser[20] found sporangia survived only 4 days at 20°C. At 30°C, sporangia remain viable for only six hours.[77] In dry air, sporangia survive for a maximum of 5 days.[11] However, there are conflicting reports about effects of humidity on sporangium survival. In most studies, viability was favored by high humidity,[20,50,77,80] but in some it was increased by low humidity.[48]

Sporangium viability is also influenced by exposure to sunlight. In Greece, Zachos[56] found that viability was permanently reduced by exposure to more than one hour of sunlight. In France, Lafon and Bulit[11] observed that sporangia collected in the evening following a hot sunny day were not viable. In India, where day temperatures reach more than 30°C, sporangia harvested between 0800 and 1800 hours failed to germinate, while sporangia collected at night from 2000 to 0600 hours germinated within one hour. Maximum germination rates were recorded for sporangia collected between 0200 and 0600 hours.[91]

Germination.   Sporangium germination and the liberation of zoospores occur only in water. Germination is also dependent on temperature and sporangium age, young sporangia germinating more readily. Germination has been observed at temperatures ranging between 2 and 9°C and 28 and 30°C, with optima between 15 and 25°C.[18,20,36] The germination process takes 30 to 60 minutes at optimum temperatures and 10 hours at extreme temperatures.[11,18,36] Germination is retarded if sporangia are exposed to sunlight prior to immersion in water. Zachos[56] found that sporangia exposed to sunlight for one hour required immersion in water for at least 24 hours before germination commenced.

4-6-5 Secondary Infection

The secondary infection process is the same as that outlined for primary infection and has similar requirements. In general, secondary infection is favored by surface wetness for at least two hours after midnight following sporulation.[40,50] Du Pisani[65] identified possible infection periods as those with 90% RH or more for four hours or longer. However, Blaeser and Weltzien[77,92] found a temperature-sum requirement of at least 45 degree hours for duration of leaf wetness before infection could occur. Blaeser and Weltzien[77,80] suggested that secondary infection is primarily associated with rain rather than dews because they found that dissemination only occurred during rain. The role of dew in relation to downy mildew epidemiology is reviewed by Populer.[48]

4-7 INOCULATION AND DISEASE ASSESSMENT

4-7-1 Culture

P. viticola can be grown on grapevine leaf discs or on cultured tissues *in vitro*.[93–95] Dual cultures of *P. viticola* and tissue-cultured grapevine plants are useful in research on grapevine resistance[96] and for fungicide evaluation.[94] To date, *P. viticola* has not been grown on artificial media.

4-7-2 Inoculum Preparation

Inoculum of *P. viticola* for experimental use is commonly prepared by the following procedure. Infected leaves with oil spots are collected from the field and surface sterilized to remove phylloplane microorganisms. Surface sterilization is essential if aseptic cultures are required. This can be achieved by treatment with an antibiotic solution (0.5g/L chloramphenicol, 0.3 g/L streptomycin, 1.33 g/L penicillin G) for one hour, 70% ethanol for 10 to 20 seconds followed by sodium hypochlorite (2.7% available chlorine) for 10 to 15 minutes, and one rinse in sterile distilled water.[96] For other purposes, surface sterilization with sodium hypochlorite (0.8% for 10 to 15 minutes) may suffice.

Leaves are incubated in moist air (SVPD ≤ 1.3 mb) in the dark for a minimum of four hours at 18 to 28 °C. Sporangia developed on these leaves are dispersed into sterile distilled water to provide suspensions ranging in concentration from $6-15 \times 10^3$ sporangia/mL for *in vitro* studies[96] to at least 1.6×10^3 sporangia/mL for field studies.[97,98]

Concentrations are usually determined using a haemacytometer. Young leaf tissues (with oil spots) have better sporangium production than older leaves[82] and usually require less surface sterilization. For best results, suspensions of recently formed sporangia should be used within 60 minutes of preparation because sporangia germinate to produce zoospores which quickly encyst. Where there is interest in the genetic variability of *P. viticola* isolates, single zoospores for culturing can be isolated by passing the sporangia-zoospore suspension through a monofilament nylon high-capacity sifting fabric with a mesh aperture of 10 μm.[99]

4-7-3 Inoculation and Incubation

Inoculation is best achieved by spraying suspensions of sporangia onto the lower surface of leaves or by spraying entire shoots (leaves, clusters, and stems) with sporangial suspensions.[97] With *in vitro* inoculations, whole shoots can be dipped in sporangial suspensions.[94,96] In bioassays, suspensions are deposited or sprayed onto leaf discs (on moistened filter paper or agar) before or after exposure to fungicides. Often a specified volume of suspension per unit area of tissue is required so the number of effective infections (resulting in lesions) per unit area is known and/or controlled. At low concentrations (20 to 100 sporangia/mL), the number of lesions produced per unit area is directly related to concentration. On leaves of *Vitis labrusca* cv. Catawba, the infection efficiency of *P. viticola* (the ratio between the number of lesions produced and the number of sporangia applied) has been estimated to be 6%.[100] Little is known of infection efficiency on more susceptible grapevine cultivars.

After inoculation, surface wetness is required for at least three hours at temperatures of 18 to 24°C to allow most rapid completion of the infection process. Several methods are used to maintain wetness. With *in vitro* studies, surface wetness is retained by enclosure within sealed containers. With *in vivo* inoculation, wetness is usually maintained either by enclosure in polyethylene bags[97,98] or by overvine misting systems.[10,52] After infection, temperatures are best maintained at 20 to 26°C for incubation. First symptoms of disease will appear four or more days after inoculation, depending on incubation temperatures and RH.

Downy mildew disease may also be increased by other methods. For example, carefully timed irrigation and/or misting treatments can be used to promote epidemics on field vines after natural or artificial inoculation.

4-7-4 Disease Assessment

A variety of methods are used for assessing downy mildew development, depending on the nature of experimental studies. Usually, assessment relates to disease incidence (i.e, presence or absence of disease), disease severity (e.g., percent area of tissue diseased), and capacity of diseased tissues to perpetuate the disease cycle (e.g., presence or absence of sporulation and/or amount of sporulation). Rating systems used vary considerably, incorporating one or more of these aspects. With severity assessments, scales of the type described by Horsfall and Barratt[101] are most appropriate, as these are logarithmic and account for visual differentiation. However, depending on the accuracy required, scales of severity are usually reduced or modified for practical reasons.

Those that are directly related to the percent area of tissues diseased are more precise and reproducible, providing that estimates of percent area are objective. Usually, these estimates are made using diagrammatic keys or image analysis.

4-8 MANAGEMENT

4-8-1 Quarantine (Regulatory)

Although downy mildew is widespread, it is apparently absent from some viticultural regions. The significance of quarantines in restricting movement and establishment of *P.*

viticola in these regions is unclear. In Western Australia, for example, strict quarantine appears to have successfully prevented the introduction of downy mildew, but recent climatic studies suggest that low temperatures during infrequent summer rainfalls are unfavorable for disease development and may be a more significant constraint.[7]

Quarantine measures used to prevent the movement of *P. viticola* include limiting the introduction of planting material to heat-treated dormant cuttings or rootlings and disease-free plants in tissue culture and making these subject to postentry scrutiny before release to industry. The introduction of soil and leaf debris, which may carry oospores and any actively growing green material in which the fungus may survive, should be prohibited. These prohibitions should also apply to grapevine produce and trash that could harbor the fungus and be introduced in packaging or in vineyard machinery such as mechanical harvesters.[102]

4-8-2 Cultural Practices

As growth of *P. viticola* is reliant on suitable climatic conditions, cultural practices can affect disease development, particularly when conditions are marginal.[9,37,103]

Microclimates in vineyards can be influenced by vineyard topography, layout, and management. Sheltered areas in vineyards with poor air drainage may develop microclimates favorable for disease when the macroclimate is mostly unfavorable.[85,104] When planning vineyards, careful attention to spacing vines, row direction, and placement of wind-breaks in relation to prevailing winds will ensure maximum air drainage and minimum wetness duration.[37] Aspects of ongoing management, such as weed control, can also contribute.[104] Efficient soil drainage is important, as the pooling of water in low-lying areas encourages the maturation and germination of oospores.[11,52]

Microclimates within grapevine canopies can also be manipulated by controlling vine vigor, number and length of vine shoots, and by changing the design of vine trellises.[105,106] The removal of leaves around berry clusters, a practice adopted by some table grape growers, increases air movement and reduces localized humidity and wetness duration. This creates a cluster microclimate less favorable for infection by downy mildew[37] and other diseases, such as Botrytis rot.[107] Trellis designs and pruning practices that reduce the development of Botrytis bunch rot[108] would also be expected to reduce downy mildew development.

Practices that maximize the distance between the soil (the source of primary inoculum) and susceptible vine foliage should also discourage early downy mildew infection.[9] These include careful disbudding and training of vines early in their development to prevent suckering, good trellis design, and, where appropriate, the trimming of vine foliage. In some European regions, growers are advised to remove and burn all low canes from vines to minimize the risk of primary infection.[52] Mid- to late-season trimming in some regions also removes young foliage that is very susceptible to infection and improves aeration of the canopy, with little or no adverse effect on cropping and vine development.[11]

Trellising and pruning practices that optimize air movement and light distribution in the vine canopy also allow good spray penetration and coverage that facilitate other forms of disease control. Mechanical and minimal pruning practices that congest vine growth should be considered with care because they increase canopy humidity and create a microclimate more favorable to disease.

In semiarid areas in particular, irrigation practices may significantly influence downy mildew development. Disease is more pronounced and crop losses are greater in overvine-irrigated vineyards[78] than in furrow-irrigated vineyards. In South Australia, in excess of 80% of crop loss from downy mildew occurred in overvine-irrigated vineyards.[26] Overvine irrigation before or after rainfall increases the risk of primary infection. Irrigation schedules and system designs that minimize the wetness duration, especially at night or in association with dews, reduce the risk of creating conditions favorable for sporulation and secondary infection.[37]

4-8-3 Host Resistance

Most cultivars of *Vitis vinifera* are susceptible to downy mildew. Other species such as *V. californica* and *V. arizonica* are also susceptible, while *V. aestivalis* and *V. labrusca* are less susceptible. In comparison, *V. berlandieri* is relatively resistant and *V. rotundifolia*, *V. cordifolia*, and *V. rupestris* are highly resistant.[11]

In China, clones of *V. pseudoreticulata*, *V. davidii*, *V. piasezkii*, *V. romanetii*, *V. flexuosa*, *V. bryonifolia*, *V. hancockii*, *V. yeshanesis*, and *V. amurensis* have some resistance to *P. viticola*.[109]

The breeding of grapevines for downy mildew resistance has already been reviewed by Matthews.[110] Nearly 300 grape varieties and species were screened,[110] and *Vitis aestivalis*, *V. labrusca*, *V. lincecomii*, and *V. riparia* were most likely sources of resistance for breeding programs. Subsequent breeding involved the intercrossing of American and European varieties followed by back-crossing to European varieties to transfer resistance from the American species. When Portugese varieties were surveyed, no resistance was observed, but as some differences in susceptibility were noted, interspecific crossing and clonal selection were undertaken.[110] It was found that resistance was expressed as a hypersensitive reaction either as limited patch resistance, mycelial growth limited beyond necroses, or as ringspot resistance (where necrotic tissues are marginal to mycelial growth). Phasic resistance in leaves related to leaf age and climate was also observed. Between 1945 and 1964, results of 200 crosses were reported, selections being made in F1, F2, and back-crosses. Out of 110 F2-resistant clones tested for production and quality, 5 were promising.

More recently, some promising resistant mutations (with limited patch or ringspot resistance) have also been induced by treating seeds with X-rays and neutrons and irradiation.[111]

In Germany, initial attempts to transfer resistance from *V. riparia* to *V. vinifera* cv. Gamay were unsuccessful.[112] Subsequently, resistance in French hybrids (Leyvre-Villard) was used in crosses with European vines. Resistant seedlings were selected after artificial inoculation with *P. viticola*. Resistant seedling hybrids derived from crosses with European vines readily transmitted resistance provided they were used as female parents. Interspecific crosses were robust, vigorous, and resistant to Botrytis as well as downy mildew. Wines produced from hybrids were considered as good as those from European vines. Hybrids were selected because of their inherent resistance.

Forms of and ways of increasing resistance to *P. viticola* have also been reported by Boubals.[113] In Vitaceae, total immunity is rare, only being found in *Cissus discolour*. Resistance to germ-tube penetration occurs in *V. lincecomii*, *V. piasezkii*, and the cultivars

Pinot Meunier and Berger and appears to be related to hairy and warty (bullate) surfaces. Resistance, involving a hypersensitive reaction associated with stomatal and mycelial death, is determined by a single dominant gene and occurs in some *Cissus* and *Tetrastigma* species. All *V. vinifera* varieties are homozygous recessive for this gene (i.e., susceptible). In other species and interspecific hybrids, resistance is polygenic and exhibited as a restriction of mycelial development in the host. In crosses with *V. vinifera*, partial dominance of resistance of susceptibility may occur depending on the *V. vinifera* cultivar used, one to four genetic factors being involved.

Boubals[113] noted that crosses between *V. riparia* and *V. vinifera* have produced highest levels of resistance in F2. Foliage of late-maturing types, however, is usually susceptible. No correlation between downy mildew and powdery mildew resistance has been observed.

In addition, programs to develop and/or evaluate hybrids with downy mildew resistance have been established at other centers.[96,114,115] With the wine industry, however, some workers believe that the prospects of developing commercially acceptable hybrids are poor because resistance is polygenic and linked with genes associated with poor wine quality.[52,116] Nevertheless, recently developed genetic engineering techniques have potential to overcome this problem.

The mechanism of disease resistance in *Vitis* species, as in other plants, involves the production of stress metabolites, some of which are fungitoxic.[117] Phytoalexins were originally conceived as antifungal compounds produced by plants in response to fungal attack,[118] but more recent and detailed studies have shown that such host-synthesised compounds are also produced in response to other forms of stress such as chemical application[61] and environmental changes.[119] In *Vitis* species, several stress metabolites with antifungal activity (i.e., ϵ-viniferin, $\propto$-viniferin, and pterostilbene) are produced together with the monomeric precursor of these compounds, resveratrol.[64] These metabolites are responsible for inhibiting growth of *P. viticola* in resistant interactions. In the age-related resistance of *V. vinifera* to *Botrytis cinerea*, $\propto$-viniferin is predominantly produced. However, despite its relatively high antifungal activity, pterostilbene does not appear to be involved in the resistance of *V. vinifera* or *V. viparia*.[64]

Other studies[120] indicate that with the exception of some *Vitis* species that exhibit resistance despite an inability to synthesize high concentrations of stilbene phytoalexins, resistance is positively correlated with capacity for resveratrol and ϵ-viniferin synthesis following induction by UV irradiation.

In recent years, knowledge of stress metabolites and their relationship to host defense mechanisms has been applied to screening procedures for resistance to Botrytis rot[121] and powdery mildew.[122] However, recent attempts to use resveratrol production as a basis for the *in vitro* screening of grapevines for resistance to *P. viticola* have been unsuccessful.[119]

4-8-4 Chemical Control

Since the discovery of Bordeaux mixture,[31] chemicals have been used to control downy mildew on vines throughout the world. The efficiency of chemical control programs depends on the efficacy of chemicals used, appropriate treatment timing, and efficient application. For economic and environmental reasons, there is increasing pressure on viticulturists to maximize the efficiency of disease control programs.

Fungicides.   Most chemicals used to control downy mildew can be classified either as preinfection (protectant) or postinfection (eradicant) fungicides, depending on their principal mode of action. The terminology pre- or postinfection is preferred because it directs attention to the activity of the fungicide in relation to key aspects of control (i.e., the infection process).

Preinfection fungicides (Protectants).   Protectant fungicides are those that either prevent healthy plant tissue from becoming infected or prevent any infected tissues already present from becoming more diseased and infectious.[123] Protectants must be applied before infection occurs. They are usually residual surface-acting fungicides that inhibit fungal development on the host surface. Generally, they are nonspecific, broad-spectrum fungitoxicants. Preinfection fungicides used for downy mildew control include copper compounds, dithiocarbamates, phthalimides, and several other types of organic fungicides (see Table 4-1).

Widely used copper compounds include Bordeaux mixture (copper sulphate neutralized with calcium hydroxide), copper oxychloride, and cupric hydroxide. Bordeaux mixture has excellent spreading properties and considerable resistance to weathering. Basic copper salts (e.g., copper oxychloride and cupric hydroxide) are less tenacious but are more readily mixed and applied. In general, copper compounds are of low hazard to humans and the environment. Some, particularly Bordeaux mixture, may retard and harden vegetative growth under some climatic conditions.[124,125] In addition, the continual use of copper compounds may lead to high copper levels in vineyard soils,[126] resulting in copper toxicity to vines.

Of the organic fungicides, mancozeb and folpet are most widely used in Europe. Although dithiocarbamates (e.g., mancozeb) tend to be less persistent than copper fungicides, they are not as phytotoxic.[125] However, some phytotoxicity to maneb and mancozeb has been reported.[127] Combinations of one or several organic fungicides and one or more copper compounds are also used as protectants in many countries. These exploit the respective advantages of each of their components while minimizing their disadvantages. Some combinations exhibit a synergistic effect (e.g., copper oxychloride with zineb).[11]

Postinfection fungicides (Eradicants).   A recent advance in the chemical control of downy mildew has been the development of fungicides that are effective when applied after infection (i.e., fungicides with eradicant or curative ability). A principal feature of postinfection fungicides is their ability to penetrate plant tissues; most are readily translocated. Fungicides with notable eradicant ability against *P. viticola* include the phenylamides, which include the acylanalines (e.g., metalaxyl) and the acylaminobutyralactones (e.g., ofurace), the phosphonates (e.g., phosethyl-Al), and other chemicals (e.g., cymoxanil) (see Table 4-1). The improved disease control provided by systemic eradicants such as the phenylamide and phosphonate fungicides is attributed not only to their ability to inhibit disease development after infection but also to their resistance to weathering and their longer residual activity.[28]

For antiresistance reasons, most systemic eradicants are now used in association with surface-acting protectants, which supplement their persistence and increase their spectrum of activity. As with protectants, some fungicide combinations (e.g., ofurace or

TABLE 4-1: SOME FUNGICIDES THAT CONTROL *PLASMO-PARA VITICOLA*

Primary Efficacy Chemical Group	Common Name
Protectants	
Copper Compounds	Bordeaux mixture
	Copper oxychloride
	Cupric hydroxide
Dithiocarbamates	Mancozeb
	Maneb
	Metiram
	Propineb
	Zineb
Phthalonitriles	Chlorothalonil
Polyhalogen-alkylthio Compounds	Captafol
	Dichlofluanid
	Folpet
Eradicants	
Acylaminobutyralactones	Cyprofuram
	Milfuram
	Ofurace
	Oxadixyl
Carbamates	Propamocarb
	Prothiocarb
Phenylamides	Benalaxyl
	Furalaxyl
	Metalaxyl
Phosphonates	Phosethyl-Al
	Phosphorous Acid
Others	Dithianon
	Cymoxanil

benalaxyl with mancozeb) also exhibit a synergistic effect.[28,128–131]

Cymoxanil is used in various formulations with dithiocarbamates, phthalimides, or copper compounds and occasionally with mixtures of these. Although cymoxanil is not systemic, it penetrates vine tissues sufficiently to inhibit mycelial development of *P. viticola* up to 3 to 4 days after infection.[132] The persistence of cymoxanil formulations is dependent on the protectant component of the mixtures (i.e., usually 10 to 12 days). This is because cymoxanil has a short half-life in plants,[129,133] and its protectant activity is lost after a few days, especially in hot weather.[125]

Metalaxyl is more frequently used as a mixture with folpet or copper oxychloride. Apart from being slightly systemic in the xylem of plants, predominantly in an acropetal direction,[28] it also diffuses as a vapor into moist air.[11] It eradicates infection and reduces sporulation at least 3 to 5 days after infection[98,134,135] and delays symptom appearance.[97,134]

However, later work has indicated efficacy up to 15 days postinfection.[97,136,137] Volatiles emitted from metalaxyl deposits also inhibit sporulation.[138]

Most mixtures containing metalaxyl persist for 2 to 3 weeks after application. In host tissue, metalaxyl inhibits the formation of secondary haustoria and mycelia inside cells, stops lesion formation and sporulation, and reduces viability of sporangia produced from treated leaves.[97,136,139,140] There is evidence to suggest that the fungitoxic mode of action of metalaxyl involves the inhibition of RNA synthesis.[28]

Phosethyl-Al (Phosethyl-aluminium), the only truly systemic fungicide for downy mildew control, is used in combination with folpet and other protectant fungicides. The phosethyl-Al component is translocated from leaves via the xylem and to some extent, the phloem, to extremities of shoots and to clusters at least until berry set. It provides downy mildew control in shoot tips up to six weeks after treatment.[11] Phosethyl-Al formulations have eradicant activity up to 3 to 4 days after infection[11] and up to a maximum of 17 days.[97,136] They will usually protect actively growing foliage and clusters for 14 to 21 days.[11] In plant tissues, phosethyl-Al breaks down rapidly to phosphorous (phosphonic) acid and carbon dioxide.[28]

The active metabolite of phosethyl-Al appears to be phosphorous acid. It acts directly on the fungus while indirect effects on host metabolism appear to be of secondary importance,[28,141] although the latter is disputed by some authors working with other pathogens.[28] Phosphorous acid alone, applied as a postinfection fungicide, has significant efficacy against *P. viticola*.[142] It reduces oil spot development when applied up to 15 days post-infection and inhibits sporulation up to 17 days postinfection. It also increases the necrosis of oil spots[97,136,137,143,144] and increases the incubation period of *P. viticola*.

The ultimate fate of phosphorous acid in plant tissues is unknown,[28] although it is readily translocated from the foliage. This accounts for the lack of preinfection control of *P. viticola* by phosphorous acid.[137,142] In avocado, actively metabolising meristematic tissue is the main sink for phosphorous acid. Thus, the sink varies with the time of application in relation to activity of shoot and root growth.[145] A similar phenomenon is likely in grapevine.

Fungicide resistance.   The mode of action of fungicides and mechanisms leading to the development of fungicide resistance in plant pathogens have been studied increasingly over recent years. These studies are the basis of present antiresistance strategies for fungicide use that are being applied to *P. viticola* as well as other foliar plant pathogens. The aim of these strategies is to prevent the development of fungicide-resistant populations and the loss of efficacy of fungicides, particularly systemics.

For decades, most protectant fungicides have been repeatedly applied to crops for disease control without any notable loss of efficacy. Generally, these surface-acting fungicides are multisite inhibitors of fungal metabolism and are potentially toxic to all fungal cells. Resistance to these fungicides appears unlikely because of the large number of mutational changes required by a fungus to achieve resistance to all sites of toxicity.

On occasions when reduced sensitivity of fungi has occurred, it is most likely due to mechanisms where fungicides have been excluded from sites of activity by reduced permeability of fungal cells or where fungicides have been detoxified as they pass through fungal membranes.[146] In contrast to protectants, most eradicant fungicides specifically inhibit one or a few sites in fungal metabolism. Most systemic fungicides are site-specific inhibitors, because this mode of action is essential for systemic movement within plants without adverse effect. Because of their specific nature, eradicant fungicides are more likely to select resistant strains with slightly different reactive sites or metabolic pathways.

Resistant races of *P. viticola* have been reported for several phenylamide fungicides. These include metalaxyl,[27, 147-150] ofurace,[149] milfuram,[149,150] and cyprofuram.[148] Herzog and Schuepp[93,151] and Clerjeau et al.[27,148] among others, identified various levels of sensitivity to metalaxyl in isolates of *P. viticola*. These were assessed on the basis of haustorium formation and/or sporulation in the presence of metalaxyl.

Metalaxyl resistance has also occurred with a number of pathogenic fungi on other crops (e.g., *Phytophthora infestans* on potatoes). With *P. infestans,* resistance has been most widespread in areas where metalaxyl was originally used without a protectant. In their review of systemic fungicides for Oomycete control, Cohen and Coffey[28] note that a disturbing feature of metalaxyl resistant isolates is their high level of resistance and their high degree of fitness. Where metalaxyl/mancozeb mixtures continue to be used, resistant isolates persist, although their frequency decreases markedly when fungicide use in the field is confined to mancozeb or stopped completely. Control of metalaxyl-resistant strains has also been achieved using other types of systemic fungicides, such as phosethyl-Al.[149,152] Studies of metalaxyl-resistant strains also suggest that they exhibit crossresistance to other phenylamide fungicides. As levels of resistance to these fungicides vary considerably among different metalaxyl-resistant isolates of *P. viticola*, several different mechanisms of resistance may exist in a fungal population.[148] The aggressiveness of metalaxyl-resistant isolates may be related to heterokaryosis in *P. viticola*.[23] Stronger fungicide synergy against metalaxyl-resistant isolates has also been observed. A possible explanation of this is that the protectant component (e.g., mancozeb) of a fungicide mixture may increase the permeability of cell membranes to increase the concentration of the site-specific inhibitor (e.g., metalaxyl) reaching the site of action in the fungus. Furthermore, since fungicides such as metalaxyl are usually fungistatic, the activity of sublethal concentrations of the protectant component may make their combined effects more fungitoxic.

The possibility of resistance to both metalaxyl and the phosphonates has been reported in some isolates of *P. viticola*,[153,154] but this has been discounted.[152,155-157] Resistance to phosethyl-Al (and phosphorous acid) has been observed with *Phytophthora* spp. in the laboratory and on one occasion in the field. The mutation rates and fitness of these isolates were similar to those of metalaxyl resistant isolates.[28] Resistant strains of *P. viticola* have also been reported for other systemic fungicides, such as oxadixyl.[158]

Antiresistance strategies.   Several theoretical models of strategies for controlling fungicide resistance have been developed in recent years. A basic assumption is that resistant mutants exist in relatively low frequencies in all fungal populations. Kable and Jeffery[159] suggested that because incomplete spray coverage occurred, the use of mixtures was a preferred antiresistance strategy over alternate applications of two fungicides. Levy et al.[160] suggested that mixtures were more efficient than alterations in delaying resistance except where the site-specific component degraded very slowly. Skylakakis[161] predicted that resistance would be delayed by mixtures or alternations because the protectant decreases the selection pressure.

The vulnerability of systemics to pathogen tolerance foreshadowed the formation of a series of Fungicide Resistance Action Committees in 1982.[162,163] As a result, a strategy for the use of phenylamide fungicides on foliar pathogens was formulated by the agrochemical industry. As part of this strategy, it was decided that (1) the fungicides would only be made available in prepacked mixtures with protectant fungicides, (2) only two to four applications per season would be recommended, and (3) postinfection use of mixtures would not be recommended.

Theoretically, the use of prepacked mixtures offered the following advantages: (1) the protectant component should control the resistant isolates, (2) the dose of the systemic component may be reduced due to additive or synergistic factors, (3) reduced concentration of the systemic component should reduce the selection pressure for resistance, and, most importantly, (4) the use of prepacked mixtures is an enforceable strategy.[163] Unfortunately, it is uncertain whether the use of mixtures will provide a long-term answer to resistance problems in the field. At present, too little is known about the basic modes of action of many systemics and the mechanisms that reduce their efficacy.[28]

Strategies for timing treatments.   Fungicide treatments for downy mildew control often are applied either in routine schedules that generally are implemented regardless of seasonal conditions or in treatment programs initiated because of assumed reasonable probability that infection has occurred or will occur. Within these generalities, various treatment strategies are adopted depending on regional conditions and resources available to the viticulturist.

Routine schedules.   A conventional strategy for downy mildew control is to apply a routine (often calendar-based) schedule of fungicide treatments each season. The aim of this is to prevent infection by maintaining a protective cover of fungicide on vines as they grow. Treatments over periods when the crop is most susceptible to infection and losses are most likely are emphasized. Most routine treatment schedules are designed for the use of protectant fungicides, most of which are less expensive than postinfection fungicides.

In most cases, postinfection fungicides are used (although not usually recommended) only in critical circumstances, such as when disease has not been satisfactorily contained because of some inadequacy in the protection provided. This often occurs during wetter seasons in semiarid regions.[164]

The timing of treatments in routine schedules is based mostly on a perception of disease risk as gained from past experience. Average season climatic conditions and associated downy mildew activity, the residual persistence of fungicides previously applied, and vine growth (or the development of unprotected susceptible foliage) are taken into account.[165] Treatment timing is usually prescribed according to calendar date or according to stages of vine growth, the latter being more effective.[52] Particular emphasis is placed on protection during the most critical period for infection (i.e. from flowering to when berries are 5 to 6 mm diameter). Intervals between treatments range from 3 to 28 days (mostly 12 to 14 days), depending on regional conditions and expected disease pressure.[52,164-166]

In overvine-irrigated vineyards (especially those irrigated at night), disease pressure is greater than in furrow-irrigated vineyards,[26,78] and the timing of treatment schedules needs to be adjusted accordingly.[164] With protectant fungicides, treatments should be applied before overvine irrigations[78] and therefore before the risk of infection increases.

Routine treatment schedules are adopted by viticulturists for a range of reasons. Where downy mildew occurs regularly, routine schedules may be used because there is little flexibility to do otherwise. They are also used where viticulturists are relatively uninformed about the risk of downy mildew infections due to a lack of knowledge about the disease or about local weather conditions. This information may be provided through downy mildew warning services or vineyard monitoring systems. A routine approach to control may also be adopted for logistic reasons. For example, when fungicides are exclusively applied by contractors (particularly when aerial spraying), spray timing is more often determined by contractor availability than disease risk. In these situations, routine schedules are employed, as there is less flexibility for reliance on the precise timing of individual applications.[52]

Dynamic strategies. Although routine schedules may provide effective downy mildew control, programs more aligned with disease risk (or expected risk) are usually more efficient, especially when vine growth is rapid and disease pressure is high. This particularly applies in more arid regions where the seasonal occurrence of downy mildew is irregular. Here, routine schedules are inefficient because fungicides are often applied on occasions when little or no disease occurs and control is not needed. In these cases, efficiency can be improved by aligning treatments with timely assessments of disease risk so controls are more likely to be applied when required.[164] This approach also offers more scope for using highly effective post-infection fungicides. Ultimately, the efficiency of treatments is dependent on the accuracy of spray timing in relation to disease risk.[165,167] For many years in some countries, regional assessments of disease risk have been provided by warning or disease management services.[11,52]

Disease management services. Systems forecasting downy mildew appearance using incubation calendar methods were initially established in France in 1912[48,52] and in Germany in 1913.[69] The basis of these systems has been to detect infection periods from daily weather records. For infections considered to be important, the incubation calendar is used to predict the duration of the latent period as computed in relation to average temperatures at that time. Spray warnings are then issued toward the end of the predicted latent period so foliage may be protected before sporulation occurs.

Susceptible stages of leaf and fruit and the ratio of unprotected to protected leaf area are also taken into account. In France, oospore maturation (an indication of potential primary infection) is also estimated on the basis of past experimental data for each region. A principal assumption of these warning systems is that successive infections give rise to separate, identifiable waves of sporulation.

In Europe, the operation of regional warning services along with the introduction of more effective fungicides and application techniques has significantly reduced the economic impact of downy mildew on European viticulture.[52] European services have also stimulated the development of improved disease management services in other countries, such as South Africa[168] and Australia.[75,169,170] The latter is based on a computerized model that simulates disease epidemiology. The model provides advice on disease risk and predicts the date of oil spot appearance in vineyards.

In 1977, Kable noted that downy mildew advisory services operating in Europe had two basic phases of activity. The first phase involved determining the time of primary infection to which the timing of the first fungicide application each season was related. The second phase involved an operational period during the growing season where the timing of subsequent treatments was determined by incubation calendar methods or by a routine schedule approach. Criteria used to identify conditions for primary infection varied considerably with different services, but there were many similarities. The most common criteria used were as follows:

1. Rainfall—Depending on locality, the minimum required varied from 6 to 20 mm (mostly 10 mm). In most cases, sufficient rain over one to three days (usually one day) was needed to cause pooling of water on the vineyard floor.

2. Temperature—The minimum temperature required during the rain varied from 7 to 12°C (mostly 10°C).

3. Host status—The extent of vine growth required varied from bud burst to 10 cm shoot length, with some systems specifying minimum leaf diameters as 2 to 2.5 cm.

4. Pathogen status—Some systems in France also accounted for oospore maturity, estimates being based on past observations in each region in relation to seasonal weather conditions.[171] Because of the high cost, monitoring of oospore samples to estimate maturity was no longer conducted each season. Depending on locality, a risk of primary infection was considered to exist when oospores germinated in 8 to 24 hours.

Currently, the method used to fix the date of the first spray varies considerably. In most cases, this spray is either applied just before or just after the appearance of primary lesions, with the aim of preventing secondary spread.

The timing of subsequent sprays also varies with different services. In most localities, sprays are applied in routine schedules with intervals between sprays varying from 3 to 17 days depending on locality, host susceptibility, weather, pathogen status, and characteristics of fungicides used. Kable[52] noted that the number of possible fungicide applications per season in Europe varied from 2 to 25, more being required in high rainfall than low rainfall areas.

In some localities, incubation calendars are also used to determine spray timing.[52,65] In Europe, fewer sprays are required for downy mildew control when the incubation calendars are used than in routine schedules, particularly in wetter regions.[52]

Recent advances in computer technology have led to the development of electronic data loggers that serve as automatic climatic monitoring devices,[172] and these have greatly extended scope for data storage, retrieval, and processing. A quantitative sampling procedure for detecting disease in vineyards has also been devised by Seem et al,[173] who found that presence or absence of oil spot symptoms can be confirmed (to 99% confidence) by carefully scanning leaves on one side of 200 to 300 vines (at about 30 seconds/vine) in areas of vineyards most likely to develop disease. This vineyard scouting is relatively efficient, especially if it is undertaken when oil spot appearance is predicted by analysis of weather events.

Simulation models of downy mildew epidemiology may provide this analysis.[75,83,169,174-177] They may also estimate disease risk and the need for fungicide treatments. Simulators and climatic data logging systems have now been incorporated in small, but to date relatively expensive, electronic devices for use in individual vineyards. Some of these devices are being marketed in Europe and elsewhere with limited success.[178] Such an approach has already been used for downy mildew control in vineyards with some success,[179] but most devices require more evaluation and refinement before they will be widely adopted in world viticulture.

The advantage of computerized systems is that they could greatly improve the accuracy of disease risk prognoses because they can monitor the variation in individual vineyard microclimates. Although more input into well-timed monitoring of the appearance of oil spots may be required for viticulturists, computer-based disease management systems are expected to provide significant savings by improving the timing of fungicide applications. More effective control with less chemical use in the vineyard is likely, particularly in semiarid regions.

Application methods.   A wide range of techniques and equipment are available for applying downy mildew fungicides (e.g., high-, low-, or ultra-low-volume spraying machines, aircraft, or irrigation systems). Their cost and effectiveness varies considerably.

Spraying techniques.   Within grapevines, the spray target varies as vines grow. In most cases, deposition of fungicides is required on stems, leaves, and clusters, particularly early in the season when the crop is most susceptible. While some machinery achieves good coverage on grape leaves, coverage on clusters may be inadequate.

The means by which fungicide spray volumes and rates are determined also vary. In vineyards, layout and canopy size affects the surface area of foliage to be treated with fungicide. These vary depending on location, vine age and cultivar, and whether or not vines are on rootstocks. Hence, in practice, fungicide recommendations (e.g., rates per hectare) are loosely defined unless related to the estimated surface area of the crop canopy to be treated.

Spray machinery.   High-volume ground spraying machines include hand spraying equipment (knapsacks and hand-directed spray guns), boom sprayers, and air-assisted sprayers. Generally, with high-volume (HV or dilute) spraying, prescribed amounts of fungicide per unit area of crop are mixed with relatively large volumes of water and applied to the point of surface runoff. Volumes of spray depend on the size of canopy but usually vary from 550 L/ha early in the growing season to more than 1100 L/ha.[180]

Low-volume or ultra-low-volume ground spraying machines utilize a wide range of nozzles or atomisers to produce sprays with small droplets that are conveyed to the vine target in a stream of air. With low-volume (LV or concentrate) spraying, the same prescribed amount of fungicide per unit area of crop is mixed with much lower volumes of water (e.g., 200 to 250 L/ha.) With ultra-low-volume (ULV or high concentrate) spraying, fungicides are mixed in only 50 to 100 L/ha. With LV and ULV techniques, good coverage is achieved by reducing droplet size.

Low-volume and ULV techniques are usually more efficient, as often up to one tenth the volume of water is required. This means less time is wasted traveling to fill spray tanks compared with high-volume spraying.[181] Also, smaller units can be used (with smaller tanks) as less volume of spray solution needs to be transported through vineyards during each application. Reduced soil compaction may be an added benefit.

Aerial spraying is also used in some viticultural regions, particularly where ground spraying operations are inefficient or limited because of vineyard size or terrain or because of the urgency for a spray application. In this way, treatments can be applied quickly despite wet or windy conditions that may prevent ground spraying. A disadvantage of aerial spraying is that spray penetration into the canopy usually is not as good as that achieved with ground spraying, particularly when vine canopies are fully developed.[52,182] Thus, there will be correspondingly less disease control later in the season.[183]

Spray deposits.   With some fungicides, the efficiency of control is influenced by droplet size, droplet density, and by chemical concentration.[181,184] Drop size appears to be particularly important. It is generally accepted that an extremely narrow range of drop sizes is desirable.[185]

For copper oxychloride, for example, there are indications that a drop size of 150 μm (volume mean diameter) and a droplet density of 50/cm^2 are required for the most efficient control of *P. viticola*.[186] With the exception of this study, efficiency parameters for other fungicides used to control downy mildew on grapevines have not been accurately defined. It is unknown whether optimal droplet densities and droplet sizes for efficient control differ for protectant and systemic fungicides.

Generally, deposits of pesticides disappear from plant surfaces in two stages. The first stage involves the rapid loss of loosely bound deposits within the first few days. In the second stage, pesticide adhering to the plant surface is gradually lost due to the effects of wind, rainfall or overvine irrigation, sunlight, and temperature.[187] Whan[188] found the copper fungicide deposits also eroded at the same rate irrespective of whether they were applied by HV or ULV techniques. While large amounts of rainfall washed fungicide deposits from the vine canopy, small amounts of rainfall redistributed copper deposits and slightly improved coverage.

Other methods.   Several alternatives to spray application of fungicides have been investigated. In Switzerland, Perret[189] reported successful control of downy mildew when fungicides were applied by a modified overhead irrigation system. In Italy and France, good downy mildew control has been achieved on vines covered by trellis-supported nets impregnated with copper fungicides or plastic nets with interwoven copper wires.[190–192] With fungicide-treated netting, care must be taken to ensure that sufficient residual fungicide is maintained.[193,194] Under netting, vines are also protected against hail damage, but in some cases the incidence of Botrytis rot is higher.[191]

4-8-5 Biological Control

Little is known of the effects of biological agents on downy mildew development. Although there are some reports of interactions between *P. viticola* and other microorganisms, these interactions have not been exploited commercially. For example, Viala[195] observed that *Trichothecium plasmopara* suppressed sporangiophore production of *P. viticola*, but this was of limited use for disease control. Bogdanova et al.[196] also reported that several fungi suppressed the development of *P. viticola*. More recently, Weltzien and Ketterer[197] reported effective control of *P. viticola* by treating vine leaves with water extracts from composted organic wastes. Such methods require further investigation with a view to exploitation.

4-9 CONCLUSION AND OUTLOOK

Downy mildew has been an important disease in most viticultural regions of the world for more than a century. Because of its destructive effects, it has been one of the most widely studied diseases of grapevines, particularly in Europe. These studies have led to the development of more effective disease management practices which use knowledge of disease epidemiology and control (particularly with fungicides). Despite these developments, there is still considerable scope for improving efficiency of downy mildew control by more precise application of available epidemiological information. This information, when combined with timely assessments of vineyard microclimates and disease status, can help viticulturists improve the accuracy of spray timing and hence the effectiveness of disease control.

The further development of new types of fungicides with curative and systemic properties, for example, would expand the range of these chemicals available for downy mildew control. In the long term, this could allow greater reliance on postinfection treatments in control programs integrated with predictive systems,[125] a shift in emphasis that could greatly improve the efficiency of control. Some studies have already highlighted advantages of using a postinfection approach to downy mildew control.[75,198] The use of curative systemic fungicides combined with a protectant offers greater flexibility with control programs, as there is less need for immediate application once infection warnings have been received. Protectants alone, however, must be applied strictly in accordance with schedules or at the time of warning, as they have no kick-back action.

The potential for applying disease management procedures based on a better use of fungicides is considerable. However, there is a need for a greater understanding of the mode of activity of fungicides, particularly the limits of their postinfection activity and strategies for their use that will enable greater precision in the timing of spray applications. This will not only reduce costs and increase the efficiency of disease control but will also preserve efficacy by delaying the development of resistant strains. An additional advantage of more precise spray timing will be less chemical use and less environmental hazard.

Regular monitoring of *P. viticola* populations in viticultural regions is advisable to identify population shifts in fungicide sensitivity and to ensure that correct antiresistance strategies are adopted. The development of safe, highly systemic fungicides which accumulate in host tissues offers other prospects. Phosphorous acid may have potential in this regard. The use of postharvest treatments to protect vines until flowering in the following season is now a possibility[199,200] and warrants further investigation.

Prospects for treating oospores or preventing their development also appear more promising than in the past. Protectant chemicals have had relatively little effect,[168] but this may no longer apply as more potent systemic fungicides are developed. More research on the mechanisms of oospore production may assist in this regard. In the past, the treatment of overwintering oospores in soils has been a daunting prospect and has received little consideration, but this also may not apply in the future. The reduction or elimination of overwintering inoculum would be worthy of study if costly treatment and monitoring programs could be avoided or reduced.

Further research into aspects of chemical application has scope for improving the efficiency of downy mildew control. Little is known about quantities of fungicides re-

quired on vine tissues for effective control and how these can be achieved most efficiently. A better understanding of these aspects will ensure that wastage of fungicides and
that needless contamination of the environment is minimized. Low-volume and ULV
techniques offer potential for reducing application rates of chemicals required for control.[201] The further development of new spraying techniques (e.g., electrostatic sprayers)
also has potential for reducing chemical use. In some regions, the application of chemicals through irrigation systems is a way of saving spraying operations. In particular, the
application of systemic fungicides through drip or other irrigation systems is a
possibility.

Recent advances in computer technology also have implications for downy mildew
management. Computerized systems are now available or are under construction for the
remote sensing and collection of climatic data.[75,202,203] The use of these systems has potential for increasing the efficiency of monitoring microclimates in vineyards and automatically processing temperature, wetness, and other data in computer-based simulators of
downy mildew epidemiology. These have the capacity to provide rapid and accurate assessments of disease risk, predictions of disease appearance, and indications of the need
for fungicide treatments. More sophisticated computer systems also provide the opportunity for creating regional data banks (on climate, etc.) which can be used to evaluate the
risk of disease in existing or new viticultural regions and ultimately classify world viticulture according to the potential impact of downy mildew.[7]

Several aspects of downy mildew epidemiology are still poorly understood, and
further knowledge of these would help to develop more precise models of epidemiology
for use in disease management systems. Geographical ecotypes of *P. viticola*, for example, appear to exist, but little is known of their significance. The same applies to alternative hosts of *P. viticola*. The susceptibility of grapevine tissues to infection by *P. viticola*
varies with age. Young leaves, flowers, and berries are highly susceptible to systemic
invasion, but as berries enlarge and leaf and stem tissues age, they become more resistant
to infection.[21,204,205] Incubation studies and incubation calendars have been developed from
research on vine leaves. But the incubation time in berries, peduncles, and pedicels
should also be considered, as this is not the same. The maturation, survival, and germination of oospores also warrants further study. In some regions, average winter temperatures appear to be related to date of appearance of disease. When monthly mean temperatures for winter months are summed (November to March in France), relatively high
values (above 40) indicate possible early appearance of disease in the following spring
(i.e., early April).[52] The significance of oospore survival for more than two seasons in
relation to seasonal disease development is also unknown. Clearly, there is a need to
develop a better understanding of overwintering inoculum potential in relation to primary
infection and epidemics.

Until now, much research has been directed toward the strategic use of chemicals
for downy mildew control, and the effects of cultural practices have received relatively
little attention. Recent studies of the influence of canopy management practices on other
diseases, such as Botrytis rot,[107,108] have shown significant controlling effects.

This suggests that carefully planned cultural practices in vineyards may contribute
to better downy mildew control, more than was appreciated in the past. When seeking
high yields, viticulturists should be encouraged to adopt canopy management techniques
that also discourage disease development. More research is required to identify the effects

of modern canopy management and cultural practices. Some canopy management approaches, such as mechanical hedging,[206] appear to favor disease development.

Recent advances in *in vitro* techniques and genetic research are also applicable to downy mildew management. This particularly applies to the breeding and screening of new grapevine cultivars with resistance to downy mildew. Techniques like those described by Barlass et al[207] enable *in vitro* screening of cultivars for resistance to downy mildew and greatly increase the efficiency of the screening process. This reduces reliance on time-consuming and costly field tests. Electrophoretic techniques (which identify protein or nucleic acid markers of parentage) have potential for predicting horticultural characteristics of newly bred hybrids. *In vitro* genetic engineering techniques[208] also offer ways of transferring gene sequences providing disease resistance into selected grapevine cultivars. The development of resistant cultivars and hybrids with high technological standards could greatly reduce disease risk and reliance on costly chemical control programs.

There is little doubt that a combination of these new areas of research and development has great potential for further improving the efficiency of downy mildew management in world viticulture.

4-10 REFERENCES

1. Viennot-Bourgin, G., History and importance of downy mildew. In *The downy mildews* (D. M. Spencer, ed.), 1, Academic Press, London, 1981.

2. Winkler, A. J., Cook, J. A., Kliewar, W. M. and Lider, L. A., *General viticulture,* University of California Press, Berkeley, p. 446, 1974.

3. Cooke, M. C., *Gardeners' Chronicle,* 15, ser. 3., 689, 1894 (after Viennot-Bourgin, 1981).

4. Commonwealth Mycological Institute, Distribution maps of plant diseases, Map 221, 4th ed., 1988.

5. Weltzien, H. C., Geographical distribution of downy mildews. In *The downy mildews* (D. M. Spencer, ed.), p. 31, Academic Press, London, 1981.

6. El Fahl, A. M., Ibrahim, A. N., El Sharkawy, T. A. and Mouard, M. Y., Susceptibility of certain grape varieties to *Plasmopara viticola*, the causal agent of downy mildew, with special reference to the effect of some meterorological factors, *Agric. Res. Rev.* 56, 119, 1978.

7. McLean, G. D., Magarey, P. A., Wachtel, M. F. and Dry, P. R., A climatic evaluation of the question: Could grapevine downy mildew develop in Western Australia? In *Proc. 5th Aust. Wine Indust. Tech. Conf.*, 1983, Perth, Western Australia (T. H. Lee and T. C. Somers, eds.), p. 249, Aust. Wine Res. Instit., Adelaide, South Australia, 1984.

8. Gregory, C. T., Studies on *Plasmopara viticola, Phytopathology*, 4, 399 (Abst.), 1914.

9. Anderson, H. W., *Diseases of fruit crops,* McGraw-Hill, New York, p. 364, 1956.

10. Rives, M. and Lafon, R., Observations sur le cycle du determinisme de la production chez la Vigne a l' occasion de la mise au point d'une methode d'infection artificielle par le mildiou (Plasmopara viticola) pour l'evaluation des fongicides au champ, *Vitis*, 11, 34, 1972.

11. Lafon, R. and Bulit, J., Downy mildew of the vine. In *The downy mildews* (D. M. Spencer, ed.), p. 601, Academic Press, London, 1981.

12. Berlese, A. N. and de Toni, G. B., in *Sylloge fungorum*, Vol. 7 (I. Pars and P. A. Saccardo, ed.), p. 239, 1888.

13. Schweinitz, L. D., *Synopsis Fungorum, Am. Boreal.*, 2663(25), 1837 (according to Wilson, 1907).

14. Wilson, G. W., Studies on North American Peronosporales, II, *Bull. Torrey. Botan. Club*, 34, 387, 1907.

15. Berkeley, J. M. and Curtis, M. A., *Botrytis viticola, Rav. Fungi Carol. Exsic.*, V., no. 90, 1848.

16. DeBary, A., Recherches sur de developpement de quelques champignons parasites, *Ann. Sci. Nat. France*, IV, 20, 5, 1863.

17. Walker, J. C., *Plant pathology*, 2nd ed., McGraw-Hill, New York, p. 224, 1957.

18. Gregory, C. T., Studies on *Plasmopara viticola* (downy mildew of grapes). In *Int. Congr. Vitic. Rept.*, p. 126, 1915.

19. Galbiati, C., Lo sviluppo del micelio di Plasmopara viticola osservato al microscopio ottico in tessuti non sezionati di Vitis vinifera (Mycelial development of *Plasmopara viticola* observed in unsectioned tissues of *Vitis vinifera* under the optical microscope), *Riv. Patol. Vegetale*, IV, 12, 35, 1976.

20. Blaeser, M., Untersuchungen zur Epidemiologie des falschen Mehltaus an Weinreben, *Plasmopara viticola* (Berk. & Burt. ex de Bary) Berl. et de Toni, Dissertation, Univ. Bonn, 1978, 127 pp.

21. Rafaila, C., Sevcenco, V. and David, Z., Contributions to the biology of *Plasmopara viticola, Phytopath. Z.*, 63, 328, 1968.

22. Santelli, V., A physiologic form of *Plasmopara viticola* found for the first time, *Phytopathology*, 47, 3, 1957.

23. Li, H., Doazan, J. P. and Clerjeau, M., Etude de la variabilute du pouvoir pathogene de Plasmopara viticola a l'egard de la vigne. I. Comparaison de clones monosporocystes; role de d'heterocaryose du parasite (Study of the variability of pathogenicity of *Plasmopara viticola* to grapevine. I. Comparison of monosporocystic clones; role of heterokaryosis of the parasite), *Agronomie*, 6, 187, 1986.

24. Renfro, B. L. and Shankara Bhat, S., Role of wild hosts in downy mildew diseases. In *The downy mildews* (D. M. Spencer, ed.), p. 107, Academic Press, London, 1981.

25. Walker, J., personal communication, 1988.

26. Magarey, P. A., and Wachtel, M. F., unpublished data.

27. Clerjeau, M., Moreau, Ch., Piganeau, B. and Malato, G., Resistance of *Plasmopara viticola* to anilide-fungicides: Evaluation of the problem in France, *Medeldeingen van de Faculteit Landbouwwetenschappen Rijksuniversiteit Gent*, 49 (2a), 179, 1984.

28. Cohen, Y. and Coffey, M. D., Systemic fungicides and the control of Oomycetes, *Ann. Rev. Phytopathol.*, 24, 311, 1986.

29. Cornu, M., *Etudes sur les Peronosporees*, 2, 1, 1882 (after Anderson, 1956).

30. Cuboni, J., La Peronospora dei grappoli, *Atti. Congr. Nazion. Bot. Crit.*, in *Parma*, 1887, 91, 1887.

31. Millardet, P. M. A., (1) Traitement du mildiou et du rot, *J. Agric. Pratique*, 2, 513, 1885. (2) Traitement du mildiou par le melange de sulphate de cuivre et de chaux, *J. Agric. Practique*, 2, 707, 1885. (3) Sur l'histoire du traitment du mildiou par le sulphate de cuivre, *J. Agric. Practique*, 2, 801 (English transl. by F. L. Schneiderhan in *Phytopathol. Classics*, 3, 1933).

32. Viala, P., *Les maladies de la vigne*, 3rd ed., Montepellier, 1893.

33. Farlow, W. G., On the American grape vine mildew, *Bussey Inst. Bull.*, 1, 415, 1876.

34. Scribner, F. L., Report on the fungus diseases of the grapevine, *U.S. Dept. Agr. Botan. Div. Bull.*, 2, 7, 1886.

35. Scribner, F. L., Report on experiments made in 1888 in the treatment of the downy mildew and black rot of the grapevine, *U.S. Dept. Agr. Botan. Div. Bull*, 10, 1, 1889.

36. Galet, P., *Les maladies et les parasites de la vigne*, Tome 1, Montpellier, Imprimerie du Paysan du Midi, 1977.

37. Palti, J. and Rotem, J., Control of downy mildews by cultural practices. In *The downy mildews* (D. M. Spencer, ed.), p. 289, Academic Press, London, 1981.

38. Chrelashvili, L. G., (About the overwintering of the mildew-causing fungus *Plasmopara viticola* Berl. & de Toni), *Soobschenija Akademii Nauk, Tbilisi*, 115, 173, 1984.

39. Barrett, J. T., Overwintering mycelium of *Plasmopara viticola* in the California wild grape, *Vitis californica* Benth., *Phytopathology*, 29, 822, 1939.

40. Gregory, C. T., Spore germination and infection with *Plasmopara viticola*, *Phytopathology*, 2, 235, 1912.

41. Royle, D. J. and Thomas, G. G., Factors affecting zoospore responses towards stomata in hop downy mildew (*Pseudoperonospora humuli*) including some comparisons with grapevine downy mildew (*Plasmopara viticola*), *Physiol. Pl. Path.*, 3, 405, 1973.

42. Ronzon-Tran Manh Sung, C. and Clerjeau, M., Techniques for formation, maturation, and germination of *Plasmopara viticola* oospores under controlled conditions, *Plant Disease*, 72, 938, 1988.

43. Ronzon, C., Modelisation du Comportement Epidemique du Mildiou de la Vigne: Etude du Role de la Phase sexuee de *Plasmopara viticola*, Ph.D. thesis, Universite de Bordeaux II, Bordeaux, 1987.

44. Gehmann, K., Staudt, G. and Grossman, F., Der Einfluss der Temperatur auf die Oosporenbildung von Plasmopara viticola (The influence of temperature on oospore formation of *Plasmopara viticola*), *Z. PflKrankh. PflSchutz.*, 94, 230, 1987.

45. Roussel, C., Etude du mildou, *Phytoma*, Mai, 19, 1971 (after Ronzon, 1987).

46. Viala, P., Une mission viticole en Amerique, *La Vigne Francaise*, 261, 1889.

47. Arens, K., Untersuchungen uber Keimung und Zytologie der Oosporen von Plasmopara viticola (Berl. et de Toni), *Jahrbuch fur wissenschaftliche Botanik*, 70, 57, 1929.

48. Populer, C., Epidemiology of downy mildews. In *The downy mildews* (D. M. Spencer, ed.) p. 57, Academic Press, London, 1981.

49. Prillieux, E., Sur les spores d'hiver du *Peronospora viticola*, *C. R. Acad. Sci.*, Paris, 93, 752, 1881.

50. Muller, K. and Sleumer, H., Biologische untersuchungen uber die Peronosporakrankheit des weinstockes mit besonderer Berucksichtigung ihrer Bekampfung nach der Inkubations kalendermethode, *Landw, Jahrb.*, 79, 509, 1934.

51. Galbiati, C. and Longhin, G., Indagini sulla formazione e sulla germinazione delle oospore di Plasmopara viticola. (Studies on the formation and germination of oospores of *Plasmopara viticola*), *Riv. Patol. Vegtetale*. 20, 66, 1984.

52. Kable, P. F., *Practical management of crop diseases*, Winston Churchill Memorial Trust Report, Department of Agriculture, New South Wales, p. 61, 1977.

53. Ravaz, L., *Traite general de viticulture*, III partie, t. III, *Le mildiou* (Coultet, ed.), Montpellier, 1914.

54. Roussel, C., Contribution a l'etude de l'evolution du mildiou de la vigne dans le Sud-Ouest de 1940 a 1952, *Bull.O.I.V.* 277, 152, 1954.

55. Strizyk, S., Modelisation en viticulture, *Viti*, 100, 67, 1986.

56. Zachos, D. G., Recherches sur la biologie et l'epidemiologie du Mildiou de la Vigne en Grece. Bases de Previsions et d'Avertissements, *Ann. Inst. Phytopath. Benaki, N.S.*, 2, 193, 1959.

57. Tsvetanov, D., Biologichni prouchvaniya v"rkhu zimnite spori na manata—Plasmopara viticola (Berk. et De Toni)—zimuvane i k lnene (Biological investigation on winter spores of mildew—*Plasmopara viticola* (Berk. & De Toni)—overwintering and germination), *Gradinarska i Lozarska Nauka*, 13, 137, 1976.

58. Gaumann, E., *Principles of plant infection*, Crosby Lockwood, London, 1950.

59. Langcake, P. and Lovell, P. A., Light and electron microscopical studies of the infection of *Vitis* spp. by *Plasmopara viticola*, the downy mildew pathogen, *Vitis,* 19, 321, 1980.

60. Arens, K., Physiologische Untersuchungen an *Plasmopara viticola*, unter besonderer Berucksichtigung der Infektions bedingungen, *Jahrbuch fur wissenschaftliche Botanik,* 70, 93, 1929.

61. Raynal, G., Ravisi, A. and Bompeix, G., Action du tris-o-ethyl phosphate d'aluminium (phosethyl d' aluminium) sur la pathogenie de Plasmopara viticola et sur la stimulation des reactions de defense de la vigne (Action of aluminium tris-o-ethyl phosphate on pathogenicity of *Plasmopara viticola* and on stimulation defence reactions of grapevine), *Annls. Phytopathologie*, 12, 163, 1980.

62. Srinivasan, N. and Jeyarajan, R., Grape downy mildew in India. III. Effect of infection on phenolics, sap concentration, organic acids and amino acids, *Madras Agric. J.,* 64, 797, 1977.

63. Srinivasan, N. and Jeyarajan, R., Grape downy mildew in India. IV. Effect of infection on sugar content and respiration of leaves. *Ind. J. Hortic*, 34, 209, 1977.

64. Langcake, P., Disease Resistance of *Vitix* spp. and the production of the stress metabolites resveratrol, ϵ-viniferin, $\propto$-viniferin and pterostilbene, *Physiol. Plant Pathol.*, 18, 213, 1981.

65. Du Pisani, A. L., Weather conditions conducive to downy mildew, *Deciduous Fruit Grower,* 20, 212, 1970.

66. Amphoux, M. and Bernon, G., La lutte control le mildiou, *Le Progr. Agr. et Vitic.*, 71, 227, 1954.

67. Magarey, P. A. and Wicks, T. J., Grapevine downy mildew, *Aust. Grapegrower Winemaker*, 22(256), 8, 1985.

68. Branas, J. and Bernon, G., Epoque des traitements du mildiai de la vigne, *Ann. Ec. Agric. Montpellier*, 23, 67, 1934.

69. Muller, K., Die biologischen Grundlagen fur die Peronospora-Bekampfung nach der Inkubationskalender-Methode, *Z. PflKrankh, PflSchutz.*, 46, 104, 1936.

70. Dry, P. R., Downy mildew of grapevines, *South Aust. Dept. Agric. Ext. Bull.*, 4, 1975.

71. Leu, L. S. and Wu, H. G., (Inoculation, sporulation and sporangial germination of grape downy mildew fungus, *Plasmopara viticola*), *Plant Protection Bull., Taiwan*, 24, 161, 1982.

72. Lipetzkaya, A. D., (On the biology of winter spores of *Plasmopara viticola*), *Pl. Prot., Lieninge,* 18, 162, 1939.

73. Magarey, P. A., Wicks, T. J. and Weir, P. C., Some new guidelines for the control of grapevine downy mildew, *Aust. Grapegrower. Winemaker*, 22(256), 13, 1985.

74. Wicks, T. J., Magarey, P. A., Wachtel, M. F. and Frensham, A. W., Effect of post-infection application of phosphorus (phosphonic) acid on the incidence and sporulation of *Plasmopara viticola* on grapevine, *Plant Disease*, 75, 40, 1991.

75. Magarey, P. A., Wachtel, M. F., Weir, P. C. and Seem, R. C., A computer-based simulator for rational management of grapevine downy mildew (*Plasmopara viticola*), *Pl. Protection Quarterly*, 6, 29, 1991.

76. Baldacci, E., Calendario d'Incubazione della Peronspora e Guida alla Lotta Antiparassitaria nella vita, *Pio agricola Vogherese*, 23rd ed., C. Gallini, Voghera, 1973.

77. Blaeser, M. and Weltzien, H. C., Epidemiologische Studien an Plasmopara viticola zur Verbesserung der Spritztermin bestimmung [Epidemiological studies to improve the control of grapevine downy mildew (*Plasmopara viticola*)], *Z. PflKrankh. PflSchutz*, 86, 489, 1979.

78. Avizohar-Hershenzon, Z. and Hochberg, N., Effet de l'irrigation sur le developpement du Plasmopara viticola en presence ou absence de fongicides (Effect of irrigation on the development of *Plasmopara viticola* in the presence or absence of fungicides), *Annls. de Phytopathologie*, I, no. hors-serie, 55, 1969.

79. Brook, P. J., Effect of light on sporulation of *Plasmopara viticola, N.Z.J. Bot.*, 17, 135, 1979.

80. Blaeser, M. and Weltzien, H. C., Die Bedeutung von Sporangienbildung, -ausbreitung und keimung fur die Epidemiebildung von Plasmopara viticola (The importance of sporulation, dispersal, and germination of sporangia of *Plasmopara viticola*), *Z. PflKrankh. PflSchutz*, 85, 155, 1978.

81. Yarwood, C. E., The relation of light to the diurnal cycle of sporulation of certain downy mildews, *J. Agric. Res.*, 54, 365, 1937.

82. Hill, G. K., Effect of temperature on sporulation efficiency of oil spots caused by *Plasmopara viticola* (Berk & Curt. ex de Bary) Berl. & de Toni, in vineyards, *Vitic. Enol. Sci.*, 44, 86, 1989.

83. Lalancette, N., Madden, L.V. and Ellis M.A., A quantitative model for describing the sporulation of *Plasmopara viticola* on grape leaves, *Phytopathology*, 78, 1316, 1988.

84. Royle, D. J., Quantitative relationships between infection by the hop downy mildew pathogen, *Pseudoperonospora humili* and weather and inoculum factors, *Ann. Appl. Biol.* 73, 19, 1973.

85. Marais, P.G., and Knox-Davies, P.S., Epidemiology of grapevine downy mildew in the western cape province of South Africa, *Phytophylactica*, 6, 135, 1974.

86. Branas, J., Etudes effectuees sur le court-nour en France et al Allemagne et conclusions qu'elles permettent, *Progr. Agric. Vitic.*, 111, 1, 1939.

87. Pinckard, J. A., The mechanism of spore dispersal in *Peronospora tabacina* and certain other downy mildew fungi, *Phytopathology*, 32, 505, 1942.

88. Corbaz, R., (Studies of fungal spores trapped in the air II, in a vineyard), *Phytopathology Z.*, 74, 318, 1972.

89. Wicks, T. J., Magarey, P. A., Emmett, R. W., Baker, G. J. and Baker, B. T., Practical management of grapevine diseases and pests—revised 1989 edition, *Aust. Grapegrower Winemaker*, 26(309), 19, 1989.

90. Zillig, H., Wie entstehen Plasmopara-Epidemien?, *Z. PflKrankh. PflSchutz*, 52, 83, 1942 (after Blaeser, 1978).

91. Srinivasan, N. and Jeyarajan, R., Viability of *Plasmopara viticola* sporangia produced at different times in a diurnal cycle, *Curr. Sci.*, 45, 106, 1976.

92. Blaeser, M. and Weltzien, H. C., Untersuchungen uber die infektion von weinreben mit Plasmopara viticola in abhangigkeit von der Blattnassedauer (Investigation of the infection of grapevine with *Plasmopara viticola* in relation to the duration of leaf wetness), *Med. Fac. Landbouww. Rijksuniv. Gent,* 42 (2), 967, 1977.

93. Herzog, J. and Schuepp, H., Three types of sensitivity to metalaxyl in *Plasmopara viticola, Phytopath. Z.*, 114, 90, 1985.

94. Lee, T. C. and Wicks, T. J., Dual culture of *Plasmopara viticola* and grapevine and its application to systemic fungicide evaluation, *Plant Disease*, 66, 308, 1982.

95. Morel, A. M., Le developpement du Mildiou sur les tissues de Vigne cultives in vitro, *Camptes rendes des adances de l'Academic des Sciences*, t. 218, 50, 1944.

96. Barlass, M., and Skene, M. G. M., Tissue culture and disease control. In *Proc. 6th Aust. Wine Indust. Tech. Conf.* (T. Lee, ed.) p. 191, Adelaide, South Australia, 1987.

97. Magarey, P. A., Wachtel, M. F. and Newton, M. R., Evaluation of phosphonic (phosphorous) acid, phosethyl-Al and several phenylamide fungicides for post-infection control of grapevine downy mildew caused by *Plasmopara viticola, Australasian Plant Pathol.*, 20, 34, 1991.

98. Wicks, T. J. and Lee, T. C., Evaluation of fungicides applied after infection for control of *Plasmopara viticola, Plant Disease*, 66, 839, 1982.

99. Schuepp, H. and Herzog, J., Reliable method to produce single-zoospore cultures of *Plasmopara viticola, Z. PflKrankh. PflSchutz.*, 93, 30, 1986.

100. Lalancette, N., Ellis, M. A. and Madden, L. V., Estimating infection efficiency of *Plasmopara viticola* on grape, *Plant Disease*, 71, 981, 1987.

101. Horsfall, J. G. and Barratt, R. W., An improved grading system for measuring plant disease, *Phytopathology*, 35, 655, 1945.

102. Anonymous, Plant Diseases Act 1914, Proclamation No. 3, *Western Australian Government Gazette,* p. 3246, 1921.

103. Butler, E. J. and Jones, S. G., *Plant pathology,* Macmillan, London, p. 833, 1949.

104. Marais, P. G., Die biologie en epidemiologie van donsskimmel in Weskaapland, M.Sc. agric. thesis., Univ. Stellenbosch, Stellenbosch, 1973.

105. Smart, R. E., Canopy microclimates and effects on wine quality. In *Advances in viticulture and oenology for economic grain* (T. H. Lee and T. C. Somers, eds.), p. 113, Proc. 5th Aust. Wine Indust. Tech. Conf., University of Western Australia, Perth, 1984.

106. Smart, R. E., Principles of grapevine canopy microclimate manipulation with implications for yield and quality. A review, *Am. J. Enol. Vitic.*, 36, 230, 1985.

107. Gubler, W. D., Marois, J. J., Bledsoe, A. M. and Bettiga, K. J., Control of Botrytis bunch rot of grape with canopy management, *Plant Disease*, 71, 599, 1987.

108. Savage, S. D. and Sall, M. A., Botrytis bunch rot of grapes: The influence of selected cultural practices on infection under Californian conditions, *Plant Disease*, 67, 771, 1983.

109. He, P. C. and Wang, G. Y., (Studies on the resistance to *Plasmopara viticola* of wild Vitis species native to China), *Acta Horticulturae Sinica*, 13, 17, 1986.

110. Matthews, P., Breeding for resistance to downy mildews. In *The downy mildews* (D. M. Spencer, ed.), p. 255, Academic Press, London, 1981.

111. Coutinho, M. P., A resistencia da videira ao mildio: Evolucao dum trabalho (The resistance of grapevine to downy mildew: The development of a program of work), *Garcia de Orta, Estudos Agronomicos*, 9, 229, 1982.

112. Becker, N. Y. and Zimmermann, H., *Wein-Wissenschaft*, 4, 238, 1976 (after Matthews, 1981).

113. Boubals, D., *Annls, Amel. Pl.*, 9, 1, 1959 (after Matthews, 1981).

114. Clingeleffer, P. R., Breeding of grapevines for hot climates, *Aust. Grapegrower Winemaker*, 22(256), 99, 1985.

115. Leppik, E. E., Gene centres of plants as sources of disease resistance, *Ann. Rev. Phytopathol.*, 8, 323, 1970.

116. Boubals, D., *Downy mildew resistance*, Thesis, Faculty of Sciences, Montpellier, 1958.

117. Stoessl, A., Stothers, J. B. and Ward, E. W. B., Sesquiterpenoid stress compounds of the Solanaceae, *Phytochemistry*, 15, 855, 1976.

118. Cruickshank, I. A. M., Phytoalexins, *Ann. Rev. Phytopathol.*, 1, 351, 1963.

119. Barlass, M., Miller, R. M. and Douglas, T. J., Development of methods for screening grapevines for resistance to infection by downy mildew. II Resveratrol production, *Am. J. Enol. Vitic*, 38, 645, 1987.

120. Dercks, W. and Creasy, L. L., The significance of stilbene phytoalexins in the *Plasmopara viticola*-grapevine interaction, *Physiol. & Molec. Pl. Pathol.*, 34, 189, 1989.

121. Stein, U. and Blaich, R., Untersuchungen uber stilbenproduktion and botrytisanfalligkeit bei Vitis-arten, *Vitis*, 24, 75, 1985.

122. Pool, R. M., Creasy, L. L. and Frackelton, A. S., Resveratrol and the viniferins, their application to screening for disease resistance in grape breeding programs, *Vitis*, 20, 136, 1981.

123. Skylakakis, G., Theory and strategy of chemical control, *Ann. Rev. Phytopathol.*, 21, 117, 1983.

124. Haeseler, C. W. and Peterson, D. H., Effect of cupric hydroxide vineyard sprays on concord grape yields and juice quality, *Plant Dis. Rep.* 58, 486, 1974.

125. Schwinn, F. J., Chemical control of downy mildews. In *The downy mildews* (D. M. Spencer, ed.), p. 305, Academic Press, London, 1981.

126. Raz, B., Schuepp, H. and Siegfried, W., (100 years of *Plasmopara* control and copper input into vineyards), Hundert Jahre Plasmopara Bekampfung und Kupfereintrag in die Rebberge, *Schweizerische Zeitschrift fur Obstund Weinbau Wadenswil*, 123, 272, 1987.

127. Egger, E. and Borgo, M., Manifestazioni fitotossiche su vite causate da antiperonosporici organici (Phytotoxic manifestations on grapevine caused by organic anti-*Peronospora* chemicals), *Informatore Fitopatologico*, 32, 55, 1982.

128. Clerjeau, M., Lafon, R. and Bugaret, Y., Etudes sur les proprietes et le mode d'action des nouveaux fongicides antimildiou chez la vigne (cymoxanil, metalaxyl, milfurame, phosethyl-Al) (Studies on the properties and mode of action of new fungicides (cymoxanil, metalaxyl, milfuram, phosethyl-Al) against downy mildew on grapevine), *Phytiatrie-Phytopharmacie*, 30, 215, 1981.

129. Gisi, U., Binder, H. and Rimbach, E., Synergistic interactions of fungicides with different modes of action, *Trans. Br. Mycol. Soc.*, 85, 299, 1985.

130. Gozzo, F., Pizzingrilli, G. and Valcamonica, C., Chemical evidence of the effects of mancozeb on benalaxyl in grape plants as possible rationale for their synergistic interaction, *Pesticide Biochem. & Physiol.*, 30, 136, 1988.

131. Samoucha, Y. and Gisi, U., Systemicity and persistence of cymoxanil in mixture with oxadixyl and mancozeb against *Phytophthora infestans* and *Plasmopara viticola*, *Crop Protection*, 6, 393, 1987.

132. Klopping, H. L. and Delp, C. J., 2-Cyano-N-[(ethylamino carbonyl)]-2-(methoxyimino) ac-teamide, a new fungicide, *J. Agric. Food Chem.*, 28, 467, 1980.

133. Douchet, J. P., Absi, M., Hay, S. J. B., Muntan, L. and Villani, A., European results with DPX 3217, a new fungicide for the control of grape downy mildew and potato late blight, *Proc. 1977 Br. Crop Prot. Conf.—Pests Dis.*, 2, 585, Br. Crop Prot. Council, Croydon, England, p. 199, 1977.

134. Lafon, R., Bugaret, Y. and Bulit, J., Essais d'un nouveau fongicide de la famille des acyla-lanines (CGA 48-988) contre le mildiou de la vigne (Plasmopara viticola (B.C.)) Berl. et de Toni (Trials with a new fungicide (CGA 48-988) of the family of acylalanines against grape-vine downy mildew (*Plasmopara viticola* (B.C.) Berl. & de Toni), *Phytiatrie-Phytopharma-cie*, 27, 263, 1978.

135. Wicks, T. J., The control of *Plasmopara viticola* by fungicides applied after infection, *Austra-lasian Plant Pathol.*, 9, 2, 1980.

136. Magarey, P. A., Wachtel, M. F., Frensham, A. B. and McHenry, W., Strategic use of acylani-lide and phosphonate fungicides against grapevine downy mildew, Abst. 61, 6th Conference, Australasian Plant Pathology Society, Adelaide, South Australia, 1987.

137. Magarey, P. A., Wicks, T. J. and Wachtel, M. F., Field evaluation of phosphorous acid for the control of grapevine downy mildew, Abstract 111, 7th Conference, Australasian Plant Pa-thology Society, Brisbane, Queensland, 1989.

138. Wicks, T. J. and Lee, T. C., Effect of fungicide volatiles on sporangial production of *Plasmo-para viticola*, *Plant Disease*, 66, 945, 1982.

139. Bruck, R. I., Fry, W. E. and Apple, A. E., Effect of metalaxyl, an acylalanine fungicide, on development stages of *Phytophthora infestans*, *Phytopathology*, 70, 597, 1980.

140. Staub, T., Dahmen, H. and Schwinn, F. J., Biological charaterization of uptake and transloca-tion of fungicidal acylanalines in grape and tomato plants, *Z. PflKrankh. PflSchutz*, 85, 162, 1978.

141. Dercks, W. and Creasy, L. L., Influence of phosethyl-Al on phytoalexin accumulation in the *Plasmopara viticola*-grapevine interaction, *Physiol. & Molec. Pl. Pathol.*, 34, 203, 1989.

142. Magarey, P. A., Wicks, T. J. and Wachtel, M. F., Phosphorous acid as Foli-R-Fos 200 con-trols downy mildew, *Aust. Grapegrower Winemaker*, 26(304), 86, 1989.

143. Magarey, P. A. and Wachtel, M. F., Control of grapevine downy mildew with phosphorous acid, Abst. 59, 6th Conference, Australasian Plant Pathology Society, Adelaide, South Aus-tralia, May, 1987.

144. Wicks, T. J., Magarey, P. A. and Wachtel, M. F., The spread of downy mildew in a vineyard, *Aust. Grapegrower Winemaker*, 27(316), 73, 1990.

145. Pegg, K. G., Whiley, A. W. and Hargraves, P. A., Phosphonic (phosphorous) acid treatment control Phytophthora diseases in avocado and pineapple, *Australasian Plant Pathol.*, 19, 122, 1990.

146. Delp, C. J., Coping with resistance to plant disease control agents, *Plant Disease*, 64, 652, 1980.

147. Bosshard, E. and Schuepp, H., Variabilitat ausgewahlter Stamme von Plasmopara viticola beziiglich ihrer Sensibilitat gegenuber metalaxyl unter Freilandbedingungen (Variability of selected strains of *Plasmopara viticola* with respect to their sensitivity to metalaxyl under field conditions), *Z. PflKrankh. PflSchutz*, 90, 449, 1983.

148. Clerjeau, M., Irhir, H., Moreau, C., Piganeau, B., Staub, T. and Diriwachter, G., Etude de la resistance croissee au metalaxyl et au cyprofurame chez Plasmopara viticola: Evidence de plusieurs mecanismes de resistance independents, In *Fungicides for crop protection, 100 years of progress*, Monogr. No. 31, 1, 303, Br. Crop Prot. Counc., Croydon, England, 1985.

149. Gay-Bellile, F., Lacouture, J., Sarrazin, J. F., Courlit, Y. and Menard, E., Le point sur le mildiou de la vigne. Les fongicides systemiques. Les souches resistantes (Downy mildew of grapevine. Systemic fungicides. Resistant races), *Prog. Agric. Vitic.*, 100, 83, 1983.

150. Leroux, P. and Clerjeau, M., Resistance of *Botrytis cinerea* Pers. and *Plasmopara viticola* (Berl. and de Toni) to fungicides in French vineyards, *Crop Protection*, 4, 137, 1985.

151. Herzog, J. and Schuepp, H., Haustorial development test to characterize metalaxyl resistance and genetic variability in *Plasmopara viticola, EPPO Bull.*, 15, 431, 1985.

152. Gaulliard, J. M., Efficacite du phosethyl-Al contre les souches de Plasmopara viticola resistantes aux anilides (acylalanines) (Effectiveness of fosetyl-Al against *Plasmopara viticola* strains resistant to anilides (acylalanines)), *EPPO, Bull.*, 15, 437, 1985.

153. Cohen, Y. and Samoucha, Y., Cross-resistance to four systemic fungicides in metalaxyl-resistant strains of *Phytophthora infestans* and *Pseudoperonospora cubensis, Plant Disease*, 68, 137, 1984.

154. Samoucha, Y. and Cohen, Y., Efficacy of fosetyl-Al in controlling metalaxyl-sensitive and -resistant isolates of *Phytophthora infestans, Phytopathology,* 75, 1384, 1985.

155. Bompeix, G., Clerjeau, M., Lafon, R. and Malfatti, P., Downy mildew: Review of the effectiveness of fosetyl-Al against anilide-resistant strains, *Phytoma-Defense des cultures*, No. 361, 1984.

156. Clerjeau, M., Moreau, Ch., Piganeau, B., Bompeix, G. and Malfatti P., Effectiveness of fosetyl-Al against strains of *Plasmopara viticola* and *Phytophthora infestans* that have developed resistance to anilide fungicides, *Proc. Brit. Crop Prot. Conf.—Pests and Diseases*, 2, 497, 1984.

157. Diriwachter, G., Sozzi, D., Ney, C. and Staub, T., Cross-resistance in *Phytophthora infestans* and *Plasmopara viticola* against different phenylamides and unrelated fungicides, *Crop Protection*, 6, 250, 1987.

158. Grabski, C. and Gisi, U., Quantification of synergistic interactions of fungicides against *Plasmopara* and *Phytophthora*, *Crop Protection*, 6, 64, 1987.

159. Kable, P. F. and Jeffery, H., Selection for tolerance in organisms exposed to sprays of biocide mixtures: A theoretical model, *Phytopathology*, 70, 8, 1980.

160. Levy, Y., Levi, R. and Cohen, Y., Buildup of a pathogen subpopulation resistant to a systemic fungicide under various control strategies: A flexible simulation model, *Phytopathology*, 73, 1475, 1983.

161. Skylakakis, G., Effects of alternating and mixing pesticides on the buildup of fungal resistance, *Phytopathology,* 71, 1119, 1981.

162. Delp, C. J., Industry's response to fungicide resistance, *Crop Protection* 3, 3, 1984.

163. Staub, T. and Sozzi, D., Fungicide resistance: A continuing challenge, *Plant Disease*, 68, 1026, 1984.

164. Magarey, P. A., Downy mildew control in 1983/84, *Aust. Grapegrower Winemaker,* 21(244), 24, 1984.

165. Magarey, P. A., A disease management system for viticulture. An insight into tomorrow's technology for disease control, *Aust. Grapegrower Winemaker*, 22(256), 6, 1985.

166. Hopkins, D. L., Fungicidal control of bunch grape diseases in Florida, *Proc. Fla. State Hortic. Soc.*, 86, 329, 1973.

167. Magarey, P. A., Systems for managing diseases and pests of grapevines, In *Proc. 7th Aust. Wine Industry Tech. Conf.* [(P. J. Williams, D. M. Davidson and T. H. Lee, eds), p. 136,] Adelaide, South Australia, 1989.

168. Matthee, F. N. and Heyns, A. J., Downy mildew (*Plasmopara viticola*), *Decid. Fruit Grower*, 19, 261, 1969.

169. Magarey, P. A., Maelzer, D. A., Kable, P. F., Woods, P., Wicks, T. J. and Wallace, H. R., A management system for grapevine downy mildew in Australia—its conception, Abst. 292, 4th Int'l. Congr. Pl. Pathol., Melbourne, Aust., 1983.

170. Magarey, P. A. and Weir, P. C., A simulator for better management of downy mildew, *Aust. Grapegrower Winemaker,* 25(292), 47, 1988.

171. Strizyk, S., Modelisation—La gestion des modeles E.P.I., *Phytoma*, 353, 13, 1983.

172. Jones, A. L., Fisher, P. D., Seem, R. C., Kroon, J. C. and Van DeMotter, P. J., Development and commercialization of an in-field micro-computer delivery system for weather-driven predictive models, *Plant Disease* 68, 458, 1984.

173. Seem, R. C., Magarey, P. A., McCloud, P. I. and Wachtel, M. F., A sampling procedure to detect grapevine downy mildew, *Phytopathology*, 75, 1252, 1985.

174. Blaise, P. and Gessler, C., Weather-based forecast of *Plasmopara viticola* epidemics of grape. In Proc. 5th Int'l. Congr. Plant Pathol. Kyoto, Japan, Poster No. 1-6, 1988.

175. Lalancette, N., Ellis, M. A. and Madden, L. V., Development of an infection efficiency model for *Plasmopara viticola* on American grape based on temperature and duration of leaf wetness, *Phytopathology* 78, 794, 1988.

176. Lalancette, N., Madden, L. V. and Ellis, M.A., A model for predicting the sporulation of *Plasmopara viticola* based on temperature and duration of high relative humidity, *Phytopathology,* 77, 1699, 1987.

177. Maurin, G., Application d'un modele d'etat potentiel d'infection a Plasmopara viticola (Application of a model of infection potential to *Plasmopara viticola*), *EPPO Bull.*, 13, 263, 1983.

178. Mandrioli, P., Brunelli, A. and Veronesi, G., I sistemi electronici per il controllo delle malattie delle piante (Electronic systems for monitoring plant diseases), *Informatore Fitopatologico*, 35, 11, 1985.

179. Gianetti, G., Mancini, G. and Scapin, I., Esperienze di lotta guidata contro la peronospora della vite con l'impiego delle centraline microclimatiche della serie Agrel (Experiences in guided control of downy mildew of grapevine using microclimatic electronic stations of the Agrel series), *Difesa delle Piante,* 9, 315, 1986.

180. Emmett, R. W., Fungicide and pesticide spray application: An overview, *Aust. Grapegrower Winemaker*, 25(297), 39, 1988.

181. Campbell, M. M., Evaluation of equipment and methods for applying agricultural chemicals. In *Proc. 6th Aust. Wine Indust. Tech. Conf.* (T. Lee, ed.), p. 180, Adelaide, South Australia, 1987.

182. Clingeleffer, P. R., Trayford, R. S., May, P. and Holt, G. E., Contrasting fungicide deposits by aircraft and ground machines in a vineyard. In *Proc. 5th Int. Agric. Aviation Congr.*, Kenilworth, England, p. 313, 1975.

183. Schmid, A., Raboud, G., Antonin, Ph. and Raymond, J.-Ch., Traitements des vignes par helicopteres—Resultats en Valais en 1979, *Rev. Suisse Vitic. Arboric. Hortic.,* 12, 25, 1980.

184. Hislop, E. C. and Baines, C. R., An analysis of some spray factors affecting the protection of foliage by fungicides. In *Symposium on Spraying Systems for the 1980's,* Monogr. No. 24 (J. D. Walker, ed.), p. 23, Br. Crop. Prot. Counc., Croydon, England, 1980.

185. Matthews, G. A., *Pesticide application methods,* Longmans, London and New York, 1979.

186. Whan, J. H., unpublished data, 1988.

187. Ebeling, W., Analysis of the basic process involved in the deposition, degradation, persistence, and effectiveness of pesticides, *Residue Rev.*, 3, 35, 1963.

188. Whan, J. H., Persistence of chemicals. In *Chemicals in the Vineyard, proceedings of a seminar,* 30 May 1985, Mildura, Vic. (D. C. Lester, R. M. Cirami and T. H. Lee, eds.), p. 39, *Aust. Soc. Viticulture Oenology,* 1985.

189. Perret, P., Erfahrungen mit der Verregnung von Pflanzenschutzmitteln im Weinbau (Results of sprinkler irrigation with incorporated plant protectants in viticulture), *Schweizerische Zeitschrift fur Obst-und Weinbau*, 115, 31, 1979.

190. Bozzini, G., Prospettive d'impiego di strutture protettive antigrandine a "dotazione rameica" in funzione antiperonosporica (Prospects for the use of hail protection structures incorporating copper for grapevine downy mildew control), *Informatore Agrario*, 41, 57, 1985.

191. Morando, A., Lotta contro la grandine e la peronospora con reti impregnate. Risultati biennali di prove (Hail and mildew control with treated nets. Results of two years' trials), *Italia Agricola*, 114, 97, 1977.

192. Olivelli, V., Protezione della vite dalla peronospora con reti antigrandine in plastica impregnate di sali rameice (Protection of grapevines from downy mildew by plastic antihail nets impregnated with copper salts), *Vignevini*, 3, 23, 1976.

193. Lafon, J. and Gouvernet, R., Un procede pour proteger la vigne contre la grele et le mildiou (A method of protecting grapevine against hail and mildew), *Prog. Agric. Vitic.*, 94, 651, 1977.

194. Lafon, J. and Gouvernet, R., Un procede pour proteger la vigne contre la grele et le mildiou (suite et fin) (A method of protecting grapevine against hail and mildew (continuation and end), *Prog. Agric. Vitic.*, 94, 686, 1977.

195. Viala, P., Un parasite du mildiou de la vigne comptes rendus, *Acad. d'Agric. de France*, 18, 19, 654, 1932.

196. Bogdanova, V. N., Marzhina, L. A. and Dima, S. G., Izuchenie antibioticheskoi aktivnosti gribov protiv mil'd'yu vinograda (Investigation of the antiobiotic activity of fungi against grapevine mildew), *Mikroorganizmy i virusy, Kishinev, USSR*, 45, 1979.

197. Weltzien, H. C. and Ketterer, N., Control of downy mildew, *Plasmopara viticola* (de Bary) Berlese et de Toni, on grapevine leaves through water extracts from composted organic wastes, *J. Phytopathology*, 116, 186, 1986.

198. Perandin, G., Ruffoni, M. and Pasqualin, G., Lotta guidata contro la peronospora della vite con l' impiego di fungicidi preventivi e curativi in pieno campo (Supervised control of vine downy mildew using preventive and curative fungicides in the field), *Informatore Agrario*, 41, 47, 1985.

199. Dunn, C. L. and Klein, S. P., Post-harvest treatment of vines: a new concept in *Plasmopara* control. In *Fungicides for crop protection*, 100 years of progress. Monogr. No. 31, 455, Br. Crop Prot. Counc., Croydon, England, 1985.

200. Wade, M., Highwood, D. P., Dunn, C. L., Moncorge, J. M. and Perugia, C., Field evaluation of post-harvest treatment to vines: A new concept in *Plasmopara* control. In *Fungicides*

for crop protection, 100 years of progress. Monogr. No. 31., 455, Br. Crop Prot. Counc., Croydon, England, 1985.

201. Whan, J. H., Smith, I. R. and Morgan, N. G., Effect of spraying techniques on brown rot of peach fruit, and on black spot, powdery mildew and the two-spotted mite of apple trees, *Pesticide. Sci.*, 14, 509, 1983.

202. Perandin, G., Moro, L., Pasqualin, G. and Christeller, G., Il computer nella difesa anti-peronosporica in viticoltura (The computer in downy mildew control in viticulture), *Informatore Agrario*, 41, 85, 1985.

203. Scapin, I., Gianetti, G. and Mancini, G., Indagini sul funzionamento di apparecchiature elettroniche per la lotta contro la peronospora della vite (Research on the working of electronic apparatus in the control of grapevine downy mildew), *Difesa delle Piante*, 9, 3, 1986.

204. Srinivasan, N. and Jeyarajan, R., Grape downy mildew in India. I. Foliar, floral and fruit infections, *Vitis*, 15, 133, 1976.

205. Srinivasan, N. and Jeyarajan, R., Influence of positions in grapevine flowers and berries on growth of downy mildew (*Plasmopara viticola*), *Madras Agric. J.*, 70, 557, 1983.

206. Clingeleffer, P. R. and Possingham, J. V., The role of minimal pruning of cordon trained veins (MPCT) in canopy management and its adoption in Australian viticulture, *Aust. Grapegrower Winemaker*, 24(280), 7, 1987.

207. Barlass, M., Miller, R. M. and Antcliff, A. J., Development of methods for screening grapevines for resistance to infection by downy mildew. I. Dual culture *in vitro, Am. J. Enol. Vitic.*, 37, 61, 1986.

208. Kerr, A., Agrobacterium: Pathogen, genetic engineer and biological control agent, *Australasian Plant Pathol.*, 16, 45, 1987.

5

POWDERY MILDEW OF GRAPE

ROGER C. PEARSON and DAVID M. GADOURY

Department of Plant Pathology
New York State Agricultural Experiment Station
Cornell University
Geneva, New York

5-1 INTRODUCTION

Powdery mildew, also called Oîdium, Mehltau, Echte Mehltau, Schimmel, äscherich, Oidio, mal bianco, and cenicilla,[1,2] is caused by *Uncinula necator* (Schw.) Burr. The fungus was first described in eastern North America in 1834 by Schweinitz, who named the fungus *Erysiphe necator*.[3] The disease caused minor damage to native American grapes and did not gain notoriety until 1845, when it was first observed in a glasshouse in Margate, England by a gardener named Tucker,[1] for whom Berkeley named the anamorph *Oidium tuckeri* in 1847.[3] The disease was first reported in California vineyards in 1859.[4] Not until the discovery of the cleistothecial stage in France by Couderc in 1892[3] was the American fungus *U. necator* and the European fungus *O. tuckeri* considered one and the same.

By 1850, the fungus had spread to all the major grape-growing areas of Europe, where it caused considerable crop loss.[5] In France, for example, production declined from 54.8×10^6 hl in 1847 to 10.8×10^6 hl in 1854, causing many ruined vineyardists to emigrate to North Africa or South America.[5] Fortunately, the discovery in 1850 that the disease could be controlled with sulfur[5] meant that crop destruction was limited to but a few years, and by 1858 production in France returned to what it had been in 1847, prior to the introduction of powdery mildew.[5]

Most pathologists consider *U. necator* to be of North American origin[2] since it appears to have been introduced into Europe on American grapevines.[1] Salmon argued strongly that *U. necator* was native to the Old World, occurring on native plants in Japan.[3] However, the relative resistance of many native American *Vitis* species to powdery mildew, as well as the relative susceptibility of many Asiatic species,[5] tends to support the idea that the fungus originated and coevolved with *Vitis* species native to North America.

Only members of the Vitaceae appear to be susceptible to *U. necator*, specifically the genera *Ampelopsis, Cissus, Parthenocissus,* and *Vitis.*[6]

5-2 DISTRIBUTION AND ECONOMIC IMPORTANCE

Today, powdery mildew can be found in most grape-growing areas of the world, including the tropics.[5,7] Uncontrolled, the disease can be very destructive, as when the disease first appeared in France.

Studies in Germany in 1974, using the cultivars Scheurebe and Portugieser, revealed that controlling powdery mildew leads to a 61 and 40% increase in yield, respectively, over unsprayed vines.[8] Sall and Teviotdale[9] indicated that powery mildew was the most widespread problem in California vineyards and that the amount of money spent on control plus the loss in yield frequently equals the value of 10% of that state's entire crop. Pool et al.[10] in New York found a 40% reduction in vine size and a 65% reduction in crop in unsprayed vines at the end of a three-year study with the interspecific hybrid cultivar Rosette. They also documented a reduction in winter hardiness, expressed as a reduction in bud survival, where fewer than one half of the buds grew on unsprayed vines following the winter of 1980–1981. In the same New York study,[10] the mechanical harvesting process resulted in substantial defoliation of unsprayed vines (42% loss in leaf area vs. 8% loss in healthy vines). This loss of leaf area can be especially significant in early ripening cultivars, where loss of postharvest photosynthesis can reduce the vines' capacity to mature fruitful, winter-hardy canes.[10]

Production of high-quality wines necessitates that fruit infection be held to very low levels. Berries infected by *U. necator* tend to be higher in acid than healthy berries.[10] Furthermore, the fungus itself produces off-flavors in wine made from infected grapes.[11] In the New York study,[10] both hydrogen sulfide and mildew-like off-aromas were detected in wines made from lots of fruit with 3% or more infected berries. In addition, infected berries tend to crack, thereby providing entry sites for *Botrytis cinerea*[12] and sour-rot organisms.[5,9,13]

5-3 SYMPTOMS

The powdery mildew fungus can infect all green tissues of the grapevine.[5,9,14] *Uncinula necator* penetrates only the epidermal cells, sending haustoria into them to absorb nutrients. Although such fungal growth is found only in epidermal cells, neighboring, noninvaded cells may become necrotic.[15,16] The presence of mycelia with conidiophores and conidia on the surface of the host tissue gives it a whitish-gray, dusty appearance (Fig.5-1). Both surfaces of leaves of any age may be infected, but susceptibility declines with leaf age.[17] Occasionally, the upper surfaces of infected leaves exhibit chlorotic or colorless spots which resemble the oil-spot symptoms of downy mildew infection.[5] Young, expanding leaves that are infected become distorted and stunted (Fig.5-1).[5,18] Petioles and cluster stems are susceptible to infection throughout the growing season,[14] and once infected they become brittle and prone to breakage as the season progresses.[19] When green shoots are infected, the

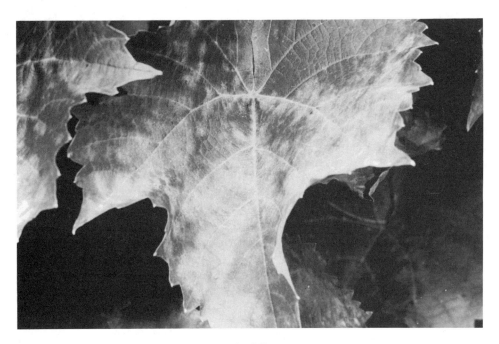

Figure 5-1 Powdery mildew on grapevine foliage.

affected tissues appear dark brown to black in feathery patches, which later appear reddish-brown on the surface of dormant canes.[5,9,14,18] Only remnants of collapsed hyphal fragments can be found at this stage of development.[19] Occasionally, white stunted shoots with small, cupped leaves can be observed early in the growing season. These shoots, which are white due to the growth and sporulation of *U.necator*, result from bud infection the previous season.[5,6,20,21,22]

Cluster infection before or shortly after bloom may result in poor fruit set and considerable crop loss.[5] Berries are susceptible to infection until their sugar content reaches about 8%,[23] although established infections continue to produce spores until the berries contain 15% sugar.[23] If berries are infected prior to attaining full size, the epidermal cells are killed, thereby preventing growth of the epidermis. Nevertheless, the pulp continues to expand and the berry splits (Fig.5-2) due to internal pressure.[5,9] Split berries either dry up or rot.[18] If berries of nonwhite cultivars are infected as they begin to ripen, they fail to color properly and have a blotchy appearance at harvest.[18] Infected berries may also develop a net-like pattern of scar tissue.[14] Such fruit is unmarketable as fresh fruit,[5,14] and wines made from it have off flavors.[10,11] In many viticultural regions, the fungus produces cleistothecia on the surface of infected leaves, shoots, and clusters during the latter half of the growing season.[9,24] They can be seen without magnification.

5-4 CAUSAL ORGANISM

Uncinula necator is an obligate parasite on genera of the Vitaceae. The superficial, but semipersistent septate, hyaline hyphae (4 to 5 μm in diameter) develop characteristic multi-

Figure 5-2   Powdery mildew on grape berries.

lobed appressoria,[21] from which penetration pegs are formed (Fig.5-3). After penetration of the cuticle and cell wall, a globose haustorium is formed within the epidermal cell. Conidiophores form perpendicularly on the prostrate hyphae at frequent intervals. The multiseptate conidiophore is attached to the mycelial hyphae by a cylindrical foot-cell (25 to 40 μm) which is flexuous at the base.[25] Conidiophore cells are generally wider than mycelial hyphae, measuring 6.2 to 7.5 μm. Conidia are hyaline, contain inconspicuous fibrosin bodies,[25] are cylindro-ovoid in shape, are formed singly but accumulate in chains, and measure 27–47 $\times$ 14–21 μm. The oldest conidium is at the distal end of the chain (Fig.5-4). Under field conditions, chains of 3 to 5 spores are common,[5] but under static, humid conditions chains of conidia may reach 8 to 10 spores in length.[19]

The fungus is heterothallic,[26,27] and most populations consist of two mutually exclusive mating types. A small percentage of isolates have the capacity to form cleistothecia in protracted associations with isolates that initially appear to be of an incompatible mating type.[19] Cleistothecia have been reported to form in nearly all viticultural regions of North America,[24,28] Europe,[29–32] Russia,[33,34] and have recently been reported from Australia[35] and Peru.[36] Cleistothecia can form on all infected tissues and are hyaline and spherical when young, but soon yellow due to the accumulation of a yellow lipid in the ascocarp.[37] Cells of the outer ascocarp wall darken as the cleistothecium approaches its mature diameter of 84 to 105 μm. The mature cleistothecium bears equatorially inserted, upwardly directed, multiseptate appendages that are one to six times as long as the diameter of the ascocarp.[7] They are brown at the base and bear the uncinate (hooked) tips for which the genus is named. While the cleistothecium is immature, it retains functional connections to the mildew colony and is roughly spherical in shape. A basal concavity forms when the ascocarp matures and hyphal connections to the mildew colony necrose; thus, the cleistothecium is concavo-

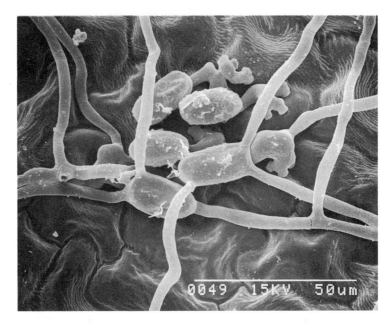

Figure 5-3 Germinating ascospores of *Uncinula necator* showing young colony and appressoria.

Figure 5-4 Epiilluminescence micrograph of conidial chains of *Uncinula necator*.

convex at maturity.[37,38] Cleistothecia contain four to six, rarely more, ovate to subglobose asci that measure 50–60×25–40 μm.[7] Asci most commonly contain four to six ovate to ellipsoid, hyaline ascospores. Ascospore cytoplasm has a low water potential,[37] and spore measurements can be significantly affected on glass microscope slides by certain mountants. In lactophenol, mean dimensions of ascospores are 22×12 μm and range from 18–29×10–15 μm.[19] Mean dimensions of unmounted ascospores, as measured by epiilluminescence microscopy, are 26×15 μm and range from 23–28×14–16 μm.[19] Similar to conidia, viable ascospores usually germinate with a short, single germ tube which terminates with a multilobed appressorium. However, multiple germ tubes may also arise from the spore.[24]

5-5 DISEASE CYCLE

Uncinula necator may overwinter as hyphae inside dormant buds of the grapevine or as cleistothecia on the surface of the vine (Fig.5-5). In greenhouses or in tropical climates, mycelia and conidia may survive from one season to the next on green tissue remaining on the vine.[5]

Infection of developing buds occurs during the growing season.[21] The fungus grows into the bud, where it remains in a dormant state on the inner bud scales until the following season.[22] Shortly after bud break, the fungus is reactivated and shoots growing from infected buds become covered with white mycelium. Conidia are produced abundantly on these infected shoots (called "flag shoots"),[6] and they are readily disseminated by the wind to neighboring vines. One study on dissemination of conidia indicated that release began

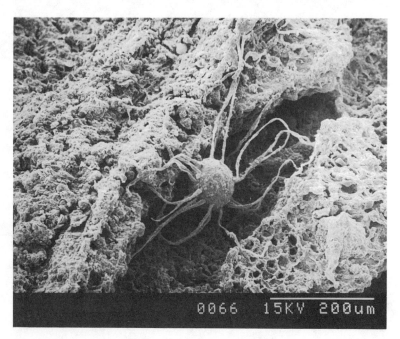

Figure 5-5   Cleistothecia of *Uncinula necator* on bark of grapevine.

two hours after sunrise, increased to a peak at mid-day, and declined sharply thereafter.[39] Little or no release occurred at night.

Cleistothecia are washed by late summer and autumn rains to the bark of the vine, where they overwinter.[24,38] The first-formed leaves of shoots growing near bark are infected first, presumably due to their proximity to the overwintered cleistothecia.[24] The concavo-convex cleistothecia swell to become spherical, dehisce circumscissilely, and discharge ascospores during rain[37] beginning shortly after bud burst, and continue to release asco-spores until the supply is depleted. In New York State, the period of ascospore release lasts from bud break until shortly after bloom.[24] Ascospores germinate and infect green tissue, resulting in colonies that produce conidia for secondary spread.[24]

5-6 INOCULATION AND DISEASE RATING

Potted and field-grown vines can be inoculated easily using a preparation of conidial sus-pensions[40] or by dusting conidia from mildewed leaves onto new host tissue. Despite the inhibitory effects of free water on conidial germination,[23] enough conidia apparently survive in most cases to achieve successful infection using spore suspensions.

For more critical inoculations involving single-spore isolates or in the inoculation of plants *in vitro*, conidia can be transferred using a human hair affixed to a Pasteur pipette.[41] This tool can be sterilized in 95% ethanol, allowed to dry for 10 seconds, and can then be used to transfer individual conidia aseptically to new host tissue. Conidia can also be mass transferred using sterile cotton-tipped swabs.[19]

Inoculations involving ascospores require a supply of physiologically mature cleisto-thecia and suitable conditions for ascospore discharge. Cleitothecia should be collected between bud burst and bloom. They may be stored at -5°C for six weeks,[42] but viability declines rapidly thereafter. Small numbers of viable cleistothecia may remain on overwin-tered leaves, but in spring the greatest numbers of viable cleistothecia are found on the bark of the vine.[38] Detached leaves of grapes have been successfully inoculated by suspending water-soaked leaf pieces bearing numerous cleistothecia over the detached leaves for 24 to 48 hours.[24] Plants may also be inoculated by harvesting cleistothecia from bark, placing the ascocarps on wet filter paper, and suspending the filter paper over host tissue.[24]

Disease assessment methods vary with the objectives of the research. Typically, inci-dence is reported as the percentage of leaves infected or the percentage of fruit clusters infected. Severity is often reported as the proportion of leaf surface infected or as the proportion of the fruit surface infected, and is often estimated from visual keys and rating scales.[43–45] Both measures of disease have strong points and limitations. Incidence is gener-ally more easily measured than is severity and is often the most appropriate measure of disease when both incidence and severity are low and large amounts of plant material must be examined. For example, in well-managed commercial vineyards, most infected leaves may bear only a single mildew colony. Severity in this case is quite uniform, and compari-sons between vineyards are more likely to be made based on the percentage of leaves infected rather than on severity. However, when disease incidence is high, as often occurs in evaluation of fungicides in research plots, severity may be the most appropriate means of comparing treatment effects. For example, two treatments may result in 70% of the fruit clusters bearing mildew colonies, but only 5% of the cluster surface may be infected in treatment A, while 30% of the cluster surface may be infected in treatment B. Mathematical relationships between disease incidence and severity can often be established that allow the estimation of severity from the more easily measured incidence.[45,46] Measurements of disease incidence made at different times during a growing season can be misleading if not corrected for host growth. For example, 50% leaf infection in early spring when there are only 10 leaves per shoot is not equivalent to 50% leaf infection in autumn when there may be 30 or more leaves per shoot. Such inaccuracies can be corrected by basing incidence measurements on the total number of leaves per shoot at the end of the season.

5-7 EPIDEMIOLOGY

The effects of various environmental factors such as temperature, moisture, and light on conidial germination and survival, as well as colony development, have been studied extensively.[47] Temperature appears to be the major limiting environmental parameter for the development of the fungus. Temperatures of 20 to 27°C are optimal for infection and disease development, although fungal growth can occur from 6 to 32°C. Temperatures above 35°C inhibit germination of conidia, and above 40°C they are killed.[23] At 25°C, conidia germinate in approximately five hours, and the time from inoculation to sporulation is five days.[23] At 7°C, incubation time is more than 32 days, whereas at 23 and 30°C it is six days.[23] Mildew colonies are reported to be killed after 10 hours exposure to 36°C or 6 hours at 39°C.[23] Sall[48] has used temperature as the independent variable in a model of disease development. Using this approach, she determined that early warm springs preceded especially severe epidemics of powdery mildew in California.

Free water causes poor and abnormal germination of conidia[23] as well as bursting of conidia, presumably due to excessive turgor pressure. Rainfall can be detrimental by removing conidia and disrupting mycelium. Atmospheric moisture in the 40 to 100% relative humidity (RH) range is sufficient for germination of conidia and infection, although germination has also been reported at less than 20% RH.[23,49] Humidity appears to have a greater effect on sporulation than on germination. For example, two, three, and four to five conidia have been reported to form during a 24-hour period at 30 to 40%, 60 to 70%, and 90 to 100% RH, respectively.[30]

Low, diffuse light favors development of powdery mildew. In fact, germination of conidia is reported to be inhibited by bright sunlight. In one study, germination of conidia was 47% in diffuse light, but only 16% in sunlight.[30]

The role of cleistothecia in the epidemiology of grape powdery mildew has recently been investigated in New York State.[24,38] Although widely reported to be of minor importance in the disease cycle, cleistothecia are the principal source of primary inoculum in New York[24] and may be additional sources of inoculum in other viticultural areas as well. In spring, cleistothecia on the bark of the vine discharge ascospores when rain falls between bud burst and bloom.[24] Ascospore discharge requires free water,[37,42] but measurable ascospore discharges have occurred after as little as 2.5 mm of rain.[37] Temperature has little effect on ascospore release in the range from 10 to 25°C, but temperatures of 4°C or less can suppress ascospore discharge.[37,42] Thus, the conditions required for an ascospore release period appear to be rainfall between bud break and bloom when temperatures are above 4°C.

Ascospores germinate equally well in free water and in saturated atmospheres.[50] Germination declines rapidly as humidity decreases, but approximately 17% of the ascospores can germinate at 54% RH at 25°C.[50] Ascospores have germinated, but have not formed appresoria or infected grapevines, at below 10°C or above 31°C.[50]

There are two principal sources of primary inoculum for grape powdery mildew: mildewed young shoots (flag shoots) emerging from infected buds,[6,20,21] and cleistothecia that have overwintered on the bark of the vine.[24] Both sources result in production of inoculum beginning shortly after bud break, but may result in different patterns of disease development. Flag shoots are intense point sources of inoculum and will cause disease foci centered on the location of the flag shoot.[24] If flag shoots are numerous, this may eventually result in a random or uniform distribution of disease throughout the vineyard. Where cleistothecia are the principal source of primary inoculum, disease is often randomly distributed throughout the vineyard. Initial infections due to ascospores are most often found on the undersides of the first-formed leaves of shoots growing in close proximity to the bark of the vine.[24] Ideal conditions for abundant primary infection will depend on the type(s) of primary inoculum present. Spread from flag shoots can be expected to be favored by optimal temperatures for sporulation and the absence of free water on susceptible tissues. However, release of ascospores is most likely to be maximized by frequent rains during the period between bud break and bloom.

Regardless of the source of primary inoculum, powdery mildew can increase rapidly under ideal conditions for secondary infection. We have recorded infection rates of 0.31 to 0.35/unit/day on the *Vitis* interspecific hybrid cultivar Rosette during the early phases of powdery mildew epidemics, which can be translated as an approximate 10-fold increase in disease incidence in one week.[19]

The magnitude and density of the overwintering pathogen population has a substantial effect on when disease reaches a detectable level in vineyards. In New York, where cleistothecia are the principal source of primary inoculum, differences in the density of ascocarp populations on bark are reflected in differences in the timing of epidemics of powdery mildew. For example, vines sprayed with a fungicide that eradicated approximately 99% of the overwintering cleistothecia were consistently two to three weeks behind unsprayed vines in disease incidence. Fruit infection at harvest was reduced from 58% on the unsprayed vines to 12% on the vines that received the eradicant treatment but no sprays during the growing season.[51] Similarly, a mathematical model developed by Sall[48] indicated that the level of infection at véraison will be reduced by 50% when initial infection is delayed more than two weeks after bud break.

5-8 MANAGEMENT

5-8-1 Regulatory

Regulatory measures have not been applied to powdery mildew because it had already spread to most viticultural areas before the concepts of quarantines and clean-stock programs were developed, and quarantine measures at the present time are probably not justified.[52] However, since the introduction of new strains in *U. necator* containing genes for fungicide resistance or virulence toward certain cultivars could be easily accomplished by transport of dormant cuttings (the buds could be infected or cleistothecia could be on the bark), perhaps sanitation procedures are justified for shipment of planting stocks. Where the cleistothecial stage is not known to be present, it might be prudent to prevent the introduction of new mating types of this heterothallic fungus. In retrospect, the movement of the pathogen to Europe and the widespread destruction it brought to that industry in the mid-1800s[5] is a lesson to be learned regarding the spread of diseased plant material into new areas.

5-8-2 Cultural

Cultural practices can reduce the severity of disease and increase the effectiveness of chemical control. Planting in sites with good air drainage and sun exposure and orienting rows to take advantage of these are helpful.[5] Moderate temperatures, high humidity, and low light intensity are generally found within dense canopies.[53] These factors favor development of powdery mildew. Therefore, the use of training systems that allow good air movement through the canopy and prevent excess shading help reduce disease.[5] An open canopy not only maintains a less favorable microclimate for development of powdery mildew, but it also allows better penetration of fungicide.[54] Savage and Sall[55] found differences in the incidence of Botrytis bunch rot in various trellis types and pointed to differences in wind penetration of the different canopy types as the most likely cause. This microclimatic effect on bunch rot was later confirmed by English et al.[56] and is the principle underlying the practice of leaf removal[57] to aid in control of Botrytis bunch rot. Excessive nitrogen fertilization, which tends to promote succulent growth, has also been associated with increased incidence of powdery mildew.[58]

5-8-3 Chemical

Control of powdery mildew in commercial vineyards is generally based on the use of fungicides. Sulfur was the first effective fungicide used for control of this disease and, due to its efficacy (both preventive and curative) and low cost, it remains the most widely used fungicide at present.[5] Sulfur is commonly applied as a dust or as a wettable powder. In dry climates, sulfur dust is preferred, whereas in regions where rainfall is plentiful during the growing season, wettable powder or flowable formulations are preferred for their retention qualities.[5] Much of the fungicidal activity of sulfur is associated with its vapor phase.[5] These toxic vapors account for the activity of sulfur observed at some distance from the site of application.[5] The production of vapors and their effectiveness depends on the type of sulfur as well as various environmental factors. Amorphous sulfur emits fungicidal vapors more quickly than crystalline sulfur and, in general, as the particle size of sulfur decreases, activity increases.[5] However, below 1 μm the persistence is reduced and the potential for phytotoxicity is increased.[5] Temperature has a major influence on sulfur activity. The optimal temperature range for sulfur activity is 25 to 30°C. Although activity is seen as low as 18°C, efficacy may be inadequate.[5] Above 30°C, the risk of phytotoxicity increases greatly and applications at 35°C or higher are not recommended. Activity of sulfur is reduced in humid air compared to dry air.[5]

Copper formulations and several organic fungicides such as dinocap, benomyl, and compounds belonging to the sterol demethylation-inhibiting (DMI) group (for example, triadimefon, penconazole, myclobutanil, and flusilazole) are used commercially, although not as extensively as sulfur. The organic fungicides maintain activity over a wider temperature range and, with the exception of dinocap, exhibit less phytotoxicity than sulfur.

One problem with the use of benzimidazole fungicides has been the selection of resistant strains of the pathogen. Benomyl resistance became widespread in commercial vineyards in New York following three to four years of intensive use.[40] Similar resistance developments have been reported in Rumania.[59] Resistance to the DMI fungicide triadimefon has been reported from California, where isolates that are tolerant of triadimefon are cross-resistant to other DMI fungicides.[60] A recent decline in the efficacy of triadimefon in New York may also be related to shifts in sensitivity of the pathogen population to triadimefon.[19]

The use of fungicides for control of powdery mildew should begin during early stages of vine development. Studies in New York[10] have documented the importance of early-season control of powdery mildew on vine vigor, yield, and crop quality. Furthermore, epidemiological studies indicate that initial infection from ascosporic inoculum occurs shortly after bud break.[24] It is especially important to begin applications of protectant fungicides, such as sulfur, prior to disease development to ensure good control.[48] In many viticultural regions, the first application is made at 15 to 20 cm of growth.[5,9] However, in light of new biological information, applications of protectant fungicides at 2.5 cm of growth are now suggested for areas where ascosporic inoculum is present. Because sulfur has poor retention qualities, application schedules of 7 to 10 days are usually required. Attempts to time sulfur applications based on ambient temperatures look promising.[61] Dinocap has been used successfully on a 10- to 14-day schedule,[40] and DMI fungicides are commonly used on a 14- to 21-day schedule.[62] Pre-bud swell, over-the-trellis sprays of lime sulfur have significantly delayed the development of powdery mildew epidemics in New York, resulting in reduced fruit infection at harvest.[51] These sprays kill the overwintering cleistothecia on the bark of the vine, thereby reducing the primary inoculum.

5-8-4 Biological

To date, biological control has not been applied to *U. necator*. The most commonly reported mycoparasites are *Ampelomyces quisqualis* Ces. (syn. *Cicinnobolus cesatii* De Bary)[3,63] and *Tilletiopsis sp.*[64] *Ampelomyces quisqualis* was found to infect naturally a large percentage of mildew colonies late in the growing season in New York,[38] and it could be an important factor in reducing the late-season spread of powdery mildew and production of cleistothecia. Although use of these fungi may have application in the controlled environment of a greenhouse, they have not been used to control grape powdery mildew commercially in the field.

5-8-5 Disease Resistance

Within *Vitis*, there are great differences in susceptibility to powdery mildew. *Vitis vinifera* and Asiatic species such as *V. betulifolia*, *V. pubescens*, *V. davidii*, *V. pagnucii*, and *V. piasezkii* are highly suceptible. By comparison, American species such as *V. aestivalis*, *V. berlandieri*, *V. cinerea*, *V. labrusca*, *V. riparia*, and *V. rupestris* are much less susceptible.[6] Although powdery mildew resistance has not been a major objective in most grape-breeding programs, considerable variation in susceptibility to powdery mildew has been observed in progeny from breeding programs in search of *Phylloxera* resistance, downy mildew resistance, cold hardiness, or other factors. *Vitis rotundifo-*

lia, the muscadine grape of the southeastern U.S.A., is resistant to powdery mildew,[65] but efforts to produce commercial interspecific hybrid fruiting cultivars from this species have so far been unsuccessful.

Cultivars within a species demonstrate considerable variation in susceptibility. For example, one French study[2] listed the following *V. vinifera* cultivars as being highly susceptible: Carignan, Piquepoul, Cinsaut, Cabernet Sauvignon, Cabernet Franc, Muscadelle, Muscat de Frontignon, Durif, Muscat d'Alexandrie, Chasselas, Frankenthal, Sylvaner, Auxenois, and Emperor. As susceptible they listed Alicante Bouschet, Clairette, Mourvedre, Chardonnay, Pinot noir, Pinot gris, Meunier, Gamay noir, White Riesling, Müller-Thurgau, Gewürztraminer, Sauvignon, Chenin blanc, Colombard, Merlot noir, Portugais bleu, Tannat, Muscadet, Petit Verdot, and Ugniblanc. Weakly susceptible cultivars listed were Aramon, Cot, Folle blanche, Meccabeu, Baroque, Grenache, and Syrah. In the eastern United States, where *V. vinifera*, native American cultivars of *V. labrusca* origin, as well as interspecific hybrids are grown in neighboring vineyards, comparisons of cultivar susceptibility have been made. A highly susceptible category would include the following *Vitis* interspecific hybrid cultivars: Aurore, Chancellor, Chelois, De Chaunac, Rosette, Rougeon, Seyval, Vidal 256, and Vignoles, as well as the *Vitis vinifera* cultivars Cabernet Sauvignon, Chardonnay, Gewürztraminer, Merlot, Pinot noir, and White Riesling. By comparison, a susceptible category would include the hybrids Baco noir, Cascade, and Foch, as well as *Vitis labrusca* Catawba, Concord, Delaware, Elvira, and Niagara.[66] A category of weakly susceptible cultivars would include the *Vitis* interspecific hybrids Canadice, Cayuga White, and *Vitis labrusca* Ives and Steuben.[66]

The presence or absence of pathogenic specialization of isolates of *U. necator* from grape is poorly understood. It has been shown that isolates from *Vitis* and *Parthenocissus* differ in pathogenicity and virulence in reciprocal inoculations.[27] Isolates from 10 *Vitis* species did not differ in virulence on the *Vitis* interspecific hybrid cultivar Chancellor.[27] However, there is circumstantial evidence that specialized forms may exist. There are inconsistencies in the relative susceptibility of various cultivars to powdery mildew. Although the North American species *Vitis labrusca* and *V. riparia* have been reported to be practically immune to infection in Europe,[5] 50% or more of the leaf surface is commonly colonized by *U. necator* on wild vines of these species in the northeastern U.S. The *Vitis* interspecific hybrid cultivar Vidal blanc was reported to be highly resistant to powdery mildew in West Germany,[67] and yet is highly susceptible to infection in New York.[66] *Vitis vinifera* White Riesling, a highly susceptible cultivar in the United States and Europe, has been reported to be resistant to mildew infection in India.[68] The inconsistency of relative susceptibility is temporal as well as spatial. In 1945, Suit[69] reported that fruit of the *Vitis* interspecific hybrid cultivars Dutchess and Missouri Riesling were uninfected and slightly infected, respectively, in New York vineyards, while *Vitis labrusca* Concord was highly susceptible and was severely diseased. In New York in 1988, Wilcox and Agnello[66] rated Missouri Riesling as extremely susceptible to powdery mildew, while Dutchess and Concord were both moderately susceptible. Finally, Doster and Schnathorst[17] reported that *Vitis labrusca* Concord and Niagara did not support sporulation of *Uncinula necator* in laboratory studies and were the most resistant to infection of all cultivars examined.

5-8-6 Integrated Control

Various cultural practices and resistant cultivars have been shown to reduce disease,[5] and biological control organisms may parasitize a significant portion of the pathogen population.[38] However, the most widespread approach to controlling powdery mildew is essentially that used in the mid-1800s: Vines are protected from infection by the repeated application of fungicides during the growing season. Thus, cultural practices, and most importantly the cultivar planted, determine the approximate rate and frequency of fungicide application.[66] While nonchemical measures are available that reduce disease, there is as yet no method to bridge the gap between these research findings and their application to control of powdery mildew without seasonal applications of fungicides.

The previously mentioned pre-bud burst, over-the-trellis sprays of lime sulfur also reduce the incidence of angular leaf scorch (*Pseudopezicula tetraspora*) and Phomopsis cane and leaf spot (*Phomopsis viticola*).[51] However, the rate of application required for effective control makes this treatment economically unattractive in many cases. Nevertheless, the strategy of dormant season eradication of primary inoculum may eventually provide control of multiple diseases.

5-9 CONCLUSION

Although powdery mildew has been a major disease problem in grapes for the last 150 years, much research is still needed in the areas of epidemiology and control. Recent research in New York has shown that the cleistothecial stage is more significant than previously thought. In that viticultural region, the cleistothecial stage is the principal source of overwintering inoculum. It would seem desirable to determine the significance of the cleistothecial stage in other viticultural regions as well. In addition to providing primary inoculum, the genetic variability made possible by the sexual stage may be exhibited as variation in the population, such as increased virulence on currently resistant cultivars or development of tolerance to fungicides.

Little is know regarding the effect of eradication of cleistothecia on subsequent development of epidemics. Specifically, the relationship between inoculum dose and disease development needs to be determined. In areas where cleistothecia are the sole source of primary inoculum, significant advances in disease control are likely if an efficient and effective means of disrupting overwintering can be found.

Questions remain to be answered concerning the survival of the fungus in infected buds. First, how widespread is bud infection and subsequent flag shoot development in various cultivars and in different viticultural regions? How does the fungus survive the dormant season in infected buds? Is the fungus itself or the infected bud susceptible to death at low temperature? What triggers reactivation of the fungus in spring? These questions need to be answered to understand how the pathogen survives the intercrop period in diverse viticultural regions.

Sulfur was developed as a control for powdery mildew relatively soon after the discovery of the disease in Europe. The effectiveness of sulfur throughout the world is probably one reason why many studies on the relationships between fungal biology and ecology, and the development of powdery mildew, have never been conducted. In our opinion, improvements in control of grape powdery mildew are as likely to be generated by research on pathogen biology and ecology as by research on chemical suppression of disease.

Many products in the DMI group of fungicides have been developed in the past ten years. Unfortunately, few experimental data on the potential for development of tolerant strains of *U. necator* are available. Nevertheless, based on experience with the use of the DMIs on other powdery mildews and the decline in efficacy of triadimefon on *U. necator*, one should be cognizant of the potential for development of tolerance to all DMI fungicides. It would be most prudent to develop base-line sensitivity data on *U. necator* in each viticultural region prior to introduction of a new DMI fungicide. This information would be invaluable in ferreting out future resistance or grower misuse problems. Also, by following population shifts, crop advisors could recommend control-strategy changes before substantial crop loss occurs. New information on the role of cleistothecia in the epidemiology of the disease raises questions concerning chemical control strategies. Various chemicals should be screened for their eradicant activity against cleistothecia. The feasibility of using chemical sprays to kill overwintering cleistothecia, or prevent their formation, may have dramatic effects on initial development of epidemics and strategies for control.

It is critical that antiresistance tactics be employed now while alternative DMI fungicides, such as myclobutanil and fenarimol, are still providing good control of grape powdery mildew. Such tactics would include eradication of overwintering inoculum coupled with reduced seasonal use of DMI fungicides, and the use of non-DMI fungicides alone, to reduce the number of applications of DMI fungicides used each year. DMI fungicides should not be used season-long, either alone or in tank mixes with other fungicides for powdery mildew control.

Biological control of grape powdery mildew has not been adequately studied. Most reports in the literature simply mention finding a mycoparasite on a colony of *U. necator* without indicating its impact on the development of the disease. We have observed significant destruction of powdery mildew colonies by *Ampelomyces quisqualis* late in the season when most of the damage by the disease has already been done, and have observed almost complete destruction of cleistothecia overwintered in soil.[19,38] Studies need to be conducted on the environmental and nutrient requirements for growth of both *Ampelomyces* and *Tilletiopsis*, as well as other potential mycoparasites. Perhaps through recent advances in biotechnology, strains of these fungi can be developed that will provide control of powdery mildew when used early in the season. Surveys also need to be conducted to find other phylloplane fungi, bacteria, or mycoviruses[70] that may have either a direct or an indirect detrimental effect on *U. necator*.

There is much need for studies on disease resistance in powdery mildew of grape. Many inconsistencies appear in the literature regarding not only varietal susceptibility but also species susceptibility. The reasons for these inconsistencies need to be determined. Perhaps races exist which have yet to be defined. A functional sexual stage may provide unlimited opportunities for recombination and development and expression of new virulence genes. Are there specific genes that confer resistance to powdery mildew in *Vitis*; if so, where are they located and how can they be transferred to desirable cultivars? Is resistance monogenic or polygenic, how are resistance genes inherited, and is the resistance durable?

Although the reported disease resistance of various species of *Vitis* has been incorporated into interspecific hybrids, acceptance of these hybrid cultivars in the commercial trade is one of reluctance and even legality in some countries. Countries such as France and Germany forbid the commercial planting of interspecific hybrid cultivars, allowing only cultivars of *Vitis vinifera*. Such restrictive laws prevent the incorporation of resistance genes from other species of *Vitis* and force cultivation of susceptible cultivars that require significant inputs of chemical sprays to control not only powdery mildew, but other diseases as well. In addition, the tradition and reputation of growing certain cultivars in specific regions of the world, the availability of effective fungicides, as well as the public's recognition of varietal wines such as Chardonnay, Cabernet Sauvignon, Riesling, and Pinot noir are the primary reasons why disease resistance as a control strategy has not been utilized.

The history of cultivar transition in the wine industry of New York State is one of shifting to more disease-susceptible cultivars to meet consumer demands. The wine industry in that state was originally based on cultivars of *Vitis labrusca*. The wines made from these grapes were generally classified as dessert wines. *Vitis* interspecific hybrid cultivars bred in France in the late 1800s and early 1900s were later planted to increase the quality and sales of table wines. Unfortunately, many of the hybrids, although equally cold-hardy, were more susceptible to diseases such as powdery mildew, downy mildew, and black rot than were the native *Vitis labrusca* cultivars. The New York wine industry, in further efforts to meet consumer preferences and to compete with wines imported from California and from Europe, has recently planted *Vitis vinifera* cultivars in regions of New York State where these cold-sensitive grapes are able to survive. Growers of these highly disease-susceptible cultivars of *Vitis vinifera* are now faced with the necessity of using intensive spray programs requiring two to three times the fungicide input required to grow the traditional American cultivars.

To circumvent the sociological and legal barriers to growing interspecific hybrid cultivars, and depending on the answers to questions of the genetic nature of resistance to powdery mildew in *Vitis*, a highly significant contribution of genetic engineering to viticulture would be the incorporation of genes for resistance to powdery mildew, as well as to other diseases, into popular and accepted cultivars such as Cabernet Sauvignon, Chardonnay, and White Riesling.

5-10 REFERENCES

1. Arnaud, G. and Arnaud, M., *Traité de pathologie vegétale*, Paul Lechevalier et Fils, Paris, 1931.
2. Galet, P. *Les maladies et les parasites de la vigne*. Tome 1. Paysan du Midi, Montpellier, 1977.
3. Salmon, E. S., A monograph of the Erysiphaceae, *Mem. Torrey Bot. Club*, 9, 1, 1900.
4. Smith, R. E., Grape mildew as viewed in the early agricultural press of California, *Plant Dis. Rep.*, 45, 700, 1961.
5. Bulit, J. and Lafon, R., Powdery mildew of the vine. In *The powdery mildews* (D. M. Spencer, ed.) Academic Press, New York, 1978.
6. Boubals, D., Étude des causes de la résistance des vitacées à l'oidium de la vigne—*Uncinula necator* (Schw.) Burr. et leur mode de transmission héréditaire, *Ann. Amelior. Plantes*, 11, 401, 1961.

7. Kappor, J. N., *Uncinula necator*. C.M.I. Descriptions of Pathogenic Fungi and Bacteria No. 160, Commonwealth Mycological Institute, Kew, Surrey, England, 1967.

8. Beetz, K. J., Ursachen fur das verbreitete Auftreten von Oidium im deutschen Weinbau und seine wirtschaftliche Bedeutung, *Weinberg und Keller* 24, 443, 1977.

9. Sall, M. A. and Teviotdale, B. L. Powdery mildew. In *Grape pest management* (D. L. Flaherty, F. L. Jensen, A. N. Kasimatis, H. Kido and W. J. Moller, eds.), Univ. Calif. Publ. 4105, 1981.

10. Pool, R. M., Pearson, R. C., Welser, M. J., Lakso, A. N. and Seem, R. C., Influence of powdery mildew on yield and growth of Rosette grapevines, *Plant Disease*, 68, 590, 1984.

11. Ough, C. S. and Berg, H. W., Powdery mildew sensory effect on wine, *Am. J. Enol. Vitic.*, 30, 321, 1979.

12. Pearson, R. C., Chemical control of *Botrytis cinerea* on grapes in New York (USA), *EPPO Bull.*, 12, 101, 1982.

13. Hewitt, W. B., Rots and bunch rots of grapes, *Cal. Agric. Exp. Stn. Bull. 868*, 1974.

14. Hewitt, W. B. and Jensen, F. L., Powdery mildew of grape, *Calif. Agric. Exp. Stn. Bull. 801*, 1973.

15. Yarwood, C. E., History and taxonomy of the powdery mildews. In *The powdery mildews* (D. M. Spencer, ed.), Academic Press, New York, 1978.

16. Lakso, A. N., Pratt, C., Pearson, R. C., Pool, R. M., Seem, R. C. and Welser, M. J., Photosynthesis, transpiration, and water use efficiency of mature grape leaves infected with *Uncinula necator* (powdery mildew), *Phytopathology*, 72, 232, 1982.

17. Doster, M. A. and Schnathorst, W. C., Comparative susceptibility of various grapevine cultivars to the powdery mildew fungus *Uncinula necator, Am. J. Enol. Vitic.*, 36, 101, 1985.

18. Pearson, R. C. and Goheen, A. C. (eds), *Compendium of grape diseases*, APS Press, St. Paul, MN, 1988.

19. Pearson, R. C. and Gadoury, D. M., unpublished, 1991.

20. Sall, M. A. and Wyrsinski, J., Perennation of powdery mildew in buds of grapevines, *Plant Dis.*, 66, 678, 1982.

21. Pearson, R. C. and Gärtel, W., Occurrence of *Uncinula necator* in buds of grapevine, *Plant Dis.*, 69, 149, 1985.

22. Bernard, A. C., Presence du mycelium de l'oidium (*Uncinula necator* (Schw.) Burril) dans les bourgeons latents de *Vitis vinifera* cv. Carignan, pendant l'hiver, *Prog. Agric. Vitic.*, 102, 360, 1985.

23. Delp, C. J., Effects of temperature and humidity on the grape powdery mildew fungus, *Phytopathology,* 44, 615, 1954.

24. Pearson, R. C. and Gadoury, D. M., Cleistothecia, the source of primary inoculum for grape powdery mildew in New York, *Phytopathology*, 77, 1509, 1987.

25. Bosewinkle, H. J., Identification of Ersiphaceae by conidial characteristics, *Rev. de Mycol.*, 41, 493, 1977.

26. Smith, C. G., Production of powdery mildew cleistocarps in a controlled environment, *Trans. Br. Mycol. Soc.*, 55, 355, 1970.

27. Gadoury, D. M. and Pearson, R. C., Heterothallism and pathogenic specialization in *Uncinula necator, Phytopathology*, 81, (in press) 1991.

28. Bioletti, F. T., Oidium or powdery mildew of the vine, *Calif. Agric. Exp. Stn. Bull. 186*, 1907.

29. Weltzien, H. C. and Weltzien, M., Cleistothecia von *Uncinula necator* in Württemberg 1961, *Z. Pflanzenkr. Pflanzenschutz*, 69, 664, 1962.

30. Aurel, T., *Ceretari privind biologia ciupercii Uncinula necator (Schw.) Burr. care provoaca*

fainarea vitei de vie si mijloacele de combatere in conditiile podgorei Dealul Mare, Ph.D. thesis, Institutul Agronomic, Bucuresti, Romania, 1974.

31. Yossifovitch, M., *Contribution à l'etude de l'Oïdium de la vigne et de son traitement*, Ph.D. thesis, Univ. Toulouse, 1923.

32. Calogne, F. D., Notas sobre el ascocarpo de los Erysiphaceae (Ascomycetes), *Acta Bot. Mal.*, 6, 17, 1980.

33. Mijušković, M., Brojno stvaranje peritecija *Uncinula necator* (Schw.) Burr. u Crnoj Gori 1962 godine, *Zašt. Bilja*, 14, 329, 1963.

34. Lemanova, N. B., O perezimovke oidiuma, *Zashch. Rast. Mosk.*, 14, 49, 1969.

35. Wicks, T. J. and Magarey, P., First report of *Uncinula necator* cleistothecia on grapevines in Australia, *Plant Dis.*, 69, 727, 1985.

36. Dongo, S. L. and Arestegui, M. E., *Uncinula necator* en vid en el Peru, *Fitopatologia*, 8, 35, 1973.

37. Gadoury, D. M. and Pearson, R. C., Ascocarp dehiscence and ascospore discharge in *Uncinula necator*, *Phytopathology*, 80, 393, 1990.

38. Gadoury, D. M. and Pearson, R. C., Initiation, development, dispersal, and survival of cleistothecia of *Uncinula necator* in New York vineyards, *Phytopathology*, 78, 1413, 1988.

39. Pady, S. M. and Subbayya, J., Spore release in *Uncinula necator*, *Phytopathology*, 60, 1702, 1970.

40. Pearson, R. C. and Taschenberg, E. F., Benomyl-resistant strains of *Uncinula necator* on grapes, *Plant Dis.*, 64, 677, 1980.

41. Coyier, D. L., Heterothallism in the apple powdery mildew fungus, *Podosphaera leucotricha*, *Phytopathology*, 64, 246, 1974

42. Diehl, H. J. and Heintz, C., Studies on the generative reproduction of grapevine powdery mildew (*Uncinula necator* Berk.), *Vitis*, 26, 114, 1987.

43. Pearson, R. C., Field evaluation of fungicides for control of diseases of grapes. In *Methods for evaluating pesticides for control of plant pathogens* (K. D. Hickey, ed), Am. Phytopathol. Soc., 1986.

44. Bouron, H., Guide-line for the biological evaluation of fungicides, *Uncinula necator* (Schw.) Burr. (powdery mildew of grapevine), *EPPO Bull. 8*, 1978.

45. Seem, R. C., Disease incidence and severity relationships, *Annu. Rev. Phytopathol.*, 22, 133, 1984.

46. Seem, R. C. and Gilpatrick, J. D. Incidence and severity relationships of secondary infections of powdery mildew on apple, *Phytopathology*, 70, 851, 1980.

47. Schnathorst, W. C., Environmental relationships in the powdery mildews, *Annu. Rev. Phytopathol.*, 3, 343, 1965.

48. Sall, M. A., Epidemiology of grape powdery mildew: A model, *Phytopathology*, 70, 338, 1980.

49. Oku, H., Hatamoto, M., Ouchi, S. and Fujii, S., Effect of temperature and humidity on the development of powdery mildew of grapevine, *The Scientific Reports of the Faculty of Agriculture, Okayama Univ.*, 45, 16, 1975.

50. Gadoury, D. M. and Pearson, R. C., Germination of ascospores and infection of *Vitis* by *Uncinula necator*, *Phytopathology*, 80, 1198, 1990.

51. Gadoury, D. M. and Pearson, R. C. The use of dormant eradicant sprays to control grape powdery mildew, *Phytopathology*, 78, 1507 (Abstr.), 1988.

52. Bovey, R. and Pelet, F., Grapevine and temperate fruits. In *Plant health and quarantine in international transfer of genetic resources* (W. B. Hewitt and L. Chiarappa, eds.), pp. 165–195, CRC Press, Cleveland, 1977.

53. Smart, R. E., Principles of grapevine canopy microclimate manipulation with implications for yield and quality, *Am. J. Enol. Vitic.*, 36, 230, 1985.

54. Sall, M. A., Mildew severity related to vineyard environment, *Calif. Grape Grower*, September, pp. 20 and 24, 1981.

55. Savage, S. D. and Sall, M. A., Botrytis bunch rot of grapes: Influence of trellis type and canopy microclimate, *Phytopathology*, 74, 65, 1984.

56. English, J. T., Thomas, C. S., Marois, J. J. and Gubler, W. D., Microclimates of grapevine canopies associated with leaf removal and control of Botrytis bunch rot, *Phytopathology*, 79, 395, 1989.

57. Gubler, W. D., Marois, J. J., Bledsoe, A. M. and Bettiga, L. J., Control of Botrytis bunch rot of grape with canopy management, *Plant Dis.*, 71, 599, 1987.

58. Bavaresco, L. and Eibach, R., Investigations on the influence of N fertilizer on resistance to powdery mildew (*Oidium tuckeri*) downy mildew (*Plasmopara viticola*) and on phytoalexin synthesis in different grapevine varieties, *Vitis*, 26, 192, 1987.

59. Nagler, M., Diaconu, V. and Alexandri, A. A., Rezistenta fainarii vitei de vie (*Uncinula necator*) si a fainarii castravetilor (*Sphaerotheca fuliginea*) la fungicide sistemice benzimidazolice, *An. Inst. Cercet. Prot. Plant*, 12, 345, 1977.

60. Ogawa, J. M., Gubler, W. D. and Manji, B. T., Effects of sterol biosynthesis inhibitors on diseases of stone fruits and grapes in California. In *Sterol biosynthesis inhibitors—pharmaceutical and agrichemical aspects* (D. Berg and M. Plempel, eds.), pp. 262–288, Ellis Horwood, Ltd., Chichester, England, 1988.

61. Sall, M. A., Wyrsinski, J. and Schick, F. J., Temperature-based sulfur applications to control grape powdery mildew, *Calif. Agric.*, July-August, pp. 4 and 5, 1983.

62. Pearson, R. C., Fungicides for disease control in grapes, advances in development. In *Fungicide chemistry, advances and practical applications* (M. B. Green and D. A. Spilker, eds.), American Chemical Society Symposium Series 304, 1986.

63. Blumer, S., *Echte Mehltaupilze (Eysiphaceae)*, Gustav Fischer Verlag, Jena, 1967.

64. Hoch, H. C. and Provvidenti, R., Mycoparasitic relationships: Cytology of the *Sphaerotheca fuliginea-Tilletiopsis* sp. interaction, *Phytopathology*, 69, 359, 1979.

65. Olmo, H. P., The potential role of (*vinifera* × *rotundifolia*) hybrids in grape variety improvement, *Experientia*, 42, 921, 1986.

66. Wilcox, W. F. and Agnello, A. M., *Grape pest control guide*, N.Y. Coop. Ext. Serv., 1988.

67. Stein, U., Heintz, C. and Blaich, R., Die in vitro-Prufung von Rebensorten auf Oidium-und Plasmopara-Resistenz, *Z. Pflanzenkr. Pflanzenschutz*, 92, 355, 1985.

68. Sohi, H. S. and Sridhar, T. S., Notes on relative resistance and susceptibility of grape varieties to powdery mildew (*Uncinula necator* Schw.), *Ind. J. Agric. Sci.*, 42, 641, 1972.

69. Suit, R. F., Field results on the control of certain grape diseases in New York, *Bull. 712, New York State Agric. Exp. Stn.*, Geneva, NY, 1945.

70. Azzam, O. I., Gonsalves, D., Namba, S., Gadoury, D. M. and Pearson, R. C., Detection of dsRNA species and virus-like particles in *Uncinula necator, Phytopathology*, 80 (Abstr.), 969, 1990.

6

BUNCH ROT
OF GRAPES CAUSED
BY BOTRYTIS CINEREA

N. G. NAIR

Biological & Chemical Research Institute
N.S.W. Department of Agriculture, Rydalmere, N.S.W., Australia 2116

G. K. HILL

Landes-Lehr und Versuchsanstalt fuer Weinbau
Zuckerberg 19, D 6504 Oppenheim, Germany

6-1 INTRODUCTION

Botrytis cinerea is perhaps the commonest and best known fungus and has been the centre of mycological research since the time of de Bary. Few, if any other fungi, have been studied so thoroughly by so many able investigators, or are the subject of so extensive a literature.

W. B. Brierley, 1921

Disease exists in natural host populations, usually in the endemic state; it becomes epidemic in uniform pioneer vegetations. Crops are just like the latter and thus favor epidemics. With effective control systems, crop diseases tend to become endemic, yet still causing losses that may be intolerable, and still necessitating conscious efforts to keep them at a low level.

J. C. Zadocks & R. D. Schein, 1979

Botrytis cinerea is a facultative parasite of aerial parts of several plants; and, among the diseases caused by this species, bunch rot of grapes (*Vitis vinifera L.*) is especially well known. It is a pathogen of significant economic importance on grapevine, causing both substantial losses in yield and lowering in quality. An idea of the importance of this disease is apparent from the number of scientific contributions that continue to be published about it; and yet there are gaps in our knowledge about the etiology and epidemiology of bunch rot disease.

Although other fungi such as *Alternaria alternata* and *Aspergillus* sp. have been associated with bunch rot,[1] it has been shown conclusively that *B. cinerea* is the causal organism of this disease.[2]

It is sometimes stated that bunch rot results from poor crop management. This is largely untrue when one considers the vineyard as an ecosystem where it is often difficult to manipulate both the crop and its environment. However, adequate ventilation of the canopy of the grapevine and good management of crop nutrition and sound plant protection will reduce the risk of *Botrytis* infection.

The dual nature of *B. cinerea* in causing the destructive bunch rot and, under certain conditions, the nondestructive noble rot, is not paralleled in plant pathology. Whereas bunch rot causes much economic loss, noble rot yields wines of a special quality that are highly economical.

6-2 SYMPTOMATOLOGY

6-2-1 Flowers

Infected flowers are symptomless. Microscopic observations of the floral parts will reveal necrosis of stamens and growth of the pathogen on the style and stigma (Fig. 6-1). During

Figure 6-1   Grape flower infected by *Botrytis cinerea.*

the period of flowering, the infected stamens dehisce, and the solitary ovary can often be seen covered with tufts of sporulating mycelia.

6-2-2 Leaves

B. cinerea can infect young and relatively older leaves and produce irregularly shaped necrotic spots (Fig. 6-2). These spots are not normally marginal; however, under certain conditions marginal necrosis occurs.

6-2-3 Berries

The most prominent symptom of the disease is found on the berries. Infected berries become dark colored and show the typical greyish, hairy mycelium all over their surface (Fig. 6-3). Often the fungus can be seen to grow along cracks or splits on the berries. Tufts of conidiophores and conidia protrude from stoma and peristomatal cracks on the skin of the berry. Under high disease pressure, all the berries in a bunch can be infected.

6-3 ECONOMIC IMPORTANCE

From an economic point of view, only infections on flowers[3] and berries are of importance in terms of lowering quantity and quality of yield. The importance of flower infection on the epidemiology of bunch rot is discussed later. At flowering, flower debris provides an excellent nutrient source for the conidia. Heavy damages of flower bunches

Figure 6-2   Grape leaf infected by *Botrytis cinerea* (see arrow).

Figure 6-3 Grape berries infected by *Botrytis cinerea.*

are reported during unusually wet and cool conditions and an elongated flowering period.[4] Under these conditions, up to 30% of the bunches can be destroyed. Dense canopies
are much more prone to this phenomenon than well-aerated trellis systems. Flower infection occurs naturally during every season in the Hunter Valley, Australia, where mild to
warm conditions prevail.

During flowering, latent infections can also occur. Such latent infections become
manifest later in the season and contribute to the bunch rot complex. During the development of berries until véraison (when the berries begin to soften), the berries are resistant
to infection.

After véraison, the berries become increasingly susceptible to infection. Then, the
degree of damage from an economic point of view depends on the severity of infection,
growth stage of the bunches, and the grapevine cultivars, and these factors are important
for the processing of the fruit.[5]

The so-called sour rot affects bunches with a sugar content under 13° Brix and
leads very often to a complete loss of attached grapes. Sour rot is common in areas with
frequent summer rainfalls.

At higher sugar content, attached berries can normally be processed, but this forces
growers to an earlier harvesting or to selective picking of molded grapes, both normally
lowering gross return from the crop.

For making good, deep-colored, red wines, an excess of rotted bunches is disastrous because *Botrytis* produces high quantities of oxidizing enzymes, which destroy the
anthocyanin.[6]

The famous noble rot represents a rare case of a rotted foodstuff becoming more
valuable than the healthy one. Under favorable dry conditions following a heavy *Botrytis*

attack, the mycelium colonizes the berry skin and kills the epidermal cells, thereby allowing abundant evaporation of water through the cuticle. The fungus utilizes some of the glucose and malic acid contents of the berry juice, but normally the drying process of the berry overcomes this loss of solid substances, and finally a raisin-like shrunken fruit is picked selectively. The famous white Auslese-type wines, the most reknown originating from the Rhine Valley or Sauterne, are made from such grapes. But if heavy rainfalls occur, the concentrated berry juice may be leached out and the grower ends up with both a reduced yield and a lowered quality of the must.[5,7,8]

Undoubtedly, the noble rot is an exception, and most of the grape growers in the world tend to control bunch rot as effectively as possible because it significantly reduces the yield. The losses are heavier if the stalks of berries are infected and the whole bunches eventually drop to the ground. In cool-climate grapevine growing areas such as Germany, a moderate *Botrytis* infection is welcome for raising the extract content of the juice; however, winemakers working in regions with high temperatures during harvest time face many problems with the rotted fruit. First, molded grapes are often colonized by fruit flies and acetic-acid bacteria. The second problem is from secondary infection of the bunches by other pathogenic fungi such as *Aspergillus* sp., *Penicillium* sp. or *Trichothecium roseum*. These fungi produce a bitter taste, an undesirable off-flavor, and mycotoxins.[9,10]

Juices from *Botrytis*-infected grapes lose much of their fruity components and tend to age soon after fermentation, altogether an undesirable feature for the production of fresh, reductive white vines.

6-4 ETIOLOGY AND EPIDEMIOLOGY

The establishment of *B. cinerea* on moribund or injured tissues normally allows the pathogen to infect the healthy tissues. Principally, *Botrytis* can attack most of the organs of grapevine even during winter. Damage to the one-year-old shoots is normally restricted to tissues exhibiting lack of maturity. Normally, the fungus forms sclerotia in the outer layers of the dead bark of shoots; whereas the inner, living tissues are protected against fungal attack by a well-formed, suberized periderm layer. The mycelium of *Botrytis* was isolated from buds mainly form the lower parts of the shoots.[11] This presumably is due to invasions of mycelium from infected grape stalks during late autumn. In the vineyard, these infections only affect seriously canes with bad maturity of the wood, leading to a lowered rate of bud burst on the basal buds of fruit canes in spring.

For bench-grafting nurseries, the use of *Botrytis*-infected scions creates serious problems. For that purpose, a careful disinfection of contaminated cuttings with oxychinoline derivatives is essential.[12]

During vegetative period, leaf infections occur occasionally during long rainy periods with continuous leaf wetness over 48 hours. Young leaves become infected from invasions starting from hydathodes on the leaf tips in spring. Sometimes floral parts from other plants are deposited on the surface of grapevine leaves by air currents. These provide a source of nutrient for germinating conidia, enabling them to penetrate and form necrotic spots even on old leaves. As soon as the leaf surface dries off, the mycelium stops its growth and no further damage occurs. Mature leaves exhibit pronounced resistance against *Botrytis* infection, resulting in restricted necrosis.[13]

6-4-1 Source of Inoculum

There are three sources of primary inoculum of importance in the epidemiology of bunch rot disease. These are mycelia, conidia, and sclerotia.

Mycelia. Nair et al.[14] have found that rapidly senescing or moribund tissues in flower bunches in cv. *Ondenc* consisted of approximately 67% calyptra and 22% stamens. At véraison, calyptra (approximately 4%) were more numerous. More calyptra were contaminated with *B. cinerea* than stamens at 80% cap-fall and véraison. These infected floral parts provide a large saprophytically based mycelial inoculum. In grape flower, calyptra and stamens dehisce at the start and end of bloom, respectively, and often these tissues adhere to the developing berries after being shed and become potent inocula for aggressive infections.

Conidia. The conidia of *B. cinerea* are dry and are largely dispersed in air currents. According to Corbaz,[15] the concentrations of conidia in the air increased as the grapevines matured. Nair and Nadtotchei[16] showed that conidia produced as a result of sporogenic germination of sclerotia infected grape berries.

Sclerotia. The sclerotia may be directly infective or sources of conidia which result in infection of grapevines. Nair and Nadtotchei[16] demonstrated that sclerotia are a source of primary inoculum and that conidia-bearing sclerotia on canes left on grapevines will initiate a further cycle of the disease. Kublitskaya and Ryabtseva[17] have shown that some sclerotia did not produce conidia but after two to five months formed apothecia. The ascospores produced from these apothecia can also initiate primary infection of grapevines. Nair and Nadtotchei[16] observed that repeated germination of sclerotia produce conidia over a relatively long period. They observed suppression of sporulation when the conidia were left on sclerotia of *B. cinerea* and the resumption of sporulation after the conidia were removed from their surface. This could extend the period of conidial production and infection. Under natural conditions in the vineyard, rainfall and splashing water would be expected to dislodge conidia from germinating sclerotia and initiate conidial production by removing the suppression on sporulation.

The optimum temperature for sclerotial germination followed by infection was between 20 and 25°C. Maximum primary infection of flowers during bloom from germinating sclerotia is likely when daily air temperature during late spring reaches 20°C. Sclerotia survived best between 20 and 25°C and in relatively dry soil. However, it is difficult to relate *in vitro* results to vineyard situations since mycoparasitism, decay, and complex interactions between soil microorganisms and the environment can decrease the survival of sclerotia. The ultrastructural features of sclerotia, like the deposition of melanin on the surface and the presence of inner cells that are equipped with electron-dense storage bodies, suggests that they are well adapted for relatively long periods of survival.[18]

Sclerotia are more likely to survive longer on canes, and these are probably more important than those in soil as a source of primary inoculum in each cycle of the disease.

6-4-2 Infection Pathway

The sequence of events beginning from primary infection and ending in bunch rot can be summarized as follows:

A **Source of primary infection** Mycelium Conidium Sclerotium → B **Primary infection** Flowers Leaves → C **Latency** → D **Secondary infection** Berries Leaves? → E Bunch rot

Müller-Thurgau observed as far back as 1888 that infection of grapes by *B. cinerea* was mainly brought about by direct penetration through lenticels. Mycelium penetrated through minute openings or cracks in the cuticle. A correlation between thickness of the epidermis and resistance to infection does not appear to exist.

After penetrating the cuticle, further passage of the hyphae through the cells of the internal tissue of the berry could be aided by pectin-hydrolyzing enzymes.

Based on natural infections of the floral parts of grapevines in the Napa Valley, California, McClellan and Hewitt[19] showed that the path of infection is through the stigma and style and then into the stylar end of the ovary. At some later stage, the fungus bridges an "abscission zone" layer of the style to the ovary. Still later, the fungus proceeds to rot the berries. Nair and Parker[3] have reported that *Botrytis* infects flowers of grapevines in the Hunter Valley, New South Wales, Australia, and the infection appears to progress from the stylar end of the ovary. These flower infections are invariably followed by a period of latency when the pathogen remains in a quiescent phase. *Botrytis* can, therefore, cause either the common berry rot following infections of mature grapes or an early *Botrytis* rot[19] and mid-season bunch rot[3] following aggressive infections of immature berries.

Stellwaag-Kittler[5] showed that the optimum temperature for infection was 20°C and the minimum period of saturation to be 16 hours at that temperature. A wet period of 12 hours at 20°C would also be sufficient for infection. It has also been shown by other workers that infection of grapes is possible during periods of not less than 13 hours at temperatures between 15 and 20°C at 100% relative humidity. Nelson[7] has reported that a wet period of 12 to 24 hours at 16°C and 72 to 84 hours at 3°C was required for infection of harvested grapes. A constant temperature of 12°C required a shorter wet period for infection than a mean temperature of 12°C fluctuating between 2 and 21°C. Nair et al.,[14] using isolates of *B. cinerea* from Hunter Valley, Australia, showed that a constant temperature between 20 and 25°C was optimum for infection of grape berries. The length of wet period necessary to produce significant infection (the infection period) at a constant temperature of 15°C was 3 hours longer than at 20 and 25°C. The effect of fluctuating temperature on infection showed that a minimum temperature difference of 8°C caused a difference of 6 to 12 hours in the length of the infection period in the cv. Cabernet Sauvignon. Infection of berries proceeded faster and to a greater extent (60% infection in 12 hours) at a constant temperature of 25°C than when the temperature fluctuated below this temperature (< 10% in 12 hours).

Previous investigators have shown that any environmental factor that reduces the period of wetness or high relative humidity reduces the chances of *Botrytis* infection. For instance, increasing the ventilation within the canopy of grapevine has led to a decrease in infection. This is discussed later in this chapter.

Using a diachronic approach, Nair (unpublished) has obtained significant correlations between rainfall, number of rain days, and bunch rot disease in the Hunter Valley, Australia. Regression analysis of climatic variables, phenology of grapevine, and incidence of *Botrytis* over a fixed lag period (0 to 54 days) showed a positive correlation between the pathogen and relative humidity at a lag time of one day. This appears to suggest that high relative humidity may result in an increased incidence of bunch rot on the following day. Relative humidity readings prior to a high incidence of *Botrytis* in grapevine were generally above 80%, indicating that relative humidity at or above this level may be conducive to *Botrytis* infection. A negative correlation of *Botrytis* infection occurred with relative humidity and rainfall at a lag time of nine days. This indicates that low relative humidity or rainfall nine days previously may result in a significantly low infection.

Based on field data (Nair, unpublished), correlations of climatic variables at a lag time of 25 to 50 days (cropping days) suggest that the degree of infection at flowering and mid- to late season are linked. The gradient of the lines on the scatter plots (Fig. 6-4) represent the rate of increasing infection of *Botrytis*. Although the gradients were different for each season, they all exhibited the same trend of increase in infection during mid- to latter part of the season.

6-4-3 Predisposition

Several factors, such as host physiology, climate, cultural factors, etc., render grapevines susceptible to infection.

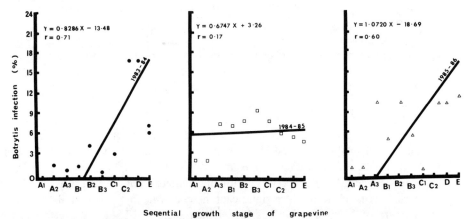

Figure 6-4 Relation between sequential growth stage of grapevine (cv. Traminer) and *Botrytis* infection during the grape seasons 1983–1984, 1984–1985, and 1985–1986. A1 to A3: Prebloom (A1—two or three leaves unfolded, A2—five to six leaves unfolded, A3—inflorescences enlarged); B1 to B3: Bloom (B1—start of flowering, B2—25% cap-fall, B3—80% cap-fall); C1 and C2: Fruit development (C1—start of fruit development, C2—bunch closure); D: Véraison; E: One week before harvest.

Moribund or senescing tissues.   Nair and Parker[3] observed that approximately 10% of flowers within an inflorescence did not set fruit and died rapidly. These flowers, if infected, could become potent inocula within developing bunches for berry rot. The importance of moribund floral tissues has been discussed earlier.

Predisposition wounding.   Grapes can be wounded by several agents, such as hail, frost, sunscald, insects, etc. (Jarvis[8]), and it has been shown that such damage predisposed them to infection by *B. cinerea*.[8,14] Wounds appear to be important avenues for entry of the pathogen. Exudations on the surface of berries as a result of natural causes or wounding could also present a nutrient source for germinating spores.[20]

Grapevine phenology.   Bunch rot has been associated with infection of mature grape berries; however, the recent demonstrations of flower infection have meant a rethinking of the classical concept of the *Botrytis* infection process. Although the rate of *Botrytis* infection is related to the sequential growth stages of grapevine, as pointed out earlier, more work is required before one could consider phenology as a predisposing factor.

6-5 HOST-PATHOGEN INTERACTION

6-5-1 Importance of Host-Pathogen Interaction

Infection of berries by *Botrytis* reveal economic damage in the field. The time of infection and the number of bunches infected vary widely depending on the stage of maturity of the host, the cultivar and, more importantly, the climate. In the presence of inoculum within the canopy of grapevine and humid conditions[5] allowing conidia to germinate and infect the host, only the defense mechanisms of the grapevine can protect the tissues from infection. Nearly all the cultivars of *V. vinifera* are more or less resistant before véraison and become susceptible as soon as the berries start the ripening process. Infections occurring before véraison are restricted to the floral parts and leaves. There are differences between the cultivars as to their resistance when fully mature.[13] Certain cultivars show a high degree of resistance over a long period and are therefore considered as *Botrytis* resistant under field conditions. Such cultivars are thought to possess both active and preformed defense mechanisms against fungal infections even at senescent stages of berry development.[21] Only a part of these defense mechanisms is known, and their role in the infection process is still not fully understood.

Knowledge of susceptible and resistant stages of the bunches makes it easier to understand the epidemiology of bunch rot and enables the choice of the right control measures.

6-5-2 Susceptibility of the Green Tissues of the Host at Different Stages of Their Development

Vegetative organs.   There exists a striking difference between vegetative and generative tissues of the grapevine in their susceptibility to *Botrytis* infection. Young leaves are known to be very susceptible, whereas fully matured ones are relatively more resistant.[13,22,23] Table 6-1 shows the relationship between age of the leaf and the resistance to infection.

Heavy infections of leaves are not very common and normally take place during spring when the duration of leaf wetness is long following rains, allowing mycelium to spread readily in the mesophyll. Under these conditions, few young shoots are infected from attached tendrils or small wounds;[4] the shoot tips are subsequently killed. Flower debris from adjacent crops blown onto leaf surfaces can enhance fungal infection by providing an excellent nutrient source for conidia of *Botrytis* to germinate.

In the field, the growth of necrotic *Botrytis* spots on the leaves stops as soon as the leaf surface dries. But these leaf-infection sites can produce conidia abundantly during wet periods later in the season. In the autumn, *Botrytis* mycelium sometimes invades the nodes of the shoots through the grape stalks[24] and occasionally colonizes the winter buds, finally killing these organs and reducing the rate of bud burst on the basal parts of the fruit canes in the next spring. Even more spectacular is a bleaching of the outer bark of attached canes accompanied by the formation of sclerotia.[4] This phenomenon is mainly restricted to canes with bad wood maturation and normally does not affect the living parenchyma of the shoot, which is protected from invasion by the large, corky periderm layer.

Generative organs.   Infections of the generative organs of the grapevine nearly always result in reduced yield and sometimes in reduced quality of fruit. Early infection can destroy flower bunches. As pointed out earlier, flower infection can also be symptomless, the infection manifesting itself at a later stage of development of the grapevine.[2,3,19] Figure 6-5 shows the effect of ontogeny on the susceptibility of the berries of cv. Riesling toward direct infection by the conidia of *B. cinerea*.

The susceptibility of berries to infection increases steadily after the start of véraison and, in particular, above a sugar concentration of 6 to 8%. In the case of relatively more resistant cultivars, only berries with a much higher sugar content are susceptible to infection, and under high disease pressure not more than 20 to 24% of all the berries are infected. Tetraploid varieties of *V. vinifera* are known to be much more susceptible than diploid ones; and some hybrids, but not all, reveal marked *Botrytis* resistance of the mature berries even at 18° Brix.[21]

Under field conditions, other factors may contribute to resistance, such as a loose grape architecture or a decreased tendency of berries to crack by excessive uptake of water during rain periods.[4,25,26] Looser bunches tend not to provide the moist chamber and high relative humidity favorable for germination of conidia or retaining flower debris. A

TABLE 6-1:   RESISTANCE OF LEAVES IN *V. VINIFERA* (CV. RIESLING) AT DIFFERENT STAGES

Leaf age	Insertion (basipetal)	Average size of necrosis (sq mm)
Young	2–4	85.7
Matured	6–10	1.5
Senescent	12–16	2.7

Note: Leaves were inoculated with mycelial plugs of *Botrytis cinerea* and incubated for 72 hours, from Hill.[13]

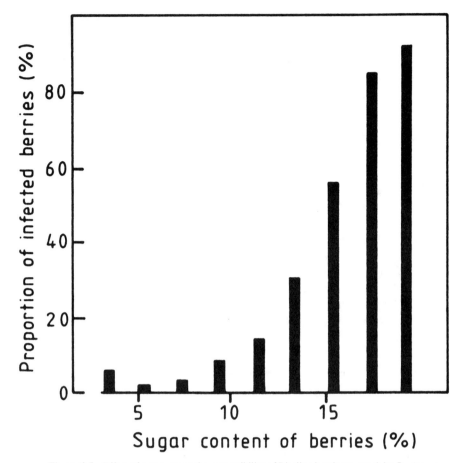

Figure 6-5   Effect of ontogeny on the susceptibility of Riesling berries to attack by *B. cinerea.*

classic example of dense clusters is seen in the cv. Pinot Noir. Sometimes, growing berries become squeezed by others of the same bunch and the weakened berries become avenues for *Botrytis* infection. Another problem may be the uneven maturation among the berries of one bunch. For instance, seedless berries exhibit an early maturity and therefore a high susceptibility. From the infected, susceptible, mature berries the fungus may invade others through the stalks.

An additional problem is the wounds on berries caused by caterpillar and light-brown apple-moth infestations. These contribute to disease severity.[5,14]

Healthy grape stalks undergo little risk from direct infection by conidia of *Botrytis* but occasionally can be invaded by mycelial material growing from flower debris or attached berries.[27]

In many cases, the problem of stalk rot is related to the incidence of grape stalk necrosis (stiellaehme), which is a physiological disease mainly based on mineral imbalance of the bunches.[28] Therefore, most of the varieties classified as susceptible toward *Botrytis* stalk rot (e.g., Riesling) also show a high incidence of stalk necrosis.

6-5-3 Defense Reactions of the Host

Botrytis resistance in grapevine.   Whatever resistance against *Botrytis* may be based on, the results can be measured by a severity rating.

We encounter many terms specifying the phenomenon of resistance. Leaf resistance, for instance, may be based on a completely different mechanism compared with the so-called bunch rot resistance. This fact is a nightmare for grapevine breeders when trying to select *Botrytis*-resistant varieties from young seedlings not yet bearing grapes. We see a situation of old-age resistance in the case of leaves and old-age susceptibility in berries. When looking closer at the infection process, we can find a resistance toward penetration of the infection hyphae, normally provided by the cuticle in the first instance and then a tissue-bound resistance to the spreading of mycelium inside the host from the primary infection site.

As mentioned earlier, many factors, morphological and physiological, contribute to the resistance against the bunch rot complex. One of the best studied features represents the stage resistance of bunches. We describe later the major resistance factors that contribute to a lowered susceptibility of grape berries. Knowing these factors may lead to a better understanding of *Botrytis* epidemics and, importantly, from the grape growers' point of view, some hints on maintaining or improving *Botrytis* resistance by cultural measures.

Host-pathogen interaction for the bunch rot complex.   *Botrytis* is well known as a versatile pathogen with complex genetical variability and good adaptation to environmental conditions. But the fungus is also considered to prefer wounded or senescent tissues. The grapevine had to cope with such a fungus during its evolution. We can therefore suppose that a similar defense system exists in the host. Figure 6-6 shows the possible mechanism for infection by the pathogen and defense by the host.

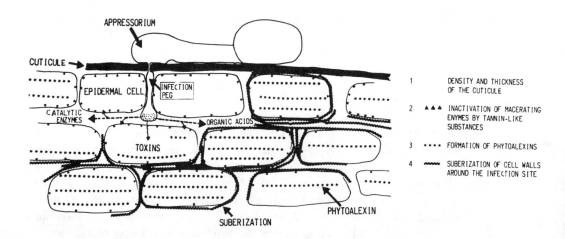

Figure 6-6   Defense mechanisms in young grape berries.

Factors and defense mechanisms in the host-pathogen interaction.

Exudate from the surface.   Fungal growth outside the cuticle is enhanced by different nutrient sources, such as pollen, floral parts, or exudates from berries. The increasing amount of the sugar exudate was considered to be partly the reason for the stage resistance of berries.[29] It is well known that strong mycelium sometimes is able to invade a few of the immature berries of a bunch.[3] Normally, as was demonstrated in inoculation experiments using conidial suspensions, the addition of nutrients to the germination sap only slightly increases the rate of infection.[13,21]

Preformed morphological barriers.   The structure and thickness of the cuticle and the epidermal layers have long been regarded as a major factor of resistance against *Botrytis* infection.[30,31]

Figure 6-7 demonstrates that removal of the cuticle followed by inoculation with conidia does not necessarily result in a much higher level of infection. Unripe, damaged berries lacking the protection of the cuticle maintain their resistance against conidial attack to a great extent, but aggressive mycelial infection overcomes the defense of the tissue. A major part of the stage-resistance mechanism can be attributed to physiological defense reactions in the tissue more than to preformed morphological barriers. The physiological defense weakens during maturation of the berries. For the mature, susceptible berry, the cuticle may play a much more important role.[32]

Mature berries without cracks, with a certain variation depending on variety, showed about a 20% less infection rate than berries with cracks. Under field conditions, the cuticle may therefore represent a resistance factor, and hardening of the cuticle may possibly reduce infection of *Botrytis*.

Preformed biochemical substances and enzyme inhibitors.   The young, resistant berries apparently do not contain preformed fungitoxic substances, but high amounts of enzyme inhibitors which act against the fungal polygalacturonase (PG)[33,34] and possibly inactivate toxins of *Botrytis*[35] as well. These PG inhibitors are tannin-like substances; in particular, proanthocyanidins in the skins of the berries. Their activity decreases steadily toward maturity by oxidation and condensation. Not all varieties exhibit the same pattern.[13] In addition, because *Botrytis* has a high potential for breaking down tannins, Stein[21] considered proanthocyanidins as a minor factor for resistance.

Preformed fungitoxic agents.   Pezet and Pont[36] isolated from whole, immature grapes a substance which acted as an inhibitor to the germination of the conidia of *Botrytis in vitro*. Using crude-pressed saps from berry skins of different maturities, Hill et al.[33] found no inhibition of conidial germination or mycelial growth. The same was observed on the surface of living, abrasive, wounded, and inoculated berries. Therefore, preformed fungitoxic substances are unlikely to be involved in the early stage of direct infection through the cuticle, although they could play a certain role in the complex of latency of flower infections.[3,19,37]

Phytoalexins are low molecular weight, antimicrobial compounds synthesized by plants against pathogens.[38] In grapevine leaves, different stilbenes (e.g., pterostilbene, resveratrol, and resveratrol polymers like the dimer ϵ-viniferine and the trimer $\propto$-viniferine) were found.[39–41] Following mechanical injuries, many substances act as inductors for resveratrol in the grapevine tissue, even monosaccharides like glucose and products of

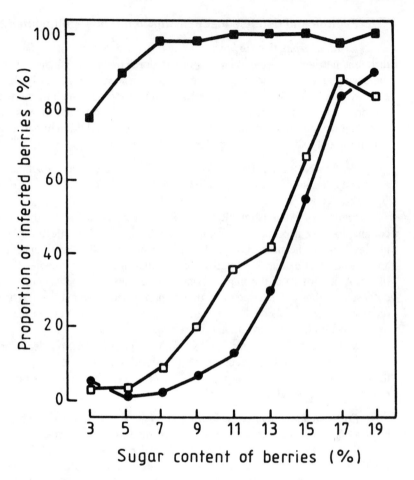

Figure 6-7 Effect of ontogeny on the role of the cuticle in the resistance of Riesling berries to *Botrytis cinerea*. (•) *undamaged berries were sprayed with B. cinerea* conidia; (□) berries were damaged by abrading the surface with glass paper and sprayed with *B. cinerea* conidia: (■) berries were damaged by abrading the surface with glass paper, inoculated with a dense mat of *B. cinerea* mycelium. All berries were held for five days at high humidity.

fungal metabolism.[42] Under UV light, stilbene emits bright blue fluorescence as it accumulates in the boundary zones around injuries of green tissues (Fig. 6-7). Unripe berries have a significant potential for the production of stilbene, whereas the response by mature grapes is less pronounced. A direct fungitoxic activity has been observed,[41,43] but their impact on the fungal growth inside the plant remains doubtful, because the water solubility of these stilbenes is low and mycelium of *Botrytis* can metabolize such compounds rather quickly *in vitro*.[44]

Hoos and Blaich[44] suggested recently that stilbenes or their oxidation products may exercise a composite action in the defense system of the grapevine, exhibiting fungistatic activity as well as being precursors of other phenolic compounds, such as lignin, formed in the cell walls around infection sites in leaves (Fig. 6-8).

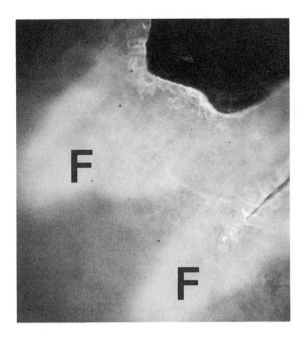

Figure 6-8 Stilbene accumulation around wounds and cracks in the epidermal tissue of immature berries (cv. Riesling) 14 hours after partial removal of the cuticles, magnification 30×. F=bluish flourescence of stilbene under incident UV incitation.

Formation of physical barriers: suberization of infection sites.   Immature grape-vine berries keep their resistance against conidial infection of *Botrytis* even when the cuticle is partly removed. The growth of the mycelium on the surface of these wounds is apparently not affected, but no rotting of the tissue underneath occurs.[33] Soon after wounding, the uninfected cells adjacent to the injured parts start suberization of cell walls. This process is followed by formation of corky barriers and leads to a complete isolation of the invaded mycelium, which is normally not able to break through this barrier[45](Fig. 6-9).

The suberization is not restricted to the presence of fungal mycelium, indicating that this process is part of the normal healing reaction of the tissue (Hill, unpublished). But the suberization process is consistently accelerated by fungal metabolic products.[27] During maturation, the potential of the berry tissue for rapid formation of mechanical barriers toward the invading pathogen could also contribute to the resistance of immature berries.

Suberization is suggested to be a successful way of stopping a pathogen. It protects tissues from both fungal enzymes and toxins, and partly from mechanical penetration. Grape stalks maintain their potential for suberization of wounds and infection sites even at full maturity of the bunches.[27]

Hardening of the grape bunches.   Lowered incidence of *Botrytis* bunch rot has been achieved by different treatments apparently exhibiting no direct fungistatic or fungitoxic activity.[5] The reason for this is not fully understood but may be due to an interaction between internal tissue-bound resistance mechanisms with external factors.

For instance, removal of leaves in the plane of bunches reduces bunch rot inci-dence.[46,47] This effect may partly be due to an improvement of the microclimate resulting

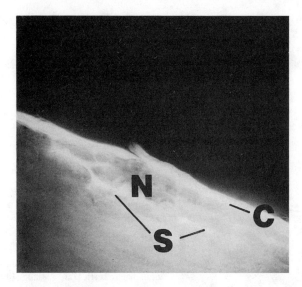

Figure 6-9 Section of the skin of immature berry (cv. Riesling) 30 hours after inoculation with conidia of *Botrytis cinerea*. C=cuticle; N=necrosis; B=bright cell walls indicating deposition of suberine.

in a better ventilation. This may facilitate quicker drying after rain and also a better spray penetration. Stellwaag-Kittler[5] found a consistently higher infection by keeping vines at high (>20% relative humidities in the field (Table 6-2).

Apart from the effects of microclimate, good exposure of the fruit to the sun generally reduces the susceptibility of berries. Stellwaag-Kittler[5] claimed that UV irradiation influenced hardening of the skin of the berries by stimulating the formation of a thick cuticle. Another explanation for the impact of UV light on disease resistance is that the phytoalexin production in grapevine leaves is stimulated by UV light.[40]

Generally, the aim to harden the bunches against *Botrytis* infection is based on good aeration of the grapes, which is provided by a training system suitable for the vineyard site. To improve these effects with "hardening" chemicals (copper compounds), it must be taken into account that slight phytotoxic effects which reduce bunch rot may also reduce the yield. Where regulations restrict the yield of grapes per hectare, these effects may even be welcome.

6-6 FUNGICIDE RESISTANCE

6-6-1 Historical Aspects and Present Situation in Chemical Control of *B. cinerea*

During the late 1950s, botryticidal phytochemicals were introduced in viticulture. Until 1968, sulphamide (dichlofluanid), the pthalimides (captan, captafol, folpet), and dithiocarbamate (thiram) were commonly used for *Botrytis* control in many countries.[48] These fungicides are multisite inhibitors. Normally, the efficacy of fungicidal treatments for bunch rot control ranged between 20 and 50% at this point in time.[5]

The release of systemic methyl benzimidazoles (MBC), such as benomyl, thiophanate-methyl and carbendazim, have consistently improved *Botrytis* control after 1968.

TABLE 6-2: THE INFLUENCE OF RELATIVE HUMIDITY ON INFECTION BY *BOTRYTIS CINEREA* ON CV. RIESLING IN A VINEYARD AT GEISENHEIM, GERMANY.

Average relative humidity	% *Botrytis* infection of the bunches
60%	6.5
80%	16.3
98%	41.6

Note: Vines were kept under plastic foils at a maturity stage of 14° Brix for six days. Disease severity was rated at harvest time.

From Stellwaag-Kittler[5].

Four years later, these fungicides surprisingly enhanced rather than suppressed attacks by *B. cinerea* in Germany. MBC acts as a mitosis inhibitor, blocking the cell division of the fungus.[49] The outbreak of tolerant strains occurred simultaneously in many temperate climatic areas.[50] An explanation for the quick outcome of resistance was the local occurrence of resistant strains among the population of *Botrytis* in the field before MBC was introduced onto the market.[50,51]

Sprays acted as a selection factor eliminating MBC-sensitive strains.[52] Sometimes, the side effects of depressing natural antagonists of *B. cinerea* increased the rate of infection in the treated blocks.[53] Tolerant strains mostly were found to be as aggressive as the sensitive strains, indicating that the mutations had not necessarily reduced the fitness of the fungus.[54] But under a mediterranean climate (e.g., in parts of Italy), satisfactory control was reported until 1977.[55] In Australia, MBC has continued to show satisfactory control of *Botrytis*.

Since 1976, the new group of dicarboximide fungicides has been available, and they exhibited good efficacy. First investigators failed to detect natural tolerant strains to dicarboximides in the vineyards and suggested a low risk in regard to fungicide tolerance in the field.[54] Dicarboximides are single-site inhibitors and disturb the synthesis of the cell wall of *Botrytis* hyphae, which burst eventually.[56] The adaptation of different strains of *B. cinerea in vitro* to dicarboximides was achieved in 1976.[57]

A few years later, resistant strains often represented more than 50% of the *Botrytis* population in German vineyards.[58,59] Since 1980, resistance has also been reported from other countries, decreasing the efficacy of the dicarboximides consistently in the northern vine-growing areas of France (e.g., in Champagne).[60] Despite abundant presence of resistant strains, the efficacy of dicarboximides did not break down as spectacularly as MBC. To delay the selection of resistant strains during the vegetative period, the use of dicarboximides was soon restricted to only two treatments after véraison in Europe.[61,62] But the residual efficacy of dicarboximides ranged under 40% and seemed to decrease with infection pressure; the latter fact made the dicarboximides uneconomical for practical use in many vineyards.[63]

Unfortunately, most of the dicarboximide-resistant strains also exhibited high resistance to MBC and seem to possess a considerable fitness, as indicated by their continuous presence even after many years without application of these fungicides.[62,64] Apparently, resistant stains exhibit a lower tolerance to high osmotic pressure; this may explain the increasing number of sensitive strains after véraison, as stated by Hartill.[65]

Control in the future may be based on mixtures of known fungicides or of promising new ones. A combination of the dicarboximide vinclozolin with the old dithiocarbamate thiram was introduced in the French market in 1986. The development of fungicides exploiting the phenomenon of negative cross-resistance is of particular interest. Some chemicals from the group of the N-phenylcarbamates specific for benzimidazole-resistant strains are now available.[66,67] Because these products act specifically on benzimidazole-resistant strains, they need to be applied in conjunction with a benzimidazole. In France, a combination of a fungicide of the new group and carbendazim gave good *Botrytis* control and has been commercially available since 1988.[68] Two to three treatments after flowering season gave about 70% efficacy. In some areas of France where the resistance is relatively low, dicarboximides are used in spray schedules alternating with diethofencarb.[69] Caution is needed because some benzimidazole-resistant strains were also found to be tolerant to diethofencarb *in vitro* in New Zealand (Beever, R. E., personal communication). In laboratory, Neskorosheniy et al.[70] obtained a strain of *B. cineria* which was resistant to all the three groups of fungicides, namely benzimidazoles, dicarboximides, and phenylcarbamates. This triple-resistant strain was as virulent as the parent strain.[70]

Another promising new botryticide in grapes is the triazole Folicur, an ergosterol biosynthesis inhibitor (EBI). Combinations with the "oldie" dichlofluanide revealed about 60 to 70% efficacy against bunch rot by three postflowering applications.[71]

6-6-2 Future Prospects of Chemical Control

It should be emphasized that until now, field resistance occurred only in single-site inhibitors. The new compound diethofencarb blocking the bypass detected by MBC-resistant strains[66,67] may undergo the same high risk as other single-site inhibitors did and should be carefully used in alternation with other botryticides.

Unfortunately, highly efficient products impose a maximum selection pressure on the residual populations.[72] A pessimist may argue that the nature of these problems is such that the future of chemical control of *Botrytis* is bleak. Nevertheless, chemical control of bunch rot is indispensable in most of the vineyards under humid climates.

A rational use of spray programs which involves alternating the new fungicide groups to delay outcome of resistance will prolong the efficacy of chemicals. In the long term, however, chemicals cannot solve problems of inefficient trellis systems or excessive vigor of the vines, both of which promote heavy *Botrytis* infections. Creating a dry microclimate on the plane of the grapes by cultural practices is therefore the most important step to achieve successful *Botrytis* control.

6-7 DISEASE MANAGEMENT

The performance of different types of chemicals on the control of bunch rot caused by *B. cinerea* was outlined earlier. Improvements on the control of the disease have occurred as a result of increased knowledge of the quantitative and qualitative epidemiology of bunch rot. Different fungicides have been used to control the disease[73] and attempts have been made to alternate or use mixtures of fungicides.[74] Work has also been carried out on timing applications of fungicides[73] according to the behavior of the pathogen in the grapevine.[75]

Chemical control involving several sprays is not only uneconomical but also bears little relation to the actual infection process. The work of Nair et al.[75] demonstrated the effectiveness of early spray applications to control the disease. Their work points to the need for fungicide application earlier than véraison, and this is consistent with the fact that *Botrytis* establishes latent infection in grape flowers.

It appears, therefore, that satisfactory control of bunch rot can be achieved with minimum use of fungicide by programming the application when the disease risk is high. The limited use of fungicides will not only conserve the efficacy of the chemical but also reduce selection pressure for resistance on the pathogen and allow reestablishment of sensitive strains. A forecasting model has been developed for bunch rot management by Strizyk.[76] It is based on the concept of disease potential and the relationship between climatic factors and phenology of the grapevine.

6-8 CONCLUSIONS

Current knowledge of the biological model of *B. cinerea* and the quantitative and qualitative epidemiology of bunch rot disease of grapes is adequate for formulating guidelines for fungicide application and for predicting disease intensity and crop loss by computer model. However, this knowledge must be consolidated to provide the necessary database. It is important in the future to use resistant grapevine cultivars. More information is also needed on latency in the life cycle of *B. cinerea*.

6-9 REFERENCES

1. Barbetti, M. J., Bunch rot of Rhine Riesling grapes in lower south-west of Western Australia, *Aust. J. Exp. Agric. Anim. Husb.*, 20, 247, 1980.

2. Nair, N. G., Fungi associated with bunch rot of grapes in the Hunter Valley, *Aust. J. Agric. Res.*, 36, 435, 1985.

3. Nair, N. G. and Parker, F. E., Midseason bunch rot of grapes: An unusual disease phenomenon in the Hunter Valley, Australia, *Plant Path*, 34, 302, 1985.

4. Gaertel, W., Uber die Eigenschaften der *Botrytis cinerea* Pers. als Rebenparasit unter besonder Bercksichtigung von Infektion und Inkubation, *Weinberg u. Keller*, 17, 15, 1970.

5. Stellwaag-Kittler, F., Moeglichkeiten der Botrytisbekaempfung an Trauben unter Bercksichtigung der epidemiologischen Grundlagen, *Weinberg u. Keller*, 16, 109, 1969.

6. Ribereau-Gayon, J., Ribereau-Gayon, P. and Sequin, G., *Botrytis cinerea* in enology. In *The biology of Botrytis* (J. R. Coley-Smith, K. Verhoeff and W. R. Jarvis, eds.), p. 251, Academic Press, London, 1980.

7. Nelson, K. E., Factors influencing the infection of table grapes by *Botrytis cinerea* Pers., *Phytopathology* 41, 319, 1951.

8. Jarvis, W. R., Epidemiology. In *The biology of Botrytis*, (J. R. Coley-Smith, K. Verhoeff and W. R. Jarvis, eds.), p.219, Academic Press, London, 1980.

9. Hewitt, W. B., Rots and bunch rots of grapes, *Calif. Agric. Exp. Stn. Bull.*, No. 868, 3, 1974.

10. Flesch, P. and Stockinger, G., Isolierung und Identifizierung von Roseotoxinen aus Kulturen des Pilzes *Trichothecium roseum, WeinWissenschaft* 42, 111, 1987.

11. Bulit, J., Bugaret, D. and Verdu, D., Sur la possibilite de la conservation hivernale du *Botrytis cinerea* et du *Phomopsis viticola* dans les bourgeons de la Vigne, *Rev. de Zool. Agric. et de Path. Veg.* 72, 1, 1973.

12. Becker, H., Botrytis und Rebveredlung, *Weinberg u. Keller,* 13, 533, 1966.

13. Hill, G. K., Early stage of pathogenesis of *Botrytis cinerea* on different development stages of vegetative and generative organs of *Vitis vinifera* L., Ph.D. thesis, Univ. Giessen (Germany), p. 81, 1977.

14. Nair, N. G., Emmett, R. W. and Parker, F. E., Some factors predisposing grape berries to infection by *Botrytis cinerea, N.Z. J. Expi. Agric.*, 16(3), 1988.

15. Corbaz, R., Etudes des spores fungiques captees dans l'air, *Phytopath. Z.*, 74, 318, 1972.

16. Nair, N. G. and Nadtotchei, A., Sclerotia of *Botrytis cinerea* as a source of primary inoculum for bunch rot of grapes in New South Wales, Australia, *J. Phytopath.*, 119, 42, 1987.

17. Kublistskaya, M. A. and Rybatseva, N. A., Biologiya Zimuyschchei stadii *Botrytis cinerea* Fv., *Mikol. i. Fitopatol.*, 4, 291, 1970.

18. Nair, N. G. and Martin, A. B., Ultrastructure and development of sclerotia of *Botrytis cinerea* Pers. *in vitro, J. Phytopath.*, 119, 52, 1987.

19. McClelland, W. D. and Hewitt, W. B., Early *Botrytis* rot of grapes, time of infection and latency of *Botrytis cinerea* Pers. in *Vitis vinifera, Phytopathology*, 63, 1151, 1973.

20. Chou, M. C. and Preece, T. F., The effect of pollen grains on infections caused by *Botrytis cinerea* F., *Ann. Appl. Biol.*, 62, 11, 1968.

21. Stein, U., Untersuchungen ueber biochemische und morphologische Resistenzmerkmale bei Vitaceen, Ph.D. thesis, Karlsruhe, Germany, University of Karlsruhe, 1984.

22. Wilhelm, A., Stand der Botrytisbekaempfung im Weinbau, *Der Deutsche Weinbau*, Wiss., Beihefte 6, 1, 1952.

23. Stellwaag-Kittler, F., Probleme des Pflanzenschutzes bei der Rebsorte Muller-Thurgau, *Wein Wissenschaft*, 31, 48, 1976.

24. Agulhon, R., Built, J., Lafon, J., Roussel, C., Pourcharesse, P. and Burgaud, L., La pourriture grise des raisins, *Vigne et Vins*, 196, 17, 1971.

25. Yamamura, H. and Naito, R., Susceptibitlity to berry splitting in several grape cultivars, *J. Japanese Soc. Hort. Sci.*, 53, 390, 1985.

26. Lang, A. and Thorpe, M., Why do berries split. In *Proc. Second Int. Cool Climate Vitic. Oenol. Smp.*, Auckland, N.Z., p. 69, 1988.

27. Hill, G. K., Suberization of cell walls: A defense reaction of grape stem tissue against invading mycelium of *Botrytis cinerea* Pers., *Quad. Vitic. Enol.*, Univ. Torino, 9, 229, 1985.

28. Theiler, R. and Mueller, H., Beziehungen zwischen Klimafaktoren und dem Stielaehmebefall bei Riesling × Silvaner, *Vitis*, 25, 8, 1986.

29. Kosuge, T. and Hewitt, W. B., Exudates of grape berries and their effect on germination of conidia of *Botrytis cinerea, Phytopathology,* 54, 167, 1964.

30. Stalder, L., Untersuchungen ueber die Graufaeule *Botrytis cinerea* (Pers.) an Trauben, I. Mitteilung, *Phytopath. Z.*, 20, 315, 1953.

31. Bernard, A. C., Resistance mechanique des baies de *Vitis vinifera* au *Botrytis cinerea* Pers., *La France Viticole*, 10, 301, 1976.

32. Blaich, R., Stein, U. and Wind, R., Perforationen in der Cuticula von Weinbeeren als morphologischer faktor der Botrytisresistenz, *Vitis*, 23, 242, 1984.

33. Hill, G. K., Stellwaag-Kittler, F., Huth, G. and Schlosser, E., Resistance of grapes in different developmental stages to *Botrytis cinerea*, *Phytopath. Z.*, 102, 328, 1981.

34. Blaich, R., Bachmann, O. and Stein, U., Causes biochimiques de la resistance de la vigne a *Botrytis cinerea*, *Revue des Oenologues*, 8, 11, 1982.

35. Stein, U., Untersuchung uber phytotoxische Stoffwechselprodukte von *Botrytis cinerea* im Zusammenhang mit der Stadien resistenz von Rebblaettern, Diplomarbeit, Univ. Mainz, Germany, 1979.

36. Pezet, R. and Pont, V., *Botrytis cinerea*—Activite antifongique dans les Jeunes Grappes de *Vitis vinifera* Variete Gamay, *Phytopath. Z.*, 111, 73, 1984.

37. Pezet, R. and Pont, V., Infection florale et latence de *Botrytis cinerea* dans les grappes de *Vitis vinifera* (Var. Gamay), *Revue Suisse Vitic. Arboricult. Hortic.*, 18, 317, 1986.

38. Paxton, J. D., Phytoalexins in plant-parasite interactions. In *Experimental and conceptual plant pathology*. Vol. 3 (R. S. Singh, U. S. Singh, W. M. Hess and D. J. Weber, eds.), p. 537, Gordon and Breach Sci. Publ., New York, London, 1988.

39. Langcake, P., A new class of phytoalexins from grapevines, *Experientia*, 33, 151, 1977.

40. Langcake, P., The production of resveratrol and viniferin by grapevines in response to ultraviolet irradiations, *Phytochemistry,* 16, 1193, 1977.

41. Langcake, P., Disease resistance of *Vitis* species and the production of the stress metabolites resveratrol, ϵ-viniferine and pterostilbene, *Physiol. Plant Path.*, 18, 213, 1981.

42. Stein, U. and Hoos, D., Induktions und Nachweismethoden fur Stilbene bei Vitaceen, *Vitis*, 23, 179, 1984.

43. Stein, U. and Blaich, R., Untersuchungen uber Stilben produktion and Botrytisanfaelligkeit bei Vitis Arten, *Vitis,* 24, 75, 1985.

44. Hoos, G. and Blaich, R., Metabolism of stilben phytoalexins in grapevines: Oxidation of resveratrol in single cell cultures, *Vitis,* 27, 1, 1988.

45. Hill, G. K., *Botrytis* control—current knowledge and future prospects. In *Proc. sixth. Aust. Wine Ind. Tech Conf.*, p. 175, 1985.

46. English, J. T., Thomas, C. S., Marois, J. J. and Gubler, W. D. Microclimates of grapevine canopies associated wtih leaf removal and control of *Botrytis* bunch rot, *Phytopathology,* 79, 395, 1989.

47. Smith, S., Codrington, I. C., Robertson, M. and Smart, R. E., Viticultural and oenological implications of leaf removal for New Zealand vineyards, *Proc. Second Int. Cool Climate Vitic. and Oenology Symp.*, Auckland, New Zealand, 1988.

48. Weiss, E., Botrytisbekampfung im Weinbau, *Der Deutsche Weinb.*, 23, 928, 1968.

49. Tripathi, R. K. and Schlosser, E., The mechanism of resistance of *Botrytis cinerea* to methylbenzimidazol-2-yl-carbamate (MBC), *Z. Pflanzenkrankh. Pflanzenschutz.*, 89, 151, 1982.

50. Ehrenhardt, H., Eichorn, K. W. and Thate, R., Zur Frage der Resistenzbildung von *Botrytis cinerea* gegenueber systemischen Fungiziden, Nachrichtenbl. *Dt. Pflanzenschutzdienst* (Braunschweig), 25, 49, 1973.

51. Bolton, A. T., Fungicide resistance in *Botrytis cinerea*, the result of selective pressure on resistant strains already present in nature, *Can. J. Plant Sci*, 56, 861, 1976.

52. Schuepp, H. and Lauber, H. P., Toleranzverhalten der *Botrytis*—population gegenueber MBC-Fungiziden (Benlate u. Enovit-M mit Wirkstoffen Benomyl u. Methylthiophanate) in den Rebbergen der Nord -u. Ostschweiz, *Schweiz. Z. Obst. Weinbau*, 114, 132, 1978.

53. Eichhorn, K. W., Erfahrungen in der Bekampfung von *Botritis cinerea* in deutschen Weinbau 1974, *Der Deutsche Weinbau*, 30, 508, 1975.

54. Lorenz, D. H. and Eichhorn, K. W., Untersuchungen zur moeglichen Resistenzbildung von *Botrytis cinerea* an Reben gegen die Wirkstoffe Vinclozolin und Iprodione, *Wein Wissenschaft*, 33, 251,1978.

55. Bisiach, M., Minervini, G., Ferrante, G. and Zerbetto, F., Ricerche sperimentali sull 'attivita antibotyritica in viticoltora di alcuni derivati della 3,5- dichloranilina, *Vignevini (Bologna)*, 5, 23, 1978.

56. Albert, G., Wirkungsmechanismen and Wirkamkeit von Vinclozolin bei *Botrytis cinerea* Pers., Ph.D. thesis, Rheinische Friedrich Wilhelms Universitat, Bonn, Germany, 1979.

57. Lerroux, P., Fritz, R. and Gredt, M., Etudes en labortatoire des souches de *Botrytis cinerea* Pers. resistantes a la dichlozoline, au dicloran, au quintozene, a la vinclozoline et au 26019 RP ou glycophene, *Phytopath. Z.*, 89, 347, 1977.

58. Hill, G. K. and Schlamp, H. A., Resistenzsituation bei *Botrytis cinerea* im Anbaugebiet Rheinhessen, *Wein Wissenschaft*, 35, 397, 1980.

59. Lorenz, D. H. and Eichhorn, K. W., Vorkommen und Verbreitung der Resistenz von *Botrytis cinerea* an Reben gegen Dicarboximid-Fungizide im Anbaugebiet der Rhinepfalz, *Wein Wissenschaft*, 35, 199, 1980.

60. Leroux, P. and Clerjeaux, M., Resistance of *Botrytis cinerea* Pers. aud *Plasmopara viticola* (Berk. et Curt) Berl. and de Toni to fungicides in French vineyards, *Crop Prot.*, 4, 137, 1985.

61. Besselat, R., Pourriture grise: Evolution des methodes de lutte, *Phytoma*, 360, 35, 1984.

62. Locher, F. J., Lorenz, G. and Beetz, K. J., Resistance management strategies for dicarboximide fungicides in grapes: Results of six years' trial work, *Crop Protection*, 6, 139, 1987.

63. Schlamp, H. A., Spritzfolgen gegen *Botrytis* im Praxis-test, *Der Deutsche Weinbau*, 43, 486, 1988.

64. Beever, R. E., Laracy, E. P. and Pak, H. A., Strains of *Botrytis coneria* resistant to dicarboximide and benzimidazole fungicides in New Zealand vineyards, *Plant Path.*, 38, 427, 1989.

65. Hartill, W. F. T., Resistance to plant pathogens in New Zealand, *N.Z.J. Agric. Res.*, 14, 239, 1986.

66. Kato, T., Suzuki, K., Takhashi, J. and Kamoshita, K., Negatively correlated cross resistance between benzimidazole fungicides and methyl N- (3,5- dichlorophenyl) carbamate, *J. Pest. Sci.*, 9, 489, 1984.

67. Gullino, M. L., Aloi, C. and Garibaldi, A., Influence of spray schedules on fungicide resistant populations of *Botrytis cineria* Pers. on grapevine, *Neth. J. Plant Path.*, 95 (Suppl. 1), 87, 1989.

68. Chamroux, I., *Botrytis,* le coup de grace?, *Viti*, 117, 46, 1988.

69. Strizyk, S., Pari tenu contre le *Botrytis* en 1987, *Viti*, 117, 39, 1988.

70. Neskorosheniy, B., Felgentreu, D. and Lyr, H., Isolation and properties of a triple resistant strain of *Botrytis cineria* to benzimidazoles, dicarboximides and *N*-phenylcarbamates. In *Proc. 9th Reinhardsbrunn Int. Sym. Systemic Fungicides and Antifungal Compounds* (H. Lyr, ed.), Germany (in press).

71. Kaspers, H., Brandes, W. and Scheinpflug, H., Verbesserte Moeglichkeiten zur Bekampfung von Pflanzendrankeiten durch ein neues Azolfungizid, HWG 1608 (Folicur, Raxil), *Pflanzenschutz-Nachrichten BAYER*, 40, 81, 1987

72. Beever, R. E. and Pak, H. E., The future of fungicides in the control of Botrytis, *Proc. Second Int. Cool Climate Vitic. Oenology Symp.*, Auckland, New Zealand, p. 85, 1988.

73. Pearson, R. C., Chemical control of *Botrytis cinerea* on grapes in New York (USA), *EPPO Bull.*, 12, 101, 1982.

74. Gullino, M. L. and Garibaldi, A., Use of mixture or alternation of fungicides with the aim of reducing the risk of appearance of strains of *Botrytis cinerea* resistant to dicarboximides, *EPPO Bull.*, 12, 151, 1982.

75. Nair, N. G., Emmett, R. W. and Parker, F. E., Programming applications of dicarboximides to control bunch rot of grapes caused by *Botrytis cinerea, Plant Path.*, 36, 175, 1987.

76. Strizyk, S., Modele d'etat potentiel d'infection—Application au *Botrytis cinerea*. In *Association de Coordination Technique Agricole*, Paris, France, 1983.

7

CITRUS CANKER

MASAO GOTO

Faculty of Agriculture, Shizuoka University
836 Ohya, Shizuoka, 422 Japan

7-1 INTRODUCTION

Citrus canker, confused with citrus scab (*Elsinoe fawcetti*) in the late nineteenth century, was recognized as a new disease in 1913 in Florida. Since that time, the etiology and ecology of the pathogen have been intensively studied in Florida, and the principal nature of the disease was elucidated by the late 1920s. A campaign to eradicate citrus canker was begun in 1915 in the southern states of America, including Florida, Alabama, Georgia, Louisiana, Mississippi, South Carolina, and Texas.[1] Citrus canker was declared to be eradicated from these areas by 1947. This achievement has been regarded as a rare instance of successful eradication of a plant disease that had once been established in certain ecosystems.

The disease continues to be one of the most important problems of citrus in Asia and South America and has stimulated much research activity on the etiological and epidemiological traits of the pathogen, as well as on potential methods of control. Several important findings have been reported since the 1960s regarding the occurrence of different types of citrus canker, in terms of symptomatology as well as host specificity. The recent widespread general interest in citrus canker was stimulated by reports of its occurrence in the Thursday islands of Australia and in Mexico and Florida in the 1980s.

7-2 ORIGIN OF CITRUS CANKER

The geographical origin of citrus canker is a matter of debate. Lee[2] reported that it may have arisen in southern China, and he assumed *Fortunella hindsii* to be the wild host plant. Natural infection was found on this plant growing on mountain tops in Kwang-tung Province, at an altitude of about 500 meters and totally isolated from commercial citrus plantings. However, Fawcett and Jenkins[3] reported that citrus canker originated in India and Java, rather than in other regions of the Orient, because they detected canker lesions on the oldest citrus herbaria kept at the Herbaria of the Royal Botanic Gardens in Kew, England (i. e., *Citrus medica* collected in India in 1827–1831 and *C. aurantifolia* in Indonesia in 1842–1844). These findings suggest the origin of disease in the tropical areas of Asia, such as South China, Indonesia, and India, where *Citrus* species are presumed to have originated and to have been distributed to other citrus-growing areas in the form of budwood.

7-3 DISTRIBUTION AND ECONOMIC IMPORTANCE

Citrus canker is found in Africa, Asia, Australasia, Oceania, and South America (Fig.7-1). Although the disease was once reported to be eradicated from Australia, New Zealand, South Africa, and the United States, during the 1980s it was reported in the Thursday Islands of Australia, as well as in Mexico and Florida. The disease found in Mexico in 1981 and in Florida in 1984 appeared to be different from that identified in Asia (A form canker). The form of citrus canker found in the Thursday Islands in 1984, however, was exactly the same as A form canker. It is noteworthy that before these outbreaks, cold weather swept these citrus-growing areas, causing severe frost damage. It was also surprising that these forms of citrus canker occurred at almost the

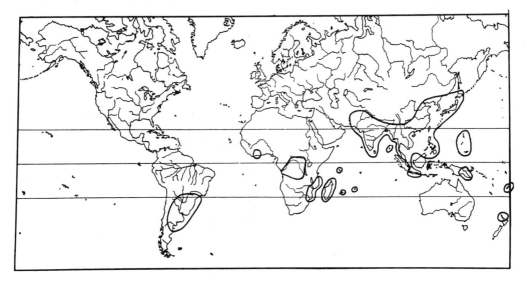

Figure 7-1 Distribution map of citrus canker (*Xanthomonas campestris* pv. *citri*) (Hasse) Dye.

same time in different regions of the world, despite the enforcement of strict international quarantine procedures. It can be suggested that the pathogen is able to maintain long periods of quiescence under certain conditions without causing the development of any visible canker lesions, or that the pathogenic bacterium may mutate under the unusual physiological stress of citrus trees subjected to unusually cold weather. There is also the possibility that the disease was introduced merely by the illegal transfer of citrus budwood, stimulated by the global trend to germplasm collection.

The economic importance of citrus canker can be analyzed from several different points of view. Loss assessment has not been determined clearly, as in the case of diseases of annual crops. When citrus infection occurs in the early growing stage, the fruits crack or become malformed as they grow, and the heavily infected ones fall prematurely. Light infection in later growth stages may cause only scattered canker lesions on the surface of fruits but makes fresh fruits unacceptable for market. The severity of fruit infection usually parallels that of foliage infection. Eighty to ninety percent of fruit infection is not uncommon in susceptible citrus trees that have already sustained severe foliage infection. Such heavy foliage infection often causes severe defoliation, leaving only bare twigs. Severe infection in susceptible young citrus trees just after transplanting is particularly critical because it may be fatal or may induce a delay of several years in achievement of full growth.

Another aspect of the economic importance of citrus canker lies in the fact that citrus fruits produced in canker-endemic regions are placed under strict international regulation of plant quarantine, which restricts the region's market.

7-4 SYMPTOMS

Citrus canker occurs on leaves, twigs, and fruits, forming necrotic brown spots with a coarse surface. The pathogen usually invades young tissues through stomata and wounds, or mature tissues only through wounds. On young leaves, lesions at the very early stage appear as small white specks that are rather difficult to recognize with the naked eye. The lesions soon become discernible on careful observation, and after several weeks they develop into brown necrotic spots 1 to 2 mm in diameter.

Although primordial lesions on young leaves may be too small to be seen by reflected light without a lens, they may be recognized by transmittable light as translucent yellow dots between oil glands when the leaves are held against light. The lesions then become raised, forming a spongy white eruption with no yellow halo. As the lesions further enlarge, the spongy eruptions begin to collapse, and brown depressions appear in their central portion, forming a crater-like shape. The margins of lesions remain to be raised above the surface of the host tissue and are characterized by a greasy appearance. As the disease advances, the central portions become grayish-white, hard, and appear as corky, dead tissues with a rough surface, surrounded by tan to dark-brown raised margins with a greasy appearance. Yellow haloes usually form around such advanced lesions (Fig.7-2). When lesions form on mid-rib or petiole, the leaf is more likely to defoliate than when scattered lesions form only on leaf blade.[4]

Canker lesions vary in size, depending on the susceptibility of the host plants. Lesions as large as 5 to 10 mm in diameter are not uncommon on susceptible cultivars

Figure 7-2 Symptoms of canker on *Citrus natsudaidai*. 1) Erupted spongy lesions developed on leaf under dry conditions. 2) Enlarged lesions with collapsed center and greasy margin developed under humid conditions. 3) Enlarged twig lesions developed on seedlings. 4) Fruit lesions.

such as grapefruit, but may be only a few millimeters in diameter on moderately resistant cultivars such as Unshu orange. The difference in resistance is greater in stems than
in leaves. On susceptible cultivars, the lesions rapidly enlarge and coalesce, encircling
entire twigs. On resistant cultivar, however, the lesions cease to enlarge at 1 to 2 mm in
diameter, and finally drop off in the form of dry brown scales, leaving a healed corky
surface (Fig.7-3).

7-5 HISTOPATHOLOGY

Citrus canker lesions are characterized by overdevelopment of parenchymatous tissues,
consisting of a large number of hypertrophic cells and a limited number of hyperplastic
cells.[5] The spongy cells near the infection site become dedifferentiated during the early
stage, displaying increased size and easier staining of the nuclei and nucleoids as well
as an increase in the amount of cytoplasm, followed by rapid enlargement. The hypertrophic cells occupy the intercellular spaces. As these cells further increase in size, the
callus tissue expands, lifting the epidermis above the leaf surface and finally causing
disruption of the epidermis, disclosing the internal callus tissues. In canker lesions,
hyperplasia usually occurs only in a few cells adjacent to healthy tissue. These hyperplastic cells develop into hypertrophic cells without continuous cell division. Hyperplasia then takes place on other new cells at the periphery of the lesion. Thus, the canker
lesion consists primarily of hypertrophic cells, with a small number of hyperplastic

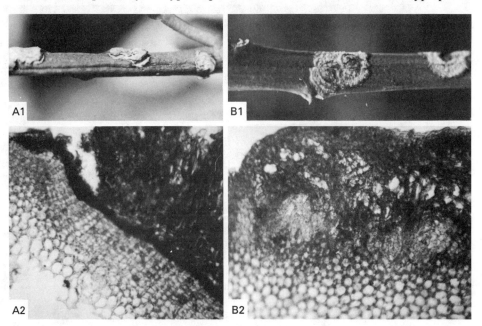

Figure 7-3 Citrus canker on twigs. A1) Canker lesions on resistant *Citrus unshu*. A2)
Transverse section of A1 showing development of cork layers between diseased and
healthy tissues. B1) Canker lesions on susceptible *Citrus natsudaidai*. B2) Transverse
section of B1 showing no defense layer developed between diseased and healthy tissues.

cells in its margin. This is the major histological difference between citrus canker and other bacterial diseases that induce hyptertrophy or bacterial gall. If infected plants are maintained in a greenhouse at low humidity, the calli continuously proliferate until the lesions become approximately 1 to 2 mm in diameter. Under natural conditions, however, the calli are contaminated by saprophytic epiphytes such as *Erwinia herbicola* and *Fusarium* spp. during rainfall. These secondary invaders soon collapse the calli from the center of the lesion, and the population of *Xanthomonas campestris* pv. *citri* rapidly declines in the collapsed area. The calli continue to develop outward at the periphery of lesions, where the pathogenic bacterium predominates.

7-6 ETIOLOGY

7-6-1 Pathogenic Bacteria

Strains.   The causal agent of citrus canker is *Xanthomonas campestris* pv. *citri* (Hasse 1915) Dye 1978.[6,56] Three distinct forms of this bacterium, A, B, and C, have been reported, mainly on the basis of geographical distribution and host range of the pathogen.[7] The pathogen for canker A (A form canker, Asiatic canker, cancrosis A, or true canker) is the authentic bacterium originally named *Pseudomonas citri* by Hasse,[6] and includes the neopathotype strain of *X. campestris* pv. *citri* (PDDCC 24; NCPPB 409). The canker A bacterium has a broad host range, producing the most severe form of canker, and is widely distributed throughout the world (i. e., Asia, Oceania, Africa, and South America). The pathogen of canker B (B form canker, cancrosis B, or false canker) has a limited host range, primarily affecting lemon and to some extent sweet orange, sour orange, lime, and citron, but not grapefruit. It is distributed in Argentina, Uruguay, and Paraguay.

The pathogen of canker C (Mexican lime canker or Galego lemon canker) was once named *Xanthomonas citri* f. sp. *aurantifolia*. It is distributed in Brazil and primarily attacks Mexican lime and, to a lesser extent, lemon. Its virulence in Mexican lime is most severe to Galego lemon, but with only slight infection on Tahiti lemon.[8]

A particular strain has also been found along the Pacific coast of Mexico in 1981. This bacterium causes pustules on leaves and young shoots of Mexican lime but not on fruits and other *Citrus* spp. It is suspected to be identical to the strain that causes canker C.[9]

In 1984, another citrus bacterial spot disease called nursery-form canker occurred in a nursery in Florida and was subsequently detected in many other nurseries in nine counties.[10] This disease was found on leaves and shoots of nursery plants but not on fruits and adult trees, although fruit can be experimentally infected. The host range of this pathogen is wide, including grapefruit, sweet orange, tangelo, tangerine, tangor, trifoliate orange, citrange, citromelo, and mandarin. The bacterium produces flat, or sometimes sunken, necrotic lesions with an extremely water-soaked appearance and chlorosis, but never produces the typical erupted canker lesions. Routine bacteriological characterization and fatty acid analyses indicate that the bacterium belongs to *Xanthomonas campestris*, but serological testing, bacteriophage sensitivity testing, and chromosomal DNA analysis have shown no relationship to the existing three strains of

X. campestris pv. *citri*. The nursery disease found in Florida was referred to as Xanthomonas leaf spot,[10] or citrus bacterial spot.[11] The pathogen was named as *Xanthomonas campestris* pv. *citrumelo*. The pathogen of canker B and C, including the strains found in Mexico, was designated as *Xanthomonas campestris* pv. *aurantifolia*. The differential characteristics of the three forms of citrus canker and of Xanthomonas leaf spot are given in Table 7-1.

Bacteriological Properties.   The morphological, physiological, and biochemical properties of Asiatic form of *Xanthomonas campestris* pv. *citri* are as follows. The cells are Gram negative, straight rods measuring $1.5–2.0 \times 0.5–0.75\ \mu$m; they are motile by means of a polar flagellum, and predominantly single (Fig. 7-4). They do not accumulate poly-β-hydroxybutyrate granules, and do not form endospores. They are encapsulated, chemoorganotrophic, and are aerobe with the oxidative metabolism of glucose. Colonies on agar plates are circular with an entire margin; they appear convex and creamy yellow on potato-glucose medium or straw yellow on nutrient medium. The pigment is characteristic xanthomonadins. Oxidase test is negative or weak, whereas

TABLE 7-1:   COMPARISON OF THREE DIFFERENT FORMS OF CITRUS CANKER AND XANTHOMONAS LEAF SPOT OF CITRUS

Characteristics	Citrus canker			Xanthomonas leaf spot
Canker form	A	B	C	
Pathogen	*X. campestris* pv. *citri*	*X. campestris* pv. *aurantifolia*	*X. campestris* pv. *aurantifolia*	*X. campestris* pv. *citrumelo*
Distribution	Asia, Africa South America Oceania	Argentina Paraguay Uruguay	Brazil Mexico	America (Florida)
Host range	Wide	Limited	Limited	Wide
Major host plant	Citrus spp.	Lemon	Mexican lime	Citrus spp. (nursery)
Symptoms	Spongy erupted lesions at first; corky rough leions with a rasied, greasy margin later; rarely water-soaked appearance			Flat or sunken lesion; extreme water soaking

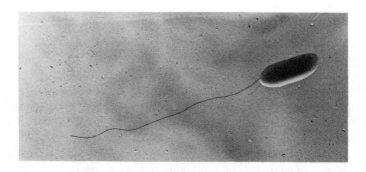

Figure 7-4   Flagellated cell of *Xanthomonas campestris* pv. *citri*.

catalase is positive. Litmus milk turns blue and becomes peptonized without being coagulated. Asparagine is not used as the sole source of nitrogen and carbon. Growth is inhibited by 0.02% triphenyltetrazolium chloride. Growth requires methionine or cysteine and is inhibited by serine. The cells are positive for hydrolysis of starch, aesculin, casein, and Tween 80, liquefaction of gelatin and pectate gel, and production of tyrosinase, lecithinase, reducing substance from sucrose, and hydrogen sulfide. The bacterium is negative for arginine dihydrolase, nitrate reduction, production of 2-ketogluconate, acetoin, urease and amino acid decarboxylases, and for methyl red test. Growth occurs in broth containing 3% NaCl but not 4%. Optimal growth temperature is 28°C, minimal temperature 6 to 7°C, and maximal temperature 36 to 38°C. The doubling time is 79 minutes. Xylose, glucose, fructose, galactose, mannose, maltose, lactose, trehalose, sucrose, glycerol, dextrin, starch, malonate, citrate, succinate, and malate are utilized as the sole sources of carbon. L-arabinose, rhamnose, raffinose, α-methylglucoside, salicin, sorbitol, inositol, dulcitol, inulin, gluconate, oxalate, acetate, and tartrate are not utilized. Potato slice is macerated when heavily inoculated. Utilization of mannitol varies among the isolates, with a close relationship to the host plants from which they are isolated. The mannitol utilization-negative isolates are mostly isolated from Unshu orange and the positive ones from other citrus, such as sweet orange and lemon. The strains of cancrosis B and C share many common characteristics with the Asiatic strain, the differences being found only in a few biochemical properties (Table 7-2).[9,12,13]

7-6-2 Susceptibility to Bacteriophages

Virulent phage.   Virulent phages propagate in host bacterial cells and are liberated through lysis of the cells after a certain latent period. These phages can be isolated either from canker lesions or from soil under citrus trees infected by canker. Three phages, Cp1, Cp2, and Cp3, have been isolated and characterized. Close relationships have been detected between susceptibility to these phages, the ability to utilize mannitol, and the host plants from which the bacteria were isolated (Table 7-3).[12-15] Therefore, the phage susceptibility can be used for rapid identification of the strains of X. campestris pv. citri, although the specific phages have not been isolated for the canker C organism.

Temperate phage.   The temperate phage is liberated through lysis of the lysogenic strain which carries phage DNA on its chromosome as the prophage. The pro-

TABLE 7-2: DIFFERENTIAL CHARACTERISTICS AMONG THE PATHOGENIC BACTERIA OF THE FORM A, B, AND C.

	Form		
Characteristics	A	B	C
Utilization of Maltose	+	−	−
Lactose	+	−	−
Mannitol	D	+	−
Malonate	+	D	−

TABLE 7-3: CHARACTERISTICS OF VIRULENT BACTERIOPHAGES OF *XANTHOMONAS CAMPESTRIS* PV. *CITRI*.

	Phages		
Characteristics	Cp1	Cp2	Cp3
Host range: A strain Mannitol +	+	−	−
Mannitol −	−	+	−
B strain	−	−	+
C strain	−	−	−
Shape	Tadpole	Spherical	Tadpole
Size (nm) Head	68	70	71
Tail	160 × 15	−	172 × 20

phage is freed from chromosomal DNA spontaneously or through induction with ultraviolet light or mitomycin C, otherwise propagating together with the bacterial chromosome through generations. Some phenotypic properties of the bacterium may be changed through transduction or phage conversion when the phage DNA is integrated into the host chromosome, becoming a prophage (lysogenization). The presence of lysogenic strains of *X. campestris* pv. *citri* was first proved by Okabe,[16] who demonstrated that the most strains of citrus canker organism are usually lysogenic, producing various temperate phages. Genetic alteration of the bacterium through lysogenization was subsequently studied, and it has been shown that temperate phages affect important traits of the bacterium, including virulence, colony form, and other properties.[17,18]

Filamentous phages.   Some temperate phages of the citrus canker organism may be filamentous. Filamentous phages thus far isolated from *Xanthomonas campestris* pv. *citri* enter a prophage state, and the infected cells become lysogenic.[19] They have an average length of 800 to 1200 nm and do not affect the growth of host bacteria. It remains to be elucidated whether all temperate phages of citrus canker organisms are filamentous.

7-6-3 Plasmids

Strains of *Xanthomonas campestris* pv. *citri* carry indigenous plasmids that were differentiated into seven with approximate molecular sizes of 12 to 53 megadaltons. No functions or phenotypes have yet been elucidated for any of these plasmids. Citrus canker organisms of different forms have different plasmid profiles (i. e., canker A strain contains major plasmids having molecular sizes 43 to 44, 45 to 46, or 53 megadaltons, canker C strain having major plasmids of 21 to 23 and 31 to 32 megadaltons, and canker B strain having a plasmid of 31 to 32 megadaltons).[20] It is interesting to note that differentiation based on the plasmid profiles is consistent with that observed on the basis of serology and of phage susceptibility. It remains for future study, however, to determine whether some of the aforementioned temperate phages are indigenous plasmids. Two insertion sequences ISXC4 and ISXC5 of 5.55 and 6.95 Kbp in size, respectively, were isolated from a particular strain XC45 of *X. campestris* pv. *citri*. These insertion sequences seemed to originate from different indigenous plasmids of the bacterial strain and were transposed, inducing a drug-resistance mutation.[21]

7-6-4 Genomic Fingerprints (Restriction fragment-length polymorphism: RFLP)

A genomic fingerprint consists of the complex genomic DNA fragment pattern as revealed after complete digestion by a specific restriction endonuclease, separation on a polyacrylamide gel, and visualization by ethidium bromide under ultraviolet rays. Because restriction endonuclease cleavage sites of a genome are uniquely conserved and stably inherited in the bacterial clone, the gemomic fingerprint can be used as an approach for evolutional and epidemiological studies of bacterial plant pathogens. The fingerprints of canker A and B strains are distinctive from each other, but indistinguishable within each strain.[22] It is noteworthy, however, that some similarities have been found in the fingerprints of canker B, a single strain of canker C from Brazil, and a single strain of Mexican citrus bacteriosis. It is interesting that the interrelationships of the fingerprints of canker B, C, and Mexican bacteriosis seem to reflect serological relationships, at least in part.

7-6-5 Serology

Although antigen-antibody interaction can be manifested in many ways, the agglutination test, the agar-gel diffusion test, the fluorescent antibody technique, and the enzyme-labeled immunosorbent assay technique have been widely used in plant bacteriology. These techniques, however, are not efficient enough to differentiate the microheterogeneity of bacterial proteins, mainly because of the heterogeneity of antisera obtained by conventional methods stemming from the polyclonality of antibodies produced by many antibody-producing B cell clones. Monoclonal antibodies are produced by hybridomas produced by fusion of a single B lymphocyte to a malignant plasma cell (myeloma) line. The benefits of monoclonal antibodies include (1) recognition of only one of many antigens on a bacterial cell, (2) generation of pure antibodies from impure antigens, and (3) availability of the identical antibodies to different laboratories, thereby eliminating variations in results caused by possible heterogeneity of antisera. Serology in the study of citrus canker has been applied mainly to diagnosis and/or detection of the causal bacterium in natural habitats. By conventional techniques, serological differentiation has been made between the pathogens of canker A and B and cankers A and C, respectively.[8] These studies have demonstrated the presence of independent antigen(s) in each strain which are not shared by others.

The monoclonal antibody technique has further elucidated the serological relationship among the strains of citrus canker. Strain-specific antibodies have been obtained for A, B, and C-form pathogens. However, there are quite complicated serological relationships among the strains of B, C, and Mexican bacteriosis, which have been differentiated on the basis of host ranges, phage susceptibility, and the capacity to grow on agar media (Table 7-4).[23] The response to monoclonal antibodies seem to vary, depending on the history of each strain or culture. This implies the possibity of antigenic alteration of the bacterium through adaptation to certain host plants, as well as to culture media.

TABLE 7-4: REACTIVITY OF *XANTHOMONAS CAMPESTRIS* PV. *CITRI* AND OTHER CITRUS BACTERIA WITH MONOCLONAL ANTIBODIES (MCA).

MCA	Citrus strains[a]			Mexican citrus strains[b]			Argentine citrus strains[c]		*Xanthomonas campestris* pv. *campestris*	*Erwinia herbicola*
	A	B	C	3	1–41	T	A-types	B-types		
A1	+	−	−	−	−	−	+	−	−	−
A2	+[d]	−	−	−	−	−	+[e]	−	−	−
B1	−	+	−	−	−	−	−	−	−	−
B2	−	−	+	+	−	−	−	+	−	−
X1	+	+	+	+	−	+	+	+	+	−
X11	+	+	+	+	−	+	+	+	+	−

[a]formalin-killed preparations of known strains of *Xanthomonas campestris* pv. *citri*. Strains were designated as A (62, 63), B (64, 69), and C (70) based on host range and phage type. B strains were readily cultured on Wakimoto medium containing Difco Bacto Agar.

[b]Most of the preparations received from Mexico were isolated from leaf lesions having symptoms of "bacteriosis." Forty-one strains (1–41) were originally thought to be pathogenic, but pathogenicity was not confirmed for any strain except No. 3. Five cultures designated T were received from Salvador Rodrigues, SARH, Tecoman, Colima.

[c]B strains were isolated from lemon by D. Zagory on modified Wilbrink's medium containing Difco purified agar. Mexican citrus strains and Argentine citrus strains were recently designated as *X. campestris* pv. *aurantifolia*.

[d]Only strain XC62.

[e]Only 3 of 12 strains. (From Alverez et al., 1985)[23]

7-6-6 Host Range

Plants of the genus *Citrus* are known to be susceptible to citrus canker in varying degrees. Plants of the family *Rutaceae* have also been reported as host plants for *X. campestris* pv. *citri* by artificial inoculation. However, most except *Poncirus trifoliata* are not sufficiently susceptible under normal conditions to warrant attention.

a. Susceptibility of plants of the suborder *Geraniineae* except those of the genus *Citrus*.   Table 7-5 shows a list of plants differentiated according to their degree of susceptibility in greenhouse and field inoculation tests.[24] However, the results of Peltier[25] on *Toddalia asiatica* and *Aegle malmelos* were inconsistent with those of Lee.[2] *Balsamocitrus paniculata*, *Feroniella obligata*, and *Matthiola incana* var. *annua* were also listed as host plants.

b. Susceptibility of species, hybrids, and cultivars of the genus *Citrus*. Foliage.   Peltier[25] used the following criteria for differentiating the susceptibility of *Citrus* spp. to citrus canker:

1. Resistant (R) or susceptible minus: No infection on twigs, but scattered small lesions on leaves; epidermis of leaf spots remains intact and does not rupture to form canker.

2. Intermediate (I): Lesions are commonly found on leaves, but occasionally on twigs; leaf spots develop to typical canker lesions with a ruptured surface.

TABLE 7-5:   SUSCEPTIBILITY OF THE PLANTS OTHER THAN *CITRUS* SPP. TO CITRUS CANKER*

A. Infection was positive both in greenhouse and field inoculation tests.

Casimiroa edulis	*Chaetospermum glutinosum*	*Eremocitrus glauca*
Feroniella lucida	*Fortunella hindsii*	*Fortunella japonica*
Fortunella margaria	*Hesperethusa crenulata*	*Microcitrus australasica*
Microcitrus australasica var. *sanguinea*		*Microcitrus australis*
Microcitrus garrowayi	*Poncirus trifoliata*	*Xanthoxylum clava-hercules*
Xanthoxylum fagara		

B. Infection was positive in greenhouse inoculation tests but not in field tests.

Aegle marmelos	*Aeglopsis chevalieri*	*Atalantia ceylonica*
Atalantia citrioides	*Atalantia disticha*	*Chalcas exotica*
Claucena lansium	*Citropsis schweinfurthii*	*Evodia latifolia*
Evodia ridleyei	*Feronia limonia*	*Fortunella crassifolia*
Lansium domesticum	*Melicope triphylla*	*Paramignya monophylla*
Paramignya longipedunculata		

C. Infection was negative both in greenhouse and field inoculation tests.

Balsamocitrus dawei	*Balsamocitrus gabonensis*	*Glycosmis pentaphylla*
Melia azerdarach	*Severinia buxifolia*	*Toddalia asiatica*
Triphasia trifolia	*Xanthoxylum bungei*	*Xanthoxylum rhetsa*
Xanthoxylum sp.		

[a]Peltier 1920 [24,25] and Lee 1918[2].

3. Susceptible (S) or susceptible plus: Lesions formed on leaves and twigs enlarge to form typical canker lesions that encircle the stems or cause defoliation.

Pelter classified the susceptibility of *Citrus* spp. as shown in Table 7-6.

Fruits.   The susceptibility of fruits to citrus canker was arbitrarily classified as follows, on the basis of results of inoculation tests made by Kawakami[26] and by Fulton and Bowman,[27] as listed in Table 7-6. Where results with the same plant do not agree between two experiments, the more susceptible indices were adopted.

1. Susceptible (S): Fifty percent or more of fruits were infected either by spray inoculation or wound inoculation.

2. Intermediate (I): Twenty to forty percent of fruits were infected.

3. Resistant (R): Less than 19% of fruits were infected.

An extensive study was recently conducted on the susceptibility of leaves of early,-intermediate-, and late-maturing cultivars of orange (*Citrus sinensis*), revealing considerable differences in canker resistance (Table 7-7).

Hybrids.   The susceptibility of foliage of hybrids to citrus canker was studied by Peltier[25] (1920) (Table 7-8). The indices are the same as those in Table 7-5.

TABLE 7-6: SUSCEPTIBILITY OF CITRUS PLANTS TO CITRUS CANKER*

Scientific name	Common name or form	Foliage	Fruit
Citrus histrix	Pointed-leaf form	R	—
	Round-leaf form	S	—
Citrus aurantifolia	Sour lime	I	R
Citrus limon	Lemons	I	I(S)
Citrus grandis	Buntan	I	S
	Hassaku	I	I
Citrus paradisi	Grapefruit	S	I
Citrus natsudaidai	Natsumikan	I	S
Citrus aurantium	Daidai	I	I
Citrus sinensis	Oranges	I	S
	Navel	I	S
Citrus temple	Temple	I	R
Citrus tankan	Tankan	I	S
Citrus mitis	Calamondin	R	R
Citrus yunos	Yuzu	R	R
Citrus unshu	Satsuma	R	S
Citrus reticulata	Tangerine	R	R
Citrus rehni	Cleopatra	R	—

*Adapted from Peltier (1920),[25] Kawakami (1921),[26] Fulton and Bowman (1929),[27] and Koizumi (1978).[49] Results are not necessarily consistent with the authors and plant organs.

TABLE 7-7: SUSCEPTIBILITY OF ORANGE CULTIVARS (*CITRUS SINENSIS*) TO CITRUS CANKER, EXPRESSED AS THE NUMBER OF LESIONS PER CM^2 OF LEAF SURFACE ON TREES SPRAYED WITH TRIBASIC COPPER SULPHATE

	No. of lesions per cm^2 of leaf surface ($\times 10^{-4}$)				
Cultivars	1–10 Highly Resist.	11–30 Mod. Resist.	31–100 Slightly Suscept.	101–300 Mod. Suscept.	301–700 Highly Suscept.
Early	Cocco Salustiana	Navelina Vive Sucral Vive	Hamlin Thomson 390 Atwood E. Navel Omb. Lentijo Westin	Trovita USA Navelate SAR Bahianina	Marrs Early
Intermediate		Sanguinelli V. Cadenera Nja. Lima	Pineapple Barao Nja. del Cielo Criolla Mejorada	Criolla Tempr.	Petropolis China
Late	Valencia Camp.	Hughes Valen. Velencia LPC Valencia Wood Valencia Late Valencia CN 35	Valencia Sp. Lue Gim Gong Pera 1743-82 Pera 1743-27 Pera 1743-52	Natal S. Pablo Valencia Frost Valencia CN 44 Enterprise	

From Zubrzycki and de Zubrzycki, 1981.[55]

TABLE 7-8:   RELATIVE SUSCEPTIBILITY OF HYBRIDS TO CITRUS CANKER*

	Parent			
Citrus aurantifolia	×	*Microcitrus australasica*	Faustrime	I
Citrus limonia	×	*Microcitrus australasica*	Faustrimon	I
Citrus mitis	×	*Microcitrus australasica*	Faustrimedin	I
Citrus sinensis	×	*Poncirus trifoliata*	Citrange	S
Citrus grandis	×	*Poncirus trifoliata*	Citrumelo	S
Citrus aurantium	×	*Poncirus trifoliata*	Citradia	I
Citrus noblis	×	*Poncirus trifoliata*	Citrandarin	S
Citrus unshu	×	*Poncirus trifoliata*	Citrunshu	I
Citrange	×	*Poncirus trifoliata*	Cicitrange	S
Citrus mitis	×	Citrange	Citrangedin	I
Citrus deliciosa	×	Citrange	Citrangarin	R
Citrus unshu	×	citrange	Citranguma	I
Fortunella margarita	×	Citrange	Citrangequat	R
Fortunella japonica	×	*Citrus aurantifolia*	Limequat	S
Citrus grandis	×	*Citrus aurantifolia*	Limelo	S
Citrus mitis	×	*Citrus aurantifolia*	Bigaraldin	S
Citrus grandis	×	*Citrus sinensis*	Orangelo	I
Foreunella margarita	×	*Citrus sinensis*	Organequat	I
Citrus noblis (Clementine)	×	*Citrus grandis*	Clemelo	I
Citrus unshu	×	*Citrus grandis*	Satsumelo	I
Citrus deliciosa	×	*Citrus grandis*	Tangelo	I
Citrus noblis (King of Siam)	×	*Citrus sinensis*	Siamor	I
Citrus deliciosa	×	*Citrus mitis*	Calarin	I
Citrus unshu	×	*Citrus mitis*	Calashu	R

*Peltier, 1920[25]

7-7 HOST-BACTERIA INTERACTION

Citrus canker research has been primarily oriented toward the ecological behavior of the causal bacterium. Studies from a physiological, biochemical, or molecular standpoint are therefore very limited.

7-7-1 Extracellular Polysaccharides (EPS) and Longevity of the Citrus Canker Organism

Xanthomonas campestris pv. *citri* produces abundant EPS, either in culture media or in host tissues. The bacterial cells in canker lesions are embedded in a dense matrix of EPS and are dispersed, together with EPS, by rain splash. However, the cells of *X. campestris* pv. *citri* are very unstable when highly diluted in water (i. e., they are rapidly killed when suspended at concentrations lower than 10^6 cells/ml). This trait is also found in most xanthomonads as well as some bacteria of other genera. I call this phenomenon lethal dilution effect.[28] The lethal dilution effect is overcome by addition of EPS at a concentration as low as 0.05 mg/ml. Bacterial suspensions at high density seem to contain enough EPS to protect bacterial cells from the lethal dilution effect (Fig.7-5). Therefore, it should be favorable for the pathogen to be splashed in such a way that bacterial exudate is not greatly diluted by rain. The citrus canker organism is

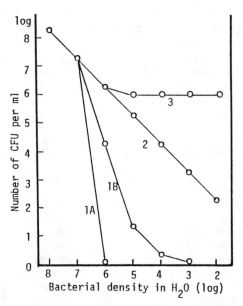

Figure 7-5 Lethal dilution effect on *X. campestris* pv. *citri* diluted in distilled water. 1A) Survival curve of pv. *citri*. 1B) Survival curve of *P. syringae* pv. *lachrymans*. 2) Survival curve of *Clavibacter michganensis* subsp. *michiganensis*. 3) survival curve of *P. solanacearum*.

also very liable to desiccation when smeared on a slide glass after dilution in water, whereas it is very stable when colonies on agar plates are directly smeared. However, the washed cells become tolerant to desiccation when smeared together with EPS. Thus, the EPS molecules exhibit great protective effects against both the lethal dilution effect in water and desiccation in air, providing substantial benefits for the bacterial ecology.

7-7-2 Extracellular Polysaccharides and Adhesion of Bacterial Cells onto Wounded Host Cells

Takahashi and Doke[29,30] reported that bacterial cells which enter the intercellular space adhere to the host cell walls through an interaction between bacterial EPS and citrus agglutinins. The citrus agglutinins consist of 96% protein and 4% carbohydrate, with estimated molecular wights of 63,000 and 16,000, respectively. The agglutinins are active with EPS of various xanthomonads at pH lower than 6.0, forming a fibrous precipitate. The precipitate formation with EPS of *X. campestris* pv. *citri* is specifically prevented by D-glucosamine. At pH 7.0, however, the agglutinins associate with the EPS without visible precipitation. Interference with bacterial adhesion to host cell walls by D-glucosamine resulted in a significant reduction of infection. Because cells of *X. campestris* pv. *citri* do grow in the intercellular space and induce symptoms, however, this bacterial adhesion must differ from the attachment of bacterial cells to host cell wall followed by the immobilization which has been demonstrated in the incompatible host-bacteria combination.

7-7-3 Extracellular Polysaccharides and Water Congestion

When the leaf water potential was increased by allowing detached leaves to absorb water from the cut end of leaf petioles, water congestion was recovered on citrus leaves into which EPS had been injected three weeks before, and the water-soaked appearance disappeared. The same phenomenon was observed under field conditions during rain. These facts imply that EPS introduced into the intercellular space remains intact and induces localized water congestion under a high water potential, enhancing the *in planta* growth of the citrus canker organism.[31]

7-7-4 Ethylene Production in Infected Tissues

Ethylene production by citrus leaves inoculated with *X. campestris* pv. *citri* occurs at three different stages (i. e., initial, intermediate, and later stages). The host cells, in association with EPS, manifest ethylene biosynthesis for several hours after inoculation. Ethylene production at this stage is not demonstrated by lipopolysaccharides (LPS) extracted from *X. campestris* pv. *citri* and is the earliest physiological response of citrus leaves inoculated with the bacterium. On the basis of the time sequence, this response seems to be related to the aforementioned bacterial adhesion and/or to "early induced selective protection."[32]

Ethylene production at the intermediate stage occurs during the period of 10 to 24 hours after inoculation. This response is induced only by the inoculation of the virulent strains and not by the infiltration of EPS and is followed by leakage of electrolytes and amino acids. Ethylene production at this stage, therefore, seems to reflect the host-bacterium interaction that causes substantial damage to the host cell membrane. The auxin produced by bacteria may play a significant role at this stage.

A large amount of ethylene is also produced after canker symptoms develop. Ethylene production at the later stage originates in the hypertrophic host cells within the canker lesions as well as in the cells in the peripheral zones, which appear to be under the influence of auxin and sometimes form yellow haloes. The high level of ethylene produced at this stage triggers the formation of an abscission layer at the base of the leaf blade or leaf petiole, thereby inducing defoliation.[31,33]

7-7-5 Changes of Membrane Permeability and Amino Acid Composition in Infected Tissues

The conductivity of the intercellular fluid rapidly increases 24 to 48 hours after inoculation. In parallel with the increase in conductivity, amino acids are released into the intercellular space, although the results of bioassay with *Erwinia herbicola* as the indicator organism reveal that amino acids are released as early as 9 hours after inoculation.[34]

Some amino acids that significantly increase in the intercellular fluid of infected citrus leaves show the following biological functions in relation to the growth of *X. campestris* pv. *citri*: (1) methionine, cystine, cysteine-growth factor; (2) aspartic acid and aspargine-nutrients; (3) serine, hydroxylysine, and glycine-growth inhibition; and (4) proline-reversal of growth inhibition. The virulent strains are more sensitive to growth inhibition by serine than the less virulent or avirulent strains. The same trend is observed in the recovery of growth by proline.[35]

7-7-6 Auxin Metabolism

Indoleacetic acid (IAA) is detected in culture filtrates of synthetic media and in the Natsudaihai leaf extract medium in which *X. campestris* pv. *citri* has been grown for two days. The concentration in the latter medium incubated for three days reached $0.042\mu g/ml$, which was about 25 times greater than that contained in the uninoculated control medium. The IAA concentration in the leaves eight days after inoculation by the infiltration method was 0.38 $\mu g/g$ dry weight, about 1.5 times greater than that of the uninoculated leaves. However, the concentration in the leaves eight days after inoculation by the prick method was 0.5 $\mu g/g$ dry weight, estimated to be about two times greater than that of uninoculated leaves. Because healthy citrus leaves contain IAA at the level of about 0.24 $\mu g/g$ dry weight, a substantial change in the endogeneous IAA metabolism in host cells must be involved in the initiation of hypertrophy, in addition to the exogeneous IAA supplied by the pathogen.[28]

7-8 DISEASE CYCLE

The disease cycle of citrus canker is described here as it occurs in the climate of the Northern Hemisphere. Citrus canker used to occur at any season on susceptible young trees that produce many angular shoots. Its occurrence on mature plants, however, is sporadic (i. e., the disease occurs in severe form every several years to a decade, even in citrus groves where it has rarely been observed). These facts indicate that survival forms or inoculum sources of the pathogenic bacterium are not as simple as those described earlier (Fig.7-6).

7-8-1 Spring Shoot

The first flush in spring is infected by the pathogenic bacterium splashed by rain from the canker lesions onto overwintered shoots. Because citrus leaves become resistant to citrus canker infection through their stomata as they mature, the disease cycle on the spring shoots may be limited to a rather short period of time, unless the leaves are injured by storms.[12]

7-8-2 Autumn Shoot

The disease cycle on angular shoots that develop from summer to autumn may continue for several months because of the continuous supply of young, susceptible shoots. Injuries caused by leaf miner (*Phyllocnistis citrella*) strongly enhance canker infection on angular shoots. Although heavily infected leaves defoliate in winter, lesions on the stems or on slightly infected attached leaves become the major inoculum source in the following spring. Therefore, citrus canker occurs in its most severe form on younger trees that flush many angular shoots rather than on mature trees bearing heavy fruit and fewer angular shoots. Infection that occurs late in autumn often results in a latent infection.

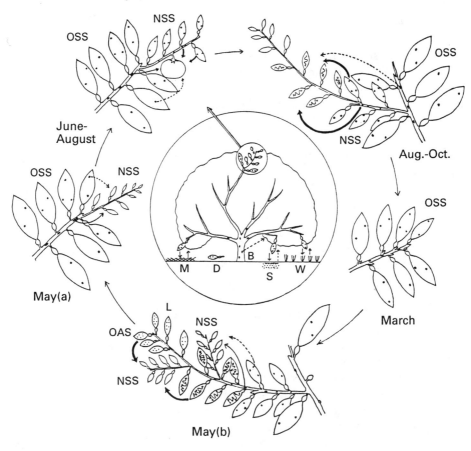

➡ (heavy arrow)	Heavy infection	OSS	Overwintered spring shoots
→	Slight infection	NAS	New autumn shoots
- - ->	Negligible infection	OAS	Overwintered autumn shoots
●	Holdover canker lesions	F	Fruits
o	New canker lesions	W	Weeds
(leaf miner symbol)	Injuries of leaf miner	S	Soil
(canker on leaf miner symbol)	Canker occurred on leaf miner injuries	D	Defoliated leaves
		M	Mulch
NSS	New spring shoots	L	Latent infection

Figure 7-6 Disease cycle of A-form citrus canker caused by *Xanthomonas campestris* pv. *citri*.

7-8-3 Fruit

Fruits suffer from canker infection for several months, beginning just after flower dropping and continuing to the early stage of maturation or until just before yellowing. Infection on fruit continues for that long because the fruits continue to grow and to provide stomata suitable for infection.[12]

7-9 SURVIVAL OF CITRUS CANKER ORGANISMS

7-9-1 Detection Method

The development of a technique for investigating the ability of *X. campestris* pv. *citri* to survive in nature is critical, particularly under saprophytic conditions. Selective media may be the most useful tool if the selection efficiency is high enough to allow only the citrus canker organism to grow out of the mixed microbial population. However, such media have not yet been developed for the bacterium. Although the fluorescent antibody technique, ELISA technique, and bacteriophage technique may be applicable, inoculation methods, especially the leaf infiltration technique, are the most simple and reliable. This technique uses young leaf tissue as a sort of selective medium. The causal bacterium, injected together with various microorganisms, may be able to grow exclusively in the leaf tissue, finally leading to formation of canker lesions. The population of the pathogen in the samples can be calculated from the number of lesions developed.

In this technique, the samples to be tested, soils or weeds, are washed in water, and the washings are centrifuged first at a low speed (e. g., 2000 rpm for 10 minutes) to remove coarse soil particles or plant residues, and then at high speed (e. g., 9000 rpm for 20 minutes) to precipitate bacterial cells. The sediments are suspended in a small amount of sterilized water and infiltrated into the mesophyll of the leaf blade with a syringe. From the number of lesions developed on the inoculated leaves, the population of the causal bacterium can be calculated on the basis of an average water infiltration rate of 7 μl/cm^2. The detection efficiency or minimal detection level of the technique is about 10^2 bacterial cells/g soil, although it takes one to two weeks before visible lesions develop on the infiltrated leaf areas. It is, however, important to use leaves as young as possible, because of the ease of infiltration into mesophyll and the likelihood that the casual bacterium may grow from single cells. The ecological behavior of *X. campestris* pv. *citri* in nature has been greatly elucidated by use of this technique.[36]

The phage technique is usually difficult to apply for detection of the citrus canker organism in a mixed microbial population because (1) the threshold of the bacterial population for phage-bacteria interaction is around 10^5 cells/ml and bacterial cells below this level can be left without being infected by phage particles; and (2) a low population of the host bacterium can rarely be raised to the level of 10^5 cells/ml by the quick multiplication of saprophytic microorganisms that are present in samples.[36] As far as the detection efficiency of *X. campestris* pv. *citri* from mixed microbial samples prepared from soils, weeds, or mulching materials is concerned, the leaf infiltration technique has demonstrated its superiority over any other methods, including serological techniques such ELISA and fluorescent antibody techniques.

7-9-2 Survival in Parasitic Form

Like other plant pathogenic bacteria, the citrus canker organism can survive for a certain period of time in the parasitic form in diseased plant tissues, in the epiphytic form on host and nonhost plants, and in the saprophytic form on straw mulch or in soil.[37,38] These survival forms, however, differ in significance as the source of citrus canker epidemics. Citrus canker lesions differ in their potentials as inoculum sources, depending on citrus cultivars, specific plants parts, or the age of the plant.

Survival in holdover cankers

***Spring shoots.* Leaf lesions.**   In canker lesions on spring leaves developed a year before, the population of *X. campestris* pv. *citri* usually becomes very low by late autumn before the plant enters dormancy, and almost extinct by the time of new flush in next spring. However, the bacterium may sometime survive at a low density in aged lesions of susceptible cultivars throughout a winter season when weather conditions are mild. These bacteria may develop new greasy spots along the periphery of old lesions in spring. In moderately resistant cultivars, such as Unshu orange, the possibility of overwintering in aged leaf lesions is negligible.

Twig lesions.   In aged twig lesions formed on susceptible cultivars, such as grapefruit and sweet orange, a cork layer is formed underneath the lesion, separating the canker tissue from healthy bark tissue. However, development of such a barrier is poor in the periphery of lesions, where the causal bacterium may be alive. In moderately resistant cultivars, however, a thick cork layer quickly develops beneath canker lesions along their entire length during summer, and the disease completely heals, dropping canker lesions as dried scales (Fig. 7-3).

Autumn shoots.   The citrus canker organism can easily overwinter in lesions formed on autumn shoots, both on leaves and twigs. The population is generally greater in the lesions developed in late autumn than in those developed in late summer.[12]

Leaves.   Heavily infected leaves usually fall in winter, so the potential of these leaves as an infection source is small. Leaves with small numbers of scattered lesions are left intact in winter, serving as the infection source in the next spring. In these leaves, new spots with a greasy surface often appear along the margin of the overwintered lesions.

Twigs.   Lesions formed on the twigs of autumn shoot are important for overwintering of the citrus canker organism, because they have a greater capacity to hold the living bacteria than do leaf lesions. In susceptible cultivars, the borders between the diseased and healthy tissues are left intact without developing any defense layer harboring a large number of the pathogenic bacteria. In resistant cultivars such as Unshu orange, the development of a defense layer is also poor during the low temperatures of autumn because of reduced metabolic activity before dormancy. The hypertrophic cells directly contact healthy ones in the cortical layer. Significantly high levels of the pathogen are dispersed from these lesions by rain. The importance of latent infection on twigs as the source of infection is the same as that in leaves.

Bark tissues.   *X. campestris* pv. *citri* can survive for long periods in the discolored bark tissue of trunks, low scaffold limbs, and lateral branches. These areas are considered to be the old canker lesions formed on the twigs of younger plants. The isolation frequency, as well as the bacterial population detected from the bark tissues, is correlated with the degree of susceptibility of the plants to citrus canker. This has been confirmed by artificial inoculation tests in which the bacterium survived at high density for three years or more in dead bark tissues which had been completely covered by a cork layer, and no canker symptoms were visible. In resistant Unshu trees with light infection, the bacterium was also detectable with high frequency but low population from six-month to three-year-old healthy-appearing green twigs, although visible symptoms were not subsequently produced on them. No direct evidence has been obtained concerning actual foliage or fruit infection resulting from citrus canker organism living in such discolored bark areas. The bark lesions harboring the pathogenic bacterium may, however, play a role as an infection source under favorable weather conditions, such as heavy rain storms. This type of survival may also be responsible for the sporadic nature of the disease.[39]

Latent infection.   In young trees developing many angular shoots, latent infection often affects several young leaves on top of twigs if infection occurs just before plants enter dormancy in late autumn. These leaves do not develop any symptoms until the following March, or exhibit tiny black dots which resemble Melanose spots. With rainfall in early spring, development of new spots or enlargement of black dots with a water-soaked appearance takes place, dispersing large numbers of bacterial cells. Latent infection may also occur on aged spring shoots as a result of injuries caused by late autumn windstorms.

7-9-3 Survival in Epiphytic Form

There is much evidence that plant pathogenic bacteria can survive in the form of microcolonies on plant surfaces without causing any pathological effects. In citrus canker, however, it is difficult to detect the epiphytic phase of the pathogen on phylloplane or gemmisphere. The microorganisms found on healthy, mature leaves are usually fungi and bacteria, such as *Fusarium* spp., *Penicillium* spp., and/or *Bacillus* spp. Artificial infestation experiments indicate that *X. campestris* pv. *citri* can epiphytically survive on a limited number of leaves for a few months during the rainy season, but it quickly declines under sunny weather to a low level undetectable by the leaf print method (Goto, unpublished data). Unlike most vegetables, the surfaces of mature leaves with a well-developed cuticle seem to be unfavorable to epiphytic colonization by the citrus canker organism, and the epiphytic form of survival should have only minor importance from the epidemiological viewpoint of citrus canker.

7-9-4 Survival in Saprophytic Form

Defoliated leaves.   The citrus canker organism usually dies quickly in tissues of diseased leaves and twigs which fall to the ground and become moist. The causal bacterium dies within three weeks in lesions of leaves that are wetted either by placing them on the soil surface or by burying them in soil at a depth of 3 to 6 cm.[40] However, the bacterium can survive for two to three months in lesions if the diseased leaves are maintained under dry conditions, either *in vitro* or *in situ*, so leaf moisture quickly decreases to less than 20%. When smeared on a glass surface, bacterial cells suspended in water are more liable to desiccation than those taken directly from agar. Within canker lesions, cells of the bacterium are usually burried in a thick matrix of EPS. These facts imply that the bacterium can survive for fairly long periods if the EPS around them dry and are left intact. When the leaves became wet, various kinds of microorganisms may invade the habitat of the pathogen and destroy favorable environments for survival.

Soil.   *X. campestris* pv. *citri* can survive or even multiply in sterilized soil but usually dies quickly in natural soil.[41,42] The survival of the citrus canker organism in soil is of interest because the disease often occurs on young citrus trees reintroduced after previous eradication of diseased plants. However, experimental evidence resulting from either germination of citrus seeds in infested soil or isolation of the bacterium from infested soil indicates that the bacterium quickly becomes extinct in natural soil. Such a short longevity has been explained on the basis of microbial interaction, and especially by the predatory effect of protozoas.[43,44] These findings are generally true when the temperature is warm enough to allow the soil microorganisms to compete with *X. campestris* pv. *citri*. When the bacterium is introduced into soil late in autumn or early in winter and the soil becomes partially frozen, the population decreases quickly for the first few weeks and then becomes stable, capable of survival at low levels for three to four months. The bacteria finally die as the soil temperature rises and active multiplication of soil microorganisms begin.[38] Therefore, it can be said that heavily infested *X. campestris* pv. *citri* can maintain its population at levels sufficient for infection for no more than a few weeks during the citrus-growing season. There is no evidence for multiplication of the bacterium in natural soil.

Weeds.   *X. campestris* pv. *citri* can be detected on the surfaces of various kinds of weeds growing in citrus groves.[37,38,45] In most cases, the bacterium appears to rest on the plant surface without active multiplication after being splashed by rain. However, in certain weeds, such as *Zoysia* spp. or *Calistegia japonica*, the longevity of the bacterium is much longer than that expected from the casual or transit form of survival of the splashed bacterium. These weeds are all immune to *X. campestris* pv. *citri*. No such infestation has been observed with weeds growing in canker-free citrus groves or those from noncitrus areas. Therefore, it is clear that the bacterium detected in weeds originates in lesions on citrus trees and splashed from them by rain. On these perennial weeds, the bacterium is usually detected from rhizomes or root systems. In the case of *Zoysia japonica*, three phagovars—A and B, which are susceptible to phage Cp1 and Cp2, respectively, and C which is resistant to both phages—could be detected, the isolation frequency of phagovar C being higher than that from diseased citrus plants. Phagovar A is also detected from weeds growing in Unshu groves where only phagovar B is distributed on plants. Phagovar B can also be isolated from the weeds late in winter or early in spring, when the pathogen is hardly isolated from holdover cankers on Unshu trees. These findings indicate that regardless of the phagovars, the pathogen can survive on the grass for at least several months and that phagovar C may be more adapted to the habitat than other phagovars.

Straw mulch.   In intensive citrus cultivation, as performed in Japan, mulch is used to protect soil from drought and to keep weeds out. Common mulch materials are rice straw or other grasses. When *X. campestris* pv. *citri* is splashed from lesions onto straw, it can survive for several months under dry and cool weather conditions, and infection originating from the inoculum on the habitat may be found on spring shoots growing close to mulch. However, when the mulch is repeatedly wetted by rain, the density of the causal bacterium quickly declines.[38] On artificially infested soil, weeds, and straw mulch, *X. campestris* pv. *citri* shows a particular type of survival curve. The number of viable cells quickly declines for the first two to three weeks to a level of 10^2 to 10^3 per gram samples, and then tends to maintain this level for a long time.[45] Such a survival curve suggests that part of the bacterial cells in the inoculum can grow to some extent under such saprophytic conditions, and that growth may be effective enough to produce EPS for protecting bacterial cells from detrimental microenvironments. The low population may have another advantage in survival, because interaction with other microorganisms should be less intense at low density. Under such conditions, the bacterium may be able to escape the unfavorable effects of competition for nutrition and living sites, or of antibiotic compounds, enabling it to coexist with other microorganisms in the habitat. These findings regarding saprophytic survival of *X. campestris* pv. *citri* also suggest that two ecotypes, parasitic and saprophytic, may be differentiated in this bacterium. In particular, phagovar C may be the ecotype more preferably adapted to saprophytic environments, although further investigations are necessary to confirm the presence of these ecotypes.

7-10 INOCULATION AND DISEASE RATING

Xanthomonas campestris pv. *citri* can infect citrus trees through stomata and wounds. Therefore, inoculation test can be done by either spray method, the rubber-block press method, the pin-prick method, or the leaf infiltration method.

7-10-1 Spray Inoculation

Bacterial growth on an agar slant is suspended in sterilized water at a concentration of about 10^7 to 10^8 cells/ml, depending on the susceptibility of the host plants. With high inoculum doses, lesions may develop so densely that they soon coalesce with each other, making them difficult to count and leading to defoliation. With low inoculum doses, the infection may be very slight or even absent on young shoots, depending on the physiological conditions of the plants or on environmental conditions. Therefore, preliminary inoculation tests with different doses are useful to determine the adequate inoculum concentration for the particular experimental conditions under consideration. Minimal infection dose through stomata is around 10^5 cells/ml.[12] No infection occurs at population below this level, the infection rate being increased in parallel with the density above this level. An inoculum size of 10^5 cells/ml roughly corresponds to the ratio of one bacterial cell to the capacity of the front cavity of a stoma.

The most susceptible zone moves from the base to the top of a shoot as it elongates; the susceptibility of a given young shoot to citrus canker is different for each leaf and each part of the stem, depending on the growth stage of the shoots. Therefore, selection of the shoots according to their growth stage becomes important for successful spray inoculation. It is recommended that the inoculated plants or plant parts be maintained under humid conditions, either by placing plants in a humid chamber or by covering shoots with plastic bags overnight. High humidity greatly enhances the infection rate, although it is not essential for establishment of infection. The disease rating in spray inoculation is determined by counting the number of lesions formed on a leaf blade and also the size of the lesions.

7-10-2 Rubber-block Press Inoculation

The press inoculation method, a modification of spray inoculation, can be applied to mature leaves that are resistant to stomatal infection.[12] In this technique, a drop of bacterial suspension is placed underneath the surface of leaves. This part of the leaf blade is placed between two pieces of rubber stoppers and pressed with the thumb and little finger so the bacterial cells are forced to enter into the stomata.

7-10-3 Leaf Infiltration Technique

Because water can be injected into mesophyll of citrus leaves at the ratio of approximately 7 μl/cm^2, irrespective of cultivars, bacterial suspensions at the concentration of 10^3 cells/ml deliver seven bacterial cells/cm^2 of leaf lamina, allowing development of seven lesions if the microenvironment of the intercellular space permits the successful growth of the causal bacterium.[36] In this technique, a definite number of bacterial cells can be introduced into the intercellular space, so the number of lesions developed should be proportional either to the susceptibility of the tissues or to the virulence of the pathogen. The technique is best applicable to young leaves because of the easier injection into mesophyll and the faster development of lesions.

7-10-4 Wound Inoculation

This method can be universally used, regardless of differences in ages, organs, susceptibility, or growth stages of host plants. A sharp needle is dipped in a bacterial suspension at about 10^8 cells/ml and then is used to prick the epidemis. A bundle of several needles can also be used. Various modification can be applied (e. g., the bacterial suspension can be smeared with a soft brush, sprayed on a leaf surface that is previously injured, or dropped on the leaf surface before puncturing). A shallow prick reaching only to the mesophyll is adequate to establish infection. Large wounds, such as holes made through a leaf blade, sometimes result in necrosis of marginal tissue that is unfavorable for initial multiplication of the causal bacteria, so infection is unsuccessful. This is particularly true in early spring when plants are still dormant at a low temperature. More successful infection is achieved when the inoculated organs are covered overnight with a polyethylene bag. Disease rating is determined on the basis of the infection ratio and size of the lesions developed around punctures. In wound inoculation, the minimal concentration of bacterial cells is around 10^2 cells/ml, approximately 1000 times smaller than that required for stomatal infection.[12] This difference may be explained simply by the physical size of the wound as compared with the size of stomata.

7-11 EPIDEMIOLOGY

As Peltier and Frederick[46] pointed out, citrus canker is severe in regions where temperature and rainfall ascend and descend together during the year. Therefore, the disease occurs in severe form in seasons and/or areas characterized by warm and humid weather conditions.

7-11-1 Spring Canker

The term *spring canker* is used for the disease that develops in early spring before first flush as a result of enlargement of a holdover lesion or as a result of new infection. When rainstorms come late in autumn, new lesions often develop in late March to early April, before first flush on overwintered leaves. Spring canker may be caused by either latent infection or new infection occurring in spring because of unhealed injuries sustained during storms.[47] Spring canker is an important source of inoculum to the new spring shoots.

7-11-2 Dispersal of the Causal Bacterum by Rain

X. campestris pv. *citri* appears in rainwater running over the surfaces of lesions and splashing onto the new shoots. The concentration of bacterial cells is largely dependent on the age of the lesions. In new lesions, such as those developed from latent infection, bacterial density often reachers 10^7 to 10^8 cells/drop, and this concentration continues from about 15 minutes after the beginning of rainfall to its end.[12] These bacterial cells seem to be supplied from the dense bacterial resorvoir contained in the abundant EPS as well as those newly multiplied in lesions in rain.

7-11-3 Epidemic under Storm Conditions

Strong wind causes many injuries to leaves and twigs of citrus trees. Such injuries are particularly severe when the trees possess thorns. The nature of the wounds varies from easily visible large injuries to small, invisible ones, such as small scratches or removal of the cuticle edges extending over stomata. From the standpoint of citrus canker infection, differences in the size of injuries have little significance in terms of the infection potential of the bacterium (i. e., a single cell is enough to cause infection). Therefore, storms such as typhoons or hurricanes encourage outbreaks of citrus canker where active inoculum sources are available. Rainstorms also aid greatly in distant dispersal of the pathogen.[47] However, the distance over which effective infection occurs is no greater than that over which the bacterium itself can be dispersed in small raindrops or aerosols. Although dispersal is usually restricted in the crown of a tree under gentle rainfall, the bacterium may cause infection on citrus trees at the distance of a few rows downwind, but rarely more than that.

7-11-4 Susceptibility of Citrus Tree to Citrus Canker in Relation to Growth Stages

Leaves and twigs show great changes in susceptibility to citrus canker during shoot development. These changes originate mainly in the structure of stomata, or in cuticle edges that extend over the front cavity.[12] In young organs such as leaves, stems, and fruits, the front cavity of stomata has a wide opening because the thin cuticular layer of the epidermis is not enough to elongate the edges. As organs approach maturity and the tissues become harder, however, the cuticular layer of epidermis becomes thicker so the edges develop over the stoma, leaving a narrow opening between them (Fig. 7-7). The slit between the two edges is so narrow that surface tension prevents entrance of rainwater carrying the pathogen into the opening of the mature stoma. In contrast, water easily enters into young stomata with wide openings. Therefore, availability of such young stomata determines the susceptibility of leaves, stems, and fruits. In very young leaves just after emergence, most stomata are still in an immature stage with no openings, so only slight infection occurs. As the organs approach maturity, they become resistant to stomatal infection because of a rapid increase in the number of stomata with only a narrow slit. The greatest susceptibility to infection through stomata is found on leaves of 0.5 : 0.8 expansion (the ratio of leaf length at a given time to the length of the fully expanded leaf), whereas the greatest incidence of infection through wounds occurs on leaves with ratios of 0.8 : 0.9 or greater. In the early days of citrus canker research,

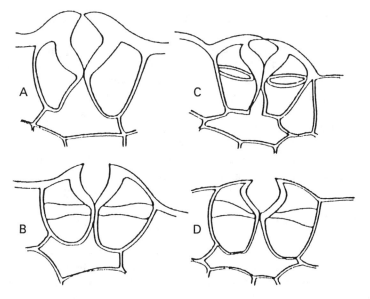

Figure 7-7 Stomata of citrus leaves. A) mature leaves of *C. natsudaidai;* B) young leaves of *C. natsudaidai;* C) mature leaves of *C. unshu;* D) young leaves of *C. unshu.* (From Goto, 1962).[12]

such structural differences among stomata were reported to be responsible for the differences in citrus canker resistance among cultivars,[48] and this idea has been referred to in many textbooks. However, all cultivars actually suffer heavy infection when inoculated at the proper growth stage, with almost no difference in the number of lesions per unit area. The true degree of resistance is shown by the size of lesions imposed by physiological, or active defense mechanisms. Fruits have the longest infection period, and stems the shortest, as a result of differences in the period of stomata development. Stomata with wide openings are continuously formed for several months on fruit. In stems, however, they disappear soon after longitudinal shoot elongation ceases. Stomatal infection of stems, therefore, can take place only for a few weeks.

The comparative tolerance of mesophyll to citrus canker seems to change somewhat during the course of growing season (i. e., tolerance becomes somewhat lower in early spring before first flush). However, such physiological tolerance may be only a part of the resistance of mature leaves to stomatal infection, because infection does occur when the causal bacterium is forced to enter the mesophyll by the rubber-block press method or through wounds.

7-11-5 Curing of Wounds

Wounds of leaves and stems heal by forming several layers of wound cork underneath the injured cells. The length of time required for healing varies widely, depending on environmental conditions, especially temperature. It takes 23 days at 10°C, 10 days at 15°C, 5 days at 20°C, and a few days at 25°C or higher.[49] Therefore, wounds sustained early in spring or late in autumn take longer to heal and therefore expose the injured tissues to infection for a longer period of time.

7-11-6 Effect of Weather on Entrance of the Pathogen

In artificial inoculation, the infection ratio increases significantly when the inoculated plant or plant part is kept in humid conditions overnight. This is particularly true in wound inoculation, where desiccation of the wound surface immediately after inoculation seems to provide an unfavorable environment for initial establishment of the pathogen. After settlement of the bacterium in wound tissue, however, humidity has no significant effect on subsequent development of lesions. In contrast, stomatal infection appears to take place as soon as the bacterial suspension makes contact with stomata, because a fairly high level of infection occurs during sunny summer weather when bacterial suspension sprayed on leaves evaporates in several minutes and the leaves receive no more moisture until after the disease has developed.[12]

7-11-7 Effect of Weather on Disease Development

The latent period of citrus canker on all young tissues is usually four to five days. As the organs approach maturity, the latent period may be extended by a few days. Generally speaking, the latent period is affected by temperature but not by humidity after infection is once established.[25] After the first symptoms or callus pustules develop, prolonged wetting by rain greatly enhances quick enlargement of lesions, the development of typical white calli being repressed.

7-11-8 Effect of Fertilization on Disease Development

Three major components—nitrogen, phosphate, and potassium—should be balanced in fertilization. An excess of nitrogen usually accelerates canker development, as occurs in most plant diseases. Under imbalanced conditions of excess nitrogen, citrus trees flush more shoots, allowing an increased number of holdover cankers. Leaves under such nutritional conditions tend to be larger and more tender, so lesions become larger than those formed on citrus trees fertilized properly. The longevity of the causal bacterium may be increased in such lesions. The physiological and biochemical details of such disease enhancement under conditions of excess nitrogen, however, remain to be determined.

7-12 RESISTANCE

7-12-1 Resistance to Aggressiveness or Structural Resistance

This resistance may be explained by the density of stomata on the leaf blade. McLean and Lee[50] have reported that the greater the number and size of stomata per unit area, the greater the susceptibility. McLean and Lee[50] calculated the number of stomata of susceptible grapefruit to be $50/mm^2$ and 20.0×19.0 μm in diameter, whereas those of resistant mandarin were $38/mm^2$ and 17.5×10.0 μm. However, there is actually no significant difference in the number of canker lesions between grapefruit and mandarin when infection occurred on leaves at the most susceptible growth stage. The essential differences are found in the size of lesions, which reflects the degree of resistance to disease development.

7-12-2 Resistance to Disease Development or Active Defense Reaction

In the processes of canker development, physiological resistance can be expressed as the degree to which growth of the causal bacterium is depressed, resulting in a lesser extent of callus formation. The mechanisms underlying such differences are not yet clear. However, it is possible that some antimicrobial substance present in healthy tissues, as well as those produced after infection, are involved in disease resistance. Citrus leaves can be heavily infected, and yet the intercellular fluids will permit growth of the causal bacterium even if the leaves contain high levels of antimicrobial substances. This implies that such substances are retained in cells and rarely leaked into the intercellular spaces so their cytotoxic effects cannot be exerted during the early stage of infection. When the hypertrophic cells are destroyed in young canker lesions, the antimicrobial substances may be released from cells, therby exerting their toxic effects against the causal bacterium and *E. herbicola* as well. However, the triggering factor for host cell destruction has not been clarified, and the nature of these antimicrobial substances remains to be studied. Some components may be the derivatives of cinnamyl alcohol, one of them being identified as citrusnin-A[51] (Fig. 7-8). The postinfectional antimicrobial substances are called phytoalexins and may vary in different *Citrus* spp.

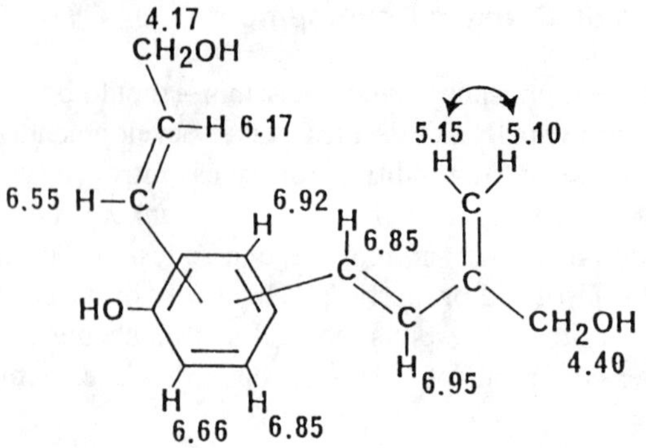

Figure 7-8   The substitution pattern of citrusnin-A (I) and the structure of its C_4 substituent. The numerals indicate the ^{1}H NMR chemical shift in CD_3OD (TMS as internal standard.)

The differences of *Citrus* spp. in resistance to citrus canker may be based on differences in defense reactions involving such phytoalexins.

7-13 DISEASE MANAGEMENT

7-13-1 Regulatory Control

The eradication of citrus canker in the Gulf states of the United States, and the subsequent restrictions on importation of citrus fruits from regions where citrus canker is endemic, comprise a typical example of regulatory control of plant diseases. Although North America was orginally free of citrus canker, it is believed that the disease was introduced into Texas in 1910 on infected trifoliate orange seedlings shipped from Japan, and became widespread in Florida by 1914. An eradication campaign was then conducted, at a cost of $6 million, requiring the destruction of 257,745 orchard trees and 3,093,110 nursery plants.[52] In 1947, the United States was declared canker free. A similar eradication campaign has been conducted in Sao Paulo, Brazil, New Zealand, Australia, and South Africa, with successful results.

Importation of citrus fruits into the United States from Japan was prohibited for many years on the basis of this historical background. In 1968, however, importation of Unshu orange was permitted in the northwestern states of Alaska, Idaho, Montana, Oregon, and Washington under certain safeguards to prevent introduction of the citrus canker organism. The area allowed for importation of Unshu orange was extended to 38 non-citrus-growing states in 1987, under the same conditions. The safeguards include the following:

1. Unshu orange to be imported to the United States shall be grown and packed in isolated canker-free export areas, where it shall be isolated from other citrus groves by a buffer zone (400 m wide) which is itself free of all non-Unshu citrus.

2. Inspection of the Unshu orange shall be performed jointly by plant pathologists of Japan and the United States in the groves prior to and during harvest, and in the packing houses during packing operations.

3. Before packing, such oranges shall be given a surface sterilization with bactericidal dip.

4. The identity of the fruit shall be maintained in several ways, including accompanying certificate from the Japanese Plant Protection Service.

5. Just prior to export, fruit from each shipment shall be tested by recognized bacteriophage methods to ensure freedom from the citrus canker organism.

No A-form canker originating in Unshu fruits from Japan has occurred in the United State since 1968, indicating that regulatory control has worked satisfactorily.

7-13-2 Cultural Control

In contrast to the communicable diseases of humans and animals, it is striking that effective chemicals are rarely available for controlling bacterial plant diseases, including citrus canker. Although copper compounds are widely applied for chemical control of citrus canker these days, their effectiveness is not sufficient to suppress the disease when it occurs in epidemic form. Therefore, cultural control methods, including field sanitation, are still the most important approach to the management of citrus canker.

Use of canker-free nursery stock.   Where citrus canker is endemic, it is most likely to occur in nurseries where many young plants are raised with flourishing abundant shoots throughout almost the entire growing season. Therefore, unless proper control procedures are applied, nursery plants usually become infected, carrying various numbers of canker lesions when leaves are carefully inspected one by one. Small lesions may sometimes be found at the basal portion of the leaf blade at the junction to the leaf petiole, or on stems. Such small numbers of lesions are a sufficient inoculum source to trigger an outbreak of citrus canker after the plants are planted in the field, because young plants flush many tender shoots. Thus, raising canker-free nursery plants is the first essential step in citrus canker management.

Windbreaks.   Because wind greatly facilitates wound infection, the establishment of windbreaks around citrus orchards is effective in preventing infection. The protective effect of a windbreak covers a distance six times greater than its height. With windbreaks, wind inside citrus groves should preferably be maintained below 6 m/sec.

Pruning.   Angular shoots holding canker lesions should be pruned in winter as long as pruning does not inhibit the growth of citrus trees. Removal of holdover cankers by pruning is useful to reduce the inoculum density within a grove.

Control of leaf miner (*Phyllocnistis citrella*).   Periodic spraying of insecticides, such as nicotine sulfate (Black Leaf) or isoxathion (Karphos), is effective for control of insect damage to newly unfolded leaves, as such damage facilitates citrus canker infection.

7-13-3 Forecasting

The outbreak of citrus canker may be forecasted on the basis of the following factors: (a) number of overwintered lesions on angular shoots: (b) temperature, frequency and amount of precipitation, and number of windy days in October and November, which determines the severity of latent infection, and also in March, which facilitates the development of spring canker; and (c) temperature in winter, which facilitates the population of latently infected bacteria. In addition, concentrations of the citrus canker organism that appear in rainwater running on the leaf surface may be monitored in spring. On this basis, an epidemic can be predicted one to two months in advance so citrus growers can be warned and take subsequent steps toward effective chemical control.

7-13-4 Chemical Control

Chemical sprays are not efficient enough to control citrus canker when it occurs in epidemic form. Therefore, prevention of primary infection on spring shoots must be emphasized in chemical control. The effective chemicals include Bordeaux mixture and other copper compounds, with or without calcium carbonate, for reducing the plant toxicity of the copper ion and antibiotics such as streptomycin and agrimycin.[47] In citrus-growing areas of Japan, the following schedules are recommended for chemical sprays against citrus canker:

1. An application of Bordeaux mixture (4-4 formula; 4 g copper sulfate and 4 g lime in 1 liter water) or copper compounds (Cu content 32 to 50%) with $CaCo_3$ in late March just before the first flush.

2. In May, applications of copper compounds with $CaCo_3$ or antibiotics every week. Application of Bordeaux mixture at this growth stage may cause chemical damage to the surface of young leaves and fruits.

3. In addition to the foregoing, an application of Bordeaux mixture (formula 2-6) in June and August, or whenever strong winds and heavy rain are predicted. To protect leaves and fruits from chemical damage, copper sulfate is reduced to one half and lime is increased to 1.5 times the formula for overwintered leaves.

The intervals for spraying of copper compounds may also be determined by the cumulative amount of precipitation, in which spraying is scheduled every 200 to 250 mm of precipitation. Spraying of antibiotics can be scheduled every 7 to 10 days in May. Although three to seven sprays are necessary for effective control of citrus canker, depending on the weather conditions, the most important sprays are those to the first flush in May and to overwintered shoots in March.

Such frequent spraying impose a great burden on citrus growers. To reduce spraying frequency, all the aforementioned control methods must be combined functionally into an integrated control system, with emphasis on (a) field sanitation, (b) building of windbreaks, (c) establishment of a disease forecasting or prediction system, and (d) spraying of copper compounds at the optimal time, based on the information from the prediction system.

7-13-5 Biological Control

Studies on the biological control of citrus canker are still in a preliminary stage. Ota[53] isolated a strain of *Pseudomonas syringae* from a haloed lesion. This bacterium was antagonistic *in vitro* to the citrus canker organism and also prevented enlargement of canker lesions as well as subsequent defoliation of infected leaves. Such effects *in vivo* were explained by the ability of the bacterium to stimulate production of phytoalexin "citrusnins" in the infected tissues. However, the study was not continued mainly because of the possible plant pathogenicity of the antagonist.

As mentioned earlier, when canker lesions are wetted by rain, *Erwinia herbicola* invades the callus tissues and multiplies, quickly collapsing hypertrophic cells. At this stage, isolation of the causal bacteria by conventional methods becomes very difficult because of the overwhelming growth of *E. herbicola* on agar plates. Studies have therefore been conducted on the interaction between the citrus canker organism and *E. herbicola* for seeking new possibilities of biological control of citrus canker with the latter organism.[28,54] Both *in vitro* and *in vivo*, *E. herbicola* grew more rapidly than *X. campestris* pv. *citri*, causing eventual extinction of the pathogen *in vitro* and quick population decline of the pathogen *in vivo*. However, *E. herbicola* grows only in the area where hypertrophic host cells are established, but never in the front boundaries at which pv. *citri* attacks healthy tissues, inducing development of hypertrophic cells. In the initial stage of infection, host cells may incite the early induced selective protection (ESP),[32] preventing growth of bacteria other than the citrus canker organism. The function of ESP based on incompatibility as well as the limited number of hypertrophic cells may prevent proliferation of *E. herbicola* in the peripheral region of canker lesions. Therefore, *E. herbicola* may become a useful agent for biological control of citrus canker if compatibility with citrus tissues can be furnished to the bacterium in some way by genetic manipulation.

7-14 FUTURE PROSPECTS

7-14-1 Citrus Canker Organism

The pathogenesis of plant pathogenic bacteria can be simply illustrated, as in Fig. 7-9. Incompatibility (I) may be derived from the interaction between bacterial surface components, such as lipopolysaccharides or outer membrane proteins, and host cell wall, leading to the resistance response. Growth capacity (G) is the ability of bacteria to initiate multiplication in a compatible combination. Virulence factors (V) are substances that are produced by the bacterium thus multiplied. When all of these processes are performed successfully, the symptoms commence to develop. On the basis of the foregoing assumption of plant pathogenesis, the A and B form strains of the citrus canker organism may be different, primarily in their degree of incompatibility and in

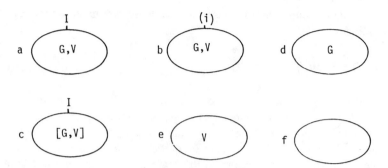

Figure 7-9 Schematic illustration of plant pathogenesis. a) Common pathogens e.g., *Pseudomonas syringae* pv. *syringae*. b) Pathogens with wide host range e.g., soft rot *Erwinia*. c) Bacteria with the capacity to induce tobacco hypersensitivity reaction, but unknown plant pathogenicity e.g., epiphytic strains of *Pseudomonas syringae* and saprophytic xanthomonads, d) Endophytic bacteria i.e., those may grow in plant tissues without inducing harmful effects. e) Bacteria with the capacity to produce virulence factor only e.g., *Escherichia coli* cloned with pectic acid lyase. f) Saprophytic bacteria.

their capacity to grow in citrus tissues, because the B form strain has a much narrower host range and is less virulent than the A form strain.

The B form strains displaying different growth capacities on agar media may have originated from nutritional adaptation. The antigenic differences between them, as demonstrated by the monoclonal antibody technique, imply possible differences in the outer membrane protein F (Omp F), which is responsible for the transfer of low molecules. Analysis of the outer membrane composition should be conducted in the future in relation to plant pathogenesis, not only in the citrus canker organism but also in plant pathogenic bacteria in general.[57]

The Florida strain of Xanthomonas leaf spot obviously lacks the capacity to produce auxin, which is the virulence factor of citrus canker organism A and B. Therefore, it can be hypothesized that the Florida strain achieved only the capacity to grow in citrus tissues but not the ability to produce auxin in the evolutionary steps, or that the A form strains lost the latter ability. It is possible that mobilizing genetic determinants, such as insertion sequences (IS) or transposons (Tn), are involved in these traits. This may be proven by genetic manipulation if this is the case.

7-14-2 Host-Bacteria Interaction

It is interesting to note that most strains of B form canker have a limited capacity to produce extracellular polysaccharides (EPS). Abundant EPS in the A form strain may prevent the attachment of bacterial cells to the host cell wall and subsequent incompatibility response. If this is the case, what is the difference between this attachment and the adhesion of bacterial cells to the host cell wall that was observed by Takahashi and Doke? Ethylene production of citrus leaf tissues at significant levels soon after EPS injection implies that host cells suffer considerable stress through the interaction between EPS and the host cell wall. The study of these sequences in detail may contribute to elucidation of the pathogenesis of *X. campestris* pv. *citri*. In some plant pathogenic bacteria, certain compounds of plant origin are essential for induction of pathogenic functions in the early stage of infection. For example, flavons and flavonoids induce the capacity of *Rhizobium* spp. to produce a substance necessary for curling of root hairs, which is the primary step of *Rhizobium* infection. Phenol compounds such as acetosyringone induce the virulence region of the Ti plasmid of *Agrobacterium tumefaciens*, to initiate processing of T DNA and its transfer to plant cells. Pectin lyase of *Erwinia chrysanthemi* is believed to be induced by some plant components through SOS responses. These phenomena might be common in plant pathogenic bacteria, including the citrus canker organism. The citrus canker organism may be affected by some compounds natural to the citrus plant that induce the capacity to overcome the resistance response of host cells, to initiate the metabolism of cell proliferation, and subsequently to synthesize auxin. These sequences in the primary steps of citrus canker infection must be substantiated in the future by experimental evidence.

Another point of interest is why the citrus canker organism primarily causes hypertrophy, with only a limited extent of hyperplasia. In many gall-forming bacteria, such as *Agrobacterium tumefaciens* and *Pseudomonas syringae* pv. *savastanoi*, production of cytokinins has been confirmed in addition to auxin, and these two compounds coordinatively induce gall tissues that are composed of both hyperplasic and hypertrophic cells. Although production of cytokinins was not tested, the histopathogenesis of *X. campestris* pv. *citri* strongly suggests that it produces only auxin but not cytokinins. The calli in citrus canker may thus cease their development even under aseptic conditions and soon deteriorate.

No accurate explanations are available for the post-infectional resistance exhibited by the resistant citrus species or cultivars. Although the complex of antibacterial substances produced in the infected tissues may be responsible for decreased multiplication of the causal bacterium, further evidence must be obtained to support this hypothesis.

7-14-3 Ecological Behavior of the Citrus Canker Organism

Citrus canker is characterized by the sporadic occurrence of epidemics. Epidemics used to occur every decade or so, and the disease would rarely occur with the same severity in the following year. Outbreaks occur in citrus groves where canker lesions have rarely been detected for several years. Therefore, the sources of inoculum have been investigated for a number of years, and a variety of habitats have been discovered. The role of autumn weather in inoculum buildup has also been clarified. However, these results are still not sufficient to explain the sudden occurrence of citrus canker epidemics. It is possible that we have missed some habitats of the citrus canker organism in which it may survive stably at low population, as in the case of the bacterium living deep in bark tissues. Our techniques may not be efficient enough to detect the pathogen living in a complex with saprophytic microorganisms. In this respect, it is interesting to note that *X. campestris* pv. *citri* usually coexists in bark tissues in association with certain kinds of slow-growing, white bacteria (possibly pseudomonads), but rarely exists by itself.[39] The effect of such association of the citrus canker organism with other microorganisms on long-term survival in nature must be studied in detail. The recent occurrence of Xanthomonas leaf spot in Florida after severe cold waves suggests that frost damage to citrus trees may directly or indirectly play an important role in release of the bacterium that is otherwise concealed in certain microhabitats without taking the parasitic form. The development of efficient detection techniques might disclose entirely new ecological behavior of *X. campestris* pv. *citri* and its related pathogens.

7-14-4 Control of Citrus Canker

The development of effective chemicals for control of citrus canker has been long claimed by citrus growers and pathologists. However, these efforts have actually been unsuccessful, as has been the case with other plant bacterial diseases in general. Most chemicals with great effectiveness *in vitro* do not necessarily show satisfactory effects. The gaps found between effectiveness *in vitro* and *in situ* may stem in part from the mode of bacterial infection. Under rainy conditions, some bacterial cells may achieve direct access to the front cavity of stomata or to wounds without being exposed to the protective chemicals left on the leaf surface. This direct ingress, even if the number of cells is very limited, is essentially different from most fungal pathogens, in that spores germinate and germination tubes elongate on the leaf surface and come into contact with chemicals. Therefore, for development of effective bactericides, emphasis must be placed on the effectiveness of chemicals reaching at least to the depth of the stomatal cavity.

For biological control, it seems difficult to find antagonistic bacteria that reside stably on the smooth surfaces of mature citrus leaves. Although some fungi, such as *Fusarium* spp., easily take on a resident form on the surfaces of citrus leaves, the density of the mycelium is usually not very high. Recent findings have demonstrated that the plants usually carry the internal resident microbes (endophyte) in vascular systems or even in parenchymatous tissues. These microbes seem to have a great capacity to associate with plant tissues (compatibility) without inciting any harmful effects, although the plant-microbe interactions have not as yet been clarified in detail. There is a substantial possibility that an antagonistic microbe may be found among these endophytes. These microbes may be more useful than those found in soil because of the unique traits of the former as stable inhabitants within the plant tissues. Even if these microbes do not have the capacity to prevent growth of the citrus canker organism, they may be useful gene sources for plant compatibility in breeding microbial agents to be used for biological control of citrus canker.

7-15 REFERENCES

1. Sinclair, J. B., Eradication of citrus canker from Louisiana, *Plant Dis. Rept.*, 52, 667, 1968.
2. Lee, H. A., Further data on the susceptibility of Rutaceous plants to citrus canker, *J. Agr. Res.*, 15, 661, 1918.
3. Fawcett, H. H. and Jenkins, A. E., Records of citrus canker from herbarium specimens of the genus *Citrus* in England and the United States, *Phytopathology*, 23, 820, 1932.
4. Goto, M. and Yaguchi, Y., Relationship between defoliation and disease severity in citrus canker, *Ann. Phytopath. Soc. Japan*, 45, 689, 1979.

5. Koizumi, M., Behavior of *Xanthomonas citri* (Hasse) Dowson and histological changes of diseased tissues in the process of lesion extension, *Ann. Phytopath. Soc. Japan*, 43, 129, 1977 (in Japanese with English summary).

6. Hasse, C. H., *Pseudomonas citri*, the cause of citrus canker, *J. Agr. Res.*, 4, 97, 1915.

7. Stall, R. E. and Seymour, C. P., Canker, a threat to citrus in the Gulf-coast states, *Plant Disease*, 67, 581, 1983.

8. Namekata, T., Estudos comparativos entre *Xanthomonas citri* (Hasse) Dow., agente causal do "cancro citrico" e *Xanthomonas citri* (Hasse) Dow., N. F. Sp. *Aurantifolia*, agente causal da "cancrose do limoeiro galego." Tese apresenta da a Escola Superior de Agricultura "Luiz de Queiroz." Para a Obencao do Titulo De Doutor Em Agromonomia, 1971 (in Portuguese with English summary).

9. Rodoriguez, S. G., Stapleton, J. J. and Civerolo, E. L., *Xanthomonas campestris* involved in Mexican lime bacteriosis in Colima, Mexico. *Plant pathogenic bacteria*, Proc. 6th Int'l Conf. Plant Path. Bact., Maryland, pp. 658–662, 1985.

10. Schoulties, C. L., Civerolo, E. L., Miller, J. W., Stall, R. E., Krass, C. J., Poe, S. R. and DuCharme, E. P., Citrus canker in Florida, *Plant Disease* 71, 388, 1987.

11. Gabriel, D. W., Kingsley, M. T., Hunter, J. E. and Gottwald, T. Reinstatement of *Xanthomonas citri* (ex. Hasse) and *X. phaseoli* (ex. Smith) to species and reclassification of all *X. campestris* pv. *citri* strains, *Int. J. Syst. Bact.*, 39, 14, 1989.

12. Goto, M., Studies on citrus canker, I., *Bull. Fac. Agr. Shizuoka Univ.*, 12, 3, 1962 (in Japanese with English summary).

13. Goto, M., Takahashi, T. and Messina, M. A., A comparative study of the strains of *Xanthomonas campestris* pv. *citri* isolated from citrus canker in Japan and cancrosis B in Argentina, *Ann. Phytopath. Soc. Japan,* 46, 329, 1980.

14. Obata, T., Distribution of *Xanthomonas citri* strains in relation to the sensitivity to phages Cp1 and Cp2, *Ann. Phytopath. Soc. Japan*, 40, 6, 1974.

15. Wakimoto, S., Some characteristics of citrus canker bacterium, *Xanthomonas citri* (Hasse) Dowson, and the related phages isolated from Japan, *Ann. Phytopath. Soc. Japan,* 33, 301, 1967.

16. Okabe, N., Studies on the lysogenic strains of *Xanthomonas citri* (Hasse) Dowson. "Papers in Comm. Dr. T. Matsumoto." Coll. Agr. Nat. Taiwan Univ. Spec. Publ. 61-73, 1961 (in Japanese with English summary).

17. Wu, W. C., Phage-induced alteration of colony type in *Xanthomonas citri.*, *Ann. Phytopath. Soc. Japan*, 38, 146, 1972.

18. Goto, M. and Starr, M. P., Lysogenization of *Xanthomonas phaseoli* and *X. begoniae* by temperate *X. citri* bacteriophages: effects on virulence, phage susceptibility, and other biological properties, *Ann. Phytopath. Soc. Japan*, 38, 267, 1972.

19. Dai, H., Tsay, S. M., Kuo, T. T., Lin, Y. H. and Wu, W. C., Neolyzogenization of *Xanthomonas campestris* pv. *citri* infected with filamentous phage Cf16, *Virology* 156, 313, 1985.

20. Civerolo, E. L., Indigenous plasmids in *Xanthomonas campestris* pv. *citri, Phytopathology*, 75, 524, 1985.

21. Tu, J., Wang, H.-R., Chang, S.-F., Gharang, Y.-C., Lurz, R., Dobrinski, B. and Wu, W.-C., Transposable elements of *Xanthomonas campestris* pv. *citri* originating from indigenous plasmids, *Mol. Gen. Genet.*, 217, 505, 1989.

22. Hartung, J. S. and Civerolo, E. L., Genomic fingerprints of *Xanthomonas campestris* pv. *citri* strains from Asia, South America and Florida, *Phytopathology*, 77, 282, 1987.

23. Alverez, A. M., Benedict, A. A., Mizumoto, C. Y. and Civerolo, E. L., Mexican lime bacteriosis examined with monoclonal antibodies. *Plant pathogenic bacteria*, Proc. 6th Int'l Cong. Plant Path. Bact., Maryland, pp. 847–852, 1985.

24. Peltier, G. L., Relative susceptibility to citrus canker of different species and hybrid of the genus *Citrus*, including the wild relatives, *J. Agr. Res.*, 19, 339, 1920.

25. Peltier, G. L., Influence of temperature and humidity on the growth of *Pseudomonas citri* and its host plants and an infection and development of the disease, *J. Agr. Res.*, 20, 447, 1920.

26. Kawakami, K., On citrus canker, *Papers in Plant Pathology*, pp. 1–114, 1921 (in Japanese).

27. Fulton, H. R. and Bowman, J. J., Infection of fruit of citrus by *Pseudomonas citri.*, *J. Agr. Res.*, 39, 403, 1929.

28. Goto, M., The role of extracellular polysaccharides of *Xanthomonas campestris* pv. *citri* in dissemination and infection: A review, Abstracts of Fallen Leaf Conference on the Genus *Xanthomonas*, September 20–23, p. 15, 1985.

29. Takahashi, T. and Doke, N., A role of extracellular polysaccharides of *Xanthomonas campestris* pv. *citri* in bacterial adhesion to citrus tissues in a preinfectious stage, *Ann. Phytopath. Soc. Japan*, 50, 565, 1984.

30. Takahashi, T. and Doke, N., Purification and partial characterization of agglutinins in citrus leaves against extracellular polysaccharides of *Xanthomonas campestris* pv. *citri.*, *Physiol. Plant Path.*, 27, 1, 1985.

31. Goto, M. and Hyodo, H., Role of extracellular polysaccharides of *Xanthomonas campestris* pv. *citri* in the early stage of infection, *Ann. Phytopath. Soc. Japan*, 51, 22, 1985.

32. Burgyan, J. and Klement, Z., Early induced selective inhibition of incompatible bacteria in tobacco plants, *Phytopath. Medit.*, 18, 153, 1979.

33. Goto, M., Yaguchi, Y. and Hyodo, H., Ethylene production in citrus leaves infected with *Xanthomonas citri* and its relation to defoliation, *Physiol. Plant Path.*, 16, 343, 1979.

34. Goto, M., Takemura, I. and Yamanaka, K., Leekage of electrolytes and amino acids from susceptible and resistant citrus leaf tissues infected by *Xanthomonas citri.*, *Ann. Phytopath. Soc. Japan*, 45, 625, 1979.

35. Goto, M. and Yamanaka, K., Growth inhibition of *Xanthomonas campestris* pv. *citri* and its reversal by amino acids found in the intercellular fluids of citrus leaves, *Ann. Phytopath. Soc. Japan*, 47, 618, 1981.

36. Goto, M., Serizawa, S. and Morita, M., Studies on citrus canker II. Leaf infiltration technique for detection of *Xanthomonas citri* (Hasse) Dowson, with special reference to comparison with phage method, *Bull. Fac. Agr. Shizuoka Univ.*, 20, 1, 1971 (in Japanese with English summary).

37. Goto, M., Ohta, K. and Okabe, N., Studies on saprophytic survival of *Xanthomonas citri* (Hasse) Dowson. 1. Detection of the bacterium from a grass (*Zoysia japonica*), *Ann. Phytopath. Soc. Japan*, 41, 9, 1975 (in Japanese with English summary).

38. Goto, M., Ohta, K. and Okabe, N., Studies on saprophytic survival of *Xanthomonas citri* (Hasse) Dowson. 2. Longevity and survival density of the bacterium on artificially in-

fested weeds, plant residues and soils, *Ann. Phytopathol. Soc. Japan*, 41, 141, 1975 (in Japanese with English summary).

39. Goto, M., Survival of *Xanthomonas citri* in the bark tissues of citrus trees, *Can. J. Bot.*, 50, 2629, 1972.

40. Peltier, G. L. and Frederich, W. J., Further studies on the overwintering of *Pseudomonas citri.*, *J. Agr. Res.*, 32, 335, 1926.

41. Fulton, H. R., Decline of *Pseudomonas citri* in the soil, *J. Agr. Res.*, 19, 207, 1920.

42. Lee, H. A., Behavior of the citrus canker organism in the soil, *J. Agr. Res.*, 19, 189, 1920.

43. Hino, I., Studies on soil protozoas. 3. Extinction of plant pathogenic bacteria in soil and water, *Nogaku-Kaiho*, 289, 528, 1926 (in Japanese).

44. Hino, I., Fate of the citrus canker organism in soil and water, *Kankitsu-Kenkyu*, 4, 167, 1931 (in Japanese).

45. Goto, M., Toyoshima, A. and Tanaka, S., Studies on saprophytic survival of *Xanthomonas citri* (Hasse) Dowson. 3. Inoculum density of the bacterium surviving in the saprophytic form, *Ann. Phytopath. Soc. Japan*, 44, 197, 1978 (in Japanese).

46. Peltier, G. L. and Frederich, W. J., Effects of weather on the world distribution and prevalence of citrus canker and citrus scab, *J. Agr. Res.*, 32, 147, 1926.

47. Serizawa, S., Control of citrus canker with copper compounds, *Shokubutsu Boeki (Plant Prot.)*, 30, 280, 1976 (in Japanese).

48. McLean, F. T., A study of the structure of the stomata of two species of citrus in relation to citrus canker, *Bull. Torrey Bot. Club*, 48, 101, 1921.

49. Koizumi, M., Resistance of citrus plants to citrus canker, *Shokubutsu Boeki (Plant Prot.)*, 32, 207, 1978 (in Japanese with English summary).

50. McLean, F. T. and Lee, H. A., The resistance to citrus canker of *Citrus noblis* and a suggestion as to the production of resistant varieties in other citrus species, *Phytopathology*, 11, 109, 1921.

51. Watanabe, K., Miyakado, M., Ohno, N., Ota, T. and Nonaka, F., Citrusnin A—a new antibacterial substance from leaves of *Citrus natsudaidai.*, *J. Pesticide Sci.*, 10, 137, 1985.

52. Knorr, L. C., Citrus. In *Plant health and quarantine in international transfer of genetic resources* (H. B. Hewitt and L. Chiarappa, eds.), ch. 9, CRC Press Inc., Florida, 1977.

53. Ota, T., Interaction *in vitro* and *in vivo* between *Xanthomonas campestris* pv. *citri* and antagonistic *Pseudomonas* sp., *Ann. Phytopath. Soc. Japan*, 49, 308, 1983.

54. Goto, M., Tadauchi, Y. and Okabe, N., Interaction between *Xanthomonas citri* and *Erwinia herbicola in vitro* and *in vivo.*, *Ann. Phytopath. Soc. Japan*, 45, 618, 1979.

55. Zubrzycki, H.M. and de Zubrzycki, A.D., Resistance to *Xanthomonas campestris* pv. *citri* (Hasse) Dowson in oranges (*Citrus sinensis* (L.) Osbeck). Proc. Int. Soc. Citriculture, pp. 405-409, 1981.

56. Young, J. M., Bradbury, J. F., Garden, L., Gvozdyak, R. I., Stead, D. E., Takikawa, Y. and Vidaver, A. K., Comment on the reinstatement of *Xanthomonas citri* (ex Hasse 1915) Gabriel et al., 1989 and *X. phaseoli* (ex Smith 1897) Gabriel et al., 1989: Indication of the need for minimal standards for the Genus *Xanthomonas, Int. J. Syst. Bact.* 41, 172, 1991.

57. Uesugi, C. H., Tsu chiya, K., Tsuno, K., Matsuyama, N. and Wakimoto, S., Sds-poly-acrylamide gel electrophoretic profiles of membrane proteins associated with host of origin in *Erwinia chrysanthimi* strains, *Ann. Phytoath. Soc. Japan,* 56, 597, 1990.

8

CITRUS BLIGHT

STANLEY NEMEC

U.S. Department of Agriculture, ARS
2120 Camden Road, Orlando, Florida

D. L. MYHRE

Soil Science Department, University of Florida
Gainesville, Florida

8-1 INTRODUCTION

Blight is recognized as one of the most important diseases of citrus in Florida and has been present in the state since it was first described in the late 1800s.[1] It is a disease which causes a number of nonspecific visual above-ground symptoms in its hosts, and for that reason it is difficult to distinguish from other decline-type diseases. The symptoms collectively are described as water stress symptoms and include leaf wilt and cupping, stem dieback, and a hardened appearance of leaves. Water sprouts develop on the lower trunk and, at times, zinc deficiency patterns occur in leaves. The disease occurs on all known rootstocks,[1,2] but rough lemon (*Citrus limon* Burm. f.) probably has the highest level of susceptibility.[2] Extensive plantings on rough lemon in Florida during the past 40 to 50 years have accounted for a large proportion of the citrus acreage in Florida. These plantings have been a reason why blight ranks as the most common citrus disease in the state.

Numerous investigations have been made to determine the nature of the disease. The first, by Swingle and Webber[3] of the U.S. Department of Agriculture, began in the 1890s and has been followed by a wide range of investigations of the disease. The most intensive efforts to resolve the etiology of blight began in the 1960s and by 1981 had resulted in over 135 publications and reports on blight.[4] In spite of the efforts of many scientists, technicians, and members of the private sector, no single cause and effect has been determined for blight. Among the literature on blight are several reviews. The history of blight and its early occurrence in Florida is well documented in the review by

Smith.[5] Additional summaries on blight appeared in literature by Burnett et al.,[6] Cohen,[7] and Reitz.[8]

A recent review[9] gave considerable attention to symptoms, scion, and rootstock relationships; diagnosis and detection; and physiological aspects. This chapter emphasizes potential causal factors and the relationship of soils and nutrition to the disease. Only limited emphasis is given to other aspects of blight highlighted in previous reviews.

8-2 DISTRIBUTION AND ECONOMIC IMPORTANCE

Blight is known to occur in Florida as well as Louisiana[10] and possibly Texas (H. Wutscher, personal communication). Blight has not been detected in California and Arizona, both of which are semiarid, citrus-producing states. In Florida, blight has appeared in almost all citrus-producing areas of the state. Prior to the December 1983 and January 1985 freezes that killed most of the citrus grown north of State Highway 50 Florida, blight was confirmed in plantings around Citra, the northernmost citrus-producing area. Blight occurs as far south as Miami and Immokalee but has not been confirmed in plantings in the Homestead area.[9] In new groves, the earliest occurrence of blight is in groves with shallow soils on the east and west coasts in Florida.

Development of blight in groves depends on a variety of factors. The site characteristics of a grove have a strong influence on the development of the disease in trees. Where groves are planted on soil with a subsoil pan within 0.3 to 1.5 m of the soil surface, blight frequently appears earlier and more severely in the shallowest areas.[1,11] Within groves, other patterns of blight development occur which are nonrandom. In some central Florida groves, the incidence of blight is most intense along highways that border the groves. The term *roadside decline* was applied to these diseased trees and is another synonym for blight.[12] In the flatwoods, where small pockets of high organic matter occur in groves, the incidence of the disease is significantly less in the organic portion than in the surrounding sandy soil.[13] More commonly, on the ridge where citrus is grown in deep, sandy soils, the development of blight is nonrandom after its initial random appearance in groves. Often a clustering of declined trees develop around a center of diseased trees.[14,15] However, the pattern of this disease development is not predictable. A healthy tree adjacent to one developing symptoms may remain healthy for many years.[1]

Losses caused by blight have been difficult to determine accurately. Trees can decline as early as three to four years of age, before they begin to fruit, or they may not develop symptoms until they are 10 to 50 years old. Fruiting trees with symptoms are seldom picked because the fruit size is too small to be commercially acceptable. Tree loss has been as high as 86% in one grove over a four-year period.[16] In any one year, loss may be less than 1% to as much as 22% in a grove.[14] In 1974, the estimate of tree loss due to blight was 50,000 acres,[17] and in 1975 that estimate was revised to 75,000 acres.[18]

8-3 SYMPTOMS

Visible above-ground blight symptoms appear related to water-stress-like reactions in the tree (Fig.8-1). Thus, blight symptoms are nonspecific and can therefore resemble those

Figure 8-1   Tree on the right with moderate symptoms of blight. The tree on the left is healthy.

occurring on trees with other diseases. Zinc deficiency symptoms in leaves, which are one of the earliest blight symptoms, are transient and may not appear at all on some trees with early blight symptoms. If zinc deficiency does occur, it usually begins in young leaves on one branch. Eventually, adjoining branches develop these symptoms until sometimes the entire tree appears zinc deficient. Blight can also begin by a cupping and hardening-off appearance to new leaves. Affected leaves also exhibit wilt, and severely wilted branches will shed leaves and develop terminal dieback. Spring bloom is delayed, but flower production is abundant. Fruit formed is smaller than normal and may be irregular in shape. Water sprouts eventually develop on the trunk from latent buds. Blighted trees have fewer and smaller leaves, less leaf area per tree, lower stomatal conductances, and lower diurnal transpiration rates than healthy trees.[19]

Below ground, earliest impressions of root distribution and health on diseased trees were made by Rhoads.[1] Later observers[5,7,20] reported that root systems on blight-diseased trees did not differ from those on healthy ones. These observations were primarily based on casual examination of roots on trees lifted from groves. Later studies identified root rot symptoms on the root system and demonstrated differences in root distribution between healthy and diseased trees (Fig. 8-2). Childs[12] described a necrosis of scaffold roots on diseased trees, and this symptom was later evaluated by Graham et al.[21] and Nemec.[22] Nemec[22] recognized two types of scaffold root symptoms. One symptom consists of a black rot originating at the most distal end, and the other is a similar but brown root rot. Both symptoms are more common on diseased rather than healthy trees.

In addition to scaffold root symptoms, fibrous root rot symptoms are present on trees. Fibrous root rot is more transient and therefore is not always evident on trees. The most common fibrous root rot is caused by the facultative parasite *Fusarium solani* (Mart.) Sacc.[23] Dry root rot of fibrous roots, another symptom, causes roots to dry and

Figure 8-2 Blight tree removed from the soil showing a noticeable absence of fibrous roots on scaffold roots below the first 15–20 cm of soil.

break apart when they are removed from the soil. *Fusarium solani* can also be isolated from these roots, but it is not certain if it causes these symptoms.

The most noticeable internal symptoms in diseased trees are located in the vessels of the wood. Vessel plugging significantly increases in wood of diseased trees compared to healthy trees.[24–28] Two types of plugs are present in vessels. One plug, designated as a resin plug because it contains a diverse number of lipids,[25] is predominately found in older wood of the scion but is more uniformly distributed in a cross-section of wood from the root system. The other plug is described as gum[25] or amorphous in nature.[27,28] Gum plugs occur uniformly in old and new wood of both the scion and rootstock.[25,28] Nemec et al[25] also described bands of narrow vessels primarily along the cambium as a third potential resistance to waterflow in diseased trees.

8-4 PATHOLOGY

8-4-1 Research on Potential Causal Agents

Many investigations have been made to discover the cause of blight. The earliest researchers to initiate studies of this type were Swingle and Webber[3] and Underwood.[29] These investigators believed the disease to be contagious. Underwood thought bacteria were the causal agent, but he did not provide supporting proof for this claim. Rhoads,[1] in his exhaustive studies on blight, attempted to isolate a causal agent, but could not detect one in diseased trees. Between the end of Rhoads's studies in 1936[1] and the beginning of Childs's[12] research in the 1950s, no reports of blight research appeared in the literature. In 1953, Childs[12] cultured *Diploidia, Fusarium,* and *Phoma* from the necrotic scaffold roots. Citrus seedlings he inoculated with these fungi did not exhibit wilt unless they were weakened, poor-quality plants. Later, Childs[30] indentified the fungus *Physoderma citri* sp. nov. in vessels of diseased trees and suggested that it may be the causal agent of blight. The structure of this fungus was reevaluated[24,26] and found to be nonliving, host-derived, vessel-plugging material identical to the resinous vessel plugs previously mentioned.

Cohen[7] reported a similar microbial flora from healthy and blight-diseased tree roots; and Nemec[31] isolated a wide range of soil fungi and bacteria from roots of healthy and diseased trees. No potential pathogen was exclusively associated with roots of diseased trees in that study, but Nemec found *F. solani* the major fungal component of the root-associated flora. Florida citrus soils contain moderate to high populations of *F. solani*,[32] and it was not until 1980 that the pathogenicity of this fungus was proven on citrus in Florida.[33] It is considered a pathogen of opportunity, causing the previously described fibrous root rot and possibly the scaffold root rot.

The association of *F. solani* with scaffold root rot was reported in two studies.[21,22] In both studies, the number of necrotic roots increased as severity of above-ground symptoms increased. In trees exhibiting diseased canopy sectors, impaired water uptake and scaffold rot were usually associated with the symptomatic side of the tree.[21,22] Graham and Timmer[21] were unsuccessful in reproducing scaffold root rot symptoms and concluded that rot was not the cause of blight but was a response to xylem blockage in the trunk and depletion of starch reserves in the tree. Graham et al.[34] later found scaffold root rot on trees with tangerine collapse and on trees with dry root rot symptoms in California, and concluded that it was a nonspecific symptom. Nemec[22] concluded that the noticeable absence of rotted scaffold roots on some blight-diseased trees and those with sectors of blight indicated that this symptom is not a prerequisite for blight. His histological studies of diseased wood demonstrated host-defense mechanisms present in areas colonized by the fungus.

The invasion of scaffold roots by *F. solani* is probably by means of infected fibrous roots. Fibrous root rot caused by *F. solani* was more common on trees with blight than healthy ones,[23] and when root samples were collected to score root rot, root weight was usually significantly less on diseased trees. Root weight differences between blight-diseased and healthy trees were greater in the first 0.3 m of soil than in the 0.3 to 0.6 m depth.[35] Albrigo et al.[36] reported that trees with early decline did not have significant feeder root loss compared to healthy trees, and claimed that major feeder root loss occurs sometime after visible above-ground symptoms appear. However, data collected earlier[23] on healthy trees indicate that fibrous root rot is evident on them several years before they develop symptoms.

Recent studies have shown that *F. solani* is an early colonizer of citrus root epidermal cells[37] and can penetrate directly as well as through broken cells. *Fusarium solani*, besides causing wilt and water stress symptoms in inoculated plants,[38] induces vessel-plugging symptoms in wood, the most common of which is the resinous plug.[22]

The involvement of nematodes with blight was examined in two studies,[39,40] but no particular nematode was constantly associated with the disease. They, as general feeders on roots, can cause wounds that may be colonized by other organisms pathogenic to roots. They many even serve as vectors of virus or virus-like agents.

Considerable interest in transmissible biological agents as potential causes of blight was shown by researchers during the early 1970s. In one investigation, Purcifull et al.[41] did not find mycoplasma-like agents in phloem of diseased trees. Rickettsia-like bacteria were recovered from wood of blight-diseased trees by Feldman et al.,[42] but Feldman and Hanks[43] were unable to culture from xylem fluid a fastidious rickettsia-like bacterium on agar. About the same time, Hopkins et al.[44] reported the transmission of the Pierce's disease rickettsia-like bacterium from citrus with blight to grape by sharpshooters. Hopkins and Adlerz[45] and Hopkins[46] later reported that this bacterium produced dieback-type symptoms on citrus seedlings inoculated in the greenhouse. These results could not be confirmed in extensive tests conducted by Feldman;[47] and Timmer and Lee[48] were not able to obtain xylem-limited bacterial (XLB) infections in grape, periwinkle, peach, plum, or rough lemon caged with sharpshooters collected from nine blight-affected citrus groves. More recently, Hopkins[49] demonstrated that rough lemon seedlings and Pineapple sweet orange grafted on rough lemon inoculated with *Xylella fastidiosa*, the XLB that causes Pierce's disease, caused elevated Zn in wood and phloem and a reduction of water uptake. Tristeza virus has also been mentioned as a possible blight causal agent, but the evidence provided so far implies that it is not causally involved. Tristeza, to be a viable causal agent, must be present in all diseased trees, and Young et al.[50] failed to detect it in all the diseased trees they tested. Furthermore, tristeza has not been detected in Cuba,[51] where blight is prevalent.

Bacterial flora other than rickettsia-like organisms have been investigated for their possible role as causal agents of blight. Feldman and Hanks,[43] in their attempt to culture a rickettsia-like bacterium from roots and twigs of diseased trees, instead cultured a variety of other xylem-resident bacteria. The majority of bacteria isolated from citrus causing hypersensitive reactions in inoculated tobacco leaves were nonfluorescent pseudomonads.[52] Rough lemon plants inoculated with hypersensitive-positive *Pseudomonas* isolates caused a significant suppression of root growth.[53] In subsequent studies, Gardner et al.[54] demonstrated that infiltration of citrus stem cuttings with nonfluorescent pseudomonads elicited fibrous (resinous) occlusions in xylem vessels, one of the symptoms observed in wood of blight-diseased trees. Additional research by Gardner, Chandler, and Feldman (personal communication) revealed that growth suppression of citrus, and in some instances vascular plugging, also can be elicited by root surface-colonizing, hypersensitive-positive nonfluorescent pseudomonads. These studies with rhizobacteria demonstrated the significant plant growth depression they can cause and their potential role as low-grade incitants of disease.

Mycoplasma-like agents, viruses, viroids, and even an XLB may cause blight;

however, their participation as etiological agents should be demonstrated by grafting experiments. Many attempts have been made to bud and graft transmit blight to healthy trees. All studies in which this has been attempted have failed,[1,3,4,44–57] except one.[58] In this one exception, blight occurred on seven of eight receptor trees that were root-grafted to known blight-diseased trees. The major criticism of this study was that soil was moved with the blighted donor trees to the study site. The soil may have contained the causal factor(s) for the development of blight in the receptor trees. This study recently was repeated to eliminate the possibility of soilborne agents causing the disease, and within two years after approach-grafting roots of trees on Carrizo citrange on rough lemon roots of trees with blight, blight symptoms developed in the receptor trees (personal communication, R. Lee). No causally related, transmissible biological agent has yet been isolated or detected in these root-grafting studies.

8-4-2 Potential Epidemiology

The development of blight in the field has epidemiological characteristics of both a slowly moving, possibly vectored, contagious cause, as well as a soilborne origin. The early random but later clustered pattern of disease development[14,15] that occurs in most groves could suggest an etiology by either of the aforementioned causal relationships. One explanation for this pattern of disease development has been observed in groves with shallow pans. In Rhoads's[1] studies and those of Nemec et al.,[11] the subsurface formations varied from as shallow as 0.3 m to more than 3 m deep, over short distances down the rows. Usually, the first cases of blight appeared on the shallowest sites and where adjacent tree sites were usually also shallow. Trees in these locations also exhibited blight early in the development of the disease in the grove. Shallow soils affect root development and cause roots to be confined to soil above the pan, where they are exposed to greater fluctuations in stress[59] which increases their susceptibility to facultative parasites. However, this explanation is not valid in deep soil where pans do not influence root development. Another explanation suggests that a slow-moving causal agent is involved and that this agent is most likely soilborne. Partial success in stopping spread of blight with subsoil barriers between trees has been presented to support this reason.[60] However, in the flatwoods where wide middles separate citrus on beds, spread of blight from a central locus is as similar down rows as it is across beds, an observation not supporting causal agent movement through the soil because wide middles are themselves a barrier to root contact and disrupt the continuity of soil between beds.[61] Blight spread in groves is also not logarithmic, which is a characteristic of airborne vectored causal agents.[15,61] This feature of blight does not support hypotheses which suggest that viruses and XLBs are causal agents. Limited attempts have been made to cage citrus trees in the field (personal communication, D. Tucker) to exclude insect vectors which may possibly spread virus or virus-like agents of blight, but these studies have been inconclusive.

Other patterns of blight development in the field strongly indicate soils as the source of the factor(s) that causes blight. Cohen[13] reported on the nonrandom distribution of blight in groves; and in one grove, he reported that most tree loss due to blight occurred adjacent to roads and ditches in 3- to 8-ha blocks of a flatwoods site. The occurrence of roadside decline in central Florida[12] has led to speculation that leaching of material from roadbeds or possibly excess fertilizer spilled as the spreader changes rows may induce blight, or that some etiological agent was spreading into the grove from the road. Additional discussion of soils and their relationship to blight will be discussed later in this chapter.

Some epidemiological information has been derived from disease development in the field. Earliest symptoms are usually expressed only on trees growing in shallow soils. Disease does not affect some trees until they have become 50 to 60 years of age, which implies that causal factors may be absent from sites for a long time in the life of a tree, or there is a considerable lag time between infection and expression of symptoms. Isolation has not prevented the occurrence of blight, because symptoms have appeared on trees isolated from other plantings by as much as 15 to 16 km. Annual loss anticipated by growers even as far back as the 1920s was 1 to 2% per year[5], a linear pattern of tree loss recently confirmed by Yokomi et al.[15] Historically, the first records of blight during the late 1800s and subsequent reports of the disease provide us with some perspective of disease expression. Smith[5] reported that blight appeared to have intensified in the 1880s, early 1900s, 1920s, 1940s, and 1960s, or in about 20-year disease cycles. Even if these disease cycles were not influenced by cultural practices, such as rootstock usage, a case for an etiological agent would be difficult to support on such a wide cycle. No other factors can be related to these cycles.

The data presented in studies with the Pierce's disease bacterium as a causal agent of blight are conflicting and are a subject requiring more research. Even though this bacterium increased bark and wood Zn levels and reduced waterflow in one of several inoculated plants,[62] very few bacteria introduced into the plant survived even very short periods of time.[47] Furthermore, results of treating trees with antibiotics have been inconclusive. In addition, Pierce's disease is endemic in grape in California,[63] where citrus is also grown, but blight has not appeared in these trees. Pierce's disease has not been detected in Queensland, Australia (personal communication, John Chapman), but blight is common on citrus in that state.

Nonfluorescent pseudomonads also stimulate the formation of vessel plugs in wood and reduce uptake of water. They would probably cause symptoms by producing toxic metabolites in the rhizosphere or in the host, but the concentration of bacteria of all types is so low in the vessels as to cast doubt on their role as pathogens in wood.[31,64] However, future studies should examine the potential of these bacteria to produce metabolites which may be produced on the root surface and translocated throughout the plant.

The recent observation that blight can be root-graft transmitted,[58] in addition to suggesting that a mechanically transmissible biological agent may incite blight, also implies that other causal factors may be translocated across the graft. If blight is caused by nutritional phenomena or by microbiological synthesized metabolites, these factors may be moved in this way from tree to tree. A root graft between trees essentially creates a split root system, and the possibility exists for movement of substances through the union.[65] Therefore, valuable future information on blight may be gained by examination of xylem fluid for hormonal and nutritional changes, bacterial and fungal toxic metabolites, as well as host-synthesized stress metabolites.

8-4-3 Management

Earliest control of blight was directed at minimizing losses to the disease. At first, cultural practices were used to diminish disease, and in Rhoads's[1] early work on blight, he recommended seven cultural practices to reduce tree loss. The best practice, use of tolerant rootstocks, is still used today.

During the 1970s, when an extensive research program on blight was developed by federal, state, and commercial personnel, a large number of studies were directed toward disease control. Several foliar-applied commercial products were used then and reported to restore trees to normal health. One of these, a cytokinin extract of seaweed,[66] and another, a technical grade of sodium erythobate,[67] did not fullfill the claims made for them by their suppliers.

Several researchers initiated studies to determine if amendments applied to soil around trees could restore trees to a healthy condition. Childs[55] applied a montmorillonite clay to soil and found a decrease in the number of blight trees in treated sites compared to untreated sites. Humus applications around trees with moderate to severe symptoms were reported to suppress disease symptoms,[68] but the initial success with this product could not be repeated in subsequent trials. Likewise, similar positive results of disease control were cited for applications of basic slag around trees.[69] Other similar tests with a mixture of calcium humates and trace elements applied to soil were claimed to reduce the severity of blight by changing the buffering capacity of the soil to make elements, such as Ca, more available to the root.[70] Although these claims of disease control with amendments seem to offer promise for a control, little or no follow-up work has been conducted with them, nor have all of the experimental tests been designed adequately to provide convincing data.

More work has been conducted with pesticides as potential controls for blight than any other group of chemical compounds. Antibiotics in initial tests to cure blight were applied to soil around trees, and in one test,[71] a remission of symptoms was reported in three of five trees treated with tetracycline. These results could not be confirmed in tests conducted by Lee et al.[72] and Timmer et al.,[73] nor in other tests with antibiotics (P. F. Smith and S. Nemec, unpublished data). However, Timmer and Lee[74] assumed that because oxytetracycline treatments reduced zinc levels in trunk wood for one year and a combination of oxytetracycline and benomyl treatment reduced zinc in wood for two years, a bacterium was the most likely cause of blight. In studies using fungicides alone for potential control, no improvement in tree health occurred,[75-77] except one, where benomyl used as a soil drench around trees appeared to improve tree health slightly compared to controls.[78]

As stated earlier, cultural practices have been used for a long time to reduce the ravages of blight, and in view of the poor results obtained with pesticides, cultural practices will continue to be used when possible to diminish disease loss. One of the cultural practices proposed even as recently as the 1970s was a process to bypass the possibly malfunctioning root system by scion-rooting. However, extensive long-term studies with scion-rooted trees have failed to demonstrate the effectiveness of this procedure.[5] Rootstocks tolerant of blight continue to be recommended as a method to reduce blight, and various investigators have attempted to define the tolerance of rootstocks to the disease. In one such study,[79] blight incidence was high in trees on rough lemon and trifoliate orange, and lowest on sour orange. Carrizo citrange was considered intermediate in its susceptibility.

8-5 SOILS AND NUTRITION

Blight occurs on most soils planted to citrus in Florida; however, the disease appears earlier and more severely on some sites, and on others, soils suppress the development of the disease.[1] The flatwoods soils are generally shallow with various subsurface marl, limestone, or shell, and organic pan materials.[80] These shallow soils have a higher incidence of blight than the quartzipsamments soils on the central ridge.[81] Rhoads[1] and Nemec[11] showed that underlying rock and clay formations, respectively, in grove soils caused root systems to be essentially confined to the soils above the subsurface formations. Blight incidence was more common in groves where these subsurface formations occurred.

Areas of organic muck are present in groves in the flatwoods part of Florida. These muck pockets contain a low incidence of blight.[13] Even on the ridge soils, areas of relatively high organic matter are present, and on these soils blight has not been as extensive as on surrounding sandy soil. Organic soils offer the root an environment which has a higher cation exchange capacity, increased water retention, and a higher biological buffering characteristic. Rooting may even occur deeper in these organic sites. Whatever the benefit the tree derives from these soils, it extends the life of the tree well beyond that of one on sandy soil.

The variable occurrence of blight in numerous sites has generated some in-depth studies on characteristics of certain locations with high and low blight incidence. Nemec[59] reported that sand over shallow clayey horizons experienced a greater fluctuation in water potential and oxygen through the year than deep, sandy soils. The shallow sands contained nutrient conditions that were unsatisfactory for root growth.[82] In a low-blight muck soil, Ca, Mg, K, and P and most minor elements exceeded levels in adjacent high-blight sands.[83] Soil location in groves and soil management have also resulted in differences in blight development. As mentioned earlier, rows along roadsides on the ridge have often developed blight (roadside decline) earlier than the remainder of the grove.[12] Also, in the flatwoods, Cohen et al.[84] reported that fewer trees showed blight symptoms on deep tillage soil than on deep tillage and lime or surface tillage treatments of the soil-water-atmosphere-plant (SWAP) project located near Ft. Pierce. Deep tillage of soil with supplemental use of dolomite in a commercial grove resulted in a higher blight incidence than in an adjacent older block of trees on soil that was surface tilled but not amended at planting with dolomite.[85] Similarly, on the ridge, blight-like symptoms appeared in plots treated with excess dolomite and lime at the Short Research Grove near Clermont.[86]

Commercial fertilizers containing nitrogen (N) have been considered as factors that may be involved in blight development. Results of a survey of 357 Florida citrus growers indicated that growers using 112 kg N/ha or less reported a lower percentage of blight than growers using over 224 kg N/ha.[87] Growers have cited (personal communication) instances of blight-like symptoms occurring on trees where excess N-containing fertilizer was accidentally spilled on the soil around healthy trees.

Excess NH_4^+ as a cause of blight-like symptoms was originally hypothesized by

Wander and Sites.[88] They suggested that some nutritional difficulties observed during recent years in commercial groves on the acid sands of Central Florida may be explained by the hypothesis that relatively large applications of NH^+_4 source materials tend to interfere with the calcium nutrition of citrus. It is possible that, even where dolomite or limestone is used for pH control, NH^+_4 sorption may interfere with calcium nutrition. The fact that dolomite or limestone remains relatively immobile in the topsoil, where NH^+_4 can readily be leached into the subsoil, makes the problem of adequate CA^{+2} nutrition even more difficult under field conditions. A leaf pattern which might be described as a chlorosis resembling Fe and Mn deficiencies occurred on both the pH-controlled and noncontrolled, 100% NH^+_4 treatments. This pattern was also observed in the NH_4 NO_3 treatments, being much more pronounced on the non-pH-controlled treatment. Analysis by Fudge (unpublished data) of leaves having a similar type of pattern indicated a low leaf Ca content.

Rhoads[1] observed that the incidence of citrus blight increased markedly with inadequate fertilization in groves where it was prone to occur. In those days, no diagnostic tests existed for blight, and various sorts of nutritionally and otherwise induced tree decline could have been confused with what we know today as blight. The only other similar observation in the literature was made by Calvert.[89] In a fertilizer experiment with young Valencia orange on rough lemon in a bedded grove on a flatwoods Spodosol, Calvert found a greater number of trees with young tree decline (YTD) symptoms, similar to citrus blight, in the low-fertility plots than in the higher-fertility plots on the interior two rows of four-row beds. It was not clear whether the blight-fertilizer relationship resulted from variations in the supply of an essential element, from interaction between the fertilizer materials and soil components, or from other unknown causes. This experiment was reported before diagnostic tests for blight had been developed; thus, the trees with YTD may or may not have had citrus blight. These low fertility treatments involved applying N, P, Mg, K, Mn, and Cu at one half the recommended rates for trees of similar age.[90] In the outer rows where much less YTD occurred, much of the A horizon soil had been removed during the formation of the beds, forcing the trees to rely on E horizon material (highly leached and nearly devoid of organic matter) as topsoil; thus, these trees responded well to additions of N and Mg fertilizer. In the inner rows where YTD occurred, all of the A horizon soil was intact and was buried under mixed fill material containing E, Bt, and C (coquina) horizon material. This mixed fill material was relatively fertile, being high in clay and $CaCO_3$, and complemented the relatively high fertility level of the underlying A horizon soil. No consistent relationship occurred between soil pH patterns in the profiles of declining versus healthy trees. Calvert[89] speculated that future research might find a link between YTD and extremely acid subsoil horizons (with pH values below 4.5).

In a more recent study reported by Anderson and Bistline,[91] no evidence was obtained from three experiments on different soils to indicate a cause-and-effect relationship between rate of N fertilization and incidence of citrus blight. The authors attempted to discount the observations of Calvert[89] as being merely coincidental. They stated that no relationship between N rates and blight was observed in any other of the many N-rate experiments carried out on citrus in Florida during the past 25 years. Ammonium nitrate, either granular or liquid, was used as an N source in most of these experiments. Blight either did not occur in these experiments or was present at insignificant levels, because no mention of it was made in any of the reports.[91] Perhaps the relationship between fertilizer rate and blight suggested by Rhoads[1] and observed by Calver[89] involved fertilizer elements other than N.[91]

In other studies, Wutscher and Hardesty[92] reported that the soil levels of NH^+_4-N and NO^-_3-N did not differ in the root zone of healthy and diseased trees over a two-year period on a deep soil, and levels of each ion were present at concentrations less than toxic to citrus roots. Wutscher[93] also reported that the cation/anion ratio was significantly higher under blight-affected trees, but that the cation exchange capacity, pH, organic matter content, moisture content, K, Ca, Mg, and Na did not differ under blight-diseased and healthy trees.[94] Other work has revealed a higher incidence of blight in soil containing well tailings (D. L. Myhre, unpublished) and in soils that receive compaction.[95] Significantly higher penetrometer resistance values occurred near blighted trees than near healthy trees in one grove. Resistance values also averaged three times higher in traffic versus nontraffic areas of the grove.

8-6 CONCLUSION

Blight remains an unresolved citrus disease problem in Florida. The immense effort of many scientists, agricultural workers, and other investigators to determine its cause(s) have not been wasted. During the last two decades, many approaches have been taken to explore avenues of its etiology or to examine control by one means or another. We have considerable knowledge of blight in the sense that information has been gathered about certain causal relations that do not fit the disease, and that standard control practices are ineffective on blight. However, several features about blight appear from the vast amount of information reported on the disease. One feature is that factors that affect the root and its environment cause trees to develop blight early, and the second is that blight can be induced in healthy trees by root-grafting diseased trees to them.

This information, by itself, rules out many possible causal factors and allows future investigations to focus attention on root health, etiological agents common to roots, and tree nutrition as factors in the causal relations of the disease. During the course of investigations on blight, studies have shown that constraining the root system to a shallow soil area initiates blight earlier.[1,11] The fact that soils amended with excess levels of dolomite and lime may cause blight earlier is significant. This information may not be just a simple case of modified nutrition, but could involve the restructuring of the soil microflora in the rhizosphere; or it may involve aspects of root-growth inhibition and root health. The fertilizer stress leading to blight or blight-like symptoms reported by growers was reproduced in studies by Nemec et al.[96] They found that NH^+_4-N stresses on the tree stimulated certain symptoms of blight but not all of the symptoms that were required to prove the trees had blight.

Each stress on the root system, whether soil compaction, fluctuating water tables in shallow soils, changes in soil pH, or other factors due to liming or excess fertilizer, probably do not cause blight alone. However, each exerts effects on the tree and when multiple stresses occur, the effects may be additive leading to earlier expression of blight symptoms. Facultative parasites, such as *F. solani*, are opportunistic on plants weakened by stress. The root rot it causes on blight trees may not be enough damage alone to cause blight, but the toxins it produces may cause enough wood cell damage remote from the site of infection to induce symptoms. Napthazarin toxins *F. solani* produces are toxic to citrus seedings[97], and have recently been detected in xylem fluid of roots and branches. Concentrations were higher in roots of blight-diseased trees than in healthy-appearing trees[98]. Future blight root-grafting research should evaluate toxin translocation, as well as virus-like agents in the etiology of blight.

8-7 REFERENCES

1. Rhoads, A. S., Blight—a nonparasitic disease of citrus trees, *Fla. Agric. Expt. Sta. Bull.*, No. 296, 64, 1936.

2. Young, R. H., Albrigo, L. G., Tucker, D. P. H. and Williams, G. L., Incidence of citrus blight on Carrizo citrange and some other rootstocks, *Proc. Fla. State Hort. Soc.* 93, 14, 1980.

3. Swingle, W. T. and Webber, H. J., The principal diseases of citrus fruits in Florida, *USDA Bull.*, No. 8, 1896.

4. Sauls, J. W., Bibliography of young tree decline, sandhill decline, and blight, *Memo on blight*, Univ. Fla., Gainesville 1981.

5. Smith, P. F., History of citrus blight in Florida, *The Citrus Ind.* 55(9), 13, 14, 16, 18, 19; (10) 9, 10, 13, 14; (11) 12, 13, 1974.

6. Burnett, H. C., Nemec, S. and Patterson, M., A review of Florida citrus blight and its association with soil edaphic factors, nutrition, and *Fusarium solani*, *Trop. Pest Management*, 28, 416, 1982.

7. Cohen, M., Citrus blight and blightlike disease, *The Citrus Ind.*, 49(7), 12, 13, 16, 26, 1968.

8. Reitz, H. J., Preliminary exploratory research, *Fla. Agric. Expt. Sta. Annu. Rept.*, 155, 1970.

9. Nemec, S., Citrus blight in Florida, *Rev. Trop. Pl. Path.* 2, 1, 1985.

10. Childs, J. F. L., Control of citrus blight disease, *Proc. Fla. State Hort. Soc.*, 94, 25, 1981.

11. Nemec, S., Fox, A. N. and Horvath, G., The relation of subsurface hardpan to blight of citrus and development of root systems, *Proc. Soil and Crop Sci. Soc. Fla.*, 36, 141, 1976.

12. Childs, J. F. L., Observations on citrus blight, *Proc. Fla. State Hort. Soc.*, 66, 33, 1953.

13. Cohen, M., Nonrandom distribution of trees with citrus blight. In *Proc. 8th Conf. Int. Organ Citrus Virol.* (E. C. Calavan et al., eds.), pp.260–263, Riverside, CA, 1980.

14. DuCharme, E. P., Tree loss in relation to young tree decline and sandhill decline of citrus in Florida, *Proc. Fla. State Hort. Soc.*, 84, 48, 1971.

15. Yokomi, R. K., Garnsey, S. M. and Young, R. H., Spatial and temporal analysis of blight incidence in a Valencia grove on rough lemon rootstocks in central Florida. In *Proc. 9th Conf. Int. Organ. Citrus Virol.* (Iguazu, Argentine, S. M. Garnsey et al., eds.), pp.260–269, Riverside, CA, 1983.

16. Norris, J. C., Young tree decline from a grower viewpoint, *Proc. Fla. State Hort. Soc.*, 83, 46, 1970.

17. Hanks, R. W. and Feldman, A. W. Results of initial indexing on citrus trees affected with young tree and sandhill decline, *Plant Dis. Reptr.*, 58, 35, 1974.

18. Anonymous, *Comprehensive review syllabus.* Part I. *Plant Pathology Dept., Inst. of Food and Agric. Sci., Univ. of Fla., Gainsville*, p.248, 1975.

19. Syvertsen, J. P., Bausher, M. G. and Albrigo, L. G., Water relations and related leaf characteristics of healthy and blight affected citrus trees, *J. Am. Soc. Hort. Sci.*, 105, 431, 1980.

20. Reitz, H. J., Unexplained citrus declines. In *Frist int. citrus short course, citrus rootstocks,
Univ. of Fla., Gainesville*, pp. 38–43, 1973.

21. Graham, J. H., Timmer, L. W. and Young, R. H., Necrosis of major roots in relation to citrus
blight, *Plant Dis. Rep.*, 67, 1273, 1983.

22. Nemec, S., Characteristics of *Fusarium solani*-infected pioneer roots on blight-diseased and
healthy citrus, *Proc. Fla. Soil and Crop Sci. Soc.*, 43, 177, 1984.

23. Nemec, S., Burnett, H. C. and Patterson, M., Observations on a citrus fibrous root rot in
blight-diseased groves, *Proc. Fla. Soul and Crop Sci, Soc.*, 37, 43, 1978.

24. Nemec, S. and Kopp, D., Extent of lipid vessel plugs in citrus with and without sandhill and
young tree decline symptoms, *Proc. Fla. State Hort. Soc.*, 87, 107, 1974.

25. Nemec, S., Constant, S. and Patterson, M., Distribution of obstructions to water movement
in citrus with and without blight, *Proc. Fla., State Hort. Soc.*, 88, 70, 1975.

26. Vandermolen, G. E., Electron microscope observations of vascular obstructions in citrus
roots affected with young tree decline, *Proc. Fla. State Hort. Soc.*, 87, 121, 1974.

27. Vandermolen, G. E., Gennaro, R. N. and Peeples, T. O., Chemical nature and statistical
analysis of the distribution of plugging in blight/YTD-affected citrus trees, *Proc. Fla. State
Hort. Soc.*, 88, 76, 1975.

28. Cohen, M., Pelosi, R. R. and Brlansky, R. H., Nature and location of xylem blockage structures in trees with citrus blight, *Phytopathology*, 73, 1125, 1983.

29. Underwood, L. M., Diseases of the orange in Florida, *J. Mycol.*, 7, 27, 1891.

30. Childs, J. F. L., Kopp, L. E. and Johnson, R. E., A species of *Physoderma* present in citrus
and related species, *Phytopathology*, 55, 681, 1965.

31. Nemec, S., Microorganisms associated with healthy and sandhill-declined citrus roots, *Plant
Dis. Reptr.*, 59, 210, 1975.

32. Nemec, S. and Chaddock, A., *Fusarium* populations in Florida citrus soils, *Phytopathology*,
4, 116, 1977.

33. Nemec, S., Baker, R. and Burnett, H. C., Pathogenicity of *Fusarium solani* to citrus roots
and its possible role in blight etiology, *Proc. Fla. State Hort. Soc.*, 93, 36, 1980.

34. Graham, J. H., Brlansky, R. H., Timmer, L. W., Lee, R. F., Marais, L. J. and Bender, G. S.,
Comparison of citrus tree declines with necrosis of major roots and their association with
Fusarium solani, Plant Disease, 69, 1055, 1985.

35. Nemec, S., Burnett, H. C. and Patterson, M., Root distribution and loss on blighted and
healthy citrus trees, *Proc. Fla. Soil and Crop Sci. Soc.*, 41, 91, 1982.

36. Albrigo, L. G., Syvertsen, J. P. and Young, R. H., Stress symptoms of citrus trees in successive stages of decline due to blight, *J. Am. Soc. Hort. Sci.* 111(3), 465, 1986.

37. Nemec, S., Achor, D. S. and Albrigo, L. G., Microscopy of *Fusarium solani* infected rough
lemon citrus fibrous roots, *Can. J. Bot.*, 64, 2840, 1986.

38. Nemec, S., Syvertsen, J. and Levy, Y., Water relations of rough lemon (*Citrus jambhiri*
Lush.) citrus seedlings infected with *Fusarium solani, Pl. Soil*, 93, 231, 1986.

39. Feldmesser, J., Childs, J. F. L. and Rebois, R. V., Occurrence of plant-parasitic nematodes in
citrus blight areas, *Plant Dis. Rep.*, 48, 95, 1964.

40. Tarjan, A. C., An evaluation of plant nematodes as causal factors in citrus young tree decline,
Proc. Soil and Crop Sci. Soc. Fla., 32, 176, 1973.

41. Purcifull, D. E., Garnsey, S. M., Storey, G. E. and Christie, R. G., Electron microscope examination of citrus tress affected with young tree decline (YTD), *Proc. Fla. State Hort. Soc.*, 86, 91, 1973.

42. Feldman, A. W., Hanks, R. W., Good, G. E. and Brown, G. E., Occurrence of a bacterium in YTD-affected as well as in some apparently healthy citrus trees, *Plant Dis. Rep.*, 61, 546, 1977.

43. Feldman, A. W. and Hanks, R. W., Recovery of rickettsialike bacterium from xylem of healthy and young tree decline-affected citrus trees. In *Proc.8th Conf. Int. Organ. Citrus Virol.* (E. C. Calavan et al., eds.), p.264, Riverside, CA, 1980.

44. Hopkins, D. L., Adlerz, W. C. and Bistline, F. W., Pierce's disease bacteria occurs in citrus trees affected with blight (young tree decline), *Plant Dis. Rep.*, 62, 442, 1978.

45. Hopkins, D. L. and Adlerz, W. C., Similarities between citrus blight and Pierce's disease of grapevine, *Proc. Fla. State Hort. Soc.*, 93, 18, 1980.

46. Hopkins, D. L., Relation of Pierce's disease to a wilt-type disease of citrus in the greenhouse, *Phytopathology*, 72, 1090, 1982.

47. Feldman, A. W., Young tree decline (blight) not reproduced in citrus inoculated with Pierce's disease bacterium, *Proc. Soil, and Crop Sci. Soc. Fla.*, 43, 81, 1984.

48. Timmer, L. W. and Lee, R. F., Survey of blight-affected citrus groves for xylem-limited bacteria carried by sharpshooters, *Plant Disease*, 69, 497, 1985.

49. Hopkins, D. L., Production of diagnostic symptoms of blight in citrus inoculated with *Xylella fastidiosa*, *Plant Disease*, 72, 432, 1988.

50. Young, R. H., Wutscher, H. K., Cohen, M. and Garnsey, S. M., Citrus blight diagnosis in several scion variety/rootstock combinations of different ages, *Proc. Fla. State Hort. Soc.*, 91, 56, 1978.

51. Hannon, C. I., Notes from a visit to Cuba and the Isle of Pines, *Citrus and Veg. Mag.*, 59, 8, 1978.

52. Gardner, J. M., Chandler, J. L. and Feldman, A. W., Growth promotion and inhibition by antibiotic-producing fluorescent pseudomonads on citrus roots, *Plant Soil*, 77, 103, 1984.

53. Feldman, A. W. and Gardner, J. M., Root suppression of citrus budlings by citrus tristeza virus and by *Pseudomonas* isolates from xylem of healthy and young tree decline-affected citrus trees, *Proc. Fla. Soil and Crop Sci. Soc.*, 41, 25, 1982.

54. Gardner, J. M., Feldman, A. W. and Stamper, D. H., Role and fate of bacteria in vascular occlusion of citrus, *Physiological Plant Pathology*, 23, 295, 1983.

55. Childs, J. F. L., Florida citrus blight. Part I. Some causal aspects of citrus blight, *Plant Dis. Rep.*, 63, 560, 1979.

56. Feldman, A. W. and Hanks, R. W., Young tree decline and sandhill decline; status of indexing investigations, *Proc. Fla. State Hort. Soc.*, 87, 101, 1974.

57. Wutscher, H. K., Youtsey, C. O., Smith, P. F. and Cohen, M., Negative results in citrus blight transmission tests, *Proc. Fla. State Hort. Soc.*, 96, 48, 1983.

58. Tucker, D. P. H., Lee, R. F., Timmer, L. W., Albrigo, L. G. and Brlansky, R. H., Experimental transmission of citrus blight, *Plant Disease,* 68, 979, 1984.

59. Nemec, S., Oxygen, temperature, and water potential in shallow and deep soils of a citrus grove with blight, *Proc. Fla. Soil and Crop Sci. Soc.*, 42, 85, 1983.

60. Albrigo, L. G., Tucker, D. P. H. and Jackson, J., Barriers against possible soil-root movement of citrus blight, *Citrus Blight Workshop Proc., Lake Alfred, FL*, March 1987 (Abstract).

61. Sonoda, R. M., Chellemi, D. O. and Pelosi, R. R. Spatial patterns of citrus bight in a grove with five rootstock/scion combinations, *Citrus Blight Workshop Proc. Lake Alfred, FL*, March 1987 (Abstract).

62. Hopkins, D. L., Role of plant growth regulators and xylem-limited bacterial disease as a cause of citrus decline, *Proc. Sym. Trop and Subtrop. Agric. Res.*, III-7 to III-12, 1986.

63. Raju, B. C., None, S. F., Docampo, D. M., Goheen, A. C., Nyland, G. and Lowe, S. K., Alternative hosts of Pierce's disease of grapevines that occur adjacent to grape growing areas in California, *Amer. J. Enology and Viticulture*, 31, 144, 1980.

64. Timmer, L. W., Lee, R. F. and Brlansky, R., Relationship of xylem-limited bacteria to citrus blight, *Citrus Blight Workshop Proc., Lake Alfred, FL*, March 1987 (Abstract).

65. Shannon, L. M. and Zaphrir, J., The relative influence of two citrus rootstock species upon plant growth and upon the inorganic composition of the scion, *Proc. Am. Soc. Hort. Sci.*, 71, 257, 1958.

66. Plimpton, R. S., The use of cytex cytokinin hormone concentrate against citrus tree decline, *Citrus and Veg. Mag.*, 39(5), 12, 14, 16, 29, 1976.

67. Biesada, A. A., Citrus growers respond to Myco-shield use for YTD, *The Citrus Ind.*, 57, 30, 31, 32, 34, 1976.

68. Pinckard, J. A., Suppression of citrus young tree decline with humus, *Plant Disease*, 66, 311, 1982.

69. Wutscher, H. K., Positive effect of basic slag application on citrus blight-affected 'Hamlin' orange trees, *Proc. Fla. State Hort. Soc.*, 908, 1, 1985.

70. Toy, R., A new approach to solving YTD, *Citrus and Veg. Mag.*, 63, 62, 1982.

71. Tucker, D. P. H., Bistline, F. W. and Gonsalves, D., Observations on young tree decline-affected citrus trees treated with tetracycline, *Plant Dis. Rep.*, 58, 895, 1974.

72. Lee, R. F. Timmer, L. W. and Albrigo, L. G., Effect of oxytetracycline and benzimidazole treatments on blight-affected citrus trees, *J. Am. Soc. Hort. Sci.*, 107, 1133, 1982.

73. Timmer, L. W., Lee, R. F. and Albrigo, L. G., Distribution and persistence of trunk-injected oxytetracycline in blight-affected and healthy citrus, *J. Am. Soc. Hort. Sci.*, 107, 428, 1982.

74. Timmer, L. W. and Lee, R. F., Chemotherapy of blighted citrus trees, *The Citrus Ind.*, 64, 47, 1983.

75. DuCharme, E. P., Feldman, A. W. and Hanks, R. W., Nature causes, and control of blight and young tree decline. In *Ann. Res. Rept. Inst. Food and Agric. Sci., Univ. of Fla.*, p.325, 1970.

76. Patterson, M. W. and Nemec, S., Evaluation of fungicides for control of citrus bight, *Am. Phytopathol. Soc. Fungicide and Nematicide Tests*, 34, 152, 1979.

77. Nemec, S., Bustillo, B., O'Bannon, J. H. and Patterson, M., Effect of fungicides and nemati-cides on citrus blight in Florida, *Phytopathology,* 72, 360, 1982 (Abstract).

78. Burnett, H. C., Nemec, S. and Gonsalves, D. G., Attempts to control young tree decline, *Proc. Int. Soc. Citriculture,* 3, 891, 1977.

79. Young, R. H., Albrigo, L. G., Tucker, D. P. H. and Williams, G. L., Incidence of citrus blight on Carrizo citrange and some other rootstocks, *Proc. Fla. State Hort. Soc.*, 93, 14, 1980.

80. Caldwell, R. E., Major land resource areas in Florida, *Proc. Soil and Crop Sci. Soc. Fla.*, 39, 38, 1980.

81. Reitz, H. J., Preliminary exploratory research, *Fla. Agric. Expt. Sta. Annu. Rept.*, 155, 1970.

82. Nemec, S., Calvert, D., Allen, L. H. and Fiskell, F., Feature of shallow soils with clay pans that are conducive to citrus blight, *Proc. Int. Soc. Citriculture*, Sao Paulo, Brazil, Vol. 2, p.398, 1984.

83. Nemec, S., Patterson, M. and Calvert, D., Factors associated with citrus blight-suppressive Florida flatwoods organic soils, *Phytopathology* 73, 504, 1983 (Abstract).

84. Cohen, M., Calvert, D. V. and Pelosi, R. R., Disease occurrence in a citrus rootstock and drainage experiment (SWAP) in Florida, *Proc. Int. Soc. Citriculture*, Japan, Vol. 1, 366, 1981.

85. Wutscher, H. K., Comparison of soil, leaf and feeder root nutrient levels in the citrus blight-free and citrus blight-affected areas of a 'Hamlin' orange grove, *Proc. Fla. State Hort. Soc.*, 99, 74, 1986.

86. Iley, J. R. and Guilford, H. E., Excess dolomite and lime plots display conditions very similar to YTD, *Proc. Fla. State Hort. Soc.*, 91, 62, 1978.

87. Phillips, R. L., Preliminary blight survey, *Univ. Fla. Institute of Food and Agric. Sci.*, 2, 1979.

88. Wander, I. W. and Sites, J. W., The effect of ammonium and nitrate nitrogen with and without pH control on the growth of rough lemon seedlings, *Proc. Am. Soc. Hort. Sci.*, 6, 211, 1958.

89. Calvert, D. V., Effects of rate and frequency of fertilizer applications on growth, yield and quality factors of young 'Valencia' orange trees, *Proc. Fla. State Hort. Soc.*, 82, 1, 1969.

90. Reitz, H. J., Leonard, C. D., Stewart, I., Koo, R. C. J., Calvert, D. V., Anderson, C. A., Smith, P. F. and Rasmussen, G. K., Recommended fertilizers and nutritional sprays for citrus, *Fla. Agric. Expt. Sta. Bull.*, 536B, 24, 1964.

91. Anderson, L. A. and Bistline, F. W., Rate of nitrogen fertilization and incidence of blight in three orange groves on the ridge, *Proc. Fla. State Hort. Soc.*, 91, 59, 1978.

92. Wutscher, H. K. and Hardesty, C. A., Ammonium, nitrite, and nitrate nitrogen levels in the soil under blight-affected and healthy citrus trees, *Commun. Soil Sci. and Plant Anal.*, 10, 1495, 1979.

93. Wutscher, H. K., Seasonal levels of water-extractable cations and anions in the soil under blight-affected and healthy citrus trees, *Commun. Soil Sci. and Plant Anal.*, 12, 719, 1981.

94. Wutscher, H. K., Smith, P. F. and Bistline, F., Zinc accumulation in trunk wood, water extractable ions in the soil and the development of visual symptoms of citrus blight, *Citrus and Veg. Mag.*, 63, 22, 24, 26, 28, 1982.

95. Myhre, D. L., Vasquez, L., Porter, K. M., Gardner, J. R. and Mansell, R. S., Penetrometer resistance for soil in a Florida grove affected with citrus blight, *Proc. Int. Soc. Citriculture*, Sao Paulo, Brazil, 1, 134, 1984.

96. Nemec, S., Myhre, D., and Burnett, H. 1989. Stress responses in citrus induced with soil applications of urea, ammonium nitrate, and ammonium hydroxide. *Proc. Fla. Soil and Crop Sci. Soc.* 48:111-117.

97. Nemec, S., Baker, R. A., and Tatum, J. H. 1988. Toxicity of dihydrofusarubin and isomarticin from *Fusarium solani* to citrus seedlings. *Soil Biol. Biochem.* 20:493-499.

98. Nemec, S., Jabajii-Hare, S., and Charest, P. M. 1991. ELISA and immunocytochemical detection of *Fusarium solani*-produced naphthazarin toxins in citrus trees in Florida. *Phytopathology,* 81 (in press).

9

CITRUS TRISTEZA VIRUS

R. F. LEE AND M. A. ROCHA-PEÑA

University of Florida
Citrus Research and Education Center
Lake Alfred, Florida

9-1 INTRODUCTION

Tristeza, caused by citrus tristeza virus (CTV), has been one of the most destructive diseases of citrus worldwide. The disease, first recognized as decline of citrus scions propagated on sour orange rootstock, was first reported in South Africa about 1910[1,2] and subsequently reported in Java in 1928,[3] in Argentina where it was called "podredumbre de las raicillas" (root rot)[4] about 1931, and in Brazil where it was first called "Tristeza" (sadness)[5] in 1937.[6] Quick decline of citrus on sour orange rootstock was first noted in California in 1939.[7] Similar diseases were later reported in New Zealand,[8] Australia,[9] West Africa,[10] Ceylon,[11] and Hawaii.[12] The cause of decline of the citrus scion on sour orange was unknown for many years and was thought to be a graft incompatibility between the scion and the sour orange rootstock or a nutrition problem.[13] The tristeza disease was first recognized in 1946 to be caused by a virus and to have aphid vectors.[7,14] The virus is transmitted through budding and grafting but not seed transmitted.[15] It has been speculated that tristeza originated in the Orient and was distributed worldwide by the movement of citrus budwood and plants, the result of humans' quest for new and novel citrus varieties.[16] The first report in South Africa of decline on sour orange rootstock[1,2] and the first report of stem pitting disease, now known to be part of the symptoms induced by certain strains of CTV, in South Africa[17] provide some evidence that South Africa was the first country to have imported the disease from the Orient. From South Africa the disease was probably distributed around the world to other citrus areas.[18] Tristeza disease in now present in most of the major citrus areas of the world except in certain Mediterranean and Central American countries.[15] Several recent review articles on tristeza have been published.[15,19,20,21] This review will emphasize the recent expansion of severe CTV strains into new areas, the movement of the efficient aphid vector into new areas, the CTV situation in Florida and Central America, methods for management of the disease, and future research needs.

9-2 DISTRIBUTION

CTV occurs in most of the citrus areas of the world except in certain parts of the Mediterranean, some Central American countries, and some isolated islands (Fig. 9-1a). Severe CTV strains which stem pit grapefruit are present in scattered locations around the world (Fig. 9-1b) and are more common than the severe strains which stem pit sweet orange (Fig. 9-1b). The known geographical area where the efficient aphid vector, *Toxoptera citricida*, is present is shown in Fig. 9-1c.

9-3 SYMPTOMS

Citrus tristeza virus exists as many strains having different biological activities. These biological activities belong to one or more of the following five major categories. With the exception of mild strains, the other categories of biological activity may occur alone or in any combination in a given CTV isolate.

Mild.   Producing no noticeable effect on most commercial and commonly grown citrus varieties. Mild strains will cause only slight stem pitting, little to no vein clearing, and flecking on Mexican lime (*Citrus aurantifolia*), the most commonly used indicator plant.

Seedling yellows (SY).   Inoculated seedlings of sour orange (*C. aurantium*), lemon (*C. limon*), and grapefruit (*C. paradisi*) have severe chlorosis and dwarfing. While the SY reaction is usually found on indicator plants under greenhouse conditions, it has also been found in the field where top-worked trees infected with a SY strain of CTV are grafted or budded with a susceptible variety.[21]

Decline on sour orange.   Inoculated budlings of sweet orange (*C. sinensis*) on sour orange rootstock will be dwarfed, often chlorotic, and decline. In the field, quick decline (QD) of sweet orange, grapefruit, or tangerine scions budded on sour orange rootstock can occur within three to six weeks. At the onset of QD, the leaves turn yellow or golden color, then wilt and fall from the tree, leaving only fruit hanging on the dead tree. Often the trunk will have an overgrowth immediately above the bud union. If a bark piece is removed from the bud union, there often will be needle-like pegs in the sour orange xylem with corresponding pinholes in the phloem (bark). The presence of the pinholes and pegs in the sour orange is not necessarily diagnostic, however, because it is often common for trees on sour orange to have a slight overgrowth at the bud union when CTV strains are present. In Florida, many QD trees will form a brownish line at the bud union within a few minutes of removing a bark patch. Not all decline inducing strains of

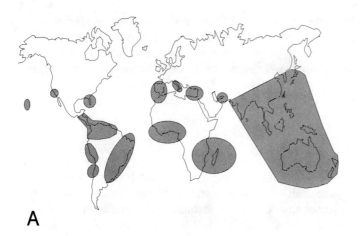

A

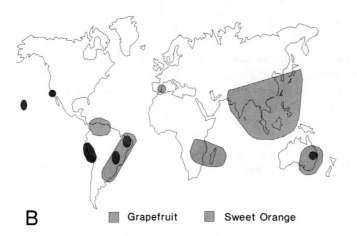

B

▨ Grapefruit   ▨ Sweet Orange

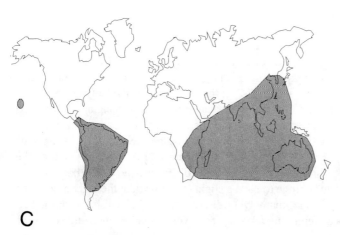

C

Figure 9-1 Distribution maps of (A) citrus tristeza virus; (B) grapefruit and sweet orange stem-pitting strains; and (C) *Toxoptera citricida*, the most effecient aphid vector of citrus tristeza virus.

CTV cause QD, but many will cause a decline over a period of several years whereby the trees will stop growing as well and become less productive. In Florida, many budwood sources have become infected with decline-inducing strains of CTV.[22] When such budwood is propagated on sour orange liners, the resultant plants often do not decline but are severely stunted and fail to come into production.[22]

Stem pitting on grapefruit.   Inoculated indicator seedlings of grapefruit show stem pitting on twigs when the bark is removed. The plants also show stunting and chlorosis. In grapefruit or pumello (*C. grandis*) trees in the field, the stunting can become very pronounced, often producing large longitudinal ridges or depressions running up and down the trunk, giving the trunk a rope-like appearance. The trees usually have a mushroom shape with occasional branches growing away from the rest of the tree. The stem pitting occurs on the scion regardless of the rootstock. Fruit size and productivity is considerably reduced.

Stem pitting on sweet orange.   Inoculated indicator sweet orange seedlings show stem pitting when peeling the bark from the wood; there is reduced growth or stunting, and often chlorosis. In the field, the trees have reduced fruit size and production, and reduced size. The twigs tend to be brittle and break easily. The pitting in field trees is obvious when the bark has been peeled from the wood. These strains can also cause stem pitting on rootstocks, such as Rangpur lime. A given CTV isolate causing stem pitting on sweet orange may or may not cause stem pitting on grapefruit, and vice versa.

A standardized host range (SHR) test using a selected battery of indicator plants has been suggested for a uniform comparison of CTV strains on a worldwide basis.[23] The battery of indicator plants includes Madam Vinous sweet orange, Mexican lime, sour orange and grapefruit seedlings, and a sweet orange scion grafted onto sour orange. All major disease syndromes, such as SY, stem pitting, and decline on sour orange rootstock, are evaluated. A world collection of CTV strains has been established in the USDA plant quarantine facility in Beltsville, Maryland, where numerous CTV strains are undergoing the SHR test under a common environment.[23,24] The results of these tests emphasize that a given strain of CTV may have any combination of the major biological activities[23,24] and the need for indexing CTV isolates on a battery of indicator plants rather than on only one indicator.

9-4 CAUSAL AGENT

CTV is a phloem-limited closterovirus having flexuous filamentous particles about $11 \times 2,000$ nm in size. Several purification procedures have been reported which result in highly purified virus for characterization of the virion components.[25-29] The virus is readily transmitted by budding and grafting.[15,21] Mechanical transmission is possible by slash inoculation of partially purified virus preparations into the stem of a host such as citron and Mexican lime.[30]

CTV virions contain a single-stranded RNA with an estimated size of 6.5×10^6 or about 20 kilobases.[15,25,31] The only gene product which has been identified has been the coat protein, which would account for about 3% of the total coding capacity of the CTV genome. The RNA extracted from purified virions appears to be plus-sense as it has been translated *in vitro* resulting in products of 65,000, 50,000, 33,000, and 26,000 daltons in size.[28] A product of about 26,000 daltons was identified as the coat protein by immunoprecipitation using CTV-specific polyclonal antisera.

Cloned cDNA libraries have been prepared from CTV RNA.[31,32] The random primer method of cloning was used by Rosner et al.[32] to prepare a library against an Israeli CTV strain. These clones ranged in size from 200 to 2000 base pairs and were shown to be viral specific. Some clones were capable of hybridizing with double-stranded (ds) RNAs of definite size. cDNA clones from this library were used to demonstrate the feasibility of diagnosing CTV infections, and two selected clones were able to differentiate selected virus strains by differential hybridization assays.[33,34] The poly-A tailing method was used to prepare libraries against Florida CTV strains.[31] These clones averaged about 800 to 1000 bases in length. However, two clones were selected which had a size 7350 and 8500 bases, with an overlap of 2,200 bases, which represents 70% of the total CTV genome. Further characterization and sequencing of these clones are underway (R. F. Lee and C. L. Niblett, unpublished).

A dsRNA of about 13.3×10^6, the size expected for the full-length replicative form, was reported from CTV-infected tissue.[35] In addition to the full-length replicative form, a number of less than full length dsRNAs, termed subgenomic, are also present. These subgenomic dsRNAs vary in size depending on virus strain and host.[36-40] The subgenomic dsRNAs are consistent for a given CTV strain in a given host and can produce a diagnostic pattern when electrophoresed on gels which can identify specific CTV strains.[38,40,41] Two of the subgenomic dsRNAs of 1.9×10^6 and 0.8×10^6 are present in all CTV strains.[38] The subgenomic dsRNAs have been isolated, ^{32}P labeled, and demonstrated to hybridize with the full-length replicative form[38] (S. Acikgoz and R. F. Lee, unpublished). Cloned cDNAs hybridize with the 13.3×10^6 dsRNA, and to the subgenomic dsRNAs.[34] The dsRNAs from CTV-infected plants have been purified and then translated in an *in vitro* reticulocyte lysate system.[42] Several denatured subgenomic dsRNAs were translated, but only the 0.8 and 1.7 kbp dsRNAs were efficiently translated. The major product of the 0.8 kbp dsRNA was a 27,000 polypeptide which immunoprecipitated with CTV antisera, whereas the major product of the 1.7 kbp dsRNA was a 21,000 polypeptide which was also immunoprecipitated by CTV antisera. Also a dsRNA of about 2.1×10^6 daltons from a mild Florida isolate has also been demonstrated to code for CTV coat protein by *in vitro* translation and immunoprecipitation using CTV-specific antisera.[43]

The CTV has been reported to have a coat protein with an estimated molecular weight of 25,000 +/− 1000 daltons.[44] More recent studies have indicated the presence of two coat proteins, a larger coat protein (CP1) with a molecular weight of about 23,000 and a smaller coat protein (CP2) with a molecular weight of about 21,000.[28,45] The CP1:CP2 ratio is about 5:1. Some Spanish CTV strains have a larger CP1 with an estimated molecular weight of 28,000 daltons, and two minor bands of 27,500 and 26,000 daltons.[46] *In vitro* translation of RNA extracted from purified virions yielded a product of about 26,000 daltons, which was immunoprecipitated using CTV-specific antisera,[28,47] and *in vitro* translation of the 1.7 and 0.8 kbp dsRNAs resulted in polypeptide products of about 21,000 and 27,000 daltons, respectively, which were immunoprecipitated.[42] Treatment of CP1 with trypsin and V8 protease resulted in the production of a polypeptide the same size as CP2.[28,45] These results suggest that the coat proteins of CTV are probably translated as part of a larger message, and after translation *in vivo*, processing takes place probably by proteolytic enzymes.

As with a number of different virus groups, aggregates of virus particles and related proteins occur in CTV-infected tissue and are usually called inclusion bodies or chromatic bodies.[48] Tristeza inclusion bodies occur in CTV-infected phloem and occasionally in the ground meristem of newly forming stems.[48,49] The inclusion bodies may be seen easily by light microscopy after Azure A staining.[50] Immunofluorescent techniques have been applied for detection of CTV inclusion bodies *in situ*.[51–53] A procedure for the purification of CTV inclusion bodies has been reported.[54]

9-5 EPIDEMIOLOGY

CTV is transmitted in a semipersistant manner by several aphid species.[15] Costa and Grant[55] found that a 24-hour acquisition feeding was the most effective for virus transmission and also reported that aphids did not remain infective after feeding on healthy plants for 24 hours or longer. The first vector reported for CTV was the oriental citrus aphid, *Aphis citricidus*, now called *Toxoptera citricida* Kirk.[13] Since this first report of aphid transmission, other aphid species which vector CTV have been reported.[21,55] These other aphid vectors of CTV are *Aphis gossypii*, *A. citricola,* and *T. aurantii*.[15] Of all the aphid vectors of CTV, *T. citricida* is the most efficient with up to 20% transmission with single aphid transfers as compared to 0.5 to 1.1% rates reported for *A. gossypii*.[56,57] Early studies of CTV transmission using *A. gossypii* consistently reported low rates of transmission even when using groups of aphids in the transmission tests. More recently, *A. gossypii* has become a relative efficient vector of CTV.[15,58] There appears to be a lag period of about 30 years when CTV is first introduced into a new area before *A. gossypii* becomes a relative efficient vector of CTV.[59] A similar lag period was observed in California[58] and in Florida.[60] This observation needs to be better explained and merits further research. Perhaps there is a helper factor which becomes widespread during the lag period and which enables the aphid to become a more efficient vector. Helper factors have been reported for other semipersistently transmitted viruses.[61]

The efficient aphid vector of CTV, *T. citricida*, was present in Brazil and Argentina when tristeza destroyed the citrus industry which at that time was on sour orange rootstock.[62] The relative effectiveness of *T. citricida* as compared to *A. gossypii* has recently been demonstrated by Yokomi & Damsteegt.[63] In side-by-side comparative tests conducted simultaneously in the quarantine facilities at Ft. Detrick, Maryland, *T. citricidus* and *A. gossypii* were used to transmit five CTV isolates from Brazil, Israel, Spain, China, and Hawaii. The probability of single aphid transmission for *T. citricidus* vs. *A. gossypii* for the five CTV isolates, respectively, were Brazil 25% vs. 0.6%; Israel 24.1% vs. 1.5%; Spain 18% vs. 2.7%; China 11.3% vs. 2.3%; and Hawaii 1.4% vs. 0%. Other research using the quarantine facilities also demonstrated the ability of Florida colonies of *A. gossypii* to transmit stem-pitting strains of CTV which are not present in Florida.[64]

The geographical area occupied by *T. citricida* has been expanding over the years. *T. citricida* has gradually been moving northward in South America. Its presence was first noted in Venezuela in 1976, while CTV has been known to be present in Venezuela since 1960.[65] Within a few years, *T. citricida* had become widespread, and this spread was tied with the subsequent rapid spread of CTV. By the mid-1980s, most of the citrus in Venezuela, originally mostly on sour orange rootstock, had been devastated.[66] The northward movement of *T. citricida* has continued. Identification of *T. citricida* was made from collections in Honduras by K. Andrews (personal communication) in 1984. In a recent survey in Central America,[67] *T. citricida* was found in Panama, Costa Rica, and El Salvador, but not in Nicaragua. *T. aurantii* and *A. gossypii* were present in all locations. Severe strains of CTV were detected in the major citrus region in the northern part of Costa Rica. In Nicaragua and El Salvador, the CTV was restricted to collections of citrus germplasm in experiment stations. In Panama, CTV was not detected but only one location was tested.

Citrus acreage has been expanding considerably in the last decade in several countries in Central America and the Caribbean Basin Islands. Most of the citrus industries in these areas, and those of Mexico and Texas, are predominantly on sour orange rootstock, which make these citrus areas extremely vulnerable to CTV. The confirmed presence of both severe CTV strains and *T. citricida* in Central America, combined with the predominance of sour orange as a rootstock, poses a serious threat to the continued citrus production in this geographical area of the world, including the citrus industries in Florida and California, where CTV already exists but *T. citricida* does not.

In the past decade, there have been several new destructive isolates of CTV which have appeared around the world. Strains of stem pitting on grapefruit were found in South Africa that severely reduce fruit size and shorten tree life.[68,69] In California, strains of CTV which carry seedling yellows and cause severe stem pitting on sweet orange were found to be spreading in the variety collection.[70,71] In Peru, severe strains of CTV were found which stem pit and reduce fruit size and vigor on Navel sweet orange scions.[72] In Israel, a very destructive decline strain of CTV has been found near Morasha which often kills the tree before CTV can be diagnosed by enzyme-linked immunosorbent assays, even when as many as 10 samples are taken from around each tree.[73,74] In the Southeast and Southwest Flatwoods of Florida, an epidemic of CTV quick decline started in the early 1980s and has not yet abated.[75] In Australia, there has been a small area in Queensland affected by a strain of CTV which severely stem pits navel sweet orange.[76] The continued appearance of new strains and development of decline epidemics serve as a reminder that tristeza is constantly on the move and becoming economically important in more places worldwide.

9-6 MANAGEMENT

There are several options for managing citrus tristeza virus involving regulation, cultural management, and biological control measures including mild strain cross-protection, genetic engineering for virus resistance, and breeding for virus resistance in commercially acceptable scions and rootstocks. The management options vary depending on whether CTV is absent or present in low or high incidence. These options are summarized in Table 9-1 and are discussed in greater detail here.

TABLE 9-1: SUMMARY OF METHODS AVAILABLE FOR MANAGEMENT OF CITRUS TRISTEZA VIRUS ACCORDING TO PREVALENCE OF THE DISEASE.

Management Measures for Citrus Tristeza Virus (CTV)[a]	
CTV prevalence	Management measure available
CTV absent	Quarantine
	Budwood certification and/or clean-stock programs
CTV present	
Low incidence	Eradication/suppression programs
	Budwood certification and/or clean-stock programs
High incidence	Tolerant rootstocks
	Varietal tolerance to stem pitting
	Mild strain cross-protection
	Genetically engineered resistance
	Budwood certification programs

[a]Modification of similar table previously published by Gonzalez-Garza.[140]

9-6-1 Regulatory

In citrus areas where tristeza is not already present, the first line of defense is quarantine. Clean-stock and certification programs for citrus serve a useful purpose in providing propagative materials and nursery stock which are true to type and free as possible from viruses and graft-transmissible diseases. Effective operation of clean-stock and certification programs also depends on strict enforcement of quarantine laws which prohibit the importation of plants or budwood infected with viruses, graft-transmissible diseases, or other pests. However, the regulations need to be flexible enough to allow for importation of virus-free material and provide the means by which virus-infected cultivars and varieties can be freed of disease agents by chemotherapy and/or shoot-tip grafting, followed by indexing for certification of the freedom of graft-transmissible agents. Without the flexibility and means to import desired cultivars which may be contaminated, illegal importations of budwood and plants will occur. Importation of CTV-infected budwood and plants is the means by which CTV has been spread around the world.

Clean-stock and certification programs are very effective for viruses which do not have a vector. Such programs are less useful against diseases which are vector transmitted, such as tristeza, stubborn, and greening, but even then they are useful to limit and prevent damage to citrus. In some areas, suppression programs have been carried out to contain tristeza and prevent its spread. For clean-stock, certification, and eradication-/suppression programs to function effectively and adequately to do the task intended, it is important that they have grower cooperation and the resources needed to accomplish their mission. It is informative to review briefly the tristeza situation and programs which have been used to control tristeza in several major citrus areas, namely Florida, California, and Spain. The tristeza situation and history of the eradication program for CTV in Israel has recently been reviewed.[21]

The history of CTV in Florida and the role of the Florida Budwood Registration Program.   The Florida Budwood Registration Program (FBRP) was started

because of the increasing grower awareness of the losses incurred due to virus diseases.[77] Following a symposium on certified nursery material and a discussion of the Texas budwood program, The Florida Horticultural Society passed a resolution requesting the establishment of a program for budwood certification in 1951. The State Plant Board took the leadership and, working with interested nurseryowners and growers, formulated the Statement of Policy under which the FBRP would operate.[77] Participation in the FBRP is voluntary. In late 1952, the FBRP began operation and within 10 years became the world's largest program for the registration of citrus budwood.[78] The diseases which were indexed for were CTV, psorosis, cachexia-xyloporosis, and exocortis; trees are observed to ensure freedom from Florida gummosis, blight, decline, leprosis, and bud mutations and trueness to type.[77]

Almost concurrent with the formation of the FBRP was the official finding of CTV in Florida by Grant in 1952.[79] However, there is circumstantial evidence that CTV was introduced much earlier. Swingle[80] noted in 1909 that over a five-year period in Florida he never saw a vigorous or productive Satsuma budded on sour orange. The Meyer lemon was introduced from the Orient into California, Texas, and Florida in 1908.[81] Indexing the clonal propagations from these introductions resulted in consistent recovery of seedling yellows CTV,[82] and it is generally accepted that the seedling yellows were present in the Meyer lemon in the original introductory material.[81] Some comments by Rhoads[83] in 1936 about blight could cause one to wonder if the blight as he was describing the disease included tristeza decline as well. For instance, he observed that blight is undoubtedly of most frequent occurrence in Lake County. In the Ridge section, where rough lemon stock was used almost exclusively for budded trees, blight was very rare. In a footnote, Rhoads comments on the difference in symptoms by the trees on rough lemon as compared to those on sour orange, and later he notes that "blight attacks the tree literally overnight, and affected branches die and the fruit withers and drops before attaining maturity."

While CTV was reported to be present in Florida, based on Mexican lime indexing, there were few tree losses. Knorr[84] conducted experiments during the 1950s whereby Mexican lime plants were placed in various locations throughout the citrus area, then later collected and then observed for CTV symptoms. The rate of infection of these trees was low, and most of the isolates produced mild to moderate symptoms on the Mexican lime. In 1956, tree loss in Orange County was reported,[85] and quick decline was first noted in the Fort Pierce area (M. Cohen, personal communication). In 1957, CTV was reported in Orange, Lake, and Seminole Counties.[86] In the 1960s, only isolated losses due to CTV were reported except in the Fort Pierce area, where natural spread of quick decline occurred.[87] Because of the increasing spread of CTV by the early 1960s, trees were no longer removed from the list of registered bud sources if they were CTV infected beginning in 1964. In 1975, a destructive outbreak of CTV loss was reported in Western Orange and southern Lake Counties.[88] Beginning in the 1980s, the prevalence and severity of CTV became much more obvious.[22] It became obvious that many trees which had been propagated on sour orange were not growing well. Several thousand trees in nurseries and planted in groves were destroyed because of severe stunting. The source of severe CTV was in budwood source trees; many of the registered trees were in fact infected with severe CTV strains. In 1984, a voluntary program was started whereby budsticks from bud source trees were collected under supervision of inspectors (C. O. Youtsey, personal communication). Using sour orange liners donated in commercial nurseries, 10 sour orange liners were budded from each source. Each clonal selection was indexed together in a block, and several replications of virus-free and mild-strain-infected bud sources of each clonal selection were included as a control from which to be able to judge the degree of dwarfing which might be caused to budlings on the sour orange rootstocks by the CTV strains in the budwood source trees. The liners were budded in February and March 1984. In September, tree performance was judged by visual evaluation and caliper measurements of the stem diameter. There were obvious differences in rate of growth. In general, about one third of the budwood source trees indexed contained CTV strains which dwarfed tree growth to the extent that the average stem diameter was one half or less of the average for the control plants. While this biological indexing method worked well, disaster struck in September 1984 when a citrus leaf spot disease[89] was thought to be citrus canker, *Xanthomonas campestris* pv. *citri* (Hasse) Dye.[90] One biological index plot was destroyed by the nursery cooperator before even a visual evaluation had been made for stunting on sour orange from the bud source trees. The resultant quarantine on movement of plant material for fear of transporting canker resulted in many propagations coming from well-performing trees often on CTV-tolerant rootstocks located in the same area as the nursery, a practice which may have further spread severe strains of CTV. A greenhouse method was developed for indexing the budwood sources in 1985 (S. M. Garnsey, unpublished). Fifteen sour orange liners in the greenhouse were budded with each budwood source and proper controls of virus-free and mild-strain-infected material. The buds were forced, and ten were selected after forcing for uniformity. The length of the growth from the bud was indirectly correlated to the severity of the CTV isolate present. This methodology has never been incorporated in the FBRP because the funds for needed facilities and personnel were not provided.

By the late 1980s, the FBRP was faced with the increasing reservoir of severe strains of CTV in budwood sources which were being spread throughout the state by propagation on CTV-tolerant rootstock.[91] There is also concern about the possible introduction of exotic CTV isolates into Florida which cause stem pitting to sweet orange and/or grapefruit scions, and the possible introduction of *T. citricida*, the most efficient vector of CTV.[22,91] The problem of severe CTV strains in budwood sources is being dealt with by making fundamental changes in the operation of FBRP.[91] Briefly, trees harboring severe strains of CTV which affect growth on sour orange are dropped as registered budwood source trees. Foundation trees will be maintained under protected conditions. Budwood-increase nurseries will be used to increase bud sources obtained from the protected foundation trees. These increase nurseries will be maintained for a short period of time, up to 18 months, for cutting of buds. In the budwood-increase nurseries, some propagations will be made on sour orange as well as a CTV-tolerant rootstock for growth comparison purposes. Budwood sources which do not grow well on sour orange would be identified and eliminated. This system will allow for better monitoring of mother trees for severe CTV isolates, and also allows for more rapid introduction of new scion varieties.

California.   CTV has been widespread and endemic in the southern California citrus-growing areas of Los Angeles, Orange, Riverside, San Bernardino, and Ventura counties. The Coachella Valley and the San Joaquin Valley remain as areas with very little tristeza. In 1979, a severe seedling yellows strain of CTV which caused stem pitting on sweet orange and grapefruit was discovered spreading in the citrus variety collection at the University of California, Riverside.[18] While these strains were transmitted very poorly by *A. gossypii* in trials 30 to 40 years previously, they were presently very efficiently transmitted by *A. gossypii*.[58] An extensive indexing and eradication program was conducted using indicator plants to detect the severe seedling yellow strains.[92] In all, over 20,000 trees were indexed over five years, and a total of 262 trees were found positive for seedling yellows and stem pitting (D. J. Gumpf, personal communication). Valuable citrus varieties were freed of viruses by shoot-tip micrografting, and the variety collection was protected under screen. The California Department of Food and Agriculture has now expanded the biological indexing to most of the southern California citrus area, where trees infected with seedling yellows are eradicated once they are detected, based on the results of a biological index.

A major objective in California has been to maintain the San Joaquin Valley as a virus-free area.[92] The Central California Tristeza Eradication Agency is charged with this responsibility and has support and aid from the California Department of Food and Agriculture, the University of California, and the citrus industry. All CTV-infected trees are removed from this area regardless of the severity of the infecting strain. Routine indexing using ELISA is used for CTV diagnosis, confirmed by biological indexing if necessary. This program began in 1963, and over 4 million trees have been indexed. This program has so far managed to keep tristeza from becoming established in the San Joaquin Valley. While CTV has not been eliminated from the San Joaquin Valley, this eradication program, perhaps the longest-running program of its kind in the world, has been effective at maintaining the Valley as essentially virus free.

Spain. CTV was first recognized to be present in Spain after an outbreak of decline on sour orange rootstock in the Ribera Alta area in Southern Valencia. An estimated 10 million trees have been lost in the provinces of Alicante, Castellon, and Valencia since then.[93] Since the original outbreak of CTV decline on sour orange, other foci of decline have appeared, probably originating from uncontrolled movement of infected budwood.[94] A severe CTV strain causing seedling yellows and stem pitting on grapefruit recently was discovered in a Satsuma cultivar which had been illegally imported into Spain, most likely from Japan.[95,96] An eradication program was initiated in the Valencia citrus area. Growers who had planted propagations of the illegally imported Satsuma infected with SY CTV causing stem pitting on grapefruit were required to have their propagations evaluated by ELISA. If CTV was identified by the ELISA assay, the planting was destroyed and the grower compensated. Plantings of citrus surrounding the Satsuma plantings are being surveyed by ELISA, and selected strains will be biologically indexed.[92]

9-6-2 Cultural

Areas which have been using sour orange as a rootstock often switch to CTV-tolerant rootstocks once CTV isolates that cause decline on sour orange become prevalent. There are some limitations to choosing an alternative rootstock. In areas where citrus blight is endemic, most alternative rootstocks are highly susceptible to blight. In Venezuela, citrus was replanted predominantly on Volkamer Lemon after CTV decline destroyed most of the trees on sour orange rootstock. When these plantings reached about five years of age, citrus blight began occurring in epidemic proportions.[97] Many trifoliate and trifoliate hybrids commonly used as rootstocks do not perform well horticulturally on calcareous or high pH soils.[98]

Areas having a severe problem with stem pitting on grapefruit will revert to planting more CTV-tolerant sweet orange varieties. Areas having stem pitting on sweet orange scions will revert to planting more tolerant sweet orange varieties or mandarins which are very tolerant of CTV stem-pitting strains.

Temple orange trees are attractive to large aphid populations presumably because of their constant flush patterns. In Florida, a block of Temples almost always is close to an area having CTV decline. Indexing on CTV isolates from Temple blocks has not indicated that extraordinary severe isolates of CTV are present in Temple trees, and passage of CTV isolates through Temple trees has not resulted in a change of biological activity of the CTV isolates.[99] The evidence suggests that the higher aphid populations around the Temple blocks probably play the greatest role in the observed association between CTV decline areas and Temple blocks.

Proper water management of citrus may be important in reducing tree loss due to CTV QD in some instances. Lee and Brlansky[100] recently collected several CTV isolates from QD trees from eight different locations in the citrus area in Florida and biologically indexed them on a battery of indicator plants.[23] Only three of the eight isolates induced decline on the sweet orange on sour orange indicator plants upon indexing, and the remaining five isolates little or no dwarfing. These five isolates were collected from locations affected by fluctuating water tables where the root systems were occasionally water logged for longer than 48 hours, conditions which would damage or kill citrus roots, making the trees more susceptible to CTV decline.[100]

9-6-3 Chemical.

No effective chemical control of CTV has been reported. It has been suggested that insect control over a large area may reduce spread of CTV by reducing aphid populations, but sufficient research has not been done to demonstrate the effectiveness or to show that this approach is cost effective.

9-6-4 Biological

Disease resistance.   Many citrus cultivars are tolerant of CTV infection, meaning that CTV replicates in the host but that no symptoms are expressed in infected plants. Some citrus relatives are immune to CTV infection, meaning that CTV does not replicate in the host. Examples include *Poncirus trifoliata, Swinglea glutinosa*, and *Severinia buxifolia*.[21] *P. trifoliata* can be hybridized to citrus, and many hybrid cultivars have been obtained which are immune to CTV replication. These cultivars are widely used as CTV-tolerant rootstocks; for example, Carrizo. The plant breeding unit at the USDA, Orlando, Florida[101,102] has been able to incorporate the immunity which exists in *P. trifoliata* into hybrid lines, which offers the potential of incorporating this immunity into scions having commercially acceptable fruit quality. Serological analysis of progeny of first-generation hybrids of *P. trifoliata* with CTV-susceptible parents finds segregation for CTV immunity, indicating that immunity was not homozygous.[101,102] Crosses between

CTV-immune hybrids of *P. trifoliata* and susceptible host also yielded segregating populations.[103] Recent breakthroughs in development of the means to obtain somatic hybrids between protoplasts of non-sexually compatible citrus relatives have resulted in the production of *Severinia* spp. and Hamlin sweet orange hybrids.[104] These plants, however, are not immune to CTV but may prove useful as CTV-tolerant rootstocks.

Mild strain cross-protection.   Cross-protection is the phenomenon whereby infection of a plant with a mild strain of a virus protects against superinfection and damage by a related and usually more severe strain.[105] Some considerations should be made before employing mild strain cross-protection as a control strategy for tristeza.[22] The disease must be endemic with no possibility of control by eradication or suppression. The mild isolate used must be mild in all citrus cultivars and susceptible hosts grown in the area. Consideration must be given to the idea that plants infected with a mild virus isolate might develop a synergistic reaction if infected with other viruses. The mild virus isolate selected for use must be stable. Artificially attenuated or mutated viruses might revert back to the original severe form. The benefit received should justify the work involved in implementing mild strain cross-protection. The possible slight reduction in tree yield and performance by the protecting mild isolate should be acceptable in terms of the potential loss which could incur without mild strain cross-protection.

Cross-protection against CTV in citrus usually deals with either decline on sour orange or protection against stem pitting. In areas where stem-pitting strains are a production problem, the strains inducing decline on sour orange are so prevalent that sour orange is not seriously considered for use as a rootstock. Most research has been done on evaluation of cross-protection against stem pitting. This has proved to be an effective control strategy. The first evidence that mild strains of CTV could cross-protect against severe strains was presented by Grant and Costa in 1951.[106] Since that time, CTV isolates which offer cross-protection against stem pitting have been selected empirically and are used commercially in many areas.[107] Over 15 million trees are cross-protected against CTV-induced stem pitting on Pera sweet orange in Brazil,[108] and cross-protection against stem pitting on grapefruit is being used on a large scale in South Africa.[109] Mild strain cross-protection is also used for grapefruit in Australia,[110–112] in Japan,[113–116] *Citrus hystrix* in Reunion,[117] limes in India,[118] and lime and grapefruit in Venezuela.[119]

Because most countries abandon sour orange as a rootstock when CTV quick decline occurs, relatively little research has been done or published on the effectiveness of cross-protection against quick decline. However, in the past decade some progress has been made to demonstrate that cross-protection against decline on sour orange is feasible and possible. Thornton et al.[112] were among the first to report on cross-protection experiments where trees were established on sour orange rootstock. VanVuuren et al.[120] evaluated 11 mild CTV isolates from Florida, Israel, and South Africa in Valencia on sour orange rootstock planted in the field in South Africa, where they were exposed to a natural challenge of severe field CTV isolates by *T. citricidus*. Quick decline appeared after three years; one of the isolates appeared to have good cross-protection against quick decline and may have commercial value. In a field plot in Florida,[120] 14 mild isolates were evaluated in grapefruit and Hamlin scions on sour orange rootstocks exposed to natural challenge in an area where severe decline strains are prevalent. Trees were monitored for decline and reactivity to a monoclonal antibody (MCA13), which reacts preferentially for severe CTV isolates,[122] after three years in the field.[121] There was good agreement between reactivity with MCA13 monoclonal antibody and visual tree decline. The CTV decline was more severe in the Hamlin scion than in the grapefruit. Four mild strains in Hamlin had none or very low rates of tree decline and reactivity to MCA13; only 6% of the grapefruit trees were infected with severe CTV, suggesting that 11 of the 14 mild strains being evaluated might have some cross-protection for grapefruit on sour orange. A similar survey[123] was made of two additional field experiments in Florida, where 14 mild isolates were being evaluated in Hamlin on sour orange and smooth flat Seville rootstocks and grapefruit, and on sour orange challenged by aphids forced to feed on plants infected with a severe CTV isolate and by graft challenge, respectively. No trees were declining in either experiment after five years and four years, respectively. MCA13 monoclonal antibody reacted to five trees in the aphid challenge plot, but these were not the trees which had been intentionally challenged. There were no trees which reacted to MCA13 monoclonal antibody in the graft-challenged plot. Thus, it would seem that with proper selection, mild CTV isolates could be selected which would cross-protect against decline on sour orange rootstock.

There are some problems involved in using cross-protection as a control strategy for CTV:

1. Budwood source trees have to be biologically indexed on a frequent basis to ensure that a severe strain has not been introduced.[69]

2. Grapefruit trees in Florida often have an erratic distribution of CTV within the scion which can result in as much as 50% of the buds being propagated virus free.[124]

3. Rootstock liners can be naturally infected with severe CTV before being budded, resulting in the apparent breakdown of cross-protection when planted into the field.[125]

4. Hot weather often results in a thermotherapy to prevent mild CTV from moving into new flushes, thus exposing virus-free, nonprotected tissue.[22,126]

5. Mild strains of CTV which have cross-protection ability must be empirically selected from a multitude of strains.[108]

6. The symptom expression and the degree of cross-protection of mild strains is influenced by the environment, and some isolates may express more severe symptoms and/or fail to cross-protect as well under different climatic conditions than those in which the isolate was originally selected[112] (L. J. Marais and R. F. Lee, unpublished).

Selection of cross-protecting CTV isolates.   The selection of mild isolates which have cross-protection potential has thus far been an empirical process. While mild isolates of CTV may be common, relatively few isolates have any potential for cross-protection.[108] In Brazil, only about 15 isolates out of 300 isolates tested had any cross-protection ability.[108] Muller and Costa's results[127] suggest that selection of CTV isolates from outstanding trees in areas with heavy losses due to CTV are the best means to find isolates for cross-protection. Other approaches which have been used include attenuation of severe isolates by selection from grapefruit recovered from seedling yellows isolate infection, vector transmission, and passage through *Passiflora* spp.[128] The empirical screening of large numbers of isolates for cross-protection is a time-consuming process. There have been some evaluation procedures developed recently which help eliminate unuseful isolates relatively quickly. Aided by glasshouse evaluation, VanVuuren and Moll[129] selected mild strains for cross-protection on limes. Ten plants were inoculated with each mild isolate, and five of these were later challenged by graft inoculation. The plants were kept to a single stem and evaluated by determining the length of growth, mass (wt) of growth, and mass of roots. ELISA was used to verify the presence of mild strains. Roistacher et al.[128] used a similar method for greenhouse selection of CTV strains protected against stem pitting on sweet orange and grapefruit seedlings. Yokomi et al.[130] evaluated limes previously inoculated with mild strains under greenhouse conditions using *A. gossypii* fed on a plant infected with a severe challenge isolate. The plants were challenged twice using aphids. The mild isolates evaluated were rated as protective, nonprotective, or inconclusive based on symptoms which developed, and infections were verified by ELISA. Yokomi's results demonstrate the need to challenge with more than one severe strain. Rocha-Peña et al. (unpublished) used similar methodology as of VanVuuren and Moll[129] but evaluated mild strains for cross-protection of Valencia on sour orange against decline. The preliminary screening of mild strains under glasshouse conditions greatly reduces the time needed to screen for isolates which may have cross-protection ability and eliminates the possibility that a freeze can prematurely end the evaluation before results are obtained.

Genetically engineered protection.   An alternative approach to mild strain cross-protection of citrus against CTV is the use of genetic engineering. There are several possible approaches to using genetic engineering to achieve virus resistance. First, satellite RNAs of cucumber mosaic virus and tobacco ringspot virus have been cloned and these cDNAs inserted into vectors which enable transformation of plant tissue.[131,132] This approach has afforded some degree of virus resistance in the transgenic plants. Second, antisense RNAs of cucumber mosaic virus,[133] tobacco mosaic virus,[134] and potato virus X[135] when expressed in transgenic plants have expressed a small degree of viral resistance. Third, the incorporation of the viral coat protein gene and expression of the viral coat protein in transgenic plants has been demonstrated to afford a high degree of virus resistance in transgenic plants.[135,136] The plants transformed with the viral coat protein gene result in plants showing the benefits of cross-protection without the virus being present.[136-138] Any of these genetic engineering strategies to obtain virus resistance are applicable to CTV, but the coat protein mediated protection in transgenic citrus may provide the best possibility of protecting against CTV decline on sour orange and/or CTV-induced stem pitting on scions. The use of transformed citrus for CTV resistance would avoid most of the problems that occur when using conventional mild strain cross-protection as a control strategy. This approach also should provide a more broad-based protection against the diverse strains of CTV.

9-7 CONCLUSION

There are several gaps in the knowledge of citrus tristeza virus. An urgent need is the ability to differentiate decline inducing from naturally occurring mild strains of the virus on a timely basis. At present, the only reliable method to do this is by biological indexing on a battery of indicator plants,[23] a procedure which requires up to a year and a lot of greenhouse space. Some progress has been made on strain differentiation in Florida for detection of CTV decline on sour orange strains by the production of strain-specific monoclonal antibodies.[122] A selected monoclonal antibody specifically reacts with severe CTV strains but unfortunately does not recognize all decline-inducing strains.[139] There is a need to continue to produce and screen for monoclonal antibodies which might be capable of differentiating the different major biological activities associated with CTV (e.g., stem pitting on grapefruit, stem pitting on sweet orange, SY, decline on sour orange, and mild strains). Selection of cDNAs capable of strain differentiation would be another approach to this problem.

There is a need to characterize the CTV genome for strains representing the different biological activities. Complete sequencing of the CTV genome would reveal the presence of open reading frames and make possible the discovery of nonstructural proteins which are encoded by the virus. At present, the only genome product which has been identified is the coat protein. Differences in sequence information between strains differing in biological activity may be of help in selecting fragments useful for strain differentiation. The success of genetically engineered virus resistance depends on characterization of the CTV genome.

There is a dearth of knowledge of virus-vector relationships. Why are some aphid species vectors and others are not? Why are some species more efficient in transmitting the virus, and how are the CTV particles carried by the aphids? Why is there a lag period between the introduction of severe CTV strains and the ability of *A. gossypii* to transmit these strains efficiently?

Of special concern is the northward movement of the efficient aphid vector of CTV, *T. citricidus*, into Central America. Concurrent with this northward movement from Brazil has been the rapid spread of severe CTV strains in neighbouring contiguous countries. Most of the citrus industries in Central America and the Caribbean Islands and those from Mexico and Texas are on sour orange rootstock. Florida also has an estimated 40 million producing trees on sour orange. These areas are all threatened and prone to suffer the catastrophic consequences that Brazil and Argentina did in the 1930s and Venezuela in the 1980s, when CTV practically eliminated their citrus industries. There is an urgent need to slow the northward movement of severe CTV strains by implementation of quarantine, budwood certification, biological indexing, and clean-stock programs. Also urgent is the need to determine the present geographical distribution of *T. citricida* in Central America, to try to implement biological control measures to slow its northward movement, and to determine what horticultural and chemical control measures might reduce populations. There is a need for a coordinated effort in the Caribbean Basin and Central America to address this problem.

9-8 REFERENCES

1. Marloth, R. H., The citrus rootstock problem: Citrus tree propagation, *Farming So. Africa*, 13, 226, 1938.

2. Webber, H. J., The "Tristeza" disease of sour-orange rootstock, *Proc. Am. Soc. Hort. Sci.*, 43, 160, 1943.

3. Toxopeus, H. J., Stock-scion incompatibility in *Citrus* and its cause, *Jour. Pom. Hort. Sci.*, 14, 360, 1937.

4. Zeman, V., Una enfermada nueva en los naranjales de Corrientes, *Physis*, 19, 410, 1930.

5. Moreira, S., Observacoes sobre a "tristeza" dos citrus ou "podridao das radicelas," *O. Biologico*, 8, 269, 1942.

6. Bitancourt, A. A., A Podridao das radicelas dos citrus na provincia de Corrientes, Argentina, *O Biologico*, 6, 285, and 356, 1940.

7. Fawcett, H. S. and Wallace, J. M., Evidence of the virus nature of citrus quick decline, *California Citrograph*, 32, 50 and 88, 1946.

8. McAlpin, D. M., Parsai, P. S., Roberts, R. and Hope, R. H., "Bud union decline" disease of Citrus, *Victoria Dept. Agr. Jour.*, 46, 25, 1948.

9. Cameron, A. E., "Tristeza" of Seville rootstock, *Citrus News*, 21, 20, 1945.

10. Hughes, W. A. and Lister, C. A., Lime disease in the Gold Coast, *Nature*, 164, 880, 1949.

11. Richards, A. V., Stock-scion influence in citrus, *Trop. Agr.* (Ceylon), 101, 61, 1945.

12. Giacometti, D. C. and Storey, W. B., Citrus quick decline discovered in Hawaii, *California Citrograph*, 37, 357, 1952.

13. Camp, A. F., Report of scientific observations in Brazil, Mimeo, *Citrus Experiment Station, Lake Alfred, Fl,* 19 pp., 1943.

14. Meneghini, M., Sobre a natureza e transmissibilidade da doenca "Tristeza" dos Citrus, *O Biologico*, 7, 285, 1946.

15. Bar-Joseph, M., Garnsey, S. M. and Gonsalves, D., The closteroviruses: A distinct group of elongated plant viruses, *Adv. Virus Research*, 25, 93, 1979.

16. Wallace, J. M., Oberholzer, P. C. J. and Hofmeyer, J. D. J., Distribution of viruses of tristeza and other diseases of citrus in propagative material, *Plant Dis. Reptr.*, 40, 3, 1956.

17. Oberholzer, P. C. J., Mathews, I. and Stiemie, S. F., The decline of grapefruit trees in South Africa. A preliminary report on so-called "stem-pitting," *Union South Africa Dept. Agr. Sci. Bull.*, No. 297, 1949.

18. Roistacher, C. N., A blueprint for disaster—part one: The history of seedling yellows disease, *Citrograph*, 67, 4, and 24, 1981.

19. Bar-Joseph, M., Roistacher, C. N., Garnsey, S. M. and Gumpf, D. J., A review on tristeza, an ongoing threat to citriculture, *Proc. 1st Int. Soc. Citric.,* pp. 419–423. ISC, Okitsu, Japan, 1981.

20. Davino, M. and Catara, A., La tristeza degli agrumi, *Inf. Fitopathologico*, 36, 9, 1986.

21. Bar-Joseph, M., Marcus, R. and Lee, R. F., The continuous challenge of citrus tristeza virus control, *Ann. Rev. Phytopathol.*, 27, 291, 1989.

22. Lee, R. F., Brlansky, R. H., Garnsey, S. M. and Yokomi, R. K., Traits of citrus tristeza virus important for mild strain cross protection of citrus: The Florida approach, *Phytophylactica*, 19, 215, 1987.

23. Garnsey, S. M., Gumpf, D. J., Roistacher, C. N., Civerolo, E. L., Lee, R. F., et al., Towards a standardized evaluation of the biological properties of citrus tristeza virus, *Phytophylactica,* 19, 151, 1987.

24. Garnsey, S. M., Civerolo, E. L., Gumpf, D. J., Yokomi, R. K. and Lee, R. F., Development of a worldwide collection of citrus tristeza virus isolates. In *Proc. 11th Conf. Int. Organ. Citrus Virol.* (R. H. Brlansky, R. F. Lee and L. W. Timmer, eds.), p. 113, IOCV, Riverside, CA, 1991.

25. Bar-Joseph, M., Gumpf, D. J., Dodds, J. A., Rosner, A. and Ginzburg, I., A simple purification method for citrus tristeza virus and estimation of its genome size, *Phytopathology,* 75, 195, 1985.

26. Gonsalves, D., Purcifull, D. E. and Garnsey, S. M., Purification and serology of citrus tristeza virus, *Phytopathology*, 68, 553, 1978.

27. Lee, R. F., Garnsey, S. M., Brlansky, R. H. and Goheen, A. C., A purification procedure for enhancement of citrus tristeza virus yields and its application to other phloem-limited viruses, *Phytopathology*, 77, 543, 1987.

28. Lee, R. F., Calvert, L. A., Nagel, J. and Hubbard, J. D., Citrus tristeza virus: Characterization of coat proteins, *Phytopathology*, 78, 1221, 1988.

29. Tsuchizaki, T., Sasaki, A. and Saito, Y., Purification of citrus tristeza virus from diseased citrus fruits and the detection of the virus in citrus tissues by fluorescent antibody techniques, *Phytopathology,* 68, 139, 1978.

30. Muller, G. W. and Garnsey, S. M., Susceptibility of citrus varieties, species, citrus relatives, and not-rutaceous plants to slash cut mechanical inoculation with citrus tristeza virus (CTV). In *Proc. 9th Conf. Int. Org. Citrus Virologists* (S. M. Garnsey, L. W. Timmer and J. A. Dodds, eds.), pp. 33–40, IOCV, Riverside, CA, 1984.

31. Calvert, L. A., Complementary DNA cloning and partial molecular characterization of citrus tristeza and citrus variegation viruses, Ph.D. thesis, Univ. Fla., Gainesville, 60 pp., 1987.

32. Rosner, A., Ginzburg, I. and Bar-Joseph, M., Molecular cloning of complementary DNA sequences of citrus tristeza virus RNA, *J. Gen. Virol.*, 64, 1757, 1983.

33. Rosner, A. and Bar-Joseph, M., Diversity of citrus tristeza virus strains indicated by hybridization with cloned cDNA sequences, *Virology*, 139, 189, 1984.

34. Rosner, A., Lee, R. F. and Bar-Joseph, M., Differential hybridization with cloned cDNA sequences for detecting a specific isolate of citrus tristeza virus, *Phytopathology*, 76, 820, 1986.

35. Dodds, J. A. and Bar-Joseph, M., Double-stranded RNA from plants infected with closteroviruses, *Phytopathology*, 73, 419, 1983.

36. Dodds, J. A., Jarupat, T., Lee, J. G. and Roistacher, C. N., Effect of strain, host, time of harvest, and virus concentration on double-stranded RNA analysis of citrus tristeza virus, *Phytopathology*, 77, 442, 1987.

37. Dodds, J. A., Jarupat, T., Roistacher, C. N. and Lee, J. G., Detection of strain specific double-stranded RNAs in citrus species infected with citrus tristeza virus: A review, *Phytophylactica,* 19, 131, 1987.

38. Dodds, J. A., Jordan, R. L., Roistacher, C. N. and Jarupat, T., Diversity of citrus tristeza virus isolates indicated by dsRNA analysis, *Intervirology,* 27, 177, 1987.

39. Dodds, J. A., Tamaki, S. J. and Roistacher, C. N., Indexing of citrus tristeza virus double-stranded RNA in field trees. In *Proc. 9th Conf. Int. Organ. Citrus Virol.* (S. M. Garnsey, L. W. Timmer and J. A. Dodds, eds.), pp. 327–329, IOCV, Riverside, CA, 1984.

40. Lee, R. F., Use of double-stranded RNAs to diagnose citrus tristeza virus strains, *Proc. Fla. State Hort. Soc.*, 97, 53, 1984.

41. Moreno, P., Guerri, J. and Munoz, N., Identification of Spanish strains of citrus tristeza virus by analysis of double-stranded RNA, *Phytopathology*, 80, 477, 1990.

42. Dulieu, P. and Bar-Joseph, M., *In vitro* translation of the citrus tristeza virus coat protein from an 0.8 kbp double stranded RNA segment, *J. Gen. Virol.*, 71, 443, 1989.

43. Niblett, C. L., Stark, D. M., Lee, R. F. and Beachy, R. N., Identification of the coat protein gene of citrus tristeza virus by denaturation and translation of double-stranded RNAs, *Phytopathology*, 81, 698, 1990.

44. Bar-Joseph, M., Loebenstein, G. and Cohen, J., Further purification and characterization of particles associated with citrus tristeza disease, *Virology*, 50, 821, 1972.

45. Lee, R. F. and Calvert, L. A., Polypeptide mapping of citrus tristeza virus strains, *Phytophylactica*, 19, 205, 1987.

46. Guerri, J., Moreno, P. and Lee, R. F., Identification of citrus tristeza virus strains by peptide maps of virion coat protein, *Phytopathology*, 80, 692, 1990.

47. Nagel, J., Hiebert, E. and Lee, R. F., Citrus tristeza virus RNA translated with a rabbit reticulocyte lysate: Capsid protein identified as one of the products, *Phytopathology*, 72, 953, 1982.

48. Schneider, H., Cytological and histological aberrations in woody plants following infection with viruses, mycoplasmas, rickettsias, and flagellates, *Annu. Rev. Phytopathol.*, 11, 119, 1973.

49. Brlansky, R. H., Inclusion bodies produced in *Citrus* spp. by citrus tristeza virus, *Phytophylactica*, 19, 211, 1987.

50. Christie, R. G. and Edwardson, J. R., Light and electron microscopy of plant virus inclusions, Fla. Agric. Exp. Stn. Mongr. Ser. No. 9., Univ. Fla., Gainesville, p. 155, 1977.

51. Brlansky, R. H., Lee, R. F. and Garnsey, S. M., *In situ* immunofluorescence for the detection of citrus tristeza virus inclusion bodies, *Plant Dis.*, 72, 1039, 1988.

52. Sasaki, A., Tsuchizaki, T. and Saito, Y., Discrimination between mild and severe strains of citrus tristeza virus by fluorescent antibody technique, *Ann. Phytopathol. Soc. Jpn.*, 44, 205, 1978.

53. Tsuchizaki, T., Sasaki, A. and Saito, Y., Purification of citrus tristeza virus from diseased citrus fruits and the detection of the virus in citrus tissues by fluorescent antibody techniques, *Phytopathology*, 68, 139, 1978.

54. Lee, R. F., Garnsey, S. M., Brlansky, R. H. and Calvert, L. A., Purification of inclusion bodies of citrus tristeza virus, *Phytopathology*, 78, 1221, 1982.

55. Costa, A. S. and Grant, T. J., Studies on transmission of the tristeza virus by the vector *Aphis citricidus*, *Phytopathology*, 41, 105, 1951.

56. Raccah, B., Loebenstein, G., Bar-Joseph, M. and Oren, Y., Transmission of tristeza by aphids prevalent on citrus, and operation of the tristeza suppression program in Israel. In *Proc. 7th Conf. Int. Organ. Citrus Virologists* (E. C. Calavan, ed.), pp. 47–49, Univ. of Calif. Press, Riverside, CA, 1976.

57. Roistacher, C. N. and Bar-Joseph, M., Transmission of tristeza and seedling yellow-tristeza virus by *Aphis gossypii* from sweet orange grapefruit and lemon to Mexican lime, grapefruit and lemon. In *Proc. 9th Conf. Int. Organ. Citrus Virol.* (S. M. Garnsey, L. W. Timmer and J. A. Dodds, eds.), pp. 9–18, IOCV, Riverside, CA, 1984.

58. Roistacher, C. N., Nauer, E. M., Kishaba, A. and Calavan, E.C., Transmission of citrus tristeza virus by *Aphis gossypii* reflecting changes in virus transmissibility in California. In *Proc. 8th Conf. Int. Organ. Citrus Virologists* (E. C. Calavan, S. M. Garnsey and L. W. Timmer, eds.), pp. 76–82, Univ. Calif. Press, Riverside, CA, 1980.

59. Bar-Joseph, M., Cross protection incompleteness: A possible cause for natural spread of citrus tristeza virus after a prolonged lag period in Israel, *Phytopathology*, 68, 1110, 1978.

60. Yokomi, R. K. and Garnsey, S. M., Transmission of citrus tristeza virus by *Aphis gossypii* and *Aphis citricola* in Florida, *Phytophylactica*, 19, 169, 1987.

61. Hunt, R. E., Nault, L. R. and Gingery, R. E., Evidence for infectivity of maize chlorotic dwarf virus and for a helper component in its leafhopper transmission, *Phytopathology*, 78, 499, 1988.

62. Roistacher, C. N. and Bar-Joseph, M., Aphid transmission of citrus tristeza virus: A review, *Phytophylactica*, 19, 163, 1987.

63. Yokomi, R. K. and Damsteegt, V. D., Comparison of citrus tristeza virus transmission efficiency between *Toxoptera citricidus* and *Aphis gossypii*. *Proc. Aphid-Plant Interactions: Populations to Molecules, an International Symposium*, p. 319, OSU, Stillwater, 1991.

64. Yokomi, R. K., Garnsey, S. M., Civerolo, E. L. and Gumpf, D. J., Transmission of exotic citrus tristeza virus isolates by a Florida colony of *Aphis gossypii*, *Plant Disease*, 73, 552, 1989.

65. Geraud, F., El afido negro de los citricos, *Toxoptera citricida* Kirkaldy en Venezuela (Resumen), 1. Encuentro Venezolano and entomologia. *UCV Fac. Agronomia Inst. Zool. Agric.*, Maracay, 1976.

66. Mendt, R., Plaza, G., Boscan, R., Martinez, J. and Lastra, R., Spread of citrus tristeza virus and evaluation of tolerant rootstocks in Venezuela. In *Proc. 9th Conf. Int. Organ. of Citrus Virol.* (S. M. Garnsey, L. W. Timmer and J. A. Dodds, eds.), pp. 95–99, IOCV, Riverside, CA, 1984.

67. Lastra, R., Meneses, R., Still, P. E. and Niblett, C. L., The citrus tristeza virus situation in Central America. in *Proc. 11th Conf. Int. Organ. Citrus Virol.* (R. H. Brlansky, R. F. Lee and L. W. Timmer, eds.), pp. 166–169, IOCV, Riverside, CA, 1991.

68. Marais, L. J., Tristeza stem pitting. The problem facing the South African citrus industry. Is it caused by new strains?, *Citrus and Subtropical Fruit J.*, March 1981, 11, 1981.

69. DaGraca, J. V., Marais, L. J. and Von Broembsen, L. A., Severe tristeza stem pitting in young grapefruit, *Citrus and Subtropical Fruit J.*, Nov. 1982, 18, 1982.

70. Calavan, E. C., Harjung, M. K., Blue, R. L., Roistacher, C. N., Gumpf, D. J. and Moore, P. W., Natural spread of seedling yellows and sweet orange and grapefruit stem pitting tristeza viruses at the University of California, Riverside. In *Proc. 8th Conf. Int. Organ. Citrus Virologists* (E. C. Calavan, S. M. Garnsey and L. W. Timmer, eds.), pp. 69–75, Univ. Calif. Press, Riverside, CA, 1980.

71. Roistacher, C. N., A blueprint for disaster—part 3: The destructive potential for seedling yellows, *Citrograph*, 67, 48, 1982.

72. Roistacher, C. N., Observation of the decline of sweet orange trees in coastal Peru caused by stem pitting tristeza, *FAO Plant Prot. Bull.*, 36(1), 19, 1988.

73. Ben Zeev, I. S., Bar-Joseph, M., Nitzan, Y. and Marcus, R., A severe citrus tristeza virus isolate causing the collapse of trees of sour orange before virus is detectable throughout the canopy, *Ann. Appl. Biol.*, 114, 292, 1989.

74. Bar-Joseph, M. and Nitzan, Y., The spread and distribution of a severe citrus tristeza virus in sour orange seedings. In *Proc. 11th Conf. Int. Organ. Citrus Virol.* (R. H. Brlansky, R. F. Lee and L. W. Timmer, eds.), pp. 162–165, IOCV, Riverside, CA, 1991.

75. Brlansky, R. H., Pelosi, R. R., Garnsey, S. M., Youtsey, C. O., Lee, R. F., Yokomi, Y. K. and Sonoda, R. M., Tristeza quick decline epidemic in south Florida, *Proc. Fla. State Hort. Soc.*, 99, 66, 1986.

76. Owen-Turner, J., Suspected severe stem pitting strain of tristeza virus discovered in Washington Navels, *Queensland Citrus Bulletin,* Autumn 1990, 2, 1990.

77. Norman, G. G., Florida state plant board program for virus-free budwood. In *Proc. 1st Conf. Intern. Organ. Citrus Virol.* (J. Wallace ed.), pp. 237–242, IOCV, Univ. of Calif. Press, Riverside, CA, 1957.

78. Calavan, E. C., Mather, S. M. and McEachern, E. H., Registration, certification, and indexing of citrus trees. In *The citrus industry,* Vol. 4 (W. Reuther, E. C. Calavan and G. E. Carman, eds.) pp. 185–222, University of California, Berkeley, CA, 1978.

79. Grant, T. J., Evidence of tristeza, or quick decline, virus in Florida, *Proc. Fla. State Hort. Soc.*, 65, 28, 1952.

80. Swingle, W. T., The limitations of the Satsuma orange to trifoliate-orange stock, U.S.D.A. Circ. 46, 10 pp., 1909.

81. Wallace, J. M. and Drake, R. J., The tristeza virus in Meyer lemon, Citrus Leaves, January, 8 and 23, 1955.

82. Olson, E. O. and McDonald, J. R., Tristeza in Satsuma varieties in Texas, *Plant Dis. Reptr.*, 38, 439, 1954.

83. Rhoads, A. S., Blight—a non-parasitic disease of citrus trees, *Florida Agr. Experiment Station Bulletin,* 64 pp., 1936.

84. Knorr, L. C. and Price, W. C., Annual report and progress report, *Univ. Fla. Citrus Expt. Sta., Lake Alfred,* 1955–1959.

85. Cohen, M., Injury and loss of citrus trees due to tristeza disease in an Orange county grove, *Proc. Fla. State Hort. Soc.,* 69, 19, 1956.

86. Knorr, L. C., Annual report and progress report: Distribution and rate of spread of tristeza, *Univ. Fla., Citrus Expt. Sta.*, Lake Alfred, 1957.

87. Bridges, G. D., Tristeza growing problem in commercial groves, *Citrus Ind.*, 47, 33, 1966.

88. Garnsey, S. M. and Jackson, J. L., A destructive outbreak of tristeza in central Florida, *Proc. Fla. State Hort. Soc.*, 88, 65, 1975.

89. Gabriel, D. W., Kingsley, M. T., Hunter, J. E. and Gottwald, T. R., Reinstatement of *Xanthomonas citri* (ex. Hasse) and *X. phaseoli* (ex Smith) to species and reclassification of all *X. campestris* pv. citri strains, *Intern. J. Sys. Bact.*, 39, 14, 1989.

90. Schoulties, C. L., Civerolo, E. L., Miller, J. W., Stall, R. E., Krass, C. J., Poe S. R. and DuCharme, E. P., Citrus canker in Florida, *Plant Disease*, 71, 388, 1987.

91. Schoulties, C. L., Brown, L. G., Youtsey, C. O. and Denmark, H. A., Citrus tristeza virus and vectors: Regulatory concerns, *Proc. Fla. State Hort. Soc.*, 100, 74, 1987.

92. Roistacher, C. N. and Moreno, P., The worldwide threat from destructive isolates of citrus tristeza virus—a review. In *Proc. 11th Conf. Int. Organ. Citrus Virol.* (R. H. Brlansky, R. F. Lee and L. W. Timmer, eds.), pp. 7–19, IOCV, Riverside, CA, 1991.

93. Cambra, M., Serra, J., Villalba, D. and Moreno, P., Present situation of the citrus tristeza virus in the Valencian Community. In *Proc. 10th Conf. Int. Organ. Citrus Virol.* (L. W. Timmer, S. M. Garnsey and L. Navarro, eds.), pp. 1–7, IOCV, Riverside, CA, 1988.

94. Moreno, P., Piquer, J., Pina, J. A., Juarez, J. and Cambra, M., Spread of citrus tristeza virus in a heavily infested citrus area in Spain. In *Proc. 10th Conf. Int. Organ. Citrus Virol.* (L. W. Timmer, S. M. Garnsey and L. Navarro, eds.), pp. 71–76, IOCV, Riverside, CA, 1988.

95. Hermoso de Mendoza, A., Ballester-Olmos, J. F. and Pina, J. A., Comparative aphid transmission of a common citrus tristeza virus isolate and a seedling yellows isolate recently introduced into Spain. In *Proc. 10th Conf. Int. Organ. Citrus Virol.* (L. W. Timmer, S. M. Garnsey and L. Navarro, eds.), pp. 68–70, IOCV, Riverside, CA, 1988.

96. DaGraca, J. V., Marais, L. J. and Von Broembsen, L. A., Severe tristeza stem pitting in young grapefruit, *Citrus and Subtropical Fruit J.*, Nov. 1982, 18, 1982.

97. Vegas, A., Ochoa, F., Beretta, J., Albanaci, N., Mendt, R., Lee, R. and Brlansky, R., A preliminary diagnosis of a sudden decline in Valencia orange trees on Volkamer lemon rootstock in Venezuela. In *Proc. 10th Conf. Int. Organ. Citrus Virol.* (L. W. Timmer, S. M. Garnsey and L. Navarro, eds.), p. 393–395, IOCV, Riverside, CA, 1988.

98. Castle, W. S., Tucker, D. P. H., Krezdorn, A. H. and Youtsey, C. O., *Rootstocks for Florida citrus*, Publication SP-42, University of Florida, Gainesville, FL, 47 pp., 1989.

99. Pelosi, R. R., Cohen, M. and Sonoda, R. M., Virulence of CTV not affected by passage through Temple orange, *Proc. Fla. State Hort. Soc.*, 99, 69, 1986.

100. Lee, R. F. and Brlansky, R. H., Variation in the severity of citrus tristeza virus isolates from groves with quick decline, *Proc. Fla. State Hort. Soc.*, 102, 1, 1989.

101. Garnsey, S. M., Barrett, H. C. and Hutchison, D. J., Resistance to citrus tristeza virus in citrus hybrids as determined by enzyme-linked immunosorbent assay, *Phytopathology*, 71, 875 (Abstr.), 1981.

102. Garnsey, S. M., Barrett, H. C. and Hutchison, D. J., Identification of citrus tristeza virus resistance in citrus relatives and its potential applications, *Phytophylactica*, 19, 187, 1987.

103. Yoshida, T., Inheritance of susceptibility to citrus tristeza virus in trifoliate orange (*Poncirus trifoliata* Raf.), *Bull. Fruit Tree Res. Stn., Okitsu*, No. 12, 1985.

104. Grosser, J. W., Gmitter, F. G., Jr. and Chandler, J. L., Intergeneric somatic hybrid plants from sexually incompatible woody species: *Citrus sinensis* and *Severinia disticha, Theor. Appl. Genet.*, 75, 397, 1988.

105. McKinney, H. H., Mosaic diseases in the Canary Islands, West Africa, and Gibralter, *J. Agric. Res.*, 39, 557, 1929.

106. Grant, T. J. and Costa, A. S., A mild strain of tristeza virus of citrus, *Phytopathology*, 41, 114, 1951.

107. Gonsalves, D. and Garnsey, S. M., Cross-protection techniques for control of plant virus diseases in the tropics, *Plant Disease*, 73, 592, 1989.

108. Costa, A. S. and Muller, G. W., Tristeza control by cross protection: A U.S.-Brazil cooperative success, *Plant Disease*, 64, 538, 1980.

109. DeLange, J. H., VanVuuren, S. P. and Bredell, G. S., Groeipuntenting suiwer sitrusklone vir die superplantskema van virusse, *Subtropica*, 2(5), 11, 1981.

110. Cox, J. E., Fraser, L. R. and Broadbent, P., Stem pitting of grapefruit; field protection by the use of mild strains and evaluation of trials in two climatic districts. In *Proc. 7th Conf. Int. Organ. Citrus Virologists* (E. C. Calavan, ed.), pp. 68–70, Univ. of Calif. Press, Riverside, CA, 1976.

111. Stubbs, L. L., Transmission and protective inoculation with viruses of the citrus tristeza complex, *Aust. J. Agric. Res.*, 15, 752, 1964.

112. Thornton, I. R., Emmett, R. W. and Stubbs, L. L., A further report on the grapefruit tristeza preimmunization against tristeza at Mildura, Victoria. In *Proc. 8th Conf. Int. Organ. Citrus Virologists* (E. C. Calavan, S. M. Garnsey and L. W. Timmer, eds.), pp. 51–83, Univ. Calif. Press. Riverside, CA, 1980.

113. Sasaki, A., Control of Hassaku dwarf by preimmunization with mild strain, *Rev. Plant Prot. Res. Jpn.*, 12, 80, 1979.

114. Leki, H., The use of cross-protection with mild strains of citrus tristeza virus (CTV) to control stem pitting disease of citrus in Japan. In *Virus diseases of citrus,* Extension Bulletin No. 284, Food and Fertilizer Technology Center, pp. 7–14, Taipei, R.O.C., 1989.

115. Koizumi, M., Citrus tristeza virus: Impact and control by preinoculation in Japan. In *Plant virus diseases of horticultural crops in the tropics and subtropics*, FFTC Book Series No. 33., pp. 148–156, 1986.

116. Koizumi, M., Kuhara, S., Leki, H., Kano, T., Tanaka, A. and Iwanami, T., A report of preinoculation to control stem pitting disease of Navel orange in fields up to 1989. In *Proc. 11th Conf. Int. Organ. Citrus Virol.* (R. H. Brlansky, R. F. Lee and L. L. Timmer, eds.), pp. 125–127, IOCV, Riverside, CA, 1991.

117. Aubert, B. and Bove, C., Mild and severe strains of citrus tristeza virus in Reunion Island. In *Proc. 9th Conf. Int. Organ. Citrus Virol.* (S. M. Garnsey, L. W. Timmer and J. A. Dodds, eds.), pp. 57–61, IOCV, Riverside, CA, 1984.

118. Balaraman, K. and Ramakrishnan, R., Strain variation and cross protection in citrus tristeza virus on acid lime. In *Proc. 8th Conf. Int. Organ. Citrus Virologists* (E. C. Calavan, S. M. Garnsey and L. W. Timmer, eds.), pp. 60–75, Univ. Calif. Press, Riverside, CA, 1980.

119. Romero, G., Gomez, K., Gruber, C., Mendt, R., Ochoa, F., Sanabria, F. and Colmenares, F., Twelve-year evaluation of citrus tristeza virus tolerant rootstocks budded with Washington Navel orange, In *Proc. 11th*. Conf. *Int. Organ. Citrus Virol.* (R. H. Brlansky, R. F. Lee and L. W. Timmer, eds.), pp. 135–136, IOCV, Riverside, CA, 1991.

120. VanVuuren, S. P., Collins, R. P. and daGraca, J. V., The performance of exotic citrus tristeza virus isolates as pre-immunizing agents for sweet orange on sour orange rootstock under natural disease pressure. In *Proc. 11th Conf. Int. Organ. Citrus Virol.* (R. H. Brlansky, R. F. Lee and L. W. Timmer, eds.), pp. 60–63, IOCV, Riverside, CA, 1991.

121. Yokomi, R. K., Garnsey, S. M., Permar, T. A., Lee, R. F. and Youtsey, C. O., Natural spread of severe citrus tristeza virus isolates in citrus preinoculated with mild CTV isolates. In *Proc. 11th Conf. Int. Organ. Citrus Virol.* (R. H. Brlansky, R. F. Lee and L. W. Timmer, eds.), pp. 86–92, IOCV, Riverside, CA, 1991.

122. Permar, T. A., Garnsey, S. M., Gumpf, D. J. and Lee, R. F., A monoclonal antibody that discriminates strains of citrus tristeza virus, *Phytopathology*, 80, 224, 1990.

123. Rocha-Peña, M. A., Lee, R. F., Permar, T. A., Yokomi, R. K. and Garnsey, S. M., Use of enzyme-linked immunosorbent and dot-immunobinding assays to evaluate two mild-strain cross protection experiments after challenge with a severe citrus tristeza virus isolate. In *Proc. 11th Conf. Int. Organ. Citrus Virol.* (R. H. Brlansky, R. F. Lee and L. W. Timmer, eds.), pp. 93–102, IOCV, Riverside, CA, 1991.

124. Lee, R. F., Garnsey, S. M., Marais, L. J., Moll, J. N. and Youtsey, C. O., Distribution of citrus tristeza virus in grapefruit and sweet orange in Florida and South Africa. in *Proc. 10th Conf. Int. Organ. Citrus Virol.* (L. W. Timmer, S. M. Garnsey and L. Navarro, eds.), pp. 33–38, IOCV, Riverside, CA, 1988.

125. Muller, G. W., Gasper, J. O., Lee, R., Brlansky, R. and Berlan, L. O. S., Determinacao da incidencia do virus da tristeza em porta-anxertos de citros pelo tetste ELISA, *Fitopatologia Brasileira*, 12, 147, 1987.

126. VanVuuren, S. P., Invloed van temperatuur op die groei en stamgleufontwikkeling van tris-teza-besmette Wes-Indiese lemmetjie-saailinge, *Subtropical*, 3(7), 13, 1982.

127. Muller, G. W. and Costa, A. S., Search for outstanding plants in tristeza infected citrus orchards: The best approach to control the disease by preimmunization, *Phytophylactica*, 19, 197, 1987.

128. Roistacher, C. N., Dodds, J. A. and Bash, J. A., Means of obtaining and testing protective strains of seedling yellows and stem pitting tristeza virus: A preliminary report, *Phytophylactica*, 19, 199, 1987.

129. VanVuuren, S. P. and Moll, J. N., Glasshouse evaluation of citrus tristeza virus isolates, *Phytophylactica*, 19, 219, 1987.

130. Yokomi, R. K., Garnsey, S. M., Lee, R. F. and Cohen, M., Use of insect vectors to screen for protecting effects of mild citrus tristeza virus isolates in Florida, *Phytophylactica*, 19, 183, 1987.

131. Gerlach, W. L., Llewellyn, D. and Haseloff, J., Construction of a plant disease resistance gene from the satellite RNA of tobacco ringspot virus, *Nature*, 328, 802, 1987.

132. Harrison, B. D., Mayo, M. A. and Baulcombe, D. C., Virus resistance in transgenic plants that express cucumber mosaic virus satellite RNA, *Nature*, 328, 799, 1987.

133. Cuozzo, M., O'Connel, K. M., Kaniewski, W., Fang, R.-X., Chua, N.-H. and Tumer, N. E., Viral protection in transgenic tobacco plants expressing the cucumber mosaic virus coat protein or its antisense RNA, *Bio/Technology*, 6, 549, 1988.

134. Powell, P. A., Stark, D. M., Sanders, R. P. and Beachy, R. N., Protection against tobacco mosaic virus in transgenic plants that express tobacco mosaic virus antisense RNA, *Proc. Natl. Acad. Sci. USA*, 86, 6949, 1989.

135. Hemenway, C., Fang, R.-X., Kaniewski, J. J., Chua, N.-H. and Tumer, N. E., Analysis of the mechanism of protection in transgenic plants expressing the potato virus X coat protein or its antisense RNA, *EMBO J.*, 7, 1273, 1988.

136. Powell-Able, P. A., Nelson, R. S., De, B., Hoffman, N., Rogers, S. G., et al., Delay of disease development in transgenic plants that express the tobacco mosaic virus coat protein gene, *Science*, 232, 738, 1986.

137. Loesch-Fries, L. S., Merlo, D., Zinnern, T., Burhop, L., Hill, K., Krahn, K., Jarvis, N., Nelson, S. and Halk, E., Expression of alfalfa mosaic virus RNA 4 in transgenic plants confers virus resistance, *EMBO J.*, 6, 1845, 1987.

138. VanDun, C. M. P., Bol, J. F. and Vloten-Doting, L. V., Expression of alfalfa mosaic virus and tobacco rattle virus coat proteins in transgenic tobacco plants, *Virology*, 159, 299, 1987.

139. Permar, T. A. and Garnsey, S. M., Comparison of biological indexing and immunological assays for identifying severe Florida isolates of citrus tristeza virus. In *Proc. 11th Conf. Int. Organ. Citrus Virol.* (R. H. Brlansky, R. F. Lee and L. W. Timmer, eds.), pp. 56–59, IOCV, Riverside, CA, 1991.

140. Gonzalez-Garza, R., Aspectos importantes de las principales virosis y enfermedades simi-lares de los citricos. In *Temas en virologia* (M. A. Rocha-Peña and R. Gonzalez-Garza, eds.), pp. 1–20, Sociedad Mexicana de Fitopatologia, 1985.

10

PHYTOPHTHORA DISEASES OF CITRUS

J. H. GRAHAM AND L. W. TIMMER

Citrus Research and Education Center
University of Florida, IFAS
700 Experiment Station Road, Lake Alfred, Florida

10-1 INTRODUCTION

Phytophthora spp. cause the most serious soilborne diseases of citrus.[1] Losses occur from damping-off of seedlings in the seedbed, root and crown rot in nurseries, and from foot rot, fibrous root rot, and brown rot of fruit in orchards. Foot rot or gummosis results from infection of the bark near the ground level, producing lesions on the trunk or crown roots that can girdle and kill the tree. *Phytophthora* spp. also decay fibrous roots, especially on susceptible rootstocks in nurseries, but also in bearing orchards where damage causes tree decline and yield losses. *Phytophthora* spp. infect fruit, causing brown rot that leads to fruit drop in the orchard and postharvest decay. The most important and cosmopolitan *Phytophthora* spp., *P. parasitica* and *P. citrophthora*, cause losses in production in Mediterranean climates where citrus is irrigated, as well as in subtropical areas receiving high rainfall. Thus, Phytophthora-induced diseases are economically important in all citrus-growing areas of the world.

10-2 HISTORICAL SIGNIFICANCE

Fawcett first determined that foot rot was caused by a fungus,[2] and detailed the early history of the occurrence and distribution of the disease.[3] Accounts of commercially important outbreaks of "gum diseases" date from 1832 in Spain and 1845 in Portugal. During the 1850s and 1860s, foot-rot–type diseases became economically important throughout the Mediterranean area and were first reported in 1860 in Australia, 1885 in New Zealand, and 1891 in South Africa. In the Western Hemisphere, reports occurred in California in 1875, Florida in 1876, and later in Cuba (1906), Paraguay (1911), Brazil (1917), Mexico (1920), and Trinidad (1935).

More than 25 years before the causal fungus was discovered, Curtis in 1888[4] reported on the extent and treatment of foot rot in Florida. Treatment consisted of exposing the base of the tree trunk, cutting away the infected bark with a knife, and applying an antiseptic. Use of sour orange rootstock, high budding of the scion, and well-drained planting sites were recommended for prevention of foot rot. In the late 1800s and early 1900s, there was a major shift from seedling trees and from trees on susceptible lemon, lime, and sweet orange as rootstocks, to trees budded on resistant sour orange rootstock.[2]

In the 1920s to 1940s, foot rot became one of the most important diseases in the ridge area of central Florida when extensive plantings were made primarily on rough lemon rootstock.[5] In Australia, major outbreaks of root rot on heavy soils in irrigated areas occurred during years of excessive rainfall. This led to the extensive replanting with Phytophthora-resistant rootstocks, such as trifoliate orange and later Benton citrange.[6,7]

10-3 DISEASE SYMPTOMS

10-3-1 Damping-Off

Damping-off can affect newly germinated seedlings of any citrus cultivar.[1] The typical symptom results when the soil or seedborne fungus penetrates the stem just above the soil line and causes the seedling to topple. *Phytophthora* spp. also cause seed rot or preemergence rot. Infected seedlings are killed rapidly when moisture is abundant and temperatures are favorable for fungal growth. Plants usually become resistant to damping-off once the true leaves have emerged and the stem tissue has matured. Symptoms of damping-off caused by *Phytophthora* spp. are similar to those caused by *Rhizoctonia* spp. and *Pythium* spp.[1]

10-3-2 Foot Rot and Gummosis

The most serious diseases caused by *Phytophthora* spp. are foot rot and gummosis.[1,8] Foot rot results from an infection of the scion near the ground level, producing lesions which extend down to the bud union on resistant rootstocks (Fig. 10-1). Scaffold root rot or crown rot may occur when susceptible rootstocks are used. Gummosis refers to a rotting of the bark anywhere on the scion of the tree.[1] Infected bark remains firm with small cracks through which abundant gum exudation occurs. Citrus gum, which is water soluble, disappears after heavy rains but is persistent on the trunk under dry conditions. Lesions spread around the circumference of the trunk, slowly girdling the tree. Badly affected trees have pale green leaves with yellow veins, a typical girdling effect. If the lesions cease to expand or the fungus dies, the affected area is surrounded by callus tissue (Fig. 10-1).

Nursery trees and young orchard trees of small trunk circumference can be rapidly girdled and killed. Large trees may be killed likewise, but typically the trunks are parti-

Figure 10-1   Phytophthora-induced foot rot lesion that has been partially callused over on the scion portion of the trunk of a low-budded sweet orange tree. Reprinted with permission of APS.

ally girdled and the tree canopy undergoes defoliation, twig dieback, and short growth flushes. On susceptible rootstocks, lesions may occur on the crown roots below the soil line and symptoms in the canopy develop without obvious damage to the trunk above ground.

10-3-3 Fibrous Root Rot

Phytophthora spp. infect the root cortex and cause a decay of fibrous roots.[1,6,9] Later, the cortex turns soft, becomes somewhat discolored, and appears water-soaked. The fibrous roots slough their cortex leaving only the white, thread-like stele, which gives the root system a stringy appearance (Fig. 10-2).

Root rot can be especially severe on susceptible rootstocks in infested nursery soil. Root rot also occurs on susceptible rootstocks in bearing orchards where damage causes tree decline and yield losses.[10,11] In advanced stages of decline, the production of new fibrous roots cannot keep pace with root death.[10] The tree is unable to maintain adequate water and mineral uptake; and nutrient reserves in the root are depleted by the repeated fungal attacks. This results in reduction of fruit size and production, loss of leaves, and twig dieback of the canopy.[10]

10-3-4 Brown Rot of Fruit

Phytophthora infection of fruit produces a decay in which the affected area is light brown, leathery, and not sunken compared to the adjacent rind.[1,12] White mycelium forms on the

Figure 10-2 Fibrous root rot of citrus caused by *Phytophthora parasitica*. The roots slough their cortex leaving only the white, thread-like stele, which gives the root system a stringy appearance.

rind surface under humid conditions. In the orchard, fruit near the ground become infected when splashed with soil containing the fungus. If favorable conditions continue, the disease spreads to fruit throughout the canopy. Most of the infected fruit soon abscise, but those that are harvested may not show symptoms until after they have been held in storage a few days. If infected fruit is packed, brown rot may spread to adjacent fruit in the container. In storage, infected fruit have a characteristic pungent, rancid odor. Brown rot epidemics are usually restricted to areas where rainfall coincides with the early stages of fruit maturity.[13] All cultivars are affected, especially lemons.

10-4 CAUSAL ORGANISMS

The most widespread and important *Phytophthora* spp. are *P. parasitica* Dast. and *P. citrophthora* (R. E. Sm. and E. H. Sm.) Leonian.[1] *P. parasitica* is more common in subtropical areas of the world and causes foot rot, gummosis, and root rot but usually does not infect far above the ground. *P. citrophthora* causes gummosis and root rot in Mediterranean climates, where seasonal rainfall occurs during the cooler winter months. It attacks aerial parts of the trunk and is most commonly the cause of brown rot.

   P. hibernalis Carne and *P. syringae* Kleb. attack citrus fruit to a limited extent in areas with cool, moist winters. *P. palmivora* (Butler) Butler and *P. citricola* Saw. are reported to attack citrus in some tropical areas.[7] An isolate of *P. arecae*, probably synonymous with *P. palmivora*, has recently been demonstrated to be highly pathogenic on roots and stems of citrus seedlings, and fruit.[14]

P. parasitica[15] and *P. citrophthora*[16] produce papillate sporangia which release large numbers of biflagellate zoospores (Fig. 10-3). The sporangia of *P. parasitica* are pear-shaped to ovoid with dimensions of 38–50 × 30–40 μm. Sporangia of *P. citrophthora* are more elongate but vary greatly in shape (45–90 × 27–60 μm). *P. parasitica* produces chlamydospores in abundance while most isolates of *P. citrophthora* do not. *P. citrophthora* rarely produces oospores, whereas *P. parasitica* commonly produces oospores 22–29 μm in diameter. The optimum temperatures for mycelial growth are 30 to 32°C for *P. parasitica* and 24 to 28°C for *P. citrophthora*.

Sporangial production by *P. parasitica* and *P. citrophthora* is favored by small deficits in matric water potential ($\psi_m = -5$ to -70 KPa) but not by saturated conditions ($\psi_m = 0$K Pa)[17] unless sporangia are produced on citrus root pieces.[18] The optimal ψ_m for sporangium formation probably represents a compromise between requirements for free water and aeration.[17] Nutrient depletion and light also stimulate sporangial production from mycelium.[19,20]

Indirect germination of sporangia to produce zoospores requires free water and is stimulated by a drop in temperature.[17,21] Under moist conditions, sporangia may also germinate directly by growth of germ tubes, but the correlation between soil saturation and severity of Phytophthora root rot suggests that indirect germination is more important in the root disease cycle.[17]

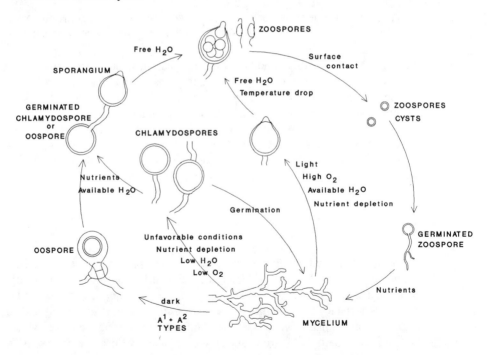

Figure 10-3 Life cycle of *Phytophthora parasitica* illustrating different spore types and some effects of environmental factors.

Chlamydospore production by *P. parasitica* in liquid culture is favored by nutrient depletion and low oxygen levels and temperatures (15 to 18 °C).[22] Water requirements for germination of chlamydospores are similar to those for sporangia. Chlamydospores of *P. parasitica* appear to become dormant below 15 °C, so exposure to temperatures of 28 to 32 °C is used to stimulate germination.[23] Nutrients, in soil extracts and from excised citrus roots, stimulate chlamydospore germination.[19] Chlamydospores germinate readily on V8® juice and cornmeal agar or in broth media.

The requirements for oospore germination are thought to parallel those of chlamydospores.[20] Oospore maturation appears to be an important factor in germinability of *Phytophthora* spp. Periods of alternating high and low temperatures may be a prerequisite for uniform germination.[20]

10-5 SOIL ENVIRONMENTAL AND MICROBIAL INTERACTIONS

10-5-1 Fungus-Soil Environmental Interactions

Phytophthora spp. are parasites but are poor saprophytes in soil.[19] The fungus grows well on nutrients obtained from the living plant and, under favorable conditions, undergoes repeated cycles of mycelium to sporangia, zoospores, and more mycelium (Fig. 10-3). *P. parasitica* is most active during the warmer seasons of the year and forms chlamydospores and oospores for survival during unfavorable periods. *P. citrophthora* is active during cooler periods and is believed to survive as resistant sporangia and possibly chlamydospores and oospores.[10] Because less is known about the life cycle of *P. citrophthora*, discussion of the spore forms in relation to soil environmental factors will be limited to *P. parasitica*.

Chlamydospores of *P. parasitica* may form when soil moisture is limiting, conditions are cool, or where the host roots are not actively growing and producing susceptible tissue (Fig. 10-3).[19,24] Formation of chlamydospores may also be stimulated by poor aeration and high carbon dioxide concentration in the soil atmosphere.[19] Chlamydospores can survive several months under unfavorable conditions.[19,24]

Optimal germination of chlamydospores occurs in well-aerated, moist environments when temperatures are favorable for root growth.[18,19] Chlamydospores require nutrients in the form of root exudates for germination.[19,20] Oospores, which occur in low numbers, are thick walled and probably resistant to desiccation and cold temperature. They may mature more slowly than chlamydospores but, once matured, they germinate in response to nutrients from roots.[20] Oospores form when the opposite mating types are present, which may occur infrequently in some citrus soils.[10,19] Thus, the importance of oospores in the disease cycle is not known, except that they provide new sources of variation in the population when sexual recombination occurs between the A^1 and A^2 mating types (Fig. 10-3).

Under well-aerated, moist conditions, chlamydospores and oospores can germinate immediately to form a sporangium that liberates zoospores (Fig. 10-3). Zoospore release is optimal for *Phytophthora* spp. in saturated soils. Diurnal temperature changes in soil may serve to synchronize the release of zoospores. The zoospores are motile and can swim short distances (cm) by flagellar movement or be carried long distances (m) by soil water.[17] The zoospores are attracted to roots[25,26] and swim to the root and encyst on contact. Zoospore cysts then germinate and penetrate the root cortex (Fig. 10-3).

10-5-2 Plant-Soil Environmental Interactions

Soil moisture.   Host susceptibility is affected when roots are stressed or damaged in saturated or in dry soil.[17] Root exudates released by living but damaged or stressed roots attract zoospores. Soils with drainage restricted by hardpans or clay layers or those with shallow water tables that temporarily rise into the root zone provide ideal conditions for fibrous root infection and rapid buildup of Phytophthora propagules.[17,18] Also, the frequency and duration of irrigations can influence the activity of the fungus and the predisposition of roots to rot.[27] If soils are saturated during an irrigation, zoospores are released and can infect roots to form more sporangia. When soils do not dry sufficiently between irrigations, sporangia can survive until the next irrigation, and zoospores will be released again. Citrus roots do not grow in saturated soil due to low O_2 availability.[28] Thus, new roots will not develop if soil is not allowed to dry adequately between irrigation cycles.[27]

Soil aeration.   Availability of oxygen in the soil atmosphere is closely related to soil moisture because the amount of pore space occupied by water directly determines the remaining pore space occupied by air. When roots are subjected to anoxia, they are damaged by reduced forms of minerals and by toxic metabolites produced by microorganisms on the roots themselves.[29] Root regeneration is restricted and root exudation increases under flooded conditions. Prolonged oxygen deprivation makes roots more attractive to zoospores and increases infection. Moreover, chronic anoxia due to flooding of the root system will damage roots outright even in the absence of *Phytophthora*.[28]

Soil temperature.   Populations of *P. parasitica* decline in the winter in Mediterranean climates, where cool temperatures ($>15°C$) induce dormancy of chlamydospores.[10,23] Populations of *P. parasitica* do not drop significantly in Florida because soil temperatures rarely fall below 15°C.[30] In California, the increase in *P. parasitica* populations coincides with a root flush in May, which occurs after the spring shoot flush in March or April.[10] Similarly, a spring flush of roots may occur in Florida when soil temperatures rise above 20 to 23°C,[31] followed by an increase in *P. parasitica* populations (Graham, unpublished). This population response is probably not temperature dependent but due to the greater amount of root substrate available for infection.

   P. citrophthora grows best at lower temperatures (24 to 28°C) than *P. parasitica* (28 to 32°C) and is probably responsible for much of the damage to roots during the winter in Mediterranean climates.[10] *P. citrophthora* could conceivably be even more damaging because this species attacks dormant roots which are incapable of regenerating during the winter months.[10]

10-5-3 Soil Microbial Interactions

Microbial antagonism resulting in the suppression of *Phytophthora* spp. in the soil, rhizosphere, or the infection court has been reported.[32] Processes such as parasitism, predation, and competition may all be operating alone or together in reducing inoculum. While the evidence for microbial suppression of citrus *Phytophthora* spp. is scant, there is substantial evidence for the existence of suppressive soils that are active against other *Phytophthora* spp.[32]

The competitive saprophytic ability of *Phytophthora* spp. is low, and rapid lysis of mycelium occurs in natural soil.[32] Extracts prepared from natural soils inhibit mycelial growth of *P. parasitica*, whereas extracts from autoclaved soils stimulate mycelial growth.[19] Amendment of natural soil with carbon and nitrogen leads to rapid growth of mycelium and then lysis.[19] In all cases, lysis is associated with intense bacterial colonization of hyphae and subsequent breakdown of the cytoplasm.[19,32] In soils associated with suppressiveness to *Phytophthora*, microbial populations are higher than in nonsuppressive or conducive soils.[32] In a soil suppressive to avocado root rot caused by *P. cinnamoni*, specific strains of *Streptomyces* and fluorescent pseudomonads occur that cause hyphal lysis and sporangium abortion of *P. citrophthora*.[33]

While specific identification of Phytophthora-suppressive citrus soils has not been made, it is possible that situations exist where populations are low even though environmental conditions are favorable. For example, the restricted occurrence of *P. citrophthora* in Florida soils but repeated outbreak of brown rot on fruit in certain orchards[13] might indicate a widespread suppression of this species in most Florida orchards.

Vesicular-arbuscular mycorrhizal fungi have been implicated in microbial antagonism of *P. parasitica* in citrus,[34] but increased tolerance of mycorrhizal plants to root rot is probably due to improved host nutrition (mainly phosphorus).[35,36]

10-6 EVALUATION OF RESISTANCE

Although most commonly planted scion cultivars are susceptible to bark infection, most rootstocks are at least moderately tolerant (Table 10-1).[1] Nevertheless, rootstocks vary in their susceptibility to diseases caused by *Phytophthora* (Table 10-1). Although the use of resistant rootstocks is the best solution for control of Phytophthora diseases, some highly resistant rootstocks are susceptible to other diseases or are horticulturally unacceptable.

TABLE 10-1: FOOT ROT SUSCEPTIBILITY OF COMMONLY PLANTED SCION AND ROOTSTOCK CULTIVARS OF CITRUS.

Scion cultivars	
Lemon, lime, sweet orange, grapefruit	Very susceptible
Mandarin, tangerine, and their hybrid	Susceptible
Rootstock cultivars	
Trifoliate orange, Swingle citrumelo	Resistant
Sour orange, Alemow	Moderately resistant
Cleopatra mandarin, Troyer and Carrizo citrange, most rough lemon selections, Rangpur, Volkamer lemon	Tolerant
Sweet orange, some rough lemon selections	Susceptible

Furthermore, there is not always a good correlation between tolerance to foot rot and to root rot.[37-39] For example, Carrizo citrange and sour orange are tolerant to foot rot, yet are susceptible to root rot when artificially inoculated.[39] The mechanisms of resistance of different citrus tissues to *Phytophthora* are poorly understood. Because of the great differences in the type of tissues affected and the response of different citrus species to infection, there are probably more than one resistance and/or tolerance mechanisms involved.

Little is known about the host specificity of *P. parasitica* and *P. citrophthora* and the relative aggressiveness of isolates within each species. *P. parasitica* has a broad host range.[40] The virulence of *P. parasitica* isolates from tomato and other noncitrus hosts toward citrus is low, but all isolates of *P. parasitica* are pathogenic on tomato.[40] Aggressiveness of *P. parasitica* on sweet orange seedlings varies widely,[41] but that does not appear to be the case for *P. citrophthora*.[6]

10-6-1 Root Rot Evaluation

Traditionally, root rot has been measured as the severity of root loss on young seedlings (> 1 year old) compared to Carrizo citrange, which is a reference rootstock with moderate tolerance. Most commonly, seedlings are inoculated in a tank filled with water or dilute nutrient solution into which swimming zoospores are released from sporangial cultures[37,42] or infested alfalfa stems.[43] The roots remain submerged for 18 to 20 hours and then are lined out in the field or the greenhouse. The soil is flooded at the time of planting and periodically thereafter to promote root rot. In other cases, the seedlings are left in the aerated nutrient solution for the duration of tests.

Disease evaluations vary considerably. In some cases, seedling mortality is used to compare tolerance.[37] More often, root loss compared to Carrizo citrange is rated.[42] Those with less root rot than Carrizo citrange are considered tolerant and are kept for further evaluation, whereas less tolerant types are discarded.

The so-called tank test has been widely used because large numbers of plants can be inoculated for mass screening and the inoculation procedures are relatively simple.[37,42] The major drawback of this technique is that nonuniform inoculum may produce erratic root rot development. This has necessitated the reinoculation or retesting of the same lots of seedlings in some cases.[37] The tank test may be too severe because the inoculum levels used are higher than those that occur in natural soils.[39]

Inoculum prepared by adding *P. parasitica* chlamydospores at levels approximating propagule densities in the field can provide conditions for uniform root rot development.[36,39] Chlamydospores produced by the method of Tsao[22] are used to infest a sandy soil.[36,39] The infested soil is incubated moist for seven days to allow for death of mycelial fragments. The inoculum density is then evaluated by dilution plating on selective media, and propagule densities are adjusted to ≥ 10 propagules/cm^3 with sterilized soil.[36,39] Young seedlings with two to four true leaves are transplanted into infested soil. Seedlings of Swingle citrumelo may be planted as a reference cultivar that is tolerant to root rot. Under warm conditions (> 25°C), the soil is flooded for the first three days and then every seven days thereafter.[44] The percentage of the root tips rotted is assessed three to four wk after inoculation. Ranking of root rot tolerance is similar with three-month-old seedlings evaluated three weeks after inoculation as with six-month-old seedlings evaluated six weeks after inoculation.[39]

The quantification of root rot susceptibility in soils infested with a range of chlamydospore densities reveals a group that is tolerant, including trifoliate orange and Swingle citrumelo, and a wide range of rootstocks that are susceptible, including sour orange, Carrizo citrange, Cleopatra mandarin, Ridge Pineapple sweet orange, and Volkamer lemon.[39] This classification of rootstocks generally agrees with the results of zoospore inoculation tests except that Carrizo citrange and sour orange, which are usually considered moderately tolerant,[37,42] are grouped with sweet orange and Cleopatra mandarin as susceptible.[39] The field tolerance of Carrizo citrange and sour orange to foot rot[1] confirms that there is not necessarily a correspondence between root rot tolerance and foot rot resistance. However, trifoliate orange and Swingle citrumelo are both tolerant to root rot and resistant to stem infections. Thus, root rot tolerance may serve as an acceptable indicator of bark resistance. This is significant because foot rot tests require larger seedlings for stem inoculations, which increases the time and cost of evaluation (see "Foot Rot Evaluation").

Previously, Carrizo citrange has been used as a standard rootstock of moderate tolerance for comparisons.[42] Swingle citrumelo, shown to be tolerant of root rot when inoculated with chlamydospores, is preferable since it provides a more rigorous standard for identifying Phytophthora-resistant germplasm.

Trifoliate orange and Swingle citrumelo, a hybrid of trifoliate orange, have the ability to regenerate roots in the presence of *P. parasitica*.[39] Their roots are not immune to infection because they support populations of *P. parasitica* in greenhouse and field soils.[45] These rootstocks apparently have the ability to grow new roots subtending infected root tips. The resistance factors that limit infection to the root tip are not known but may be related to coumarin phytoalexins found in woody tissue of citrus infected with *P. citrophthora*.[46]

Root rot tolerance to *P. citrophthora* as evaluated by zoospore inoculation appears to be similar to that of *P. parasitica*.[43] Chlamydospores are not consistently produced by *P. citrophthora* isolates; therefore, inoculation with these structures may not be applicable to this species. Other means of soil infestation such as cornmeal-infused sand cultures mixed into soil or zoospore suspensions poured onto the soil around the base of the seedling may be used.[42] Propagule levels should be assayed on selective media to determine whether inoculum levels approximate those in naturally infested soils.

10-6-2 Foot Rot Evaluation

Phytophthora infection will not occur through suberized tissues of *Citrus*.[47] Stem tissue must be wounded for assessment of foot rot resistance.[48] Most routinely, the susceptibility of citrus germplasm to foot rot has been determined by removing bark disks from the lower trunk and placing agar disks of mycelium on the exposed cambium.[42,43] The area is then sealed by wrapping with plastic to prevent drying of the inoculum and the adjacent host tissues. Several weeks later, the lesion area on the stem is measured. This inoculation method bypasses any resistance that might be due to differences in natural entry points or suberization of the bark. The inoculum potential of agar disks is not readily quantifiable and is probably much higher than under field conditions. In screening, agar-disk inoculations may be capable of discriminating high levels of resistance and eliminate germplasm which possess intermediate but potentially useful levels of resistance.[49]

Inoculations for evaluation of resistance can also be performed by placing zoospore suspensions in a water-tight collar around the base of the trunk.[47] The trunk is wounded by making a vertical cut in the bark, and moisture is maintained around the wound for six days by adding water or wet, absorbent cotton within the collar. Alternatively, washed mycelial mats[48] or blocks of chlamydospore-infested soil infused with 2% cornmeal agar[49] can be inserted under bark flaps cut like an inverted V. Moisture is maintained around trunks with moistened peat moss or pasteurized sandy soil contained in a styrofoam cup cut to fit around the trunk.

In any case, inoculation of trunks of seedlings is difficult.[47] Disease development is often nonuniform and may not be indicative of field resistance to foot rot because of the severity of the inoculation methods and the juvenility of the stem tissue at the seedling stage. Foot rot resistance can be evaluated more conveniently and rapidly using stem pieces and bark strips from greenhouse or field trees inoculated with agar disks of mycelium.[50,51] The disks are placed on the inner cambial surface of bark strips or on the exposed cambium of stems after removal of the bark. Extension of lesions from the inoculation site is measured after four to seven days incubation of the plant material in a moist chamber. Lesion length differs depending on the resistance of the cultivar. Tissue is more susceptible during periods of the year when trees are actively growing than during months when the trees are dormant.[51] Tissue to be evaluated should be carefully standardized because tissue age also affects susceptibility.

In the screening process, lesion development should be compared to a reference rootstock of at least moderate tolerance which is grown under comparable conditions to the germplasm material being tested. Also, tests should be conducted when trees are growing and the cambium is active so resistance responses dependent on rapid callusing or formation of wound tissue can be evaluated.

10-7 EPIDEMIOLOGY

Phytophthora spp. are already widespread in most citrus orchards around the world.[1] Thus, in many situations, we are dealing with an endemic pathogen rather than an epidemic situation. The disease cycle as it occurs under most conditions is portrayed in Fig. 10-4.

Fungal populations in the soil are maintained by repeated infection of the fibrous roots. Under favorable conditions of high moisture and temperature, infected roots produce sporangia which, in turn, release motile zoospores. Zoospores are probably attracted to the zone of elongation of new roots by nutrients exuded from this area.[25,26] Upon contact with the root, zoospores encyst, germinate, and then infect in the area of the zone of elongation. Once the fungus has entered the root tip, the infection may advance in the cortex resulting in rot of the entire rootlet. The cycle can repeat itself as long as conditions are favorable and susceptible tissue is available.

In the case of *P. parasitica*, the fungus most likely survives unfavorable periods in root debris. The rotted cortex is sloughed and the fungus produces chlamydospores, which may persist in soil for long periods. When favorable conditions return, chlamydo-

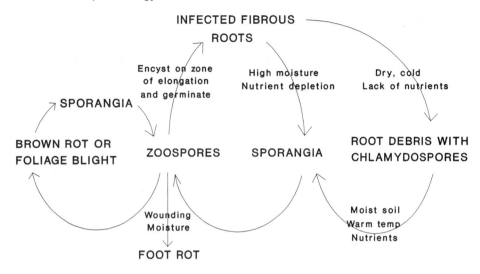

INFECTED FIBROUS

ROOTS

Encyst on zone
of elongation
and germinate

High moisture
Nutrient depletion

Dry, cold
Lack of nutrients

SPORANGIA

BROWN ROT OR
FOLIAGE BLIGHT

ZOOSPORES

SPORANGIA

ROOT DEBRIS WITH
CHLAMYDOSPORES

Wounding
Moisture

Moist soil
Warm temp
Nutrients

FOOT ROT

Figure 10-4 Disease cycle of *Phytophthora parasitica* in a citrus orchard.

spores germinate indirectly to produce sporangia and zoospores or directly to produce mycelium. When both mating types are present, oospores may also be produced and aid in survival of the fungus. With *P. citrophthora*, which often does not produce chlamydospores or oospores, the mechanism of survival is uncertain. However, both species may be able to persist as mycelium in infected, living roots or as sporangia.

Foot rot or gummosis of the trunk occurs when zoospores or other propagules are splashed onto the trunk above the bud union. A wound and moisture on or around the base of the trunk are necessary for infection, and wounds are susceptible to infection for up to 14 days.[52] Foot rot lesions do not usually produce inoculum for subsequent infections and, thus, are of no epidemiologic significance.

Even though *Phytophthora* spp. in citrus are usually endemic, epidemics can develop in certain cases. When *Phytophthora* spp. are first introduced into an orchard or nursery, spread to other plants may be epidemic. However, epidemics of foliage blight and brown rot are much more common, especially on highly susceptible hosts such as lemons. When conditions are favorable, fruit and young foliage may become infected by propagules splashed from the soil. Secondary infections may then be caused by inoculum from the above-ground parts of the plant which is dispersed by rain splash or wind-blown rain. This seldom occurs with *P. parasitica*, which does not readily produce aerial sporangia, but is more common in *P. citrophthora* and other species which produce abundant sporangia on fruit and leaf surfaces. Brown rot is most common in areas where these species are present and where citrus is grown in Mediterranean climates with prolonged winter rains.

10-8 SPREAD

The primary means by which *Phytophthora* spp. are spread through citrus orchards is by use of infested nursery stock. The pathogen may be present in soil or infected roots even though disease symptoms are not readily apparent. The fungus is also carried in soil on farm equipment when vehicles move from infested to noninfested orchards or nurseries. Propagule densities decline sharply when soil is air dried,[53] reducing the probability of spread.

Irrigation water may also move the pathogen from area to area. Within orchards, dispersal by irrigation water occurs, especially where furrow or flood irrigation is used.[18] Surface water following heavy rains may carry the fungus as it drains from the orchard. More serious problems can arise in irrigated citrus areas where run-off water carries the pathogen into canals, streams, or rivers. Use of water from these sources may then contaminate previously noninfested areas.[54,55]

Wind is not a major factor in dispersal of *Phytophthora* spp. However, windborne soil carries *Phytophthora* spp. and may recontaminate fumigated soils.[56] Wind-blown rain can disseminate sporangia produced on the surface of above-ground plant parts.

10-9 DETECTION METHODS

For regulatory purposes or for proper disease management, it is frequently desirable to know whether *Phytophthora* spp. are present and, if so, at what level. Fruit and leaf baiting methods were developed many years ago for detection of *Phytophthora* spp.[57,58] These are usually considered qualitative, but propagule densities can be quantitated by use of most-probable-number techniques.[59] They are relatively simple and require minimal equipment and supplies.

Subsequently, media have been developed for the selective isolation of *Phytophthora* spp. from soil to determine propagule densities.[60,61] Methods for sample collection and handling have been developed.[53] In studies in Florida, propagule densities are highest where fibrous root densities are greatest. Thus, populations diminish with depth and with distance from the tree.

Phytophthora populations are not uniform across the orchard. Spatial distribution studies in Florida indicate that the distribution of the fungus is random.[30] Clumping of propagules might become apparent if smaller or larger quadrant sizes are used. However, for routine propagule density determinations, samples can be collected at random in the orchard. About 20 to 40 samples per 4-ha orchard provide a reliable estimate of the propagule density when populations are moderate to high.[53] Propagule densities vary considerably between sample times,[30] and populations may show seasonality.[10,45]

Genus-specific polyclonal and monoclonal antibodies have been produced against *Phytophthora*.[62] Diagnostic kits using enzyme-linked immunosorbent assay have been developed for detection of *Phytophthora* spp. in roots and in soil debris.[63] The ELISA method is highly sensitive and can detect the presence of *Phytophthora* at lower population densities than dilution plating onto selective media.[63] Theoretically, the tissue baiting and the ELISA soil assay should be the most sensitive since they utilize larger volumes of soil. However, in one study, the leaf baiting and selective media were about equally sensitive.[64]

Although good techniques are now available to quantitate populations of *Phytophthora* spp., the significance of those populations remains to be determined. Populations may vary considerably from one time of the year to another.[30] Propagule densities in most Florida and California orchards range from 1 to 20 propagules/cm[3], but occasionally range as high as 100 to 200 propagules/cm[3].[65,66] It may be difficult to establish thresholds, but populations of less than 5 propagules/cm[3] have been considered insignificant.[66] Increases in fibrous root densities and, in some cases, yield have been observed following fungicide treatment of orchards with 10 propagules/cm[3] or higher.[11]

10-10 DISEASE MANAGEMENT

10-10-1 Nurseries

Ideally, citrus nurseries should be maintained free of *Phytophthora* spp. to avoid disseminating the fungus to the orchard. Citrus seed is treated at 50°C for 10 minutes prior to planting to avoid introduction of the fungus by this means.[67] Previously, seed was planted in raised beds in a field nursery, and consequently many problems occurred due to the damping-off of seedlings by *Phytophthora* spp. More recently, in the United States and in other citrus areas, seed is planted in sterilized, soilless media in containers in greenhouses, thus eliminating most Phytophthora problems at the seedbed stage.

Optimally, the site selected for a field nursery or seedbed has never been planted to citrus and is located at some distance from existing plantings. If sites must be used repeatedly for citrus nurseries, they should be fumigated, especially if disease problems have been observed in previous crops. Preplant application of methyl bromide at 500 to 600 kg/ha or metam-sodium at 425 liters/ha will reduce pathogen populations substantially and, in some cases, eliminate these fungi completely.[67] However, fumigation eliminates beneficial mycorrhizal fungi and may result in stunting of seedlings due to poor uptake of phosphorus, zinc, and copper.[68,69] This problem may be partially overcome by adjusting the fertilization program. Trifoliate orange and its hybrids are generally less susceptible to fumigation-induced stunting than sour orange, Cleopatra mandarin, and other rootstock species.[34,68,69]

Field nurseries and greenhouse production facilities can be maintained free of *Phytophthora* spp. by following certain sanitary practices. Plant material from sources not known to be free of *Phytophthora* spp. should not be allowed in clean facilities. Implements, machinery, and personnel which have been in the orchard should not enter nursery facilities. Care must be taken not to allow run-off water from contaminated areas to enter *Phytophthora*-free zones. If portions of nurseries become infested, trees may be destroyed and the site fumigated if the area is small. If not destroyed, they should be isolated from other areas of the nursery, and no traffic should be allowed between clean and infested areas.

The systemic fungicides, metalaxyl, and fosetyl-Al, are highly effective against *Phytophthora* spp. and provide effective control of diseases caused by these fungi in the field.[21,65,70–73] However, they are often used routinely by nursery owners to suppress *Phy-*

tophthora populations and reduce root damage in nursery stock. Application of fungicides to *Phytophthora*-free nurseries is not necessary. In infested nurseries, fungicides may suppress populations temporarily, but populations may rise sharply after nursery trees are transplanted to the orchard. *Phytophthora*-free stock should be used when planting sites not previously used for citrus. However, infested trees with only minor root loss can be used when replanting orchards with existing populations of the fungus.

10-10-2 Control of Foot Rot or Gummosis in Young Orchards

Most of the rootstocks used commercially for citrus are tolerant to bark infection by *Phytophthora* spp. Thus, gummosis usually occurs just above the bud union on the more susceptible commercial scions such as sweet orange, grapefruit, and lemons.

The majority of foot rot problems can be solved by manipulation of cultural practices. Budding the rootstock seedlings well above the soil line will help reduce foot rot incidence.[74] In areas where rainfall is high and soils are conducive to maintenance of high fungal populations, trees are often budded 50 cm or more above the soil line. In locations where frequent freezes occur, trees must be budded lower so trunks can be wrapped to protect them from freeze damage.

The soil surface under the tree must be kept clean and dry. Injuries to the trunk bark must be avoided since they provide entry points for infection. Use of herbicides in citrus orchards has helped reduce foot rot damage by reducing humid conditions under the trees and by eliminating wounds produced by hoes, mowers, and other implements. Water-absorbent trunk wraps used for freeze protection often create conditions highly favorable for foot rot development. This is especially true when irrigation wets the trunk area for long periods. Often, foot rot damage can be reduced by redirecting sprinklers so the trunk remains dry. In some situations, improvement of surface drainage to eliminate standing water is helpful in minimizing damage.

Even when appropriate cultural practices are used, foot rot may still be a problem. In previous years, foot rot lesions were treated by carving out infected bark tissues and painting the wounds with copper fungicides.[67] However, this procedure has never been highly successful and is laborious and time consuming.[21,75] With the introduction of the systemic fungicides metalaxyl and fosetyl-Al, curative treatments are much more feasible. Trunk paints and sprays with systemic fungicides are highly effective without the surgical treatments. Trunk application of the systemic fungicides as well as the copper fungicides are effective preventive treatments, and a single application lasts for many months.[21,72] Where foot rot incidence is high, it may be necessary to apply soil treatments of metalaxyl or foliar sprays of fosetyl-Al to protect remaining healthy trees. Either fungicide may be applied through drip or microsprinkler irrigation systems.

10-10-3 Control of Root Rot in Mature Orchards

Once an orchard matures and begins to bear regular crops of fruit, foot rot usually ceases to be a serious problem. Loss of fibrous roots due to Phytophthora infection may still produce mild tree decline and reduced yields. As mentioned previously, most commonly used rootstocks are somewhat tolerant to Phytophthora foot rot. Trifoliate orange and Swingle citrumelo are nearly immune to bark infection, but some root infection occurs.

Populations usually remain low in orchards on these stocks, and probably little loss is incurred on orchards on these rootstocks.[45] However, some rootstocks, such as sour orange and Cleopatra mandarin, are tolerant of bark rot but nearly as susceptible as the highly sensitive sweet orange to fibrous root rot.[39]

Prior to the development of the systemic fungicides, it was not possible to determine the importance of fibrous root loss due to Phytophthora infection on mature trees. Application of fungicides have routinely increased fibrous root densities from 20 to 40%, and in some cases increases of 100% were observed.[11] Thus, it may be necessary to control *Phytophthora* in orchards with high populations even on tolerant rootstocks to maximize yields. Fibrous root losses and Phytophthora populations may be reduced by installation of tile drainage or improvement of surface drainage. Regular foliar applications of fosetyl-Al and soil applications of metalaxyl increased yield and fruit size in some orchards.[11,65,76]

10-10-4 Brown Rot and Canopy Blight

Where *P. parasitica* is the primary species present, these diseases are seldom of much importance. Only occasional losses of fruit near the soil are incurred. However, where *P. citrophthora* and other species occur in rainy areas, these diseases may be severe. Usually, the primary problem is loss of fruit to brown rot, which may continue to develop and spread in packing boxes during storage or transport. Leaf and twig blight are seldom serious, but if the fungus infects large branches, cankers may be formed and entire branches may be girdled, especially on highly susceptible species such as lemons.

Sometimes brown rot losses can be reduced by changing from overhead to under-tree sprinklers for irrigation. However, most of the problem stems from winter rains, and fungicides must be applied. Copper fungicides or captan applied prior to the beginning of the rains are usually quite effective.[77,78] Preharvest applications of the systemic fungicides fosetyl-Al and metalaxyl to the canopy provide effective control of brown rot. Most of the standard postharvest fungicides used in packinghouses are ineffective against brown rot, but the systemic fungicides control the disease.[79]

10-11 REFERENCES

1. Timmer, L. W. and Menge, J. A., *Phytophthora*-induced diseases. In *Compendium of citrus diseases* (J. O. Whiteside, S. M. Garnsey and L. W. Timmer, eds.), p. 22, American Phytopathological Society, St. Paul, MN, 1988.

2. Fawcett, H. S., Two fungi as causal agents in gummosis of lemon trees in California, *Phytopathology*, 3, 194, 1913.

3. Fawcett, H. S., *Citrus diseases and their control*, McGraw-Hill, New York, p. 656, 1936.

4. Curtis, A. H., Report on "Foot-Rot" and "Scab" of citrus trees and the prevention of orange "Rust," *Fla. Agric. Expt. Stn. Bull.*, 2, 27, 1888.

5. Childs, J. F. L., Foot rot in Florida: Its habits and suggestions for its control, *Citrus Ind.*, 28(9), 5, 1947.

6. Fraser, L. R., Phytophthora root rot of citrus, *J. Austral. Inst. Agric. Sci.*, 8, 101, 1942.

7. Broadbent, P., Phytophthora diseases of citrus: A review, *Proc. Int. Soc. Citric.*, 3, 986, 1977.

8. Fawcett, H. S., Gummosis of citrus. I. Gummosis due to *Pythiacystis citrophthora*, *J. Agric. Res.*, 24, 191, 1923.

9. Klotz, L. J., DeWolfe, T. A. and Wong, P. P., Decay of fibrous roots of citrus, *Phytopathology*, 48, 616, 1958.

10. Lutz, A. and Menge, J., Citrus root health II: Phytophthora root rot, *Citrograph*, 72(2), 33, 1986.

11. Sandler, H. A., Timmer, L. W., Graham, J. H. and Zitko, S. E., Effect of fungicide applications on populations of *Phytophthora parasitica* and on feeder root densities and fruit yields of citrus trees, *Plant Dis.*, 73, 902, 1989.

12. Feld, S. J., Menge, J. A. and Pehrson, J. E., Brown rot of citrus: A review of the disease, *Citrograph*, 64(5), 101, 1979.

13. Whiteside, J. O., Factors contributing to the restricted occurrence of citrus brown rot in Florida, *Plant Dis. Rptr.*, 54, 608, 1970.

14. Timmer, L. W., Zitko, S. E. and Sandler, H. A., An isolate of *Phytophthora arecae* from Florida pathogenic to citrus (Abstr.), *Phytopathology*, 80, 1025, 1990.

15. Waterhouse, G. M. and Waterston, J. M., *Phytophthora nicotianae* var. *parasitica*. In *Description of pathogenic fungi and bacteria*, 35, C.M.I., Kew, England, p. 2, 1964.

16. Waterhouse, G. M. and Waterston, J. M., *Phytophthora citrophthora*. In *Description of pathogenic fungi and bacteria*, 33, C.M.I., Kew, England, p. 2, 1964.

17. Duniway, J. M., Role of physical factors in the development of Phytophthora diseases. In *Phytophthora: Its biology, taxonomy, ecology, and pathology* (D. C. Erwin, S. Bartnicki-Garcia and P. H. Tsao, eds.), p. 175, American Phytopathological Society, St. Paul, MN, 1983.

18. Feld, S. J., *Studies on the role of irrigation and soil water matric potential on Phytophthora parasitica root rot of citrus*, Ph.D. thesis, University of California, Riverside, p. 104, 1982.

19. Tsao, P. H., Studies on the saprophytic behavior of *Phytophthora parasitica* in soil, *Proc. First Int. Citrus Symp.*, 3, 1221, 1969.

20. Ribeiro, O. K., Physiology of asexual sporulation and spore germination in *Phytophthora*. In *Phytophthora: Its biology, taxonomy, ecology and pathology* (D. C. Erwin, S. Bartnicki-Garcia and P. H. Tsao, eds.), p. 55, American Phytopathological Society, St. Paul, MN, 1983.

21. Timmer, L. W., Preventive and curative trunk treatments for control of Phytophthora foot rot of citrus, *Phytopathology*, 67, 1149, 1977.

22. Tsao, P. H., Chlamydospore formation in sporangium-free liquid cultures of *Phytophthora parasitica*, *Phytopathology*, 61, 1412, 1971.

23. Lutz, A. and Menge, J. A., Breaking winter dormancy of *Phytophthora parasitica* propagules using heat shock, *Mycologia*, 28, 148, 1986.

24. Holdaway, B. F., *The survival and population of Phytophthora parasitica in soils as influenced by ecological factors*, Ph.D. thesis, University of California, Riverside, p. 77, 1971.

25. Khew, K. L. and Zentmyer, G. A., Chemotactic response of zoospores of five species of *Phytophthora*, *Phytopathology*, 63, 1511, 1973.

26. Khew, K. L. and Zentmyer, G. A., Electrostatic response of zoospores of seven species of *Phytophthora*, *Phytopathology*, 64, 500, 1974.

27. Feld, S. J., Menge, J. A. and Stolzy, L. H., Influence of drip and furrow irrigation on Phytophthora root rot of citrus under field and greenhouse conditions, *Plant Dis.*, 74, 21, 1990.

28. Stolzy, L. H., Lefey, L., Klotz, L. J. and DeWolfe, T. A., Soil aeration and root rotting fungi as factors in decay of citrus feeder roots, *Soil Sci.*, 99, 403, 1965.

29. Ford, H. W., Bacterial metabolites that affect citrus root survival in soils subject to flooding, *Proc. Amer. Soc. Hortic. Sci.*, 86, 205, 1965.

30. Timmer, L. W., Zitko, S. E., Sandler, H. A. and Graham, J. H., Seasonal and spatial analysis of populations of *Phytophthora parasitica* in citrus orchards, *Plant Dis.*, 73, 810, 1989.

31. Bevington, K. B. and Castle, W. S., Annual root growth potential of young citrus trees in relation to shoot growth, soil temperature, and soil water content, *J. Amer. Soc. Hortic. Sci.*, 110, 840, 1985.

32. Malajczuk, N., Microbial antagonism of *Phytophthora*. In *Phytophthora: Its biology, taxonomy, ecology and pathology*, (D. C. Erwin, S. Bartnicki-Garcia and P. H. Tsao, eds.), p. 197, American Phytopathological Society, St. Paul, MN, 1983.

33. Broadbent, P. and Baker, K. F., Association of bacteria with sporangium formation and breakdown of sporangia in *Phytophthora* spp., *Austral. J. Agric. Res.*, 25, 139, 1974.

34. Graham, J. H., Citrus mycorrhizae: Potential benefits and interactions with pathogens, *HortScience*, 21, 1302, 1986.

35. Davis, R. M. and Menge, J. A., Influence of *Glomus fasciculatus* and soil phosphorus on Phytophthora root rot of citrus, *Phytopathology,* 70, 447, 1980.

36. Graham, J. H. and Egel, D. S., Phytophthora root rot development on mycorrhizal and phosphorus-fertilized nonmycorrhizal sweet orange seedlings, *Plant Dis.*, 72, 611, 1988.

37. Carpenter, J. R. and Furr, J. R., Evaluation of tolerance to root rot caused by *Phytophthora parasitica* in seedlings of citrus and related genera, *Phytopathology*, 52, 1277, 1962.

38. Grimm, G. D. and Hutchison, D. J., Evaluation of *Citrus* spp., relatives and hybrids for resistance to *Phytophthora parasitica* Dastur, *Proc. Int. Soc. Citric.*, 3, 863, 1977.

39. Graham, J. H., Evaluation of tolerance of citrus rootstocks to Phytophthora root rot in chlamydospore-infested soil., *Plant Dis.*, 74, 743, 1990.

40. Matheron, M. E. and Matejka, J. C., Differential virulence of *Phytophthora parasitica* recovered from citrus and other plants to rough lemon and tomato, *Plant Dis.*, 74, 138, 1990.

41. Grimm, G. R. and Whidden, R., Range of pathogenicity of Florida cultures of the foot rot fungus, *Proc. Fla. State Hortic. Soc.*, 75, 73, 1962.

42. Grimm, G. R. and Hutchison, D. J., A procedure for evaluating resistance of citrus seedlings to *Phytophthora parasitica, Plant Dis. Rptr.*, 57, 669, 1973.

43. Broadbent, P., Fraser, L. R. and Waterworth, Y., The reaction of seedlings of *Citrus* spp. and related genera to *Phytophthora citrophthora, Proc. Linn. Soc. N.S.W.*, 96(3), 119, 1971.

44. Tsao, P. H. and Garber, M. J., Methods of soil infestation, watering and assessing the degree of root infection for greenhouse *in situ* ecological studies with citrus Phytophthoras, *Plant Dis. Rptr.*, 44, 710, 1960.

45. Agostini, J. P., Timmer, L. W., Castle, W. S. and Mitchell, D. J., Effect of citrus rootstocks on soil populations of *Phytophthora parasitica, Plant Dis.*, 75, 532, 1991.

46. Afek, U. and Sztejnberg, A., Accumulation of scoparone, a phytoalexin associated with resistance of citrus to *Phytophthora citrophthora, Phytopathology*, 78, 1678, 1988.

47. Whiteside, J. O., Zoospore-inoculation techniques for determining the relative susceptibility of citrus rootstocks to foot rot, *Plant Dis. Rptr.*, 58, 713, 1974.

48. Timmer, L. W. and Ellis, M. A., Methods for evaluating fungicides for control of Phytophthora crown and root diseases of tree crops. In *Methods for evaluating pesticides for control of plant pathogens* (K. D. Hickey, ed.), p. 163, American Phytopathological Society Press, St. Paul, MN, 1985.

49. Smith, G. S., Hutchison, D. J. and Hendersen, C. T., Screening sweet orange cultivars for relative susceptibility to Phytophthora foot rot, *Proc. Fla. State Hortic. Soc.*, 100, 64, 1987.

50. Afek, U., Sztejnberg, A. and Solel, Z., A rapid method for evaluating citrus seedlings for resistance to foot rot caused by *Phytophthora citrophthora, Plant Dis.*, 74, 66, 1990.

51. Matheron, M. E. and Matejka, J. C., Temporal changes in susceptibility of citrus phloem tissue to colonization by *Phytophthora citrophthora* and *P. parasitica, Plant Dis.*, 73, 408, 1989.

52. Whiteside, J. O., Some factors affecting the occurrence and development of foot rot on citrus trees, *Phytopathology*, 61, 1233, 1971.

53. Timmer, L. W., Sandler, H. A., Graham, J. H. and Zitko, S. E., Sampling citrus orchards in Florida to estimate populations of *Phytophthora parasitica, Phytopathology*, 78, 940, 1988.

54. Hough, A., Treatment of irrigation water for the removal of fungi and nematodes, *Citrus Subtrop. Fruit J.*, 547, 20, 1979.

55. Klotz, L. J., Wong, P. and DeWolfe, T. A., Survey of irrigation water for the presence of *Phytophthora* spp. pathogenic to citrus, *Plant Dis. Rptr.*, 43, 830, 1959.

56. Ridings, W. H., Schenck, N. C., Snell, R. R., Keen, W. M. and Cornell, J. A., Reinvasion of methyl bromide-treated soil by soil-borne fungi and their subsequent effect on citrus seedling growth, *Proc. Fla. State Hortic. Soc.*, 90, 70, 1977.

57. Grimm, G. R. and Alexander, A. F., Citrus leaf pieces as traps for *Phytophthora parasitica* from soil slurries, *Phytopathology*, 63, 540, 1973.

58. Klotz, L. J. and DeWolfe, T. A., Technique for isolating *Phytophthora* spp. which attack citrus, *Plant Dis. Rptr.*, 42, 675, 1958.

59. Tsao, P. H., A serial dilution method for estimating disease potentials of citrus *Phytophthora* in soil, *Phytopathology*, 50, 717, 1960.

60. Kannwischer, M. E. and Mitchell, D. J., The influence of a fungicide on the epidemiology of black shank of tobacco, *Phytopathology*, 68, 1760, 1978.

61. Carpenter, J. B., Burns, R. M. and Furr, J. R., Phytophthora-tolerant rootstocks for lemons, *Plant Dis. Rptr.*, 59, 54, 1975.

62. Miller, S. A., Rittenburg, J. H., Petersen, F. P. and Grothaus, G. D., Development of modern diagnostic techniques and benefits to the farmer. In *Monoclonal antibodies in agriculture,* (A. Schots, ed.), p. 15, Pudoc, Wageningen, 1990.

63. Skaria, M. and Miller, S. A., A rapid test for detecting *Phytophthora* spp. in citrus, *J. Rio Grande Valley Hortic. Soc.*, 42, 63, 1989.

64. Zitko, S. E., Timmer, L. W. and Castle, W. S., Survey of Florida citrus nurseries for *Phytophthora* spp., *Proc. Fla. State Hortic. Soc.*, 100, 82, 1987.

65. Menge, J. A., Use of new systemic fungicides on citrus, *Citrograph,* 71(12), 245, 1986.

66. Timmer, L. W., Graham, J. H., Sandler, H. A. and Zitko, S. E., Populations of *Phytophthora parasitica* in bearing citrus orchards in Florida and response to fungicide applications, *Citrus Ind.*, 69(11), 40, 1988.

67. Kotz, L. J. and Calavan, E. C., Diseases of citrus in California, *Calif. Agric. Expt. Stn. Ext. Ser.*, Riverside, 36, 1969.

68. Schenck, N. C. and Tucker, D. P. H., Endomycorrhizal fungi and the development of citrus seedlings in Florida fumigated soils, *J. Amer. Soc. Hortic. Sci.*, 99, 284, 1974.

69. Timmer, L. W. and Leyden, R. F., Relationship of seedbed fertilization and fumigation to infection of sour orange seedlings by mycorrhizal fungi and *Phytophthora parasitica, J. Amer. Soc. Hortic. Sci.*, 103, 537, 1978.

70. Davis, R. M., Control of Phytophthora root and foot rot of citrus with systemic fungicides metalaxyl and phosethyl aluminum, *Plant Dis.*, 66, 218, 1982.

71. Farih, A., Menge, J. A., Tsao, P. H. and Ohr, H. D., Metalaxyl and efosite aluminum for control of Phytophthora gummosis and root rot on citrus, *Plant Dis.*, 65, 654, 1981.

72. Timmer, L. W., Preventive and systemic activity of experimental fungicides against *Phytophthora parasitica* on citrus, *Plant Dis. Rptr.*, 63, 324, 1979.

73. Timmer, L. W. and Castle, W. S., Effectiveness of metalaxyl and fosetyl-Al against *Phytophthora parasitica* on sweet orange, *Plant Dis.*, 69, 741, 1985.

74. Whiteside, J. O., Foot rot of citrus trees—the importance of high budding as a preventive measure, *Citrus Ind.*, 53(4), 14, 1972.

75. Brodrick, H. T., Attempts to control collar rot in grapefruit trees with tree surgery, *Info. Bull. Citrus & Subtrop. Fruit Res. Inst.*, 22, 10, 1974.

76. Pond, E., Menge, J. A. and Ohr, H. D., The effect of metalaxyl and efosite-Al applied through the drip irrigation system on *Phytophthora parasitica* in the soil and on the yield of navel oranges, *Phytopathology*, 74, 854 (Abstr.), 1984.

77. Klotz, L. J., Calavan, E. C., DeWolfe, T. A., Miller, M. P., Burns, R. M., Colladay, C., Lindsay, J. and Cairns, T., Brown rot control trials with high and low volume sprays, *Citrograph*, 57(7), 267, 1972.

78. Solel, Z., Aerial versus ground application of fungicides to control brown rot of citrus fruit, *Plant Dis.*, 67, 878, 1983.

79. Cohen, E., Metalaxyl for postharvest control of brown rot of citrus fruit, *Plant Dis.*, 65, 672, 1981.

11

FUSARIUM WILT
OF BANANA
(PANAMA DISEASE)

RANDY C. PLOETZ

University of Florida, IFAS
Tropical Research and Education Center
18905 SW 280th Street, Homestead, Florida

11-1 INTRODUCTION

The first recording of fusarium wilt of banana was in Australia in 1876.[1] Reports from Hawaii (1904),[2] India (1911),[3] and numerous locations in tropical America and the Caribbean (beginning in 1890)[4-7] soon followed, and by 1932 the disease had been reported from every continent on which bananas are economically or socially important[8](Fig. 11-1). As of 1987, fusarium wilt had been found in all banana-growing regions except the South Pacific islands, including Papua New Guinea and most of the island of Borneo, Somaliland, and countries bordering the Mediterranean[9,10](Fig. 11-1). For specific locations and histories for countries listed in Fig. 11-1, see Stover[8] and Ploetz et al.[11]

Bananas are large, perennial jungle herbs in the order Zingiberales.[12] With minor exceptions, cultivated edible bananas are natural, triploid inter- and intraspecific combinations of two diploid species in the family Musaceae, *Musa acuminata* Colla and *M. balbisiana* Colla (AA and BB genotypes, respectively). Two groups of banana cultivars that are affected by fusarium wilt will be considered in this chapter: (1) those grown for export (formerly Gros Michel and now cultivars of the Cavendish subgroup, both AAA genotype), and (2) those produced and consumed locally (primarily diverse AAA, AAB, and ABB cultivars).

As a result of epidemics in export plantations, Simmonds[13] considered fusarium wilt to be one of the six most destructive plant diseases in recorded history. The first report of the disease in many countries was on Gros Michel, usually within 20 years of when the cultivar was used to initiate export production[8] (Fig. 11-1). By 1960, fusarium wilt had

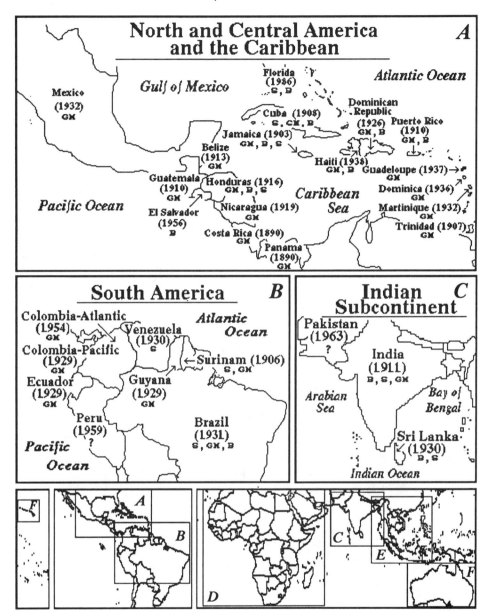

Figure 11-1 The year fusarium wilt was first reported and the banana cultivar(s) on which the disease was first noted in countries in: (A) North and Central America and the Caribbean; (B) South America; (C) the Indian subcontinent; (D) Africa; (E) Southeast Asia; and (F) Australia and Oceania. The composite global figure beneath (B) and (C) depicts the relative position of each region. Cultivar abbreviations are B, Bluggoe; DC, Dwarf Cavendish; GC, Giant Cavendish; GM, Gros Michel; PA, Pisang awak;

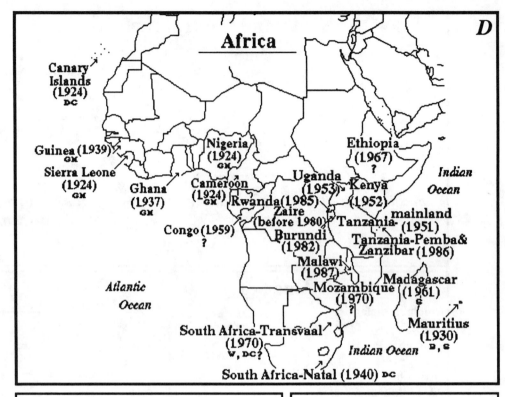

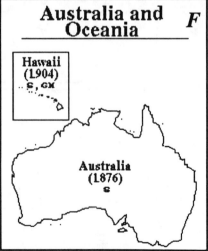

Figure 11-1 Continued. S, Silk; and W, Williams; "?" alone indicates that the cultivar on which the disease was first noted is not known, and "?" after a cultivar abbreviation notes an uncertain cultivar identity. For countries in East Africa, cultivars on which the disease was first described are Burundi (PA); Kenya (B?); Malawi (B); Rwanda (PA); Tanzania—mainland (B,GM); Tanzania-Pemba and Zanzibar (B,S); Uganda (GM,PA); and Zaire (PA and Ney Poovan). From Anonymous,[10] Ploetz et al,[11] and Stover.[8]

destroyed an estimated 40,000 ha of Gros Michel, and profitable production of the cultivar became impossible in many areas; the trades were forced to replace Gros Michel.[8]
Conversion to resistant Cavendish cultivars occurred about 30 years ago, and these cultivars remain the standards for export today.

Although Cavendish cultivars still resist the disease in tropical production areas,
they are now being destroyed by race 4 in the following subtropical production areas in
the eastern hemisphere: Australia (New South Wales and Queensland), the Canary Islands
(Tenerife and possibly other islands), South Africa (Natal and Transvaal), and Taiwan.[11]
Since adequate replacements for the Cavendish cultivars generally do not exist (for an
exceptional situation see Hwang[14]), race 4 threatens the survival of export and local trades
in the subtropics that are based on these cultivars.

Perhaps more important than the erosion of resistance in the Cavendish subgroup
are current and widespread epidemics of fusarium wilt on locally consumed cultivars of
banana. These are important sources of food throughout the humid tropics, and for native
populations in many areas they are the sole or primary source of carbohydrates.[15] Fusarium wilt causes considerable damage on locally consumed cultivars in Brazil, East Africa, Southeast Asia, and other banana-growing regions.[8,11,16–19] Ignorance of the racial
structure of the pathogen [*Fusarium oxysporum* (Schlecht.) f. sp. *cubense* (E. F. Smith)
Snyd. & Hans.] in the affected areas, the inadequate infrastructures which exist for help-
ing the large and diverse areas that are affected, the lack of adequate replacement cultivars, and the absence of economic incentives in some areas which are found in trades
based on the Cavendish cultivars make these especially difficult problems.

11-2 ECONOMIC AND SOCIAL IMPORTANCE

Combined world production of bananas and plantains (an AAB subgroup of bananas) in
1987 totalled 66 million metric tons (MMT), which ranked them first among all fruit
grown in the world.[20]

11-2-1 Fusarium Wilt on Cavendish Cultivars

Fusarium wilt currently affects production of Cavendish cultivars only in the subtropics
of the Eastern Hemisphere.[11,21,22] About 1 MMT of Cavendish fruit are produced each year
in the subtropics, most of which is sold in the producing countries.[9] Although monetary
losses caused by fusarium wilt in these situations are not available, producers in the
affected countries consider the disease a top priority.[11] Almost a third of Taiwan's production areas are affected by the disease, and the Taiwan Banana Research Institute in
Pingtung has recently devoted much time and effort to developing a replacement for Giant
Cavendish.[14] In 1988, the South African Banana Board contributed US$ 1 million toward
research, much of which was used to construct a research facility in Nelspruit which has a
primary focus of developing a replacement for the most commonly used cultivars in the
country, Dwarf Cavendish and Williams (a Cavendish cultivar). Active research programs are also underway in Australia and the Canary Islands.[11]

In the tropics, Cavendish cultivars are affected in Guadeloupe, Jamaica, the Philippines, and elsewhere, but in none of these locations is damage widespread or severe.[11] In

general, the export trades in the tropics have not been affected by fusarium wilt since the Cavendish cultivars replaced Gros Michel. However, the recent development of race 4, past experience with Gros Michel, the near total dependence on the Cavendish subgroup, and the lack of a replacement cultivar suggest that the current situation in the tropics is not cause for complacency. The international export trades produce about 7 MMT annually, and import value of the fruit in the United States was US$ 404 ton^{-1} in 1981.[23] For many producing nations, Cavendish bananas are the primary source of export revenue.[9] Race 4, or something like it, could cause great hardship if it developed in these production areas.

11-2-2 Fusarium Wilt on Locally Consumed Cultivars

In excess of 50 MMT of bananas and plantains that are produced each year are not exported.[9] Many of these bananas are consumed by subsistence producers. Accurate figures on production and economics are not available, but it is clear that locally consumed bananas and plantains are far more important to world food supplies and local food economics than are those produced for export.

About one third of the locally consumed fruit are Horn-type plantains (AAB) that are resistant to fusarium wilt.[9] Other dessert and cooking bananas in this category, however, are susceptible.[8,11] In several different banana-growing regions, the production of popular clones has been eliminated or reduced by fusarium wilt. For example, the AAB dessert cultivars that are preferred by Brazil, the world's largest banana producer, are all susceptible.[11,16,24] Production of Prata (Pome), Prata Aña, and Pacovan (a large-fruited mutant of Prata) is becoming increasingly difficult, while the area planted with the most susceptible cultivar, Maçã (Silk), has been reduced greatly because of fusarium wilt. In East Africa, several cultivars of cooking and beer bananas are damaged by fusarium wilt.[11,18,25,26] In many instances, the affected cultivars are either those whose taste is preferred (e.g., Silk) or those best adapted to seasonal rainfall in the region (e.g., Bluggoe and Pisang awak, both ABB). Since the highest per capita consumption of bananas in the world is found in some of these countries (ca. two thirds kg day^{-1} in Rwanda and Uganda[27]), it would be difficult to overestimate the importance of a lethal disease such as fusarium wilt on this staple crop.

11-3 SYMPTOMS

The first external, albeit inconspicuous, symptoms of fusarium wilt is a pale green cast of lower portions of petioles of the oldest leaves; on close inspection, discoloration of vascular strands below the epidermis in these areas is evident (Fig. 11-2). The first obvious external symptom is yellowing of leaf margins that eventually encompasses the entire leaf (Fig. 11-2). Yellowing begins with the oldest and progresses to the youngest leaves of a plant. Petioles of these leaves buckle, and eventually the central, unfurled leaf is left standing alone (Fig. 11-3). In Gros Michel, a different syndrome may develop whereby leaves remain predominantly green, but petioles buckle and leaves collapse.[8]

Splitting of the pseudostem base may occur in the yellowing and nonyellowing syndromes, and the oldest plants in a mat are affected first. Young suckers rarely develop external symptoms. Suckers in an affected mat may eventually produce bunches of fruit

Figure 11-2 Symptoms of fusarium wilt of banana: (A) internal vascular discoloration of the pseudostem; (B) marginal chlorosis of the lamina of a leaf that has also buckled at the petiole base; (C) damage ultimately caused in a mature plantation (note surviving, nonsymptomatic suckers); and (D–F) inconspicuous, subepidermal vascular discoloration of outer leaf sheaths of a pseudostem that is the first symptom evident above ground. Note dull discoloration of thin vascular strands in (D) in the area cut out of the pseudostem in (E) and (F).

that develop abnormally and ripen prematurely or irregularly. However, fruit are not infected by the pathogen and are not discolored.[9] Lack of symptoms in the latter situation distinguishes fusarium wilt from bacterial wilt (caused by *Pseudomonas solanacearum* E. F. Smith and also known as Moko disease), a disease whose symptoms often resemble those of fusarium wilt, but which affects young suckers and discolors fruit.[9] Pseudostems affected by fusarium wilt may stand for one to two months after leaves begin to collapse, but these plants ultimately fall down.

Vascular discoloration is the first internal symptom of fusarium wilt. Vascular dis-

Disease Cycle of Fusarium Wilt of Banana

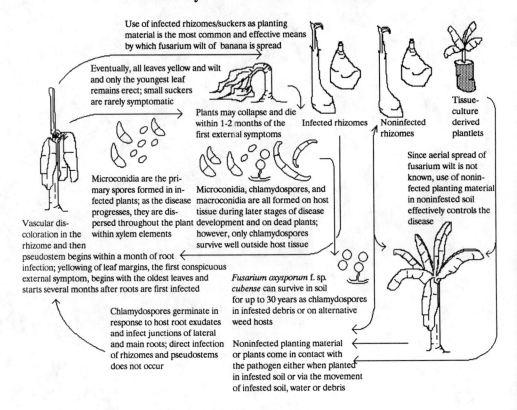

Figure 11-3 Disease cycle of fusarium wilt of banana. From Beckman[34] and Stover.[8,33]

coloration begins as a yellowing of a few vascular strands in the rhizome and progresses to a dark red or almost black discoloration of most of the pseudostem and fruit stalk (true stem) (Fig. 11-2).

11-4 CAUSAL ORGANISM AND THE DISEASE CYCLE

Smith[6] recovered the pathogen from host tissue sent from Cuba in 1910, and named it *Fusarium cubense.* Ashby[7] published the first good description of the fungus in 1913, and Brandes[5] reported the first conclusive work on its pathogenicity to banana in 1919.

11-4-1 The Pathogen

No teleomorph is known for *F. oxysporum* f. sp. *cubense;* it forms three types of asexual spores: microconidia, macroconidia, and chlamydospores (Fig. 11-3). Although the pathogen is highly variable, it cannot be distinguished morphologically from saprophytic *F. oxysporum* or other plant pathogenic formae speciales of the species. Currently, the only reliable methods for its identification are either: (1) with pathogenicity tests on differential cultivars (see section 11-5) or seedlings of *M. balbisiana* or other *Musa* species, or (2) with vegetative compatibility tests.

Thirteen vegetative compatibility groups (VCGs) have been described for the pathogen to date.[28-30] Each VCG represents a different, genetically isolated subpopulation of the forma specialis. In general, VCGs of *F. oxysporum* f. sp. *cubense* are comprised of members that share common biological, genetical, pathological, and physiological attributes[30] (Table 11-1). In addition to identifying *F. oxysporum* f. sp. *cubense,* vegetative compatibility tests have also assisted recent phylogenetic studies of the pathogen.[30-32]

11-4-2 The Disease Cycle

Chlamydospores (thick-walled resting structures) begin the disease cycle in soil by germinating in response to host root exudates and infecting lateral roots[33] (Fig. 11-3). Further development and spread of most initial infections is usually stopped in the xylem by vascular occluding responses of the host, which include the formation of gels, tyloses, and the collapse of vessels.[34-36] In susceptible cultivars, however, a few of these infections become established in the xylem and advance ahead of host defense mechanisms.

Microconidia are formed in xylem elements and facilitate spread of the fungus within the host vascular system.[33] They and mycelium of the pathogen are restricted to the xylem until later stages of the disease cycle. During advanced stages of the disease, microconidia are also formed in parenchyma and other host tissues, while macroconidia may be formed on leaves and petioles.[33] Microconidia and macroconidia can be spread by wind or water, but because they are relatively short-lived outside host tissue, they play limited roles in the spread of fusarium wilt. Chlamydospores, which are produced either on mycelia or from conidia, are formed in host tissue during later stages of the disease cycle. They are the primary survival structure of the fungus in soil or infested host tissue, and they can remain dormant for several years in soil after the hosts dies. If dormant chlamydospores are stimulated to germinate by host root exudates, they may begin the disease cycle again.

TABLE 11-1: RACIAL STRUCTURE AND ORIGINS FOR VEGETATIVE COMPATIBILITY GROUPS (VCGs) OF *FUSARIUM OXYSPORUM* F. SP. *CUBENSE*[28-30]

VCG	Race(s)	Origin(s)
VCG 0120	1, 4	Australia, Brazil, Canary Islands, Costa Rica, Guadeloupe, Honduras, Malaysia, South Africa, Taiwan
VCG 0121	4	Taiwan
VCG 0122	?	Philippines
VCG 0123	1	Philippines, Taiwan
VCG 0124–0125	1, 2	Australia, Brazil, Burundi, Honduras, Jamaica, Malawi, Nicaragua, Tanzania, Uganda, U.S.A. (Florida), Zaire
VCG 0126	1	Honduras
VCG 0127	1	South Africa
VCG 0128	2	Australia, Comores Islands
VCG 0129	4	Australia
VCG 01210	1	U.S.A. (Florida)
VCG 01211	?	Australia
VCG 01212	?	Tanzania

Fusarium wilt is spread primarily via infected rhizomes[33] (Fig. 11-3). The disease can also be spread by humans, machinery, and running water. In tropical alluvial valleys that are used often for production, periodic floods which move large volumes of soil can effectively disperse the pathogen. Figure 11-3 provides additional details on the disease cycle.

11-5 DETERMINATION OF PATHOGENICITY AND VIRULENCE OF *FUSARIUM OXYSPORUM* F. SP. *CUBENSE* AND RESISTANCE OF THE BANANA HOST

Four races of *F. oxysporum* f. sp. *cubense* are currently recognized.[9,21,37] Race 1 was the cause of the devastating epidemics on Gros Michel. It also affects the AAB dessert cultivars Silk, Pome, and Taiwan Latundan, and the first bred banana IC2 (AAAA). Race 2 is virulent on Bluggoe and other ABB cooking clones, plus some AAAA tetraploids bred in Jamaica. Race 3 affects species of *Heliconia* (close relatives of banana) but has no or a limited affect on bananas. Race 4 affects Gros Michel and Bluggoe in addition to all commercial Cavendish cultivars. Race 4's wide spectrum of virulence and its effect on Cavendish cultivars has caused concern among producers for the international export trades in the tropics who depend on these cultivars.[37,38]

Determining the racial structure of populations of *F. oxysporum* f. sp. *cubense* in a given area and the resistance or susceptibility of different banana genotypes requires reliable pathogenicity and virulence tests. Standard race evaluations are conducted with clean (noninfected) rhizomes of the following differential cultivars: Gros Michel, race 1; Bluggoe, race 2; and Cavendish, race 4. If other than naturally infested soils are to be tested, some form of organic food base, such as oat seed or millet seed, colonized with test isolates is used to infest the planting hole. Alternatively, rhizomes with roots can be dipped in suspensions of microconidia, macroconidia, or chlamydospores of an isolate. Obviously, noninfested soil is a prerequisite for such work. Tissue-culture derived plantlets have also been used instead of rhizomes,[39-42] but results with plantlets have been inconsistent among different laboratories.[39]

Unfortunately, pathogenicity and virulence tests for fusarium wilt of banana are fraught with difficulties.[37,39] Field tests may take two to nine months to complete, and results from such work may be equivocal due to inherent variability in the pathosystem, climate, edaphic conditions, and planting material used in a given location.[37,39,43] Additional difficulties arise if nonendemic, quarantined isolates are tested. Since quarantine restrictions usually prohibit the use of nonendemic isolates of the pathogen in the field, researchers must conduct such work in glasshouses, growth chambers, or some similar structure that can be secured and disinfested after these tests. The difficulties associated with using large rhizomes under confined conditions and the inconsistent results that can be obtained when plantlets are used demonstrate the need for a reliable way to test nonendemic isolates.[39] Furthermore, the current set of differential cultivars do not adequately assess virulence that probably exists among populations of the currently described races.[44] For example, IC2 was resistant to race 1 in Trinidad, but susceptible to race 1 in Honduras, and Giant Cavendish somaclone 73-53 was resistant to race 4 in Taiwan, but not in Australia.[30,37] Additional differential cultivars are needed to better evaluate the racial structure of different populations of *F. oxysporum* f. sp. *cubense*[44].

11-6 EPIDEMIOLOGY

Pathogen behavior in the host and its movement and survival in soil are discussed in sections 11-6 and 11-7. Most of the work on pathogen behavior in soil and epidemiology of the disease was conducted more than 30 years ago in response to the epidemics on Gros Michel. Comparatively little work in these areas has been conducted since that time.

Levels of *F. oxysporum* f. sp. *cubense* were usually < 200 propagules g soil^{-1} in a study in Honduras, and rhizosphere populations of the fungus were reported to be low.[45] In Taiwan, higher inoculum densities in soil (850 to 1,000 vs. 100 propagules g soil^{-1}) were required to cause 50% disease incidence in resistant somaclones of Giant Cavendish compared to the susceptible parent.[14] Since natural populations seldomly approached the highest inoculum levels tested, the somaclones performed well in the field.

Survival and growth of *F. oxysporum* f. sp. *cubense* are generally greater in acidic or light-textured soils than in clayey or alkaline soils with a high calcium content.[8] Montmorillonite clays have been associated with suppressive, "long-life" soils in which disease progress is retarded.[33]

In general, disease development is greatest when temperatures are optimal for host and pathogen growth. In Gros Michel, pathogen growth and disease development in roots occurred at temperatures of 21 and 27°C, but not 34°C[46]; symptom development declined during cooler, winter months at elevations above 700 m in Jamaica.[47] Some suboptimal conditions have also been associated with higher disease severity. Saturated or poorly drained soils have been correlated with greater disease.[47,48] Low levels of Zn and high Ca:Mg and K:Mg ratios have been associated with more severe disease in the Canary Islands; low Zn was hypothesized to reduce tylose formation; and high Ca:Mg and K:Mg ratios were thought to affect adversely pectin formation in the host.[11,49,50]

11-7 MANAGEMENT

Chemical control, use of organic amendments, flood fallowing, and other less widely tested control measures have proven ineffective in managing fusarium wilt of banana; host resistance is the most effective means for combating the disease.[9] However, since most edible bananas are triploid and infertile, breeding new, resistant cultivars is quite difficult.[51] Thirty years after initiating the effort, the Fundación Hondureña de Investigación Agrícola (FHIA) in Honduras has still not bred an acceptable, resistant alternative to the Cavendish cultivars.[38] Using a different approach, the Taiwan Banana Research Institute has recently produced somaclonal variants of Giant Cavendish that resist race 4 in Taiwan.[14]

Six improvement programs are established worldwide. The program in Taiwan is devoted to improvement via tissue culture. Traditional breeding programs exist at the Empresa Brasileira de Pesquisa Agropecuária (EM-BRAPA) in Brazil, the Institute de Recherches sur les Fruits et Agrumes (IRFA) in Guadeloupe, FHIA, the Banana Board in Jamaica, and the International Institute of Tropical Agriculture (IITA) in Nigeria. In addition, a new improvement program has just started in South Africa. In general, the programs in Guadeloupe, Honduras, Jamaica, South Africa, and Taiwan were or will be devoted, primarily, to developing replacements for the Cavendish cultivars. Only the programs in Brazil (AAB dessert bananas) and Nigeria (plantains) focus on nonexport-type bananas, and activity in the former program may now be waning (K. Shepherd, personal communication). New or renewed efforts for targets other than an export banana should be considered.[44] Much good work could be done to incorporate fusarium wilt resistance into the AAB dessert bananas and ABB cooking bananas.

Where susceptible banana cultivars are still grown, fusarium wilt can be controlled by planting noninfected rhizomes or tissue-culture derived plantlets in noninfested soil. Whenever possible, new plantings should be established with plantlets rather than rhizomes. Rhizomes infected with *Fusarium oxysporum* f. sp. *cubense* are usually not symptomatic. Thus, in most cases they cannot be distinguished from noninfected rhizomes.

11-8 CONCLUSION

Over 100 years after the first report of fusarium wilt and 30 years after the conversion of the export trades to the Cavendish cultivars, the disease is still a serious problem in diverse banana-growing regions. Research efforts are now underway on genetic resistance in the host, screening techniques, and pathogen behavior and variability in several different countries, but much additional work is needed if fusarium wilt is to be controlled adequately.[39] The recent formation of the International Network for the Improvement of Banana and Plantain (INIBAP), an assembly of regional networks that are coordinated from a central office in Montpellier, France, should enhance and assist these efforts.[15]

11-9 REFERENCES

1. Bancroft, J., Report of the board appointed to enquire into the cause of disease affecting livestock and plants. Queensland. *In Votes and proceedings 1877,* 1876 (3), 1011.
2. Higgins, J. E., The banana in Hawaii, *Hawaii Agr. Expt. Stn. Bull.,* 1904, 7.
3. Basu, S. K., Report on the banana disease in Chinsurah, *Quart J. Dept. Agric. Bengal,* 4, 196, 1911.
4. McKenney, R. E. B., The Central American banana blight, *Science,* 31, 750, 1910.
5. Brandes, E. W., Banana wilt, *Phytopathology,* 9, 339, 1919.
6. Smith, E. F., A Cuban banana disease, *Science,* 31, 745, 1910.
7. Ashby, S. F., Banana diseases in Jamaica, *Bull. Dept. Agric. Jamaica,* 2, 95, 1913.
8. Stover, R. H., *Fusarial wilt of banana and other* Musa *species*, Longman, London, p. 117, 1962.
9. Stover, R. H., and Simmonds, N. W., *Bananas,* 3rd ed., Longman, London, p. 310, 1987.
10. Anonymous, CMI, Distribution Maps of Plant Diseases. Map no. 31, ed. 4., 1977.
11. Ploetz, R. C., Herbert, J., Sebasigari, K., Hernandez, J. H., Pegg, K. G. Ventura, J. A. and Mayato, L. S., Importance of fusarium wilt in different banana-growing regions. In *Fusarium wilt of banana* (R. C. Ploetz, ed.), APS Press, St. Paul, MN, p. 146, 1990.

12. Kress, J. W., The phylogeny and classification of the Zingiberales, *Ann. Miss. Bot. Garden,* 77, 698, 1990.

13. Simmonds, N. W., *Bananas,* 2nd ed., Longmans, London, p. 512, 1966.

14. Hwang, S. C., Somaclonal resistance in Cavendish banana to fusarium wilt. In *Fusarium wilt of banana.* (R. C. Ploetz, ed.), APS Press, St. Paul, MN p. 146, 1990.

15. Anonymous, *Annual report, 1988,* International Network for the Improvement of Banana and Plantain, Montpellier, France, p. 30, 1989.

16. Alves, E. J., Shepherd, K. and Dantas, J. L. L., Cultivation of bananas and plantains in Brazil and needs for improvement. In *Banana and plantain breeding strategies* (G. J. Perseley, and E. A. De Langhe, eds.), p. 187, ACIAR Proceedings No. 21, Canberra, Australia, 1987.

17. Lakshmanan, P., Selvaraj, P. and Mohan, S., Efficacy of different methods for the control of Panama disease, *Trop. Pest Manag.,* 33, 373, 1987.

18. Sebasigari, K. and Stover, R. H., Banana diseases and pests in East Africa. Report of a survey in November, 1987, INIBAP, Montpellier, France, 15 pp., 4 appendices, and 4 tables, 1988.

19. Valmayor, R. V., Banana improvement imperatives—the case for Asia. In *Banana and plantain breeding strategies* (G. J. Perseley and E. A. De Langhe, eds.), p. 187, ACIAR Proceedings No. 21, Canberra, Australia, 1987.

20. Anonymous, *1987 FAO production yearbook,* Rome, vol. 41, 1988.

21. Su, H. J., Hwang, S. C. and Ko, W. H., Fusarial wilt of Cavendish bananas in Taiwan, *Plant Dis.,* 770, 814, 1986.

22. Pegg, K. G. and Langdon, P. W., *Fusarium* wilt (Panama disease): A review, In *Banana and plantain breeding strategies* (G. J. Perseley and E. A. De Langhe, eds.), p. 187, ACIAR Proceedings No. 21, Canberra, Australia, 1987.

23. Anonymous, *FAO commodity review and outlook* 1982–83, Rome, p. 25, 1983.

24. Ventura, J. A., Problemas fitopatologícos de bananiera cultivar Prata. In *Proc. Symp. Sobre Bananaeira Prata,* p. 17, 1983.

25. Pindji, M. M., *Rapport de mission de prospection de bananiers et des systèmes de production agricoles à base de la banane au Nord-Kivu,* INERA, Mulungu, Zaire, 1985.

26. Sebasigari, K. and Gatsinzi, F., *Rapport de mission effectuée à l'Imbo-Nord en Province de Cibitoke,* IRAZ, Gitega, Burundi, 1987.

27. Flinn, J. C. and Hoyoux, J. M., Le bananier plantain en Afrique. Estimation de son importance, rentabilité de sa recherche, suggestions économiques, *Fruits,* 31, 520, 1976.

28. Ploetz, R. C. and Correll, J. C., Vegetative compatibility among races of *Fusarium oxysporum* f. sp. *cubense, Plant Dis.,* 72, 325, 1988.

29. Ploetz, R. C., Vegetative compatibility in *Fusarium oxysporum* f. sp. *cubense:* Classifying previously noncharacterized strains, *Acta Hort.* 275, 699, 1990.

30. Ploetz, R. C., Population biology of *Fusarium oxysporum* f. sp. *cubense.* In *Fusarium wilt of banana* (R. C. Ploetz, ed.), APS Press, St. Paul, MN, p. 146, 1990.

31. Kistler, H. C and Momol, E. A., Molecular genetics of plant pathogenic *Fusarium oxysporum,* In *Fusarium wilt of banana* (R. C. Ploetz, ed.), APS Press, St. Paul, MN, p. 146, 1990.

32. Miao, V. P. W., Using karyotype variability to investigate the origins and relatedness of isolates of *Fusarium oxysporum* f. sp. *cubense*.In *Fusarium wilt of banana*, (R. C. Ploetz, ed.), APS Press, St. Paul, MN, p. 146, 1990.

33. Stover, R. H., *Banana, plantain, and abaca diseases,* CMI, Kew, Surrey, England, p. 316, 1972.

34. Beckman, C. H., Host response to the pathogen, In *Fusarium wilt of banana* (R. C. Ploetz, ed.), APS Press, St. Paul, MN, p. 146, 1990.

35. Beckman, C. H. and Halmos, S., Relation of vascular occluding reactions in banana roots to pathogenicity of root-invading fungi, *Phytopathology,* 52, 893, 1962.

36. Beckman, C. H., Mace, M. E., Halmos, S. and McGahan, M. W., Physical barriers associated with resistance to fusarium wilt of bananas, *Phytopathology*, 51, 507, 1961.

37. Stover, R. H. and Buddenhagen, I. W., Banana breeding, polyploidy, disease resistance, and productivity, *Fruits*, 41, 175, 1986.

38. Rowe, P. R., Breeding bananas and plantains for resistance to fusarium wilt: The track record. In *Fusarium wilt of banana* (R. C. Ploetz, ed.), APS Press, St. Paul, MN, p. 146, 1990.

39. Ploetz, R. C., Roundtable sessions. In *Fusarium wilt of banana* (R. C. Ploetz, ed.), APS Press, St. Paul, MN, p. 146, 1990.

40. Sun, E. J. and Su, H. J., Rapid method for determining differential pathogenicity of *Fusarium oxysporum* f. sp. *cubense* using banana explants, *Trop. Agric.,* 61, 7, 1984.

41. Ploetz, R. C. and Shepard, E. S., Fusarial wilt of banana in Florida, *Mycol. Res.,* 93, 242, 1989.

42. Ploetz, R. C., Shepard, E. S. and Crane, J. H., Current importance of fusarial wilt of banana in Dade Country, Florida and relationships among and the origins of the incitant, *Fusarium oxysporum* f. sp. *cubense,* from Florida and other banana-growing regions, *Proc. Fla. State Hort. Soc.,* 101, 256, 1988.

43. Shepherd, K., Dantas, J. L. L. and Alves, E. J., Banana breeding in Brazil. In *Banana and plantain breeding strategies* (G. J. Perseley and E. A. De Langhe, eds.), p. 187, ACIAR Proceedings No. 21, Canberra, Australia, 1987.

44. Buddenhagen I. W., Banana breeding and fusarium wilt. In *Fusarium wilt of banana* (R. C. Ploetz, ed.), APS Press, St. Paul, MN, p. 146, 1990.

45. Trujillo, E. E. and Snyder, W. C., Uneven distribution of *Fusarium oxysporum* f. sp. *cubense* in Honduras soil, *Phytopathology,* 53, 167, 1963.

46. Beckman, C. H., Halmos, S. and Mace, M. E., The interaction of host, pathogen and soil temperature in relation to susceptibility to Fusarium wilt of bananas, *Phytopathology*, 52, 134, 1962.

47. Rishbeth, J., *Fusarium* wilt of bananas in Jamaica. II. Some aspects of host-parasite relationships, *Ann. Bot. N. S.,* 21, 215, 1957.

48. Stover, R. H. and Malo, S. E., The occurrence of fusarial wilt in normally resistant 'Dwarf Cavendish' banana, *Plant Dis. Rept.,* 56, 1000, 1972.

49. Gutierrez Jerez, F., Trujillo Jacinto del Castillo, I. and Borges, Perez, A., Estudio sobre el mal de Panamá en las Islas Canarias. I. Características físcas y químicas de los suelos y su relacíon de la formedad, *Fruits,* 38, 677, 1983.

50. Borges, Perez, A., Trujillo Jacinto del Castillo, I., Gutierrez Jerez, F. and Angulo Rodridguez, D. Estudio sobre el mal de Panamá en las Islas Canarias. II. Influencia de los desequilibrios nutritivos P-ZN y K-Mg del suelo. en la alteracíon de los mecanismos de resistencia de la plantera (Cavendish enana) al mal de Panamá, *Fruits,* 38, 755, 1983.

51. Rowe, P. R., Breeding an intractable crop. Bananas. In *Genetic engineering for crop improvement* (K. O. Rachie and J. M. Lyman, eds.), Rockefeller Foundation, New York, 1981.

12

MOKO DISEASE OF BANANA

S. S. GNANAMANICKAM AND C. S. ANURATHA

Center for Advanced Studies in Botany
University of Madras, Guindy Campus, Madras 600 025, India

12-1 INTRODUCTION

Moko disease, also known as bacterial wilt of banana, plantains, and abaca, has had a long history. A bacterial wilt disease almost eliminated the Moko plantain in Trinidad about 1890. Earlier, Schomburgk, during his travels in British Guyana (1840–1844), observed a peculiar disease of banana and plantains in the island of Wakenaam. These were later investigated by Rorer in 1911 and described as the Moko disease.

Rorer[1,2] described the symptoms of Moko disease and did pioneering work on isolation and pathogenicity of the bacterial strain. It attracted attention because the disease severely attacked the Moko fig or plantain and almost eliminated this variety about 1890 in Trinidad. It is now known to have a very wide distribution in the tropics, affecting abaca and many other host plants[3] as well as bananas and plantains (Fig. 12-1).

12-2 DISEASE SYMPTOMS ON BANANA

Moko disease is manifest as a wilt, which may be confused with Panama wilt caused by *Fusarium oxysporum* f. sp. *cubense*. On young, rapidly growing plants, the youngest three leaves turn pale green or yellow and collapse near the junction of the lamina and the petiole. Within three to seven days, most leaves collapse. The most characteristic symptoms occur on young suckers that have been cut back once and begun regrowth. These are blackened, stunted, and may be twisted. If sucker leaves are present, these may turn

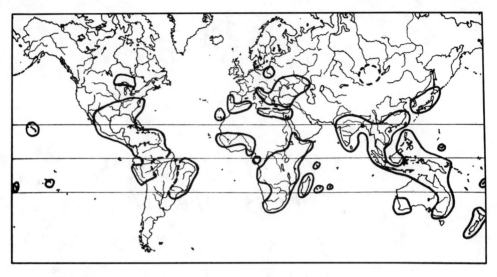

Figure 12-1   Geographical distribution of Moko disease of banana.

yellow or become necrotic. Vascular discoloration in plants that have not shot fruit is concentrated near the center of the pseudostem, becoming less apparent peripherally and centrally (Figs. 12-2a and 12-2b).

If fruit symptoms are present, it is indicative that the disease is Moko and if not, Panama disease. The presence of yellow fingers in an otherwise green bunch will often indicate the presence of Moko disease. A firm, brown, dry rot is found within fruits of infected plant.[4]

12-3 THE PATHOGEN

Moko disease is caused by a specific race (race 2) of *Pseudomonas solanacearum* (Smith) Smith, a soilborne bacterium which affects *Musa* and *Heliconia*. The race comprises several strains which differ in virulence and transmissibility.[5]

The pathogen, however, comprises a large and complex taxonomic group of strains, a fact that is reflected in the systems of classification that have been employed.

Races   *P. solanacearum* has five races which differ in host ranges, geographic distribution, and ability to survive under different environmental conditions.

1. Race 1: Solanaceous strain

2. Race 2: Musaceous strain

3. Race 3: Potato strain

4. Race 4: Ginger strain from the Philippines

5. Race 5: Mulberry strain from china.[5,6]

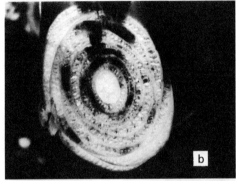

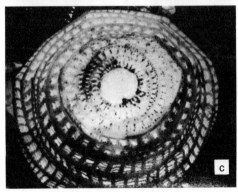

Figure 12-2a   Moko disease:
Cross-section of pseudostem of banana
showing internal discoloration of xylem
characteristic of Moko disease in India.

Figure 12-2b   Panama wilt: Cross-section
of pseudostem of banana showing
discoloration of vascular tissues toward the
periphery in India.

Figure 12-2c   Blood disease: Cross-section
of pseudostem showing internal vascular
discoloration and bacterial ooze in west
Java, Indonesia. (Courtesy: S. Eden-Green,
U.K.)

Biovars   *P. solanacearum* has also been classified into five biovars on the basis
of their ability to oxidize certain hexose alcohols and disaccharides.[7–9]

RFLP groups and serotypes   A restriction fragment length polymorphism
(RFLP) analysis involving nine DNA probes, seven of which encode information for
virulence and the hypersensitive response (HR) studied the relationship among 62 strains
of *P. solanacearum,* representing three races and five biovars. Two major divisions and
28 distinct RFLP patterns were identified: Division I contained all members of race 1, -
biovars 3, 4, and 5; division II contained all members of race 1, biovar 1, and races 2 and

3.[10] A more recent study of RFLP analysis of 150 strains of *P. solanacearum* established that several unknown strains from heliconias imported into Australia belonged to SFR group of race 2.[11] A preliminary examination of RFLP patterns of the Indonesian blood disease bacterium has shown that it differs from all strains of *P. solanacearum* race 2 isolated in Central and South America. This is strong evidence that blood disease is related to but has evolved independently from Moko.[12]

Hybridoma cell lines secreting monoclonal antibodies (mAbs) can be a very useful future tool in serological classification of *P. solanacearum*. Analysis of surface antigens of the pathogen with a panel of mAbs will provide data on the existence of subgroups (serotypes), geographical distribution of serotypes, and the relationships of surface antigens to genetic characteristics (as determined by RFLP analysis). Information available at present shows that mAbs specific to *P. solanacearum* and another mAb that delineates a banana strain from other pseudomonads have been found.[13]

Strains of race 2 of *P. solanacearum*   These strains affect musaceous hosts: plantains, bananas, and *Heliconia* spp. (a musaceous ornamental/weed). A summary of the principal characteristics reported for race 2 appears in Table 12-1. They cause symptoms ranging from stunting and distortion to rapid wilt.[14]

In culture, all strains of *P. solanacearum* produce mutants that are weakly virulent or avirulent and appear as small butyrous colonies with a distinct dark red center on 2, 3, 5 triphenyl tetrozolium chloride agar. The virulent wild type forms irregularly round, fluidal white colonies which have light pink centers. Virulent type is best maintained by storage in distilled water at room temperature.[5]

12-4 DISTRIBUTION OF MOKO

Until recently, the bacterial wilt of bananas (Moko disease) was limited in its distribution to parts of central and South America and the West Indies. It was believed that Moko does not occur in any part of Asia.[5,18] It has chiefly been found in Costa Rica, Dominican Republic, Guadeloupe, Grenada, Guatemala, El Salvador, Haiti, Nicaragua, St. Vincent, Trinidad, Argentina, Brazil, British Guyana, Colombia, Panama, Paraguay, Surinam, Venezuela, Ethiopia, French Guyana, Libia, Mauritius, Mexico, Peru, Sierra Leone, Ceylon, Malaysia, Fiji, and Hawaii[3,4,16,18] (Fig. 12-1). Peru, Brazil, Grenada, and Surinam are newer regions or countries within Latin America to have Moko disease.[19]

In Indonesia, "blood disease" of banana (Penyakit darah) has occurred since 1910 in Southern Sulawesi and caused serious damage to bananas.[20] Gaumann[21] reported that the disease was caused by *Pseudomonas celebensis*. However, Wardlaw[22] believed that the blood disease was identical to Moko disease. As discussed earlier, recent evidence from RFLP analysis indicates that blood disease may be closely related but not identical to Moko.

Collaborative research on the blood disease of banana between Tjahyono (of Bogor Agricultural University) and Eden-Green and his coworkers (of Rothamsted Experimental Station, U. K.) recently led to the belief that the disease has spread from Sulawesi to Java. These workers considered that the current designation of the causal agent as *Xan-

TABLE 12-1: CHARACTERISTICS OF STRAINS OF RACE 2 OF *PSEUDOMONAS SOLANACEARUM* [15,16]

Strain	Pathogenicity	Origin	Cultural characteristics[a]
R	Stunting and distortion of *Heliconia* only	*Heliconia* in Costa Rica	Elliptical; lace-like slime; slight formazan pigmentation towards center 2[b]
D	Leaf distortion and slow wilt of banana[c]; stunting and distortion of *Heliconia*	*Heliconia* and banana in Costa Rica	Elliptical; lace-like slime; slight formazan pigmentation towards center 2[b]
B	Rapid wilt of banana in Guyana, Venezuela, Costa Rica, Panama, Honduras, Trinidad	Probable mutation of strain D in SW Costa Rica and South America	Elliptical; lace-like slime; slight formazan pigmentation towards center 2[b]
H	Moderate wilt on plantain[d], not pathogenic to banana	Costa rica	Eliptical; lace-like slime; slight formazan pigmentation towards center 2[b]
SFR	Rapid wilt of banana and plantain; insect disseminated in Venezuela, Honduras, Guatemala, El Salvador, Nicaragua, Costa Rica, Columbia	South America (probably Venezuela)	Near round; little slime, lace-like initially; slight formazan pigmentation
SFR-C	As SFR, but a variant in Columbia	Probable mutation of SFR in Amazonas State on plantain	Near round; moderate slime deposition, lace-like initially; slight formazan pigmentation
A	Rapid wilt of plantain and banana; insect disseminated from Columbia, similar to SFR-C	Amazon Basin in Peru and neighbouring Brazil, on plantain and banana	Near round; plentiful slime, faintly lace-like; formazan in faint helical pattern

[a]Colonies on Kelman's agar medium observed with oblique light after 24 hours at 32°C.

[b]Strain R[15,17] reported to produce rapid formazan pigmentation.

[c]Banana is Musa Group AAA.

[d]Plantain is Musa Group AAB.

Reprinted with permission from *Bacterial Wilt Disease in Asia and the South Pacific* (Persley, G. J., ed.), ACIAR, 1986.

thomonas campestris pv. *celebensis* is erroneous and that the original name *P. celebensis* Gaumann should be reinstated. Symptoms of the blood disease were fully described recently by Eden-Green and Sastraatmadja.[23] In mature plants, midribs and petioles of oldest leaves become flaccid and petioles collapse near the junction with the pseudostem. Younger leaves become bright yellow and then necrotic and dry. Fruit bunches may appear outwardly normal, even when all of the leaves are discolored, but a few split, blackened, or prematurely ripened fruits may be present. Systemic wilting is rarely seen in mature plants but may affect young suckers. Internally, brown vascular streaking is seen throughout the plant extending into the roots. Characteristic vascular streaking, as it appears in a transverse section of the pseudostem, is seen in Fig. 12-2c. On mature plants this may be concentrated in the fruit stem and extend into the fruits, which are internally discolored and contain dry cavities or pockets of reddish-brown mucoid or gelatinous tissues. Characteristically, all of the fruits show internal symptoms. A white to reddish-brown "blood-like" bacterial ooze exudes slowly from cut vascular bundles.

The disease affects both dessert (AAA group; pisang Ambon, pisang Nangka) and cooking (ABB group; pisang Kepok, pisang Siam) bananas. In Sulawesi, the symptoms were most frequently observed on pisang Kepok, but dessert varieties were not widely grown, possibly due to high susceptibility.

These symptoms are similar to floral infections caused by insect-borne SFR strains of Moko disease (*P. solanacearum* race 2) in Central and South America, but bacteriological characteristics of the blood disease bacterium differ from those of *P. solanacearum*.

Blood disease shows similarities to Bugtok and Tapurok disease of bananas in the Southern Philippines,[24] where Moko disease, believed to be caused by the mechanically transmitted B strain of *P. solanacearum* race 2, has also been identified recently.[19]

True Moko now occurs also in the Philippines, with little doubt that it was introduced on banana rhizomes from Honduras in about 1968 on the island of Mindanao.[25,26] It is believed to be caused by B strain of race 2 of *P. solanacearum* rather than the SFR strain.[19] The Moko pathogen has remained confined to commercial banana plantations on the island by application of rigid control practices. A non-banana strain of race 1 of *P. solanacearum* has also been known in the Philippines since the 1930s, which is reported to be a weak pathogen on *Musa* spp. naturally infecting both bananas and abaca.[27] Symptoms are mild, and apparently only plants growing under unfavorable conditions are affected.

Occurrence of Moko disease in India has been reported from the state of West Bengal by Chattopadhya and Mukhopadhyay[28] and from Southern India by Gnanamanickam et al.[29] Stover[16] suspected that what was reported from West Bengal may refer to yellow mat disease or to symptoms caused by a solanaceous strain of *P. solanacearum* weakly pathogenic on *Musa*. The report from Southern India was from our laboratory and referred to a bacterial wilt causing severe losses on newly cleared areas planted with Robusta (AAA: a semitall mutant of Dwarf Cavendish) banana during growing seasons from 1976 to 1978. Pathogenic isolates were characterized by us as biotype/biovar II or III[29] of *P. solanacearum*. Two isolates (B8605 and B8606) we deposited at the Commonwealth Mycological Institute, U. K., were characterized as biovar 2 and 1, respectively.[30] Of the two, our banana isolate, B8606 (biovar I), was deposited with the National Collection of Plant Pathogenic Bacteria (NCPPB), Harpenden, Herts, U. K., under their accession number NCPPB 3214.[31]

Subsequently, we undertook field surveys with small financial support made available by the Indian Council of Agricultural Research (ICAR) in the states of Tamil Nadu and Kerala. Occurrence of both Panama wilt and Moko diseases was carefully investigated by making isolations. Admittedly, Panama wilt was more widespread in several districts and was quite severe (up to 90%) in some varieties such as Sirumalai (*Musa paradisiaca* Linn).[32] However, another incidence of Moko was recorded in one of the districts of Tamil Nadu on Poovan banana, and 20% of the plants showed characteristic wilt symptoms[32] (Fig. 12-3).

If scientific evidence is any guide, Moko pathogen is present in Indian bananas and none of the aforementioned reports can be discounted. It is true, however, that we do not encounter Moko disease severities to the degree and dimensions reported in the Caribbean and Latin American countries. There are two primary reasons for this. First, most of the traditional banana cultivars farmers grow in large areas under banana cultivation are resistant or moderately resistant to Moko pathogen as observed in our artificial inoculations.[33] Every farmer grows a different cultivar, and therefore, in this multiline situation, the pathogen may be kept under check. Second, the farmers practice rigorous crop rotations. Even in areas where the susceptible cv. Robusta or *Musa cavendishi* is grown, the crop is rotated with sugarcane and rice in a three-year rotation. Undoubtedly, this farmer practice of growing other cash crops and cereals becomes an efficient disease management practice for Moko.

12-5 ECONOMIC IMPORTANCE

Stover[16] observed that Moko disease is one of the major diseases of commercial plantations in Honduras and the Pacific Coast banana zones of Costa Rica and Panama. In a banana zone of 30,000 acres, less than 1% of the plants are lost annually due to Moko disease, which could otherwise increase plant loss to 5%. This loss rate of 1% is maintained by tight control measures costing nearly $400,000 annually.

Figure 12-3 Moko wilt of banana (cv. Poovan) plants in a plantation at Pudukkottai district, Tamilnadu, India.

Crop losses are quite high in peasant plantings of ABB (Bluggoe) type and AAB type bananas in Central America, Colombia, Venezuela, and Peru, particularly if the SFR strain was present.[16] French and Sequeira[15] reported that the disease affected the well-being of thousands of small banana plantings along the Peruvian head waters of the Amazon River. At present, Moko disease is under control in large banana plantations in Central America, but it adds considerably to the cost of production. For small producers of bananas and of Bluggoe plantains in particular, the problem continues to be severe. Bluggoe plantains have literally disappeared from Central America as a result of the disease and of eradication programs to protect banana plantations. In the Amazon basin, the disease is now endemic rather than epidemic as it was in the 1960s. Nevertheless, it has continued to spread along the Amazon tributaries and is still a problem for the native population.[34]

In Asia, prior to 1912 the blood disease of banana caused serious crop losses in southern Sulawesi of Indonesia, which resulted in loss of income to farmers and losses in the main staple food of the people. In recent years, the disease is present in Java and Sulawesi, but disease severity is low.[20] In Southern India, crop losses were high (up to 70%) in newly cleared areas in 1977–1978, and since then the disease has not reccurred as a threat to banana cultivation.

12-6 ORIGIN OF MOKO PATHOGEN

The origin of race 2 of *P. solanacearum* is in Latin America, according to an original hypothesis proposed by Buddenhagen[35,36] and Sequeira and Averre.[17] This remains unchallenged. Banana strains of the pathogen could have evolved from a strain attacking heliconias, a common weed found near banana fields.[17] However, it cannot be said today that Moko is not present in Asia, as it was originally believed.[19,37] In countries such as Indonesia and India, it may well be that strains evolved locally and were not introduced from outside.

12-7 EPIDEMIOLOGY

In Latin America, Moko bacteria are readily transmitted from plant to plant on pruning machetes. Open wounds formed on pseudostems of older diseased plants become exit points for bacteria, and often insects carry these bacteria and effectively transmit them onto disease-free suckers. The B strain of the Moko pathogen can persist in soil for 12 to 18 months[16] and can spread through root to root. The SFR strain has explosive epidemiological potential and is transmitted through insects.[19,38] It does not survive in soil longer than six months. Moko bacteria ooze from infected male flower bracts (of Bluggoe banana - ABB) and can be carried by visiting insects (Fig. 12-4).

The ability of our banana strain (B 8605) of *P. solanacearum* to survive in artificially infested soil was studied over a period of two years. At the end of two years, the population had declined by about three lesser orders of magnitude, but the bacterium survived.

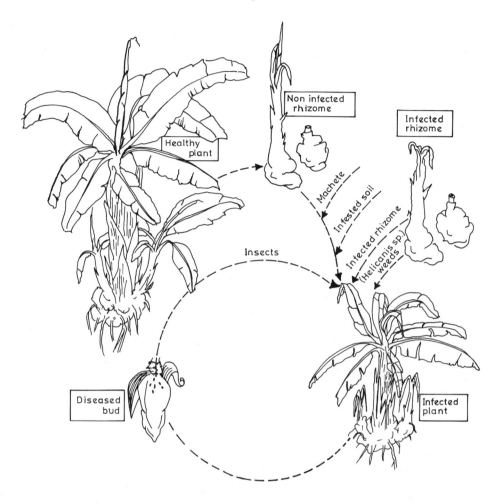

Non infected
rhizome

Infected
rhizome

Healthy
plant

Machete

Infested soil

Infected rhizome

Insects

(Helicanis sp.)
weeds

Diseased
bud

Infected
plant

Figure 12-4   Disease cycle of Moko disease of banana.

12-8 INOCULATION AND DISEASE RATING

Inoculation techniques for evaluating resistance to *P. solanacearum* have been described.[39] We have followed the method of Buddenhagen[35] with some modifications for inoculation of banana plants.[40]

In recent studies, we have used *M. balbisiana* seedlings raised from seeds when they were three months old and had three to four leaves for inoculations. The following methods of inoculation were used:

1. Injection: *P. solanacearum* cells (24 hours grown) (10^8 cfu/ml) are injected (at 1 ml/plant) into the pseudostem near the soil with hypodermic syringe.

2. Root dip: After the seedling root tips are excised, the roots are dipped in *P. solanacearum* cell (10^8 cfu/ml) suspension for 1 hour. The inoculated plants are subsequently transplanted in field soil contained in clay pots or wooden boxes.

Plants inoculated with sterile distilled water serve as controls. All plants are maintained in the greenhouse at room temperature 32° ± 2°C and natural light and are observed for the development of wilt symptoms. Inoculated plants wilted in 11 to 15 days (Fig. 12-5).

12-9 CONTROL

12-9-1 Use of Resistant Varieties

All varieties of commercial bananas and plantains were susceptible to Latin American strains (B and SFR) of *P. solanacearum*,[16] In Honduras, all *Musa* species except *M. balbisiana* were proved susceptible in artificial inoculation with the B strain. Against the SFR strain, variety Pelipita (ABB) was highly resistant and was recommended as a disease-resistant substitute for the susceptible Bluggoe (ABB), which has been devastated by the SFR strain of *P. solanacearum*.[41] In greenhouse experiments the Indian banana strain, B 8605, was artificially inoculated onto *M. acuminata* (AA), *M. balbisiana* (BB) *M. cavendish;* varieties, Dwarf Cavendish (AAA), Robusta (AAA), and also to traditional Indian varieties, Monthan (ABB), Nendran (AAB), Peyan (ABB), and Poovan (AAB) when plants were at the 4-leaf stage. The commercial cultivars, Monthan and Poovan, were resistant while all others were wilted.[33]

Figure 12-5 Test for pathogenicity of *Pseudomonas solanacearum* (NCPPB 3214) on *Musa balbisiana* seedlings.

12-9-2 Exclusion and Eradication of the Pathogen

The basis for Moko control is early detection of the infected plant and its rapid destruction along with adjacent, apparently healthy plants.[16] All pruning tools were used in areas where Moko is present should be disinfected with formaldehyde (1 part formaldehyde : 3 parts water).[42] Male flower buds are removed to avoid infection by the insect-transmitted SFR strain after the emergence of female hands. To eradicate the infested plants along with an adjacent buffer zone of healthy plants, herbicides such as 2,4-D have been used. After eradication of infected plants, the area must remain in fallow for 12 months to eliminate the B strain and 6 months to eliminate the SFR strain.[16]

Exclusion of the pathogen should involve the use of disease-free suckers.[3]

12-9-3 Crop Rotation

Use of sorghum in a nine-month rotation, and possibly also other Gramineae, has been found effective to reduce soil population of *P. solanacearum* in Honduras.[22] In parts of Tamilnadu (India), farmers use a three-year rotation of banana, sugarcane, and rice (two crops), which we believe forms an effective bacterial wilt management practice.

12-9-4 Chemical Control

Various chemicals have been used as broadcast and other applications to disinfect the soil in planting holes and as soil fumigation. However, none has prevented infection when diseased areas were replanted with bananas.[22] In Honduras, the United Fruit Company used chloropicrin to disinfect the soil in cylinders at the rate of 8 cc per square feet, and this was effective in eliminating *P. solanacearum* from soil. It should be noted, however, that Moko cannot be controlled economically or effectively by chemical means.[43]

12-9-5 Biological Control

The use of microbial antagonists has been noted as a promising control strategy for bacterial wilts. Kelman[3] reported various fungi, actinomycetes, and bacteria that exhibited antiobiotic effects against *P. solanacearum*. There are numerous reports on induced resistance to bacterial wilt of crops other than banana by prior inoculation with nonpathogens, avirulent or incompatible forms of a pathogen, heat-killed cells, etc.[44-51] There are also reports of using bacteriocin-producing strains of *P. solanacearum* for the control of bacterial wilts of tobacco, tomato, potato, and other hosts.[52-57] However, no work has been reported for biological suppression of Moko disease. In our laboratory, we have used selected strains of antagonistic bacteria of Indian origin for treatment of *Musa balbisiana* plants for suppression of Panama wilt.[58] We applied similar bacterization procedures with an efficient strain of *Pseudomonas fluorescens* (strain Pfcp) to treat *M. balbisiana* plants inoculated with *P. solanacearum* (B 8606 or NCPPB 3214) both in sterile and nonsterile soil in greenhouse. This led to increase in percent survival of plants (Tables 12-2a and 12-2b; Fig. 12-6) and to enhanced plant height and biomass (Fig. 12-7). In a field experiment, bacterized and pathogen inoculated *M. balbisiana* plants had 90% plant survival, whereas only 40% plants survived in the pathogen-inoculated plots (Table 12-3). Bacterization with *P. fluorescens* also protected the pathogen-inoculated plants from

TABLE 12-2a: SUPPRESSION OF BACTERIAL WILT CAUSED BY *PSEUDOMONAS SOLANACEARUM* (NCPPB 3214) IN BANANA (*MUSA BALBISIANA*) TREATED WITH *PSEUDOMONAS FLUORESCENS* (STRAIN PFCP) IN STERILE FIELD SOIL.

Treatment	Number of plants survived	Percent[a] survival
Inoculated (I)	24	40.0
Bacterized and inoculated (BI)	54	90.0
Bacterized (B)	60	100.0
Control (C)	60	100.0

[a]Average of two experiments with three replications for each treatment.

TABLE 12-2b: SUPPRESSION OF BACTERIAL WILT CAUSED BY *PSEUDOMONAS SOLANACEARUM* (NCPPB 3214) IN BANANA (*MUSA BALBISIANA*) TREATED WITH *PSEUDOMONAS FLUORESCENS* (STRAIN PFCP) IN NONSTERILE FIELD SOIL.

Treatment	Number of plants survived	Percent[a] survival
Inoculated (I)	0	0.0
Bacterized and inoculated (BI)	54	90.0
Bacterized (B)	60	100.0
Control (C)	60	100.0

[a]Average of two experiments with three replications for each treatment.

Figure 12-6 Bacterial wilt suppression in *Musa balbisiana* plants in nonsterile soil by bacterization with *Pseudomonas fluorescens* (Pfcp).

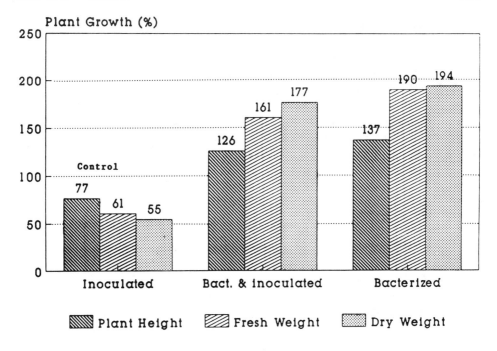

Figure 12-7   Growth enhancement due to bacterization with *Pseudomonas fluorescens* (strain Pfcp) in *Pseudomonas solanacearum* inoculated *Musa balbisiana* plants in sterile field soil.

stuntedness. The plant heights of pathogen-inoculated, bacterized + pathogen inoculated, and bacterized plants in field plots were, 38.0, 72.0, and 70.0 cm, respectively.[59]

Possible mechanisms for wilt suppression in *M. balbisiana* plants in greenhouse and field tests by *P. fluorescens* were investigated.[60] *P. fluorescens* strain Pfcp was found to produce an antibiotic in culture.   Partially purified preparations of the antibiotic had retained their toxicity toward *P. solanacearum* (B 8606) and to a large number of fungal pathogens. From its mobility on thin-layer chromatograms and from its characteristic UV absorption spectrum (major peak at 251 nm, broad peak at 364 nm, and a shoulder at 353 nm), the antibiotic appears to be a phenazine.[61] The identification, however, needs to be confirmed by further characterization.

TABLE 12-3: SUPPRESSION OF BACTERIAL WILT CAUSED BY *PSEUDOMONAS SOLANACEARUM* (NCPPB 3214) IN BANANA (*MUSA BALBISIANA*) TREATED WITH *PSEUDOMONAS FLUORESCENS* (STRAIN PFCP) IN FIELD.

Treatment	Number of plants survived	Percent survival
Inoculated (I)	12	40.0[a]
Bacterized and inoculated (BI)	27	90.0
Control (C)	30	100.0

[a]Average of three replications.

12-10 CONCLUSIONS

Moko disease of banana caused by strains of race 2 of *Pseudomonas solanacearum* occurs widely in several countries of Latin America and the Caribbean region, where the pathogen is believed to have originated. True Moko is also known to occur in the Island of Mindanao in the Philippines along with Bugtok disease of cooking bananas, which mainly affects the flowers and fruits. The disease also occurs in India and perhaps also in Indonesia, although the blood disease that occurs in Java and Sulawesi cannot yet be considered identical to Moko disease. The Bugtok disease of the Philippines and the blood disease in Indonesia show similarities. Therefore, it cannot be said the Moko disease does not occur in Asia.

It is true, however, that disease problems caused by Moko in Asia (Philippines, Indonesia, and India) are not severe. The pathogen is kept under check by tight control practices, as in the Philippines, or by cultural practices and crop rotations adopted by farmers in India.

It is important that the identity of stains present in Asian bananas is determined with clarity. It is likely that insect-transmitted SFR strains are not present in Asia, as otherwise the disease may be widely distributed. Comparative studies on Moko strains present in Asia will also establish if they were introduced or evolved locally.

Management of Moko by effective crop rotations and fallow appears feasible. Biological control by using strains that induce cross-protection or by others which produce bacteriocins or other antibiotics are promising alternate strategies for disease management.

Efficient strains of *Pseudomonas fluorescens* afford effective suppression of Moko in *M. balbisiana* seedlings, and the disease suppression is antibiotic mediated. With the use of molecular biology approaches, it should be possible to enhance the biological control capabilities. This will require the identification of gene(s) that code for antibiotic (such as phenazine) production in strains such as Pfcp, and for cloning and incorporation into banana genotypes. This will greatly contribute to bacterial wilt management by resource-poor banana farmers in India and Asian tropics.

12-11 REFERENCES

1. Rorer, J. B., A bacterial disease of bananas and plantains, *Trinidad and Tobago Agr. Soc. Proc.* 10, 109, 1910.

2. Rorer, J. B., A bacterial disease of bananas and plantains, *Phytopathology,* 1, 45, 1911.

3. Kelman, A., The bacterial wilt caused by *Pseudomonas solanacearum.* A literature review and bibliography, *N. C. Agr. Exp. Stn. Tech. Bull.,* 99, 194, 1953.

4. Buddenhagen, I. W., Bacterial wilt of bananas, history and known distribution, *Trop. Agr.* (Trinidad), 38, 107, 1961.

5. Buddenhagen, I. W., and Kelman, A., Biological and physiological aspects of bacterial wilt caused by *Pseudomonas solanacearum. Ann. Rev. Phytopathol.,* 2, 203, 1964.

6. Persley, G. J., Batugal, P., Gapasin, D. and Vander Zaag, P., Summary of discussion and recommendations. In *Proc. Int. Workshop, Bacterial Wilt Disease in Asia and the South Pacific* (G. J. Persley, ed.), p.7, Los Banos, Philippines, 1985.

7. Hayward, A. C., Characteristics of *Pseudomonas solanacearum, J. Appl. Bacteriol,* 27, 265, 1964.

8. He, L. Y., Sequeira, L. and Kelman, A., Characteristics of strains of *Pseudomonas solanacearum* from China, *Plant Dis.* 67, 1357, 1983.

9. He, L. Y., Bacterial wilt in Indonesia. In *Proc. Int. Workshop, Bacterial Wilt Disease in Asia and the South Pacific* (G. J. Persley, ed.), p. 40, Los Banos, Philippines, 1985.

10. Cook, D., Barlow, E. and Sequeira, L., Genetic diversity of *Pseudomonas solanacearum:* Detection of restriction length polymorphisms with DNA probes that specify virulence and the hypersensitive response, *Mol. Plant. Microbe Interact.,* 2, 113, 1989.

11. Barlow, E., Cook, E. and Sequeira, L., Use of restriction fragment length polymorphisms to characterize strains of *Pseudomonas solanacearum, Proc. of Amer. Phytopath. Soc. and Canadian Phytopath. Soc.,* Grand Rapids, USA, 1990.

12. Eden Green, S. J., personal communication, 1990.

13. Alvarez, A. M. and Benedict, A. A. Monoclonal antibodies for the identification of plant pathogenic bacteria: Potential applications to *Pseudomonas solanacearum.* In *Symp. Bact. Wilt of Peanut,* Kuala Lumpur, Malaysia, March 1990.

14. French, E. R., Interaction between strains of *Pseudomonas solanacearum.* its hosts and the environment. In *Proc. Int. Workshop, Bacterial Wilt Disease in Asia and the South Pacific* (G. J. Persley, ed.), p. 99, Los Banos, Philippines, 1985.

15. French, E. R. and Sequeira, L., Strains of *Pseudomonas solanacearum* from Central and South America: A comparative study, *Phytopathology,* 60, 506, 1970.

16. Stover, R. H., *Banana, plaintain and abaca diseases,* Commonwealth Mycol. Inst., Kew, Surrey, England, p. 316, 1972.

17. Sequeira, L. and Averre, C. W., Distribution and pathogenicity of strains of *Pseudomonas solanacearum* from virgin soils in Costa Rica, *Plant Dis. Reptr.,* 45, 435, 1962.

18. Meredith, D. S., Major banana diseases. Past and present status, *Rev. Plant Pathol.,* 149, 539, 1970.

19. Buddenhagen, I. W., Bacterial wilt revisited. In *Proc. Int. Workshop, Bacterial Wilt Disease in Asia and the South Pacific* (G. J. Persley, ed.), p. 126 Los Banos, Philippines, 1985.

20. Machmud, M., Bacterial wilt in Indonesia. In *Proc. Int. Workshop, Bacterial Wilt Disease in Asia and the South Pacific* (G. J. Persley, ed.), p. 30, Los Banos, Philippines, 1985.

21. Gaumann, E., Onderzoekingen over de bloedziekte der bananen op Celebes I, *Meded Instituut Plantenziekten,* 50, 1921.

22. Wardlaw, G. W., *Banana diseases including plantains and abaca,* Longman, London, p. 878, 1972.

23. Eden-Green, S. J. and Sastraatmadja, H., Blood disease of bananas in Sulawesi and Java, *FAO Pl. Prot. Bull.,* 1990.

24. Zehr, E. I. and Davide, R. G., An investigation of the cause of the 'Tapurok' disease of cooking banana in Negros Oriental, *The Philippine Phytopathology,* 5, 1, 1969.

25. Quimio, A. J., The bacterial wilt problem in the Philippines. In *Proc. of the First Int. Planning Conf. and Workshop on the Ecology and Control of Bacterial Wilt caused by Pseudomonas solanacearum* (L. Sequeira and A. Kelman, eds.), p. 103, Raleigh, NC, 1976.

26. Rillo, A. R., Differences of *Pseudomonas solanacearum* isolates in abaca and banana, *The Philippine Agriculturist,* 64, 329, 1981.

27. Zehr, E. I., Isolation of *Pseudomonas solanacearum* from abaca and banana in the Philippines, *Plant Dis. Reptr.,* 54, 516, 1970.

28. Chattopadhya, S. B., and Mukhopadhyay, N., Moko disease of bananas: A new record, *FAO Plant Prot. Bull.,* 16, 52, 1968.

29. Gnanamanickam, S. S., Lokeswari, T. S., and Nandini, K. R., Bacterial wilt of banana in Southern India, *Plant Dis. Reptr.,* 63, 525, 1979.

30. Bradbury, J. F., personal communication, 1980.

31. Bradbury, J. F., personal communication, 1983.

32. Sivamani, E. and Gnanamanickam, S. S., Occurrence of Moko wilt in "Poovan" banana in Pudukkottai district of Tamil Nadu, *Indian Phytopathol,* 40, 233, 1987.

33. Govindarajan, G., *Studies on the biology of Pseudomonas solanacearum E. F. Smith, the incitant of the bacterial wilt of banana,* Ph.D. thesis, University of Madras, Madras, India, 1983.

34. Sequeira, L., personal communication, April 1990.

35. Buddenhagen, I. W., Strains of *Pseudomonas solanacearum* in indigenous hosts in banana plantations of Costa Rica and their relationship to bacterial wilt of banana, *Phytopathology,* 50, 660, 1960.

36. Buddenhagen, I. W., The relation of plant pathogenic bacteria to the soil. In *Ecology of soilborne plant pathogens,* University of California Press, 1964.

37. Buddenhagen, I. W., *Plant diseases,* McGraw-Hill yearbook of Science and Technology, New York, McGraw-Hill, 1971.

38. Buddenhagen, I. W., and Elsasser, T. A., An insect-transmitted wilt epiphytotic of Bluggoe banana, *Nature,* 194, 164, 1962.

39. Winstead, N. N. and Kelman, A., Inoculation techniques for evaluating resistance to *Pseudomonas solanacearum, Phytopathology,* 42, 628, 1952.

40. Buddenhagen, I. W., Bacterial wilt of certain seed bearing *Musa* spp. caused by the tomato strain of *Pseudomonas solanacearum, Phytopathology,* 52, 286, 1962.

41. Stover, R. H. and Richardson, D. L., 'Pelipita,' an ABB Bluggoe-type plantain resistant to bacterial and fusarial wilts, *Plant Dis. Repr.,* 52, 901, 1968.

42. Buddenhagen, I. W. and Sequeira, L., Disinfectants and tool disinfection for prevention of spread of bacterial wilt of bananas, *Plant Dis. Reptr.,* 42, 1399, 1958.

43. Montcel, H. T., The problem of some banana diseases which cannot be controlled by chemical means. An example: Moko disease due to *Pseudomonas solanacearum, Bull. Tech. d'Information,* 409, 397 (French), 1986.

44. Goodman, R., Defenses triggered by previous invaders: Bacteria. In *Plant diseases* (J. Horsfall and Cowling, E., eds.), p. 305, Academic Press, New York, 1980.

45. Kuc, J., Plant immunization and its applicability for disease control. In *Innovative approaches for plant disease control* (I. Chet, ed.), p. 255, John Wiley & Sons, New York, 1987.

46. Matta, A., Microbial penetration and immunization of uncongenial host plants, *Annu. Rev. Phytopathol.* 9, 387, 1971.

47. Daheb, M. K., and Goorani, M. A., Antagonism among strains of *Pseudomonas solanacearum, Phytopathology,* 59, 1005, 1969.

48. Kempe, J. and Sequeira, L., Biological control of bacterial wilt of potatoes: Attempts to induce resistance by treating tubers with bacteria, *Plant Dis.,* 67, 499, 1983.

49. McLaughlin, R. J. and Sequeira, L., Evaluation of an avirulent strain of *Pseudomonas so-lanacearum* for biological control of bacterial wilt of potato, *Amer. Pot. J.* 65, 255, 1988.

50. Tanaka, H. and Maeda, H., Biological control, *Bact. Wilt News.* 4, 5, 1988.

51. Wakimoto, S., Biological control of bacterial wilt of tomato by nonpathogenic strains of *Pseudomonas glumae, Korean J. Plant Pathol.*, 3, 300, 1987.

52. Cuppels, D., Bacteriocin production by *Pseudomonas solanacearum*. In *Proc. of the First Planning Conf. and Workshop on Ecology and Control of Bacterial Wilt caused by P. solanacearum* (L. Sequeira and A. Kelman, eds.), p. 166, Raleigh, NC, 1976.

53. Cuppels, D. A., Hanson, R. S. and Kelman, A., Isolation and characteristics of a bacteriocin produced by *Pseudomonas solanacearum, J. Gen. Mycol.*, 109(2), 295, 1978.

54. Chen, W. Y. and Echandi, E., Effect of avirulent bacteriocin producing strains of *Pseudomonas solanacearum* on the control of bacterial wilt of tomato, *Plant Pathology,* 33, 245, 1984.

55. Tsai, J. W., Hsu, S. T. and Chen, L. C., Bacteriocin producing strains of *Pseudomonas solanacearum* and their effect on development of bacterial wilt of tomato, *Plant Prot. Bull.,* Taiwan, 27, 267, 1985.

56. Ren, X., Zhang, J. H., Shen, D. L. and Zhang, J., Bacteriocins of *Pseudomonas solanacearum* to control bacterial wilt of tomato. In *Proc. 7th Int. Conf. Pl. Path. Bact.,* Budapest, p. 11, 1989.

57. Aspiras, R. B. and de la Cruz, A. R., Potential biological control of bacterial wilt in tomato and potato with *Bacillus polymyxa* FU6 and *Pseudomonas fluorescens*. In *Proc. Int. Workshop, Bacterial Wilt Disease in Asia and the South Pacific* (G. J. Persley, ed.), p. 89, Los Banos, Philippines, 1985.

58. Sivamani, E. and Gnanamanickam, S. S., Biological control of *Fusarium oxysporum* f. sp. *cubense* in banana by inoculation with *Pseudomonas fluorescens, Plant and Soil,* 107, 3, 1988.

59. Anuratha, C. S. and Gnanamanickam, S. S., Biological control of bacterial wilt caused by *Pseudomonas solanacearum* in India with antagonistic bacteria, *Plant and Soil,* 124, 109, 1990.

60. Anuratha, C. S., *Biological control of bacterial wilt caused by Pseudomonas solanacearum E. F. Smith with bacterial antagonists,* Ph. D. thesis, University of Madras, Madras, India, 1990.

61. Gurusiddaiah, S., Weller, D. M., Sarker, A. and Cook, R. J., Characterization of an antibiotic produced by a strain of *Pseudomonas fluorescens* inhibitory to *Gaeumannomyces graminis* var. *tritici* and *Pythium* spp., *Antimicrob. Agents Chemother.,* 29, 488, 1986.

13

PHYTOPHTHORA ROOT ROT OF PAPAYA

W. H. KO

Department of Plant Pathology
Beaumont Agricultural Research Center
University of Hawaii, Hilo, Hawaii

13-1 INTRODUCTION

Disease of papaya (*Carica papaya* L.) caused by a *Phytophthora* was recognized in the Philippines as early as 1916.[1] The pathogen which attacked papaya fruit was identified as *Phytophthora faberi* Maubl. This species was subsequently found to be a synonym of *Phytophthora palmivora* (Butl.) Butl.[2]

Severe root rot of papaya caused by *P. palmivora* was reported first in 1957 from Australia. There, the disease frequently originated in the poorly drained portions of the plantations, and it then spread out to adjacent areas if wet weather continued.[3] It was further observed in Australia that papaya replanted in the areas with the disease in the first planting always succumbed to the disease. Subsequently, Phytophthora root rot of papaya was also recognized as the main factor contributing to the papaya replant problem in Hawaii.[4,5]

13-2 DISTRIBUTION AND ECONOMIC IMPORTANCE

In addition to Hawaii, Australia, and the Philippines, Phytophthora disease of papaya has also been reported in Ceylon,[1] Malaysia,[6] Brazil,[7] Spain,[7] and Taiwan.[8] Among these areas, Phytophthora root rot was reported only in Hawaii, Australia, Malaysia, and Taiwan.

Each year during the rainy periods, Phytophthora root rot caused heavy losses of papaya plants in the poorly drained areas infested with *P. palmivora*. For example, more than 8000 papaya plants in southeastern Queensland were destroyed by Phytophthora root rot during 1955 and 1956.[3] In central Taiwan the disease also killed more than 20% of the plants in one papaya plantation in 1975.[8]

In Hawaii, commercial cultivation of papaya is mainly for production of fresh fruit for export to the U.S. mainland and Japan. Most of Hawaii's papaya is grown on approximately 3000 acres of lava rock land in the southeastern section of the island of Hawaii.[9] Before the development of the virgin soil method, it was not possible to replant papaya in this area because of Phytophthora root rot.[4] The disease was previously considered a limiting factor for Hawaii's papaya industry since the area suitable for cultivation of papaya in Hawaii was limited.

13-3 SYMPTOMS

P. palmivora attacks fruit, stems, and roots but not the leaves of papaya in nature. Infection of young fruit is characterized by the appearance of a milky latex on the lesion. A whitish mycelial mass usually develops in the infected area on mature fruit (Fig. 13-1). Diseased fruit may fall prematurely or they may shrivel, turn dark brown and become mummified, and then fall to the ground.[10,11]

When the infection occurs on the fruit-bearing region of the stem, many fruit and leaves fall prematurely and the stem frequently snaps off at the infected area in the wind (Fig. 13-2). Older portions of stems may also become infected and develop water-soaked lesions. The infected area may enlarge and completely girdle the stem. It may also weaken the stem and cause the plant to break-off at the infected site in a strong wind.[10,11]

P. palmivora attacks lateral roots of papaya initially and, eventually, tap root in the poorly drained areas during the rainy periods. The whole root system becomes soft and

Figure 13-1   Papaya fruit infected with *Phytophthora palmivora*.

Figure 13-2 Papaya trees with the fruit-bearing portion destroyed by *Phytophthora palmivora*.

shredded. The leaves of the affected tree turn yellow, wilt, and hang limply around the stem until they fall. Only a few small leaves are left at the apex of tree. Eventually, the affected trees succumb to the disease.[3]

In infested replant fields, *P. palmivora* attacks the root system of young seedlings and frequently causes yellowing of leaves, premature defoliation, and death of plants (Fig.13-3). Sometimes only a portion of the roots is destroyed by the pathogen, when the plants reach the resistant stage.[4] In this case, disease development ceases and the plants resume normal growth. However, during the later stage of growth when plants are loaded with fruit, they are easily blown over by winds because of the unbalance in the strength of the root system.

13-4 CAUSAL ORGANISM

The causal organism of Phytophthora root rot of papaya is *P. palmivora*. In Hawaii, this papaya pathogen was originally referred to as *Phytophthora parasitica* Dast.,[11] but it was subsequently shown to be *P. palmivora*.[12] Papaya fruit attacked by *P. parasitica* or *Phytophthora capsici* Leonian were observed occasionally in isolated areas in Hawaii.[13] It is not known if these two species can also cause root rot of papaya. In Peru, *Phytophthora cinnamomi* Rands had been associated with root rot of papaya.[14] However, even though *P. cinnamomi* is commonly present in the papaya-producing area of the island of Hawaii,[15] root rot of papaya caused by this fungus has never been observed. Moreover, papaya seedlings inoculated with *P. cinnamomi* also failed to show any disease symptoms.[3]

P. palmivora produces sporangia in compact sympodial sporangiophores averaging 50×33 μm with a length/diameter ratio of 1.6.[1,16] Light is highly stimulatory to sporangial production on artificial media but not on green papaya fruit.[17] Sporangia germinate in

Figure 13-3   Papaya seedlings with Phytophthora root rot in a replant field.

nutrient solution directly by producing germ tubes, whereas in water sporangia germinate by releasing zoospores.[18] At high concentrations of zoospores, swimming zoospores aggregate, a phenomenon which has been shown to be temperature dependent with an optimum at 16°C.[19]

P. palmivora produces chlamydospores in corn meal agar that average 39 μm in diameter.[1] Most of the chlamydospores produced in infected papaya fruit and 100% papaya juice have thick walls.[20] However, the majority of chlamydospores produced in papaya juice at lower concentrations are thin-walled. On the nutrient medium, both thick- and thin-walled chlamydospores germinate by producing germ tubes which continue to grow and form mycelial masses. On the other hand, chlamydospores germinate in water and on water agar by producing short germ tubes, each with a sporangium at the tip.[20]

P. palmivora requires a hormone produced by the opposite mating type for induction of sexual reproduction.[21] In mixed cultures of opposite mating types, oospores averaging 23 μm in diameter were produced in oogonia with amphigynous antheridia attached.[1] High frequency of oospore germination of *P. palmivora* has been achieved using the method recently developed by Ann and Ko.[22]

13-5 DISEASE CYCLE

Chlamydospores, sporangia, and zoospores of *P. palmivora* can all survive in natural soil for a certain period of time.[23-25] When a naturally infested soil was sprinkled on the surface of a selective medium, 50 of the 59 colonies of *P. palmivora* obtained originated from chlamydospores, 7 from sporangia and 2 from uncertain origins but possibly from oospores, according to Chee.[23] The natural occurrence of oospores needs to be critically investigated, because sexual reproduction of *P. palmivora* requires the presence of opposite mating types and, therefore, the chance for production of oospores in nature is very slim.

In natural soil, surviving chlamydospores, sporangia, and zoospores of *P. palmivora* may attack roots of papaya seedlings, causing root rot and death of plants.[26,27] These propagules may be released into air in splash droplets by the impact of falling raindrops and become the source of aerial infection of fruit and stems.[28] On the surface of the infected fruit the pathogen produces numerous sporangia, which can be easily dispersed by wind-blown rain and cause severe infection of plants in the same and nearby plantations.[29] Diseased fruit covered with sporangia and chlamydospores of *P. palmivora* fall to the ground and serve as an important source of inoculum in the soil.[9]

13-6 INOCULATION AND DISEASE RATING

The method described by Ko and Chan[26] can be used to inoculate roots of papaya seedlings. *P. palmivora* is grown on V8® juice agar under fluorescent light for seven days at 24°C to induce sporangial formation.[17] A sporangial suspension is obtained by spraying the culture plates with distilled water using an atomizer. Zoospores are then released from sporangia by incubation at 16°C for one hour and separated from sporangia by passing the suspension through a 20 μm sieve.[19] Chlamydospores are produced by growing the fungus in papaya juice medium for one month at 24°C.[20] They are detached from mycelium by grinding the washed mycelial mats suspended in water at 3800 rpm for one minute with an Omni mixer, and separated from mycelial fragments by passing through two layers of cheesecloth and by sedimentation in a test tube.[26] Spore concentration can be determined and adjusted by the microliter syringe method.[30] Plant 25 papaya seeds in natural soil in a 2-liter plastic container. After one month, inoculate seedling roots by evenly distributing 50 ml of spore suspension over the soil surface. Water the seedlings and remove wilted seedlings twice daily. Record the disease rating based on the percentage of seedlings killed by the pathogen after two weeks.

13-7 EPIDEMIOLOGY

Rain and wind are two of the most important epidemiological factors of Phytophthora disease of papaya. Rain splash is needed for release of sporangia of *P. palmivora* from infected papaya fruit[29] and for projecting the soil inoculum into air,[28] while wind is required for dispersal of the inoculum when it reaches the air.[29] Therefore, wind-blown rain is essential for initiation of the primary infection and for the development of epidemics of the disease in papaya plantations.

Phytophthora root rot is most serious during rainy periods and is especially prevalent in poorly drained areas.[3] The waterlogging condition appears to weaken the defense mechanism of papaya roots against the invasion by the pathogen. Mobility of zoospores of *P. palmivora* under such conditions may also contribute significantly to the severity of the disease. Motile zoospores are more infective than nonmotile ones because of their positive tactic response to papaya roots.[31]

Favorable temperature is also a contributing factor to the severity of the disease because of its effect on growth and sporulation of *P. palmivora*. The optimum temperature for mycelial growth of the pathogen is 30°C, the maximum is 36°C, and the minimum is 12°C.[8] The optimum temperature for sporangial formation is 25°C, the maximum is 35°C, and the minimum is 15°C.[17]

13-8 MANAGEMENT

The chief means for controlling fruit rot of papaya caused by *P. palmivora* is by spraying the fruiting area of the plants with a protective fungicide fortnightly in dry weather and weekly in wet. Both Tribasic Copper Sulphate and mancozeb (Dithane M-45) have been shown to be effective in protecting papaya fruit from Phytophthora rot.[32]

Phytophthora root rot of papaya in replant fields has been successfully controlled by planting seeds in virgin soil placed in planting holes (10 × 30 cm).[4,9] Since roots of papaya plants three months of age and older become resistant to *P. palmivora*, all trees growing in small islands of virgin soil are as vigorous and productive as those growing in first-planting fields. Unlike chemical applications, the virgin soil method has the advantages of being relatively inexpensive, very effective, and nonhazardous.

Field sanitation is also important in the management of Phytophthora disease of papaya. Removal of infected fruit from trees and ground would greatly reduce the inoculum for aerial infection of fruit and stems and infection of seedling roots in subsequent plantings.

13-9 CONCLUSION

Currently, it is still not known how long *P. palmivora* can survive in soil in the absence of host plants. There is also no information concerning the effect of environmental factors on its survival in nature. Papaya fields in Taiwan are frequently rotated with paddy rice. Submersed conditions are detrimental to the survival of different soilborne pathogens,[33,34] including *Phytophthora*. This may explain why Phytophthora root rot of papaya is not as serious as expected in replant fields in Taiwan. Data are needed to determine if this explanation is correct.

Methods for screening papaya seedlings for resistance to Phytophthora root rot have been developed, and resistant lines have also been found.[35,36] However, these resistant lines still have not been adopted by growers in the main papaya-producing area in Hawaii because fruits produced by these lines are still not suitable for export. Development of a resistant papaya that is commercially acceptable would be valuable because of its simplicity and economy in solving the disease problem.

13-10 REFERENCES

1. Gadd, C. H., *Phytophthora faberi* Maubl, *Ann. Roy. Bot. Gdns. Peradeniya*, 9, 47, 1924.

2. Ashby, S. F., Strains and taxonomy of *Phytophthora palmivora* Butler (*P. faberi* Maubl.), *Trans. Br. Mycol. Soc.* 14, 18, 1929.

3. Teakle, D. S., Papaw root rot caused by *Phytophthora palmivora* Butl., *Queensland J. Agric. Sci.* 14, 81, 1957.

4. Ko, W. H., Biological control of seedling root rot of papaya caused by *Phytophthora palmivora, Phytopathology*, 61, 780, 1971.

5. Trujillo, E. E. and Hine, R. B., The role of papaya residues in papaya root rot caused by *Pythium aphanidermatum* and *Phytophthora parasitica, Phytopathology*, 55, 1293, 1965.

6. Thompson, A., Notes on plant disease in 1939, *Malay. Agric. J.*, 28, 400, 1940.

7. Zentmyer, G. A., Mitchell, D. J., Jefferson, L., Roheim, J. and Carnes, D., Distribution of mating types of *Phytophthora palmivora, Phytopathology*, 63, 663, 1973.

8. Huang, T. H., Chen, D. W., and Leu, L. S., Phytophthora fruit and root rot of papaya in Taiwan, *Plant Prot. Bull.*, 18, 293, 1976.

9. Ko, W. H., Biological control of Phytophthora root rot of papaya with virgin soil, *Plant Disease*, 66, 446, 1987.

10. Hine, R. B., Holtzmann, O. V. and Raabe, R. D., Diseases of papaya (*Carica papaya* L.) in Hawaii, *Hawaii Agr. Exp. Sta. Bull.*, 136, 1, 1965.

11. Parris, G. K., *Phytophthora parasitica* on papaya (*Carica papaya*) in Hawaii, *Phytopathology*, 32, 314, 1942.

12. Tokunaga, J. and Bartnicki-Garcia, S., Cyst wall formation and endogenous carbohydrate utilization during synchronous encystment of *Phytophthora palmivora* zoospores, *Arch. Mikrobiol.*, 79, 283, 1971.

13. Aragaki, M. and Uchida, J. Y., A new papaya fruit rot in Hawaii caused by *Phytophthora capsici, Plant Dis. Reptr.*, 62, 765, 1978.

14. Bazan de Segura, C., *Carica papaya*, another possible host of *Phytophthora cinnamomi, Plant Dis. Reptr.*, 35, 335, 1951.

15. Kliejunas, J. T. and Ko, W. H., Association of *Phytophthora cinnamomi* with ohia decline on the island of Hawaii, *Phytopathology* 66, 116, 1976.

16. Ho, H. H., Synoptic keys to the species of *Phytophthora, Mycologia*, 73, 705, 1981.

17. Aragaki, M. and Hine, R. B., Effect of radiation on sporangial production of *Phytophthora parasitica* on artificial media and detached papaya fruit, *Phytopathology* 53, 854, 1963.

18. Aragaki, M., Mobley, R. D. and Hine, R. B., Sporangial germination of *Phytophthora* from papaya, *Mycologia*, 59, 93, 1967.

19. Ko, W. H. and Chase, L. L., Aggregation of zoospores of *Phytophthora palmivora, J. Gen. Microbiol.*, 78, 79, 1973.

20. Kadooka, J. Y. and Ko, W. H., Production of chlamydospores by *Phytophthora palmivora* in culture media, *Phytopathology*, 63, 559, 1973.

21. Ko, W. H., Heterothallic *Phytophthora:* Evidence for hormonal regulation of sexual reproduction, *J. Gen. Microbiol.*, 107, 15, 1978.

22. Ann, P. J. and Ko, W. H., Induction of oospore germination of *Phytophthora parasitica, Phytopathology*, 78, 335, 1988.

23. Chee, K. H., Phytophthora leaf disease in Malaysia, *J. Rub. Res. Inst. Malaya,* 21, 79, 1969.

24. Chee, K. H., Production, germination and survival of chlamydospores of *Phytophthora palmivora* from *Hevea brasiliensis, Trans. Br. Mycol. Soc.*, 61, 21, 1973.

25. Turner, P. D., Behaviour of *Phytophthora palmivora* in soil, *Plant Dis. Reptr.*, 49, 135, 1965.

26. Ko, W. H. and Chan, M. J., Infection and colonization potential of sporangia, zoospores, and chlamydospores of *Phytophthora palmivora* in soil, *Phytopathology*, 63, 1307, 1974.

27. Ramirez, B. N. and Mitchell, D. J., Relationship of density of chlamydospores and zoospores of *Phytophthora palmivora* in soil to infection of papaya, *Phytopathology*, 65, 780, 1975.

28. Okaisabor, E. K., Phytophthora pod rot infections from the soil. In *Phytophthora disease of cocoa* (P. H. Gregory, ed.), p.161, Longman, London, 1974.

29. Hunter, J. E. and Kunimoto, R. K., Dispersal of *Phytophthora palmivora* sporangia by wind-blown rain, *Phytopathology*, 64, 202, 1974.

30. Ko, W. H., Chase, L. L. and Kunimoto, R. K., A microsyringe method for determining concentration of fungal propagules, *Phytopathology*, 63, 1206, 1973.

31. Kliejunas, J. T. and Ko, W. H., Effect of motility of *Phytophthora palmivora* zoospores on disease severity in papaya seedlings and substrate colonization in soil, *Phytopatholgy*, 64, 426, 1974.

32. Hunter, J. E. and Buddenhagen, I. W., Field biology and control of *Phytophthora parasitica* on papaya (*Carica papaya*) in Hawaii, *Ann. Appl. Biol.*, 63, 55, 1969.

33. Hwang, S. C. and Ko, W. H., Biology of chlamydospores, sporangia, and zoospores of *Phytophthora cinnamomi* in soil, *Phytopathology* 68, 726, 1978.

34. Su, H. J., Hwang, S. C. and Ko, W. H., Fusarial wilt of Cavendish bananas in Taiwan, *Plant Disease*, 70, 814, 1986.

35. Aragaki, M., A papaya seedling assay for Phytophthora root rot resistance, *Plant Dis. Reptr.*, 59, 538, 1975.

36. Mosqueda-Vazquez, R., Aragaki, M. and Nakasone, H. Y., Screening the *Carica papaya* L. seedlings for resistance to root rot caused by *Phytophthora palmivora* Butl., *J. Am. Soc. Hort. Sci.*, 106, 484, 1981.

14

PEAR DECLINE

ERICH SEEMÜLLER

Biologische Bundesanstalt für Land- und Forstwirtschaft
Institut für Pflanzenschutz im Obstbau
D-6915 Dossenheim, Germany

14-1 INTRODUCTION

Pear decline (PD) is a serious disease of pear induced by a mycoplasma-like organism (MLO). It causes crop loss as well as stunting and dying of trees and is most damaging on trees grown on susceptible rootstocks such as *Pyrus pyrifolia* (=*P. serotina*) and *P. ussuriensis*. The disease was first described in some detail in western North America. It was first observed in British Columbia in 1946.[1] Only a few years later, PD was discovered in central Washington and subsequently spread further south to Oregon and was found in California in 1959.[2,3] There is a clear link between the occurrence of the disease and the spread of pear psylla within the Pacific Coast pear-growing areas. For western North America, this insect was first discovered in Washington in 1939 and became a general pest in this state in the late 1940s. The occurrence of the insect has preceded in a general way the spread of PD toward the south, reaching California by 1953.[4]

The several-year lag between the observation of pest and disease is most likely due to the occurrence of the disease in some isolated places before the pest was introduced. Its spread began when these places became infested by the pear psylla.[5] A different history of PD might be true for eastern North America, where the pest was observed as early as 1832 but where the disease was identified only in 1977.[6,7] Therefore, it is thought that the disease was introduced much later than the pest. However, it is also possible that the disease escaped observation for a long time due to its nonspecific and often only mild symptoms.

A decline-like disorder called pear moria has been reported as early as 1934 in Italy and seems to occur there since at least 1908.[8-10] Histological investigations and other evidences indicate that this disorder is most likely identical with or very similar to the PD of North America.[11,12] Therefore, it appears that the disease is not of North American but rather of European or Eurasien origin and was introduced into the New World together with *P. communis* cultivars and rootstocks from Europe.

Due to the nonspecific symptoms of PD, the cause of the disease was obscure for many years. Several abiotic factors able to induce similar symptoms were thought to be involved. These include winter injuries, excessive soil moisture, summer heat, drought, and nutritional deficiency.[13,14] Moreover, scarcity of small feeder roots was observed on affected trees, and pathogenic fungi isolated from dying rootlets were considered to be the agents.[15-17] Plant pathogenic nematodes were also found in the rhizosphere of diseased trees.[18]

Since none of these factors proved to be primarily responsible for the disease, a transmissible agent (virus) was suggested to be the cause.[18] This hypothesis was supported by anatomical studies which revealed phloem aberrations similar to those found associated with the virus-induced tristeza disease of citrus.[19] Before the transmissibility of PD was eventually demonstrated, Lindner et al.[20] proposed that the disease was caused by a toxin of the pear psylla. However, only one year later, three different groups reported successful transmission of the PD agent by grafting and by psylla feeding.[4,21-23] After this discovery, the disease was thought to be induced by a virus, but in 1970 MLOs were found in diseased trees by Hibino and Schneider[24] and in 1971 in the psylla vector by Hibino et al.[25]

14-2 DISTRIBUTION AND ECONOMIC IMPORTANCE

14-2-1 Distribution

The exact distribution of PD is unknown since it does not produce disease-specific symptoms on commercially grown pear cultivars and rootstocks and is, therefore, often difficult to recognize. Characteristic symptoms are induced on the seedling Precocious, which is, however, seldom used as an indicator in indexing.[26] The most widely applied methods for diagnosis are anatomical studies and fluorescence as well as electron microscopy. With the latter two procedures, MLOs can be demonstrated, but none of the three methods allows the characterization of the causal organism. Due to this situation, it is difficult to decide if the disorders reported as pear decline from different areas of the world are identical even if the presence of MLOs has been demonstrated. The problem is increased because in several cases decline-like disorders were reported without performing microscopic examinations.

Based on microscopic evidence, PD occurs in the United States of America, Canada, Austria, Czechoslovakia, Germany, Italy, and the Netherlands.[1,12,24,27-30] A disorder called Parry's disease reported in Great Britain is in all probability identical with PD.[31] Parry's disease, as well as the disease occurring in Germany, induced symptoms in the indicator Precocious similar to those described for PD in North America.[31,32] According to reliable reports based mainly on symptom expression, the disease also occurs in France, Greece, Yugoslavia, Spain, the European part of the Soviet Union, and Switzerland.[33-38] It is likely that PD is present in all European countries where pears are grown.

Decline-like disorders have been reported in Argentinia, Australia, and Israel.[39-41] Although their relationships to PD have not yet been demonstrated, it can be assumed that the disease occurs in parts of the world other than Europe and North America, where the *P. communis*-type pear is grown.

14-2-2 Economic Importance

The severity of the disease greatly depends on the rootstock the trees are worked on. During the PD epiphytotic in the Pacific Coast of North America in the 1950s and 1960s, the disease was very disastrous. For instance, the production of cv. Bartlett dropped an average of about 30% from 1956 to 1959 in north-central Washington.[19] In California, more than a million of trees were affected and thousands of acres of mature, productive pear trees were destroyed. In some areas, pear production has been reduced by half. In the Sierra foothills, the entire pear industry was almost wiped out.[42] This heavy loss occurred mainly on trees on the highly susceptible rootstocks *P. pyrifolia* and *P. ussuriensis*, which were widely used at that time. Trees on *P. communis* and *P. calleryana* rootstocks were considerably less affected.[19,43] However, Refatti[44] reported that more than 50,000 trees on *P. communis* rootstock were killed in the Trentino-Alto Adige area of northern Italy during the epiphytotic from 1945 to 1947. In two field trials performed in Germany, tree losses from 20 to 40% were recorded on *P. communis* rootstock as well as a severe reduction in both vigor and yield.[45,46] On quince, the most important rootstock of the European pear industry, the symptoms are usually mild and transient in nature. In most cases, the trees recover after a few years, and only a small percentage is severely affected or may die.[31,46,47] The effect of the disease on the yield of trees on tolerant rootstocks such as quince is poorly understood.

The scion cultivars do not appear to have a great influence on the development of PD, although there seem to be some variations. For instance, cvs. Clapp's Favorite and Charneu are obviously less affected than Bartlett.[45] However, the resistance among cultivars is not sufficiently high for much emphasis to be given to this factor in selecting cultivars.

14-3 SYMPTOMS

Symptom expression is significantly influenced by the rootstock and by the stage of the disease. Generally, three basic forms can be differentiated. These are quick decline, slow decline, and reddening of the foliage, which sometimes includes leaf curl. None of these is highly specific for the disease.

Quick decline is characterized by a sudden wilting of the leaves, which become dry and usually turn dark. The trees may die within a few days or weeks. Before the onset of quick decline, trees may or may not have shown slow decline, reddening, or leaf curl symptoms. Quick decline may happen any time in summer or fall, but it is more common when the trees are under stress due to hot and dry weather. In some cases, mainly when the first symptoms appear late in the growing season, the trees may die over the following winter or in spring immediately after sprouting (Fig.14-1a). The quick decline syndrome

Figure 14-1 Pear decline symptoms: (A) quick decline (left); tree showed red-leaf symptoms the previous fall and died at sprouting; (B) slow decline of Beurre Hardy on *P. communis* rootstock (left), healthy tree of same age (right); (C) leaf curl of Bonne Louise d'Avranche; (D) premature leaf drop on Beurre Hardy (left and right healthy trees).

is especially prevalent on trees on the oriental rootstocks *P. pyrifolia* and *P. ussuriensis*.[2,43] Quick decline may also occur on trees grown on *P. communis* rootstocks, but there seems to be a considerable variability within this species. For instance, rapid dying is apparently rather unusual on Bartlett seedlings, but up to 40% of the trees grown on of Kirchensaller Mostbirne seedlings may be killed.[19,45,46]

Slow decline occurs on the oriental as well as some less tolerant *P. communis* stocks. It is characterized by a progressive weakening of the trees which may fluctuate in severity. In its earlier stages, the vigor is lessened but the trees usually set and mature a normal crop. Moderately affected trees may bloom heavily and have a fair set of fruit, but the size of the crop is often reduced. In advanced stages, little or no terminal growth is made (Fig. 14-1b), both fruit set and size are poor, and leaves are few, small, leathery, and light green, the margins being slightly uprolled. In fall, leaf color may turn to the normal yellow or the leaves may become reddish and may drop prematurely. The trees may live for many years or may die within a few years.[2]

Reddening of the foliage in late summer and fall is the most characteristic symptom of diseased trees on the more tolerant rootstocks, including seedlings of *P. calleryana* and *P. betulifolia* as well as the less susceptible clone and seedling rootstocks of *P. communis*. It is the mildest form of PD and was once thought to be a different disease until the importance of the rootstock on symptom expression was realized. Reddening seems to reduce both vigor and yield, but certainly less than slow decline. It is sometimes associated with leaf curl symptoms which are characterized by downward curled leaves (from the tip toward the midrip) while the margins are rolled upward along the longitudinal axis (Fig.14-1c). The leaves also may be thickened and undulated. Trees with reddening symptoms usually drop their leaves earlier than healthy trees (Fig.14-1d). They also may show slow decline symptoms in subsequent years.[5,48–51]

Reddening of the foliage is unusual for healthy trees on highly compatible stocks including all types of *P. communis*, but it occurs frequently on healthy trees on quince. However, reddening of diseased trees on quince appears several weeks earlier in the season than reddening of healthy trees.

A distinct brown line on the cambial side of the bark at the union of scion and rootstock may be observed, mainly on severely affected trees. It has some diagnostic value for trees grown on oriental rootstocks but it occurs only occasionally on trees on *P. communis* rootstock.[43,53]

Depending on the severity of the disease, the root system is also affected. On trees with quick decline symptoms, the feeder roots are dead and so are most of the smaller roots up to 4 mm diameter. Also, some of the larger roots are dead or dying. Slow-decline diseased trees have a reduced number of feeder roots and, mainly after several years of disease, a much smaller root system than healthy trees. The root system of trees with the red-leaf syndrome may also be reduced to a variable degree according to the severity of symptoms.[2,19,49,55]

14-4 CAUSAL ORGANISM

14-4-1 Agent

Although Koch's postulates are not yet fulfilled, there is very little doubt that the causal agent is an MLO. This organism belongs to a group of nonculturable, plant pathogenic procaryotes bound with a trilaminar unit membrane but lacking a rigid cell wall. Like true mycoplasmas, these organisms are spherical to oblong, elongated, filamentous or irregular-shaped bodies, 50 to 800 nm in diameter. The taxonomic position of the MLOs, which are insect vectored and known only since 1967, is still unclear. With perhaps a few exceptions in nonwoody hosts, MLOs seem to occur in plants only in intact sieve tubes.[24,56]

14-4-2 Disease Induction

The main effect of MLO infection apparently is the impairment of the sieve tube function. The first indication of damage seems to be the deposition of pathological callose on the sieve areas (Fig.14-2). Then the sieve tubes may become necrotic. As a reaction to the loss of conducting tissue, vigorously growing trees may form additional phloem by prolonged or newly commencing cambial activity. The additional phloem is generally referred to as replacement phloem. This excessive phloem, which is abnormal in that cells including sieve tubes are smaller than usual, may also necrotize. This again induces the formation of new tissue. In this way, a big layer of phloem may form which, in extreme cases, is three to four times wider than normal (Fig.14-3).[19,53]

Pathological phloem was first observed immediately below bud union on trees grown on the highly susceptible rootstocks *P. pyrifolia* and *P. ussuriensis*. The necrosis at bud union blocks translocation of nutrients to the roots. An accumulation of starch in the scion results, and the root system weakens or dies, the feeder roots being first af-

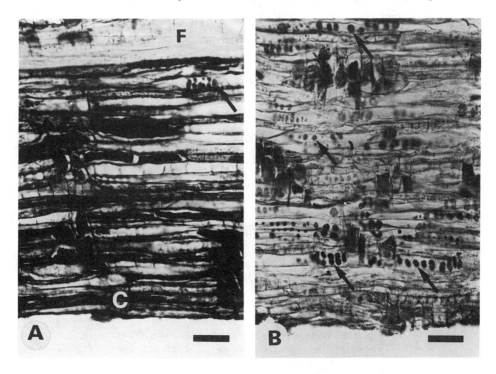

Figure 14-2 Formation of pathological callose: (A) complete current season's phloem of a healthy tree with callose in one sieve tube of the oldest portion (arrow); F=phloem fibers, C=cambial zone; (B) youngest portion of a replacement phloem of a diseased tree showing callosed sieve areas (arrows) all over the phloem, even in the youngest sieve tubes; bars 40 μm.

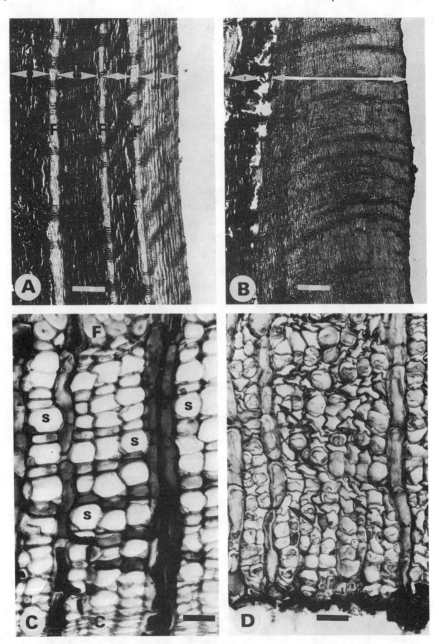

Figure 14-3   Replacement phloem and sieve tube necrosis: (A) phloem of a healthy tree; four annual rings of normal width separated by phloem fibers (F); bar 250 μm; (B) current season's phloem (arrows) of a newly diseased tree with a wide replacement portion. Previous year's phloem of normal width (arrowheads); bar 250 μm; (C) complete current season's phloem of a healthy tree in early fall, fully functional; S=functional sieve tubes, C=cambial zone, F=phloem fibers; bar 25 μm; (D) youngest portion of a replacement phloem with small, mostly collapsed sieve tubes (arrows); bar 25 μm.

fected.[19,53,57] This course of disease development led to the hypothesis that a substance produced by MLOs mainly in the leaves and translocated down the trunk is toxic to the rootstock sieve tubes but not to the sieve tubes of the scion. For that reason, PD is often considered to be a bud union disorder and, therefore, a disease of grafted trees.[19,58]

There are three main reasons why the toxin and bud-union pathology hypothesis should be revised. First, PD occurs on nongrafted trees as well. When infection occurred by psylla feeding or when the inoculating scion was removed after infection, the disease was observed on trees of several species with variable susceptibility such as *P. pyrifolia, P. ussuriensis, P. calleryana, P. communis* (including own-rooted Barlett trees), and quince. Even death was observed on such trees.[30,49,54] The second reason is that pathological phloem is not confined to bud union but may occur throughout the scion cultivar. Callose deposition, sieve tube necrosis, and replacement phloem were found in trunks, branches, shoots, and leaf veins.[26,30,59] The third evidence is that, in correspondence with the anatomical symptoms, the MLOs may be present in high numbers throughout the scion, and in most of the stocks also in the roots.[30,60]

In contrast to the foregoing hypothesis, it is much more likely that there is no long-distance effect of infection, but that the presence of MLOs is directly responsible for the phloem aberration. The reaction of a given stock-scion combination can be explained with the different susceptibility of the two parts to infection. For instance, when a tree consisting of a culinary pear cultivar and a *P. ussuriensis* rootstock becomes infected, the organisms multiply and spread in the scion and eventually reach the rootstock. In this system, the phloem of both scion and rootstock is affected, but the rootstock is much more sensitive to infection than the scion and is more severely damaged. If the scion is grafted to a *P. communis* rootstock, both parts may have similar host properties and are, therefore, similarly affected. Trees grafted onto quince give yet another reaction. In this combination, the scion is the better host and also more sensitive to infection than the quince, which is only slightly affected.[46]

However, the mode of action of MLOs is still unknown. It is not impossible that toxic metabolites which act near the place of infection are involved. Schneider[26] observed that bundle sheats and other vein-associated tissues of infected veins were necrotized. He presumed that toxic substances diffuse laterally from the sieve tubes into the vein-encasing tissue and cause pathoses.

14-4-3 Colonization Behavior

Seasonal variation   MLO colonization of the stem of decline-diseased pear trees is subject to seasonal fluctuation. Under German climatic conditions, the organisms in diseased trees are usually first detectable in May, occasionally in April. At that time, the number of colonized trees is low and so is the MLO population in these trees. In June and July, the percentage of trees showing infection of the stem, as well as the MLO population in the stem, increase. The highest number of infected trees is detected from August to November, and the highest level of MLO population is reached in September or October. In December the MLOs start to degenerate and, as verified by graft transmission experiments, no or only very few viable organisms are present in the stem in February, and none of them in March (Fig. 14-4).[61]

Since the PD MLOs in pear trees apparently depend on the physical and chemical

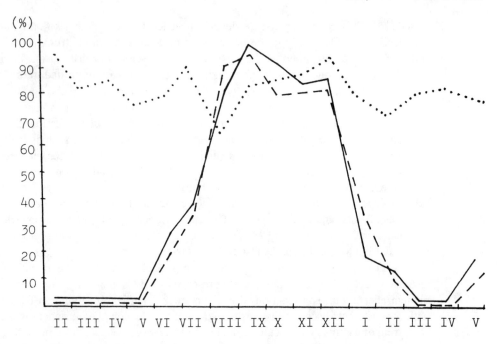

Figure 14-4 Detection of MLOs by fluorescence microscopy (in % of samples examined) in shoots (---), trunks (—) and roots(···) of diseased pear trees on *P. communis* rootstocks throughout the year.

environment of functional sieve tubes, the elimination of the organisms in the above-ground parts of the trees is closely related to the degeneration of the sieve elements in the stem (Fig.14-5a). This degeneration, first described by Evert[62] and later confirmed by Schaper and Seemüller,[59] occurs in November and December. New phloem starts to form in late winter or early spring. According to Schaper and Seemüller, sieve tube degeneration is not as complete as reported by Evert, since they always found a few intact sieve tubes in the first or second cell layer adjacent to the undifferentiated cambial zone throughout the winter. Sieve tube degeneration is also influenced by infection. In phloem that has a replacement portion, the sieve tubes may degenerate later, so a greater number of intact elements may still be present in February. In trees having this phloem arrangement, viable MLOs may be present longer than in trees with normal phloem.[59] However, according to experiments by Seemüller et al.,[61] the probability of MLO survival in the stem is extremely low since only three of 165 stem scions transmitted the disease in February and none of 115 in March. Many of the scions used in these experiments had phloem with a replacement portion. It is unclear why survival of the organisms in the few intact sieve tubes present in the pear stem phloem is not possible.

The survival of the organisms in the stem is somewhat influenced by the overwintering conditions. When diseased trees were kept in a cold room at 4°C or in the greenhouse at temperatures between 15 and 30°C, a graft transmission rate of about 10 and 20%, respectively, was obtained in February. In March, transmission grafting was unsuccessful with all trees kept under either condition, even if the scions were still in leaf at the time of grafting.[61] An explanation of these results might be that the degeneration of the

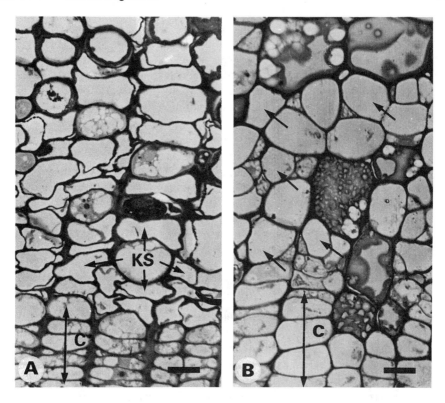

Figure 14-5 Pear phloem in winter; (A) stem phloem; all sieve tubes are collapsed (KS); (B) root phloem; all sieve tubes of this portion are functional (arrows); C=cambial zone; bars 12.5 μm.

phloem is delayed when the trees are overwintering at higher temperatures than those occurring in the field.

Overwintering of PD MLOs seems not to be directly influenced by low temperatures, although this factor is considered to be responsible for the elimination of the organisms in their host plants. Westwood and Cameron[63] shipped PD-infected trees from Oregon to Colorado to be planted and observed there. No PD developed in Colorado, while the disease occurred severely in Oregon where matched trees were grown. The authors proposed that the trees moved to Colorado were exposed to low enough temperatures, either in shipping or after planting, to eliminate the causal agent. In contrast, Seemüller et al.[61] could not find any direct effect of low temperatures on the survival of PD MLOs. In the middle of October, while the phloem was still functioning, infected pear trees were slowly chilled to a temperature of -20°C. Thereafter, PD was transmitted at a normal rate by stem grafting. The fact that culturable mycoplasmas and spiroplasmas can be stored at either -20 or -80°C is another reason to believe that MLOs are not sensitive to cold.

In contrast to the stem, no major change in the MLO population occurs in the roots throughout the year. The organisms are present in relatively high numbers in all infected trees, at least in such grown on *P. communis* rootstocks (Fig. 14-4). The survival in the roots is due to the presence of phloem rings of variable width with functional sieve tubes in winter (Fig.14-5b). In some cases, only three or four layers with intact sieve tubes (about 20% of the total) were found, while in others 90% of these elements in the annual ring were functional. As an average, about 50% of the annual root phloem was functioning throughout the year. These results show that some or most of the root phloem remains intact during winter.[59,61] The survival of MLOs in the roots of woody hosts was also found with other MLO diseases such as apple proliferation, elm phloem necrosis, and mulberry dwarf.[59,64,65]

Reinvasion of the stem   If MLOs are eliminated from the aboveground parts of the trees during the winter and persist in the roots, the stem must be reinvaded by the organisms after the winter to produce symptoms. This reinvasion was studied by Schaper and Seemüller.[66] Over a four-year period, from the middle of March to the middle of June, a narrow strip of bark was removed at regular intervals all around the trunk or the scaffold limbs of orchard pear trees that were colonized in the roots. The girdling interrupted the continuity of the phloem, which prevented the traveling of the MLOs above the girdling line.

The experiments revealed that recolonization starts as soon as newly differentiated sieve tubes are present in the stem. In southwestern Germany, this state was usually reached by mid-March. The recolonizing organisms moved upward rather slowly. Most of them passed the girdling height of 100 to 130 cm above soil level by the middle or the end of May, which is usually 10 to 30 days after full bloom. However, time and speed of recolonization varied considerably among individual trees and, corresponding to phenology, from year to year. For instance, in one year with very early bloom, recolonization occurred five to six weeks earlier than in the other three years investigated.

From the data obtained on recolonization, the velocity of MLO spread can be calculated at 7.5 to 30 mm per day. The mechanism of the upward movement is unknown. The driving force may be an acropetally oriented phloem stream from the roots to the shoots at the beginning of the growing season, when the newly forming leaves need carbohydrates and the roots are able to serve as a source. The main problem with this hypothesis is the slow rate of recolonization, compared with the translocation rates of solutes in the phloem. Also, there is no evidence that the flow in the phloem is polarized for a period of 10 weeks or longer. Another explanation for translocation is a possible motility of the causal organisms as found for some nonhelical mycoplasmas such as *Mycoplasma pneumoniae*.[67]

Latency   Symptom expression of PD-diseased trees, especially of those grown on rootstocks that induce the red-leaf syndrome, may vary considerably over the years. Symptoms usually appear during the first few years of disease. Later, there may be a symptom remission for one or several years, after which the disease may reappear. The reason for this course of disease development was investigated by Seemüller et al.[52] with trees grown on *P. communis* Kirchensaller Mostbirne seedling rootstock. This study relates the fluctuation of symptoms to the colonization behavior of the causal agent. When the trees show reddening symptoms, they are usually moderately or heavily colonized in the stem, at least in the trunk. In contrast, trees with normal foliage are in most cases not colonized in the stem, or the organisms occur in low numbers and may be confined to the trunk. Therefore, it can be concluded that symptom remission is due to the failure of the

MLOs to reinvade the stem, to the restriction of recolonization to the trunk, or to the insufficient multiplication of the organisms in an invaded stem. The reason for the failure of MLOs to recolonize the stem is not clear. It appears to be influenced by weather conditions, since there is a significant variation in symptom expression from year to year.

Although symptom remission is common for trees grown on *P. communis* rootstocks, there seems to be no true recovery on these stocks since all trees examined by Seemüller et al[52] remained colonized in the roots for a lifetime. Different colonization patterns were observed on diseased trees on quince rootstock. In this case, the causal agent is either not detectable in the rootstock or occurrs in low numbers only. Furthermore, a given population decreases over time and seems to become eliminated within a few years. The poor host properties of quince may account for the mild symptom expression and the transient nature of PD on quince rootstocks.[31,46]

14-4-4 Host Range

PD occurs on cultivars and rootstocks of *P. communis* L. (French pear), and on *P. pyrifolia* (Burm.) Nakai. (Japanese pear), *P. ussuriensis* Maxim. (Chinese pear), *P. calleryana* Decne., P. 'Variolosa' (a pear of unknown origin), *P. elaeagrifolia* Pal, and *Cydonia oblonga* Mill. when used as rootstock or grown as own-rooted trees.[19,26,43,47,48,54] In all these species, the causal agent was observed by either electron or fluorescence microscopy.[24,27,28,30,46]

Seedlings of a great number of Pyrus species were inoculated by Seemüller[30] by stem grafting with infected Bartlett scions. In addition to the above mentioned species, the disease appeared at least on some of the trees grown on seedlings of the following species: *P. amygdaliformis* Vill., *P. balansae* Decne., *P. betulifolia* Bunge, *P. bretschneideri* Rehd., *P. cossonii* Rehd., *P. fauriei* Schneid., *P. michauxii* Bosc, *P. nivalis* Jaqu., *P. pyraster* Burgsd., *P. serrulata* Rehd., and *P. syriaca* Boiss. Since the disease occurred for several years and since the agent is not able to survive in the scion cultivar, the seedling rootstock of diseased trees must have been hosts of the PD MLO.

PD was transmitted to *Catharanthus roseus* (periwinkle) by psylla feeding and dodder (*Cuscuta campestris*) bridges. The infected plants became chlorotic and stunted and gradually declined in vigor. The flowers were smaller and fewer, and MLOs were observed in diseased plants.[68,69]

14-5 TRANSMISSION

14-5-1 Psylla Transmission

Since PD followed closely in the wake of the pear psylla (*Psylla pyricola* Foerst.) in its spread down the Pacific Coast of North America, this insect was soon supposed to be the vector of the disease. Jensen and Erwin[22] and Jensen et al.[4] were the first to prove this hypothesis. They caged noninfested pear psylla adults on diseased trees and kept the colony for one to two months to acquire the causal agent. Then adults were removed and transferred to one- to three-year-old healthy trees (100 Psylla/tree), where they fed on for five to eight days. The first symptoms were observed after 56 days. By the end of the growing season, 62% of the trees showed severe symptoms or died, 21% showed mild symptoms (reddening), and only 17% appeared healthy. In a different experiment, similar-aged trees were exposed to only 25 psyllas presumed to be infectious. In this case, symptoms appeared only the following year on 15 of 20 trees exposed to psylla feeding.[50] It was shown that the pear psylla is able to acquire the PD MLO in a few hours feeding, and that the organisms persist in the vector for at least three weeks but probably the entire life of the insect.[70]

The pear psylla overwinters as adult on trunks, branches, and spurs of pear or of other trees in the vicinity of pear orchards. Eggs are laid on pear shoots and spurs from late winter to early spring. Hatching starts at about the green-tip stage and goes to about petal fall. After five nymphal instars, the adult stage is reached several weeks later. In cooler climates, there are three generations, and in warmer climates five. Eggs of the summer brood are normally deposited along the mid-vein on the upper leaf surface. Psylla adults are very active but do not move far from home trees unless conditions such as overcrowding or nutritional depletion of the foliage trigger a mass exodus from the tree or the orchard. The insect feeds mainly on pear but may also occur on quince.[71] Reproduction on other plants is not possible or very poor.[72]

In North America, *P. pyricola* is the only feeding psyllid as well as the only PD vector. In Europe, however, several *Psylla* spp. occur on pear. In Great Britain, *P. pyricola* seems to be the only pear sucker of importance. In the other pear-growing areas of northwestern Europe, both *P. pyricola* and *P. pyri* L. occur. In central and southern Europe, *P. pyri* is predominant, but P. pyrisuga Foerst. is also present and to a lesser extent P. pyricola, too.[35,73,74] The life history of *P. pyri* is very similar to that of *P. pyricola*. However, *P. pyrisuga* has only one generation which migrates in summer to conifers where overwintering occurs. The adults return to pear for oviposition in March.[71]

There are no experiments proving unequivocally that *P. pyri* and *P. pyrisuga* are vectors of PD. In field trials, Refatti[73] obtained a very low transmission rate with *P. pyri*. Since he worked with unprotected trees, uncontrolled infection cannot be excluded. However, there are circumstantial evidences for PD transmission by these two insects. In Italy and Yugoslavia, *P. pyri* is the predominant species while *P. pyrisuga* occurs less common and *P. pyricola* is found only occasionally. In both countries natural spread is common, even in the absence of *P. pyricola*.[35,75,76]

14-5-2 Graft Transmission

PD is transmitted by budding and by grafting of stem and root scions. For successful grafting of stem portions, the colonization behavior of the MLOs has to be considered. No transmission or very limited success is obtained from January to March.[54,61] Since during summer and fall not all trees are colonized in the top, the presence of the organisms should be confirmed before grafting. Stem colonization is likely to occur on trees showing reddening or leaf curl symptoms. However, it is recommended to prove the presence of the agent with an appropriate method. When the scionwood is colonized, a transmission rate of more than 80% is usually achieved by grafting of scions.[61] Budding is significantly less successful. In experiments performed by several authors, a transmission rate from 8 to 26% was obtained.[21,23,33,77] Using the same infection source, Seemüller[30] obtained 12% transmission by budding and 96% by scion grafting. The inferior results of budding might be due to the uneven distribution of the MLOs in the budwood and/or the poorer survival of the organisms in the smaller piece of tissue.

The results of root grafting depend on the nature of the donor rootstock. Using roots of fairly or heavily colonized stocks such as *P. communis*, transmission is generally successful throughout the year. However, the transmission rate is lower than by grafting of stem scions. An average of about 60% was obtained by Seemüller et al.[46,52,61] Transmission was much lower (about 6%) when roots from the unsuitable host quince are grafted to quince.[46]

The incubation period after graft inoculation depends on the time of inoculation and on the size of the test trees. When one- to two-year-old trees were inoculated in late summer or fall, symptoms usually appear in the following year. Trees of the same age inoculated in spring may develop symptoms in the same year or one year later.[21,54,61] Mature orchard trees in most cases show symptoms only two or three years after inoculation.[30]

14-6 CONFIRMATION

Since symptoms similar to those of pear decline may be produced by other factors such as incompatibility, girdling, poor drainage, malnutrition, winter injuries, and drought, the visual diagnosis has to be confirmed by other means, which include mainly microscopy and indexing. Before the causal agent was discovered, the anatomical symptoms in the current season's ring of phloem were used for diagnosis. Radial sections of the bark removed immediately below bud union were examined for the presence of pathological callose, sieve tube necrosis, and replacement phloem.[19] With this method reliable results were obtained with trees on the susceptible oriental rootstocks, but problems arose when samples from the more tolerant stocks like *P. communis* had to be examined. In the latter cases, phloem aberrations are less pronounced and often just pathological callose is detectable, sometimes even in doubtful quantities only.[19,30,53]

With the discovery of the causal agent, electron microscopy became available as a diagnostic tool. However, detection of the organisms sometimes proves to be difficult due to their low numbers in the tissue and their uneven distribution in diseased trees (Fig.14-6a). Also, electron microscopy is an elaborate procedure which is not available in many laboratories. A quick, reliable, and widely used method is the demonstration of the organisms by fluorescence microscopy. With this procedure, frozen or embedded sections are stained with the DNA-binding fluorochromes 4, 6-diamidino-2-phenylindole (DAPI) or the benzimidazole derivative Hoechst 33258.[7,8] These dyes bind preferably to A+T-rich DNA, which seems to occur in MLOs. After the treatment, the organisms appear in the sieve tubes as small, brightly fluorescent particles occurring singly or in clusters (Fig.14-6b). It is important that only the fluorescence in sieve tubes is evaluated, since other cell types of the phloem also contain fluorescent particles such as nuclei, mitochondria, and plastids. These are mostly absent in mature sieve tubes and do not, therefore, interfere with the intense MLO fluorescence.

For the investigation of trees grown on rootstocks with good host properties such as *P. communis*, best results are obtained by examining root samples. The MLO population

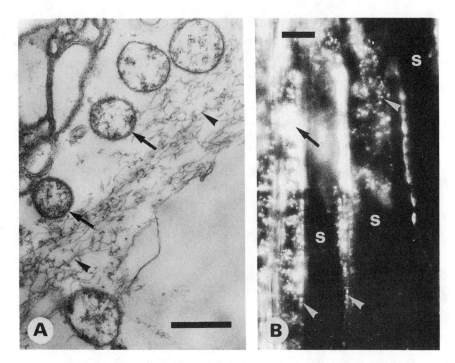

Figure 14-6 (A) Few PD MLOs (arrows) in a sieve tube; arrowheads=P-protein; bar
500 nm; (B) MLO fluorescence in sieve tubes (S) after DAPI staining; MLOs appear as
single particles or small aggregates (arrowhead) or as larger, brightly fluorescent areas
(arrow); bar 12.5 μm.

is usually highest in the roots and is also not subject to seasonal fluctuation. Stem samples
should only be used when trees show distinct red-leaf curl symptoms, or when they are
grafted to a rootstock with poor host properties such as quince.[46,52,61] There is some variation in the suitability of different stem parts for diagnosis. In the commercial cultivars, the
MLO titer usually decreases from the trunk to scaffold limbs, twigs, shoots, petioles, and
leaf veins.[60] However, Schneider[79] found in a few pear types, including Precocious, the
highest titer in the small minor veinlets, fewer MLOs in the coars veins, and none or only
low numbers in the woody parts.

In contrast to the rather nonspecific symptoms induced in commercial pear cultivars, some seedlings of *P. communis* and of some wild *Pyrus* species develop characteristic leaf symptoms. Of such plants, the *P. communis* seedling Precocious was selected and
introduced as an indicator by Schneider.[26] It first shows slightly chlorotic leaves and
broadened and swollen veins. Then, first along the adaxial side of the midrib and later of
the other framework veins, corky tissue is formed which is first whitish and later becomes
brown (Fig.14-7). Simultaneously with the development of vein symptoms, the leaves
become leathery and brittle.

For indexing work, either root or stem scions should be used according to the
colonization of the donor tree. Root scions are whip-and-tongue-grafted to the roots or
side-grafted to the stem, and stem scions are side-grafted. In both cases, the indicator
usually is top-grafted to a rootstock conductive for the disease. The plants should be kept

Figure 14-7   Pear decline symptoms on Precocious: (A) swollen veins and white, corky tissue along the midrip; healthy leaf (left); (B) later stage of (A) showing vein browning.

in the greenhouse at temperatures between 25 and 30°C, and should be observed until the end of the growing season of the year following grafting.[26,32,61] Precocious proved to be unsuitable for indexing work in the field because symptom development is unsatisfactory, at least under the conditions in Germany.[80] P. 'Variolosa' and the *P. communis* cv. *Magness* were also proposed as indicator hosts for PD by Schneider.[26] Both were supposed to exhibit a distinctive type of browning of leaf veins. However, in experiments by Kunze and Seemüller,[80] both indicators proved to be less suitable than Precocious. Symptom expression of P. 'Variolosa' was not reliable enough, and that of Magness not sufficiently specific.

14-7 MANAGEMENT

Like other MLO diseases, PD is difficult to control. It is not possible to avoid completely infection and crop loss, but damage can be reduced significantly if proper attention is given to all control measures. These include mainly rootstock selection, use of healthy planting material, orchard culture, psylla control, and chemotherapy.

14-7-1 Rootstock Selection

Due to the great influence of the rootstock on disease severity, the choice of a suitable rootstock is the most important factor to prevent loss by PD. Expression of decline in various stocks is influenced by both inherent susceptibility and natural vigor. Thus, tolerant stocks with moderate or low vigor may show more decline than moderately susceptible stocks with high vigor that often outgrow the effects of decline. Therefore, tolerant and vigorous stocks resist decline most successfully.[81] The reaction of most of the established rootstocks to PD has already been discussed in early sections. Their susceptibility was compared in field experiments under natural infection conditions with that of a great number of new and candidate rootstocks as well as wild *Pyrus* species, by Westwood and coworkers in Oregon.[82-84] The resistance of the material was rated as follows:

1. Highly resistant: *Cydonia oblonga* (quince), *P. betulifolia,* and the *P. communis* clones Anjou (self-rooted), Barlett (self-rooted), Old Home, and several Old Home × Farmingdale selections.

2. Resistant: *P. communis* seedlings of Bartlett, Nelis, and Kirchensaller Mostbirne, *P. calleryana, P. dimorphophylla, P. elaeagrifolia, P. nivalis, P. pashia,* and *P. syriaca.*

3. Slightly suceptible: Imported French *P. communis* seedlings, *P. amygdaliformis, P. caucasica, P. cordata, P. fauriei.*

4. *Highly susceptible: P. pyrifolia* and *P. ussuriensis.*

Some of these pear types were tested in graft inoculation experiments by Seemüller and coworkers[30,46] in Germany. In this study, the high susceptibility of *P. pyrifolia, P. ussuriensis,* and *P. amyadaliformis* and the resistance of quince could be confirmed. The authors also found a great variation among *P. communis* seedlings of different origin. Although there were fairly resistant seedling sources, the resistance was generally lower than rated by Westwood and coworkers.[82-84] For instance, trees on seedlings of Kirchensaller Mostbirne proved to be moderately susceptible, since there was tree death, slow decline, and long-lasting reddening and stunting. Also, the Old Home × Farmingdale (OH × F) clones were not uniformly highly resistant as claimed by the American workers. Rather, some of them, including OH × F 69 and OH × F 217 were highly susceptible, as expressed by quick decline and slow decline. A high susceptibility was also found in seedlings of the French *P. communis* cv. Feudière and, in contrast to the evaluation in Oregon, in such of *P. elaeagrifolia.* The reasons for that variation may be differences in specificity and/or virulence of the strains occurring in Oregon and Germany, or a more stringent and critical evaluation in Germany.

The reaction to infection of most of the aforementioned rootstocks and their influence on diseased trees is poorly understood. Only the resistance behavior of seedlings of *P. communis* cv. Kirchensaller Mostbirne and of Quince A was investigated in more detail.[46] The former rootstock is a suitable host for the PD agent, allowing both a fairly high reproduction of MLOs and their long-time persistence in diseased trees. Although pathological phloem aberrations do occur, the trees are in most cases only moderately affected. For that reason, Kirchensaller Mostbirne seedlings, like most *P. communis* rootstocks, were considered to be resistant by Westwood and coworkers.[82-84] A different kind of host reaction was found in quince. Trees on this rootstock were significantly less affected than such on pear seedlings. The reason for the better resistance seems to be the poor host properties of quince in which the agent occurs only in low numbers or not at all. Since the organisms have to overwinter in the root system, a low population results in a slow recolonization of the stem. A second resistance factor is the only temporary persistence of the MLOs in quince roots.

Due to this situation, the evaluation of the seedlings of Kirchensaller Mostbirne and of quince rootstocks by Westwood and coworkers has to be revised. Because of its good host properties and the damage induced, the former rootstock should be rated as moderately susceptible. Since there is no doubt that trees on quince are affected as well, this rootstock should be rated as slightly susceptible. A resistant rootstock should have better properties. Ideally, it should be an unsuitable host which cannot be colonized or in which the agent is not able to persist over the dormant period. In both cases, the pathogen would be eliminated from the tree during the winter following infection.

14-7-2 Healthy Planting Material

Healthy trees should be used for establishing new plantations. Therefore, rootstocks and scionwood should be tested with an appropriate method to ensure that they are free from infection. Although there is no perpetuation of the disease by seeds, seedlings may become infected in the nursery. When the health state of hardwood cuttings and scionwood is doubtful, there is virtually no risk in perpetuating PD if they are removed in winter and planted or grafted in spring. Also, the PD agent seems to be readily eliminated from plant material by thermotherapy.

14-7-3 Orchard Culture

Good cultural practices greatly improve the performance of decline-diseased trees. According to field observations, increasing nitrogen use has been of benefit to most decline orchards. Apparently, more nitrogen is needed to supply the weakend root system of diseased trees. In irrigation areas, a more frequent application of water in decline orchards help the trees to size fruits and maintain vigor. Experience shows that twice the frequency should be used than in healthy orchards. Severe topping and pruning is another means to maintaining sufficient vigor and fruit size.[81]

14-7-4 Psylla Control

Psyllids are not only responsible for primary transmission of PD but also seem to influence the severity of the disease. The more psyllas are feeding on a tree, the greater the showing of decline symptoms. On the other hand, little decline is observed in well-cared-for mature orchards where the psyllids are excluded, even if a high percentage of the trees is infected. If a psylla infestation is reduced from heavy to light or very light, declining trees will start to regain vigor in two or three years.[81] For these reasons, it seems that infested pear suckers reinfect the trees and thus enhance or initiate colonization of the top. In this way, an MLO titer necessary for symptom induction might be reached earlier in the season than when normal recolonization occurs from the roots. In the case where recolonization of the stem from the roots would not take place, the psyllids would be responsible for the reinfection of the crown.

The goal of a good psylla management program is, therefore, to hold psylla population to a level as low as possible throughout the season. There are several strategies to reach this goal, which depend on factors such as climate, level of population, psylla migration, site of plantation, significance of predators in the management program, and the occurrence of pesticide resistance.

One strategy is mainly based on a dormant season spray schedule. Dormant sprays are especially effective because only the vulnerable adult psyllas are present. If the over-wintering adults are severely reduced before egg-laying starts, the population can then be maintained at low levels throughout the foliage season by using sprays that would nor-mally be applied for other pests. Dormant sprays can be applied anytime from leaf fall until the start of egg-laying if there is only little migration. If migration is important, control measures are not successful until the psyllids migrate back to start egg laying. Dormant oil, dormant oil plus an insecticide, DNOC, or synthetic pyrethroids are recom-mended for dormant application.[81,85] When the psylla counts remain high, further sprays are needed, mainly until May or June. Foliage oil may be used for this purpose as well as those selective chemicals which are least damaging to beneficial insects. If a high survival of predators is insured, no further psylla sprays may be needed later in the season.[81]

Another strategy in psylla management concentrates on control of the first genera-tion. Broad-spectrum insecticides such as synthecic pyrethroids can be used if no resist-ance occurs to this group. The sprays have to be directed against the first three nymphal stages because they are most easy to kill. Insect growth regulators or growth regulators plus oil may also be applied in this program, which, like the first, is also considered to be beneficial for predators as they do not colonize the orchard at that time of the year. If further sprays are needed later in the season, pesticides should be choosen that allow greatest survival of important natural enemies of psylla, such as authocorids.[86]

A further approach, which may be appropriate for cooler climates where psylla is a less important pest, is to defer a decision on spraying until summer, when the need for a spray may be assessed more confidently by monitoring of psylla and predator abundance. However, any spray applied in summer has to be selective to minimize poisoning predators.[86]

An important question in psylla control according to the guidelines of integrated pest management is whether the current thresholds, which are based on direct psylla damage to the trees, are adequate to limit disease spread. If this is not the case, it may be necessary to reduce psyllid density thresholds, particularly in late summer when MLOs in the phloem are most prevalent and the risk of disease spread is high.

14-7-5 Chemotherapy

Soon after MLOs were recognized as plant pathogens, the therapeutic effect of tetracy-cline compounds was discovered. Nyland[87] was the first to introduce these antibiotics for the control of PD, and as early as 1973 oxytratrecycline-HCI (OTC, trade name Terramy-cin) was registered in California for postharvest injection.[42] Depending on the size of the trees and the time of application, from 0.1 to 0.8 g a. i. OTC per tree is administered into the trunk. In most cases, 0.5 to 0.6 g is optimal for a mature pear tree.[81,88,89] The antibiotic is applied either by infusion or pressure injection. In both cases, the method may vary according to the volume to be administered. The treatment is usually carried out as soon as possible after picking but not later than two weeks before anticipated leaf fall.

To apply the antibiotic in a large volume by infusion, an apparatus consisting of a 4-liter plastic jug, three plastic tubings, and plastic tube connectors is used. Holes about 6 mm in diameter and 7 to 10 cm apart are drilled horizonally 3.5 to 5 cm deep into the trunk as near the ground as practicable. The jug is filled with about 3 liters OTC solution of an appropriate concentration, and then the whole apparatus is suspended from a branch of the tree 1.0 to 1.5 m above the level of the holes. One or two apparatus are needed per tree, depending on its size. Normally, the liquid is taken up in a few hours.[81] For the application of concentrated OTC solution with the infusion method, two or more holes slanting downward are drilled around the trunk. Plastic 10-ml pipets are set firmly into the holes, and then the pipets are filled with OTC solution. Uptake of the solution requires no longer than 30 minutes.[89]

Two different methods may be applied for pressure injection. In one case, a machine is used which injects about 1 liter of OTC solution per tree with a pressure of 7 bars. Injection occurs through hollowed, tapered wood screws inserted into drilled holes (6 to 8 cm deep, 6 mm in diameter) spaced in a line around the trunk about 40 cm above soil level. Usually, three holes will be sufficient. Injection can be done in less than one minute. It takes longer in early morning and at night than in the afternoon, and varies according to weather and tree conditions. It takes longer on cloudy and humid days.[90] In the second injection method, up to eight holes 7 mm in diameter and 3 to 4 cm deep are drilled in a spiral pattern around the trunk. In each hole, 5 ml of OTC solution is injected with a hypodermic syringe.[91]

Chemical treatment is widely applied in the U.S.A., mainly on trees grown on the more sensitive rootstocks. For instance, between 140,000 and 200,000 trees were treated in California during the 1976 season.[42] Two basic effects can be prevented or decreased by the treatment: slow decline and reddening including leaf curl. Treated trees improve considerably in vigor and regain nearly normal growth. Also, there is a remission of foliage symptoms and a great increase in yield. Growers are getting double the fruit production they had before beginning the OTC treatment. This higher yield is mainly due an increased weight of individual fruits. However, there is no complete recovery since the production of treated trees seems to be only 70 to 80% that of healthy trees.[42,81,89]

The reason for incomplete recovery is the uneven distribution of OTC after trunk injection. The antibiotic is mainly translocated in the aboveground parts of the trees, while only little is distributed in the roots. Thus, the MLO population in the roots is not or only little affected by the treatment.[92] From the roots, the causal agent is able to reinvade the stem after the concentration of the antibiotic has fallen below the threshold level of activity. For that reason, the treatment should be applied anually or at least for two or three consecutive years. If, in the latter case, recovery is achieved, treatment may be withheld until trees show signs of decline again.[42,81]

Since the antibiotic treatment affects mainly the MLO population in the stem, which is naturally eliminated during winter, the question about the real effect of the OTC application arises. It seems to be a negative influence on the recolonization of the stem from the roots in spring. This may be either a slight reduction of the MLO population in the roots resulting in a slower reinvasion of the stem or, more likely, a prevention or retardation of stem recolonization by antibiotic residues still present in the stem at the time of recolonization. If the latter is true, the time of OTC application should be reconsidered. Then it would be better to apply the antibiotic at a low dose in late spring or early summer rather than relatively high amounts in fall. An application early in the season does not seem to create residue problems.[92]

Tetracycline antibiotics may be phytotoxic to pear. The detrimental effects are dose related and are more pronounced at application late in the season toward leaf fall. In contrast, there is very little phytotoxicity at application early in the foliage season. Symptoms of phytotoxicity include blackened leaves in fall immediately after application and, in the following spring, yellowing of the foliage, delayed bloom, blossom necrosis, reduced fruit set, retarded limbs, and trunk damage at the site of application.[89,92]

14-8 CONCLUSION

Present knowledge of how MLOs induce plant diseases is inadequate. Several other aspects of the disease are also poorly understood. The major obstacle to improved knowledge of the disease is that, like the other MLOs, the PD-causing agent cannot be cultured yet *in vitro*. For that reason, Koch's postulates cannot be fulfilled, and, more important, the organisms could not be sufficiently characterized thus far. Its identity is, therefore, obscure, and it is unknown if the natural plant habitat of the agent is confined to the genus *Pyrus* or if there is a wider host range comprising woody and/or herbaceous plants which might be a reservoir of the disease. Due to this situation, it is most important to develop disease-specific diagnostic tools such as monoclonal antibodies and genomic probes. The techniques to produce these tools are now available, as shown for other MLO diseases such as aster yellows and X-disease of stone fruit.[93,94] A disease-specific confirmation would also considerably improve diagnosis and would allow reliable investigations on the distribution of PD and the relatedness of the disorder in different areas of the world.

There are several unsolved problems concerning the psylla vectors. While it is well established that *P. pyricola* is an efficient vector, it is not definitely proven if *P. pyri* and *P. pyrisuga* are able to transmit the disease. The ability to transmit is very important for the stringency of control measures. For instance, if *P. pyrisuga* is not a vector, less emphasis would be necessary for its control since this pest migrates to other hosts in early summer. An important question in psylla control is whether the current thresholds based on direct damage to the trees are adequate to limit disease spread. Due to the rapid development of resistance to most pesticides, psylla control will remain a very difficult task. Therefore, more emphasis should be given for breeding psylla-resistant pear cultivars. Resistance to *P. pyricola* occurs in several *Pyrus* species, including *P. betulifolia, P. calleryana,* and *P. fauriei.*[83] Another important aspect is the persistance of the PD agent in psyllids. The three major pear suckers overwinter as adults and are present on pear in spring. If the PD agent persists in psyllids, infection of healthy trees and reinfection of diseased trees would occur. In this case, recolonization of the stem would be enhanced or even initiated, which would increase damage considerably (see section 14-7-4).

Since the PD agent seems to overwinter exclusively in the root system, the use of resistant rootstocks is the most efficient and most promising strategy to avoid damage. Resistant rootstocks should prevent either colonization by or overwintering of the organism. In this way, infection having taken place during the growing season would be eliminated in winter. At present, there is no rootstock fulfilling completely this requirement and there is no *Pyrus* species which delivers a high percentage of progenies showing this property. However, among seedlings of tolerant species there seem to be individuals showing the desirable level of resistance.[30] These plants could be used in resistance breed-

Figure 14-8   Red-leaf symptom on a Bartlett pear tree in late summer.

ing or, if they prove suitable as rootstocks, should be propagated by appropriate methods.

14-9 REFERENCES

1. McLarty, H. R., Killing of pear trees, *Ann. Rep. Canad. Plant Dis. Survey*, 28, 77, 1948.

2. Woodbridge, C. G., Blodgett, E. C. and Diener, T. O., Pear decline in the Pacific Northwest, *Plant Dis. Rep.*, 41, 569, 1957.

3. Nichols, C. W., Schneider, H., O'Reilly, H. J., Shalla, T. A. and Griggs, W. H., Pear decline in California, *Bull. Calif. Dep. Agric.*, 49, 186, 1960.

4. Jensen, D. D., Griggs, W. H., Gonzales, C. Q. and Schneider, H., Pear decline virus transmission by pear psylla, *Phytopathology*, 54, 1346, 1964.

5. Kaloostian, G. H., Oldfield, G. N. and Jones, L. S., Effect of pear decline virus and pear psylla (*Psylla pyricola*) toxin on pear trees, *Phytopathology*, 58, 1236, 1968.

6. McIntyre, J. L., Dodds, J. A., Walton, G. S. and Lacy, G. H., Declining pear trees in Connecticut: Symptoms, distribution, symptom remission by oxytetracycline, and associated mycoplasmalike organisms, *Plant Dis. Rep.*. 62, 503, 1978.

7. McIntyre, J. L., Schneider, H., Lacy, G. H., Dodds, J. A. and Walton, G. S., Pear decline in Connecticut and response of diseased trees to oxytetracycline infusion, *Phytopathology*, 69, 955, 1979.

8. Catoni, G., Casi di deperimento di peri e di meli, *Boll. Agr.*, 47, 148, 1934.

9. Catoni, G., Deperimenti e moria dei peri nella Venezia Tridentina, *Agric. Trent.*, 2, 1, 1947.

10. Baldacci, E., Catoni, G., Ciferri, R., Marimpietri, L. and Refatti, E., Relazione sui dati e le esperienze circa la moria del pero nella Venezia Tridentina (Trentino ed Alto Adige), *Not. Mal. Piante*, 4, 1, 1949.

11. Shalla, T. A., Chiarappa, L., Blodgett, E. C., Refatti, E. and Baldacci, E., The probable coidentity of the moria disease of pear trees in Italy and pear decline in North America, *Plant Dis. Rep.*, 45, 912, 1961.

12. Refatti, E., La moria del pero in Italia, *Notiz. Mal. Piante*, 68, 1, 1964.

13. Degman, E. S., Is pear decline caused by overirrigation?, *Proc. Wash. State Hort. Assoc.*, 50th Ann. Meet., 66, 1958.

14. Woodbridge, C. G., The nutrient content of Bartlett pear trees, *Proc. Wash. State Hort. Assoc.*, 54th Ann. Meet., 155, 1958.

15. Sprague, R., Fungi isolated from the roots of pear trees at Wenatchee, *Proc. Wash. State Hort. Assoc.* 51th Ann. Meet., 93, 1955.

16. Sprague, R., Fungi isolated from underground parts of pear trees, *Plant Dis. Rep.*, 41, 74, 1957.

17. McIntosh, D. L., The infection of pear rootlets by *Phytophthora cactorum*, *Plant Dis. Rep.*, 44, 262, 1960.

18. Blodgett, E. C. and Aichele, M. D., Progress report on pear decline research, *The Goodfruit Grower*, 9, 2, 1959.

19. Batjer, L. P. and Schneider, H., Relation of pear decline to rootstocks and sieve-tube necrosis, *Proc. Amer. Soc. Hort. Sci.*, 76, 85, 1960.

20. Lindner, R. C., Burts, E. C. and Benson, N. R., A decline condition in pears induced by pear psylla, *Plant Dis. Rep.*, 46, 59, 1962.

21. Blodgett, E. C., Aichele, M. D. and Parsons, J. L., Evidence of a transmissible factor in pear decline, *Plant Dis. Rep.*, 47, 89, 1963.

22. Jensen, D. D. and Erwin, W. R., The relation of pear psylla to pear decline, *Calif. Agric.*, 17 (1), 2, 1963.

23. Shalla, T. A., Chiarappa, L. and Carroll, T. W., A graft transmissible factor associated with pear decline, *Phytopathology*, 53, 366, 1963.

24. Hibino, H. and Schneider, H., Mycoplasmalike bodies in sieve tubes of pear trees affected with pear decline, Phytopathology, 60, 499, 1970.

25. Hibino, H., Kaloostian, G. H. and Schneider, H., Mycoplasma-like bodies in the pear psylla vector of pear decline, *Virology*, 43, 34, 1971.

26. Schneider, H., Indicator hosts for pear decline: Symptomatology, histology, and distribution of mycoplasmalike organisms in leaf veins, *Phytopathology*, 67, 592, 1977.

27. Blattny, C. and Vana, V., Pear decline accompanied with mycoplasma-like organisms in Czechoslovakia, *Biol. Plant. (Prague)*, 16, 74, 1974.

28. Behnke, H. D., Schaper, U. and Seemüller, E., Elektronenmikroskopischer Nachweis von mykoplasmaaähnlichen Organismen bei Birnbäumen mit pear-decline-Symptomen in der Bundesrepublik Deutschland, *Phytopath. Z.*, 97, 89, 1980.

29. Kegler, H. and Klinkowski, M., Untersuchungen zum Nachweis des virösen Birnenverfalls (pear decline), *Phytopath. Z.*, 58, 293, 1967.

30. Seemüller, E., unpublished results.

31. Davies, D. L., Clark, M. F. and Adams, A. N., A mycoplasma-like organism associated with a decline-like disease in English pears, *Acta Hortic.*, 193, 329, 1986.

32. Seemüller, E., Kunze, L. and Schaper, U., Übertragung des europäischen Birnenverfalls (pear decline) auf die amerikanischen Indikatorsorten 'Magness' und 'Precocious,' *Nachrichtenbl. Deut. Pflanzenschutzd.*, 32, 33, 1980.

33. Lemoine, J., A diebak of pear observed in France and resembling pear decline or moria, *Acta Hortic.*, 44, 131, 1975.

34. Plakidas, A. G., Pear decline on Ikaria island, *Plant Dis. Rep.*, 46, 150, 1962.

35. Grbic, V., Neke stetne vrste iz Psyllidae na plantazama krusaka in Vojvodini, *Zast. Bilja*, 25, 121, 1974.

36. Rallo, L., Decaimiento del peral en plantaciones del valle del Ebro, An. INIA, *Ser. Prot. Veg.*, 3, 147, 1973.

37. Anonymous, Pear decline. European and Mediterranean Plant Protection Organization, Data sheets on quarantine organisms, List A, 2, 8, 1978.

38. Schmid, G., Pear decline, *Schweiz. Z. Obst-Weinbau*, 110, 197, 1974.

39. Sarasola, A. A., El decaimento del peral an el Valle del Rio Negro, *Publ. téc. Inst. Pat. Veg. Argentina*, 60, 1960.

40. Pares, R. D. and Hutton, K. E., Virus diseases of pome fruit in New South Wales, *Agric. Gaz. N.S.W.*, 72, 414, 1961.

41. Gur, A., The compatibility of the pear with quince rootstocks, *Spec. Bull. Israel Minist. Agric. Res. Stat.*, 10, 9, 1957.

42. Beutel, J. A., Moller, W. J. and Cress, F. D., Research review: Antibiotic injections control pear decline disease, *Calif. Agric.*, 31(8), 12, 1977.

43. Blodgett, E. C., Schneider, H. and Aichele, M. D., Behavior of pear decline disease on different stock-scion combinations, *Phytopathology*, 52, 679, 1962.

44. Refatti, E., Pear moria in Italy, *FAO Plant Prot. Bull.*, 12, 6, 1964.

45. Spaar, D., Vater, J. and Helm, W., Ökonomische Auswirkungen des Birnenverfalls in einer Ertragsanlage der GPG "Elbaue" Rogaätz, *Gartenbau*, 19, 161, 1972.

46. Seemüller, E., Schaper, U. and Kunze, L., Effect of pear decline on pear trees on 'Quince A' and *Pyrus communis* seedling rootstocks, *Z. PflKrankh. PflSchutz*, 93, 44, 1986.

47. Kunze, L. and Seemüller, E., Nachweis des Verfalls (pear decline) bei Birnen auf Sämlings und Quittenunterlagen, *Nachrichtenbl. Deut. Pflanzenschutzd.*, 23, 170, 1971.

48. Griggs, W. H., Jensen, D. D., Iwakiri, B. T. and Beutel, J. A., Leaf curl in young pear trees, *Calif. Agr.*, 21(10), 16, 1967.

49. Griggs, W. H., Jensen, D. D. and Iwakiri, B. T., Development of young pear trees with different rootstocks in relation to psylla infestation, pear decline, and leaf curl, *Hilgardia*, 39, 153, 1968.

50. Kaloostian, G. H. and Jones, L. S., Pear leaf curl virus transmitted by pear psylla, *Plant Dis. Rep.*, 52, 924, 1968.

51. Millecan, A. A., Gotan, S. W. and Nichols, C. W., Red-leaf disorders of pear in California, *Bull. Calif. Dep. Agr.*, 52, 166, 1963.

52. Seemüller, E., Kunze, L. and Schaper, U., Colonization behavior of MLO, and symptom expression of proliferation-diseased apple trees and decline-diseased pear trees over a period of several years, *Z. PflKrankh. PflSchutz*, 91, 525, 1984.

53. Seemüller, E. and Kunze, L., Untersuchungen über den Birnenverfall (pear decline) in Südwestdeutschland, *Mitt. Biol. Bundesanst. Land-Forstw., Berlin-Dahlem*, 144, 47, 1972.

54. Schneider, H., Graft transmission and host range of the pear decline causal agent, *Phytopathology*, 60, 204 , 1970.

55. Soma, K. and Schneider, H., Developmental anatomy of major lateral leaf veins of healthy and of pear-decline diseased pear trees, *Hilgardia*, 40, 471, 1971.

56. Lee, I. M. and Davis, R. E., Phloem-limited prokaryotes in sieve elements isolated by enzyme treatment of diseased plant tissues, *Phytopathology*, 73, 1540, 1983.

57. Catlin, B. P., Olsson, E. A. and Beutel, J. A., Reduced translocation of carbon and nitrogen from leaves with symptoms of leaf curl, *J. Amer. Soc. Hort. Sci.*, 100, 184, 1975.

58. Schneider, H., Cytological and histological aberrations in woody plants following infection with viruses, mycoplasmas, rickettsias, and flagellates, *Ann. Rev. Phytopath.*, 11, 119, 1973.

59. Schaper, U. and Seemüller, E., Conditions of the phloem and the peristence of mycoplasmalike organisms associated with apple proliferation and pear decline, *Phytopathology*, 72, 736, 1982.

60. Schaper, U. and Seemüller, E., Einfluss des Besiedlungsverhaltens auf die fluoreszenzmikroskopische Nachweisbarkeit der Erreger der Apfeltriebsucht und des Birnenverfalls, *Nachrichtenbl. Deut. Pflanzenschutzd.*, 36, 21, 1984.

61. Seemüller, E., Schaper, U. and Zimbelmann, F., Seasonal variation in the colonitaion patterns of mycoplasmalike organisms associated with apple proliferation and pear decline, *Z. PflKrankh. PflSchutz*, 91, 371, 1984.

62. Evert, R. F., Phloem structure in *Pyrus communis* L. and its seasonal changes, *Calif. Univ. Publs. Bot.*, 32, 127, 1960.

63. Westwood, M. N. and Cameron, H. R., Environment-induced remission of pear decline symptoms, *Plant Dis. Rep.*, 62, 176, 1978.

64. Braun, E. J. and Sinclair, W. A., Histopathology of phloem necrosis in *Ulmus americana*, *Phytopathology*, 66, 598, 1976.

65. Tahama, Y., Studies on the mulberry dwarf disease. XXIII. Electron microscopic examination on the mycoplasmalike organisms in the overwintered mulberry trees, Bull. *Hiroshima Agric. Coll.*, 5, 151, 1975.

66. Schaper, U. and Seemüller, E., Recolonization of the stem of apple proliferation and pear decline-diseased trees by the causal organisms in spring, *Z. PflKrankh. PflSchutz*, 91, 608, 1984.

67. Radestock, U. and Bredt, W., Motility of *Mycoplasma pneumoniae*, *J. Bacteriol.*, 129, 1495, 1977.

68. Kaloostian, G. H., Hibino, H. and Schneider, H., Mycoplasmalike bodies in periwinkle: Their cytology and transmission by pear psylla from pear trees affected by pear decline, *Phytopathology*, 61, 1177, 1971.

69. Raju, B. C., Nyland, G. and Purcell, A. H., Current status of the etiology of pear decline, *Phytopathology*, 73, 350, 1983.

70. Bailey, J. B., Jensen, D. D., Nickel, J. L., Tanada, Y., Catlin, P. B., Griggs, W. H., Ryugo, K., Shalla, T. A., Dickson, R. C., Schneider, H., Tsao, P. W., Pine, T. S. and Kaloostian, G. H., Progress report on pear decline, *Calif. Agric.*, 19(7), 14, 1965.

71. Müller, H. J., Psylloidea. In *Handbuch der Pflanzenkrankheiten*, Vol. 5, 3. Lief., 1. Teil (H. Blunck, ed.), p. 306, Paul Parey, Berlin und Hamburg, 1956.

72. Kaloostian, G. H., Transitory hosts of the pear psylla, *J. Econ. Entomol.*, 63, 1039, 1970.

73. Refatti, E., Present status of the research on pear "moria" in Italy, *Tagungsber, Deutsch. Akad. Landw.wiss. Berlin*, 97, 97, 1968.

74. Rieux, R., Recherches sur les psylles du poirier et la lutte integrée dans le sud de l'Europe, *IOBC WPRS Bull.*, 9(4), 181, 1986.

75. Refatti, E., Pear decline and moria. In *Virus diseases of apples and pears, Techn. Comm. Commonw. Bur. Hort. Plant Crops, No. 30*, 108a, 1967.

76. Refatti, E., Malattie da virus e micoplasmsi delle pomacee: Situazione italiana e problemi di quarantena, *Inform. Fitopat,*, 33, 38, 1983.

77. Shalla, T. A., Carroll, T. W. and Chiarappa, L., Transmission of pear decline by grafting, *Calif. Agric.*, 18(3), 4, 1964.

78. Seemüller, E., Investigations to demonstrate mycoplasmalike organisms in diseased plants by fluorescence microscopy, *Acta Hortic.*, 67, 109, 1976.

79. Schneider, H., Distribution of mycoplasmalike organisms (MLO) in pear decline indicator trees, *Proc. Amer. Phytopath. Soc.*, 3, 320, 1976.

80. Kunze, L. and Seemüller, E., unpublished results.

81. Anonymous, Pear pest management, *Agric. Sci. Publ.,* Univ. Calif., Richmond, CA, 1978.

82. Westwood, M. N., Cameron, H. R., Lombard, P. B. and Cordy, C. B., Effects of trunk and rootstock on decline, growth and performance of pear, *J. Amer. Soc. Hort. Soc.*, 96, 147, 1971.

83. Westwood, M. N. and Lombard, P. B., Rootstocks for pear, *Proc. Oregon Hort. Sci.*, 73, 64, 1982.

84. Westwood, M. N. and Lombard, P. B., Pear rootstocks: Present and future, *Fruit Var. Jour.*, 37, 24, 1983.

85. Stäubli, A., Utilisation pratique du fenoxycarb, un regulateur de croissance des insectes, dans le lutte contre le psylle du poirier, *IOBC WPRS Bull.*, 9(4), 194, 1986.

86. Campbell, C. A. M., Pear psylla management in England, *IOBC WPRS Bull.*, 9(4), 190, 1986.

87. Nyland, G., Remission of symptoms of pear decline in pear and peach X-disease in peach after treatment with a tetracycline, *Phytopathology*, 61, 904, 1971.

88. Nyland, G. and Moller, W. J., Control of pear decline with tetracycline, *Plant Dis. Rep.*, 57, 634, 1973.

89. Lacy, G. H., McIntyre, J. L., Walton, G. S. and Dodds, J. A., Rapid method for and effects of infusing trees with concentrated oxytetracycline-HCl solutions for pear decline control, *Can. J. Plant Path.*, 2, 96, 1980.

90. Reil, W. O. and Beutel, J. A., A pressure machine for injecting trees, *Calif. Agric.*, 30(12), 4, 1976.

91. Casanova, R., Llacer, G. and Sanchez-Capuchino, J. A., Remission of symptoms of apple proliferation, after injection of concentrated tetracycline solutions, *Acta Phytopath. Acad. Sci. Hungaricae*, 15, 273, 1980.

92. Schaper, U., Untersuchungen zum Nachweis und Besiedlungsverhalten sowie zer Bekämpfung der Erreger der Triebsucht des Apfels und des Birnenverfalls, Diss. Univ. Göttingen, 1981.

93. Lin, C. P. and Chen, T. A., Monoclonal antibodies against the aster yellows agent, *Science*, 227, 1233, 1985.

94. Kirkpatrick, B. C., Stenger, D. C., Morris, T. J. and Purcell, A. H., Cloning and detection of DNA from a nonculturable plant pathogenic mycoplasma-like organism, *Science*, 238, 197, 1987.

15

PRUNUS NECROTIC RINGSPOT VIRUS

GAYLORD I. MINK

Washington State University
Irrigated Agriculture Research and Extension Center
Prosser, Washington

15-1 INTRODUCTION

15-1-1 What Is Prunus Necrotic Ringspot Virus?

The term *Prunus necrotic ringspot virus* is often used in two different but overlapping ways. As a generic term, it identifies all the serologically interrelated members of ilarvirus subgroup III.[1,2] These include not only viruses that cause various ringspot diseases of Prunus, rose, and hops but also viruses that cause mosaic diseases of apple, rose, and other perennial plants as well as viruses that cause certain line pattern diseases of plums.[2,3] As a specific term, Prunus necrotic ringspot virus (PNRSV) identifies those viruses with characteristics similar to a virus first described by Fulton,[4–6] which he initially called sour cherry necrotic ringspot virus. This name was subsequently changed to PNRSV.[7] Used in this narrow context, PNRSV identifies a virus which differs both biologically and serologically from apple mosaic virus (ApMV) but which resembles ApMV morphologically and chemically.[2,3]

While PNRSV and ApMV are often treated as distinct viruses,[7,8] they are related serologically[9] and sometimes occur in the same hosts.[7] Their separate names were retained because they can be differentiated serologically and they are usually associated with different diseases.[9] However, it seems likely that if the initial representative of these two viruses had been isolated from rose or hop, they would now be regarded only as serotypes rather than distinct viruses.

Despite differences in host ranges and symptomatology, there appear to be no greater differences between viruses regarded as PNRSV and ApMV than there are among variants or strains of either virus. Therefore, in this chapter I apply the term *Prunus necrotic ringspot virus* in its broadest sense; that is, to include both PNRSV and ApMV.

15-1-2 Historical Background

Ringspot and mosaic diseases of Prunus and Malus have been recognized since the 1930s (Table 15-1). However, a certain amount of confusion develops when one attempts to review the early literature, especially literature covering viruses which are now regarded as biological or serological variants of PNRSV. One reason for this is the changes in terminology that have been applied to these viruses as technology changed. For nearly three decades, ringspot viruses of Prunus were described primarily on the basis of the diseases they produced. A partial list of similar names that were applied during this period include cherry necrotic ringspot virus,[18] necrotic ringspot virus,[24] Prunus ringspot virus,[25] ringspot virus,[15] stone fruit ringspot virus,[25] sour cherry necrotic ringspot virus,[26] and tatter leaf virus.[27] Because most of these names were applied before techniques were available to study the causal agents, little is known about the chemical, physical, and serological properties of many of the original isolates. However, it is likely that each of these names represents a synonym for PNRSV.

A second reason for confusion in interpreting early reports of the various Prunus ring spot viruses involves the names applied to a group of diseases which later were found to be caused by a distinct virus: prune dwarf virus (PDV).[28] Prune dwarf virus was first described in 1936.[29] For many years, the relationship between PNRSV and PDV was the subject of intense debate among researchers as was the role that both viruses played in a disease of sour cherries referred to as "yellows." Evolution in understanding the relationships among these viruses and diseases they cause was reviewed in detail by Fulton in

TABLE 15-1: CHRONOLOGY OF NAMES APPLIED TO DISEASES OF PERENNIAL PLANTS THAT APPEAR TO BE CAUSED BY BIOLOGICAL VARIANTS OF PRUNUS NECROTIC RINGSPOT VIRUSES.

Year	Host	Disease	Reference
1932	Peach	Rings and spots	10
1933	Apple	Infectious variegation	11
1937	Apple	Mosaic	12
1939	Almond	Calico	13
1939	Sweet cherry	Rugose mosaic	13
1939	Mazzard cherry	Mosaic	13
1939	Rose	Mosaic	14
1941	Peach	Ringspot	15
1942	Peach	Necrotic leaf spot	16
1942	Sweet cherry	Lace leaf	17
1944	Sour cherry	Necrotic leaf spot	18
1946	Sweet cherry	Tatter leaf	19
1956	Sour cherry	Stecklenberger	20
1963	Hop	Ringspot	21
1963	Sour cherry	Prunus necrotic ringspot	22
1970	Plum	Line pattern	23

1968.[29] At the time of the review, technological changes had provided methods to examine causal agents as well as diseases.

Beginning in 1948, a series of reports described transmission of viruses from fruit trees to herbaceous plants.[30-36] During this period, Fulton[4-6] transmitted four viruses from sour cherry to herbaceous plants, determined many of their properties, back transmitted them to Prunus species where three of the viruses (A, G, and H) caused diseases of the ringspot type and one (B) caused prune dwarf disease in Italian prune.[37] The three ringspot viruses were found to be interrelated serologically, but all were serologically distinct from PDV.[38]

The name Prunus necrotic ringspot virus was first used to designate a specific virus in 1963.[22] In 1970, Fulton[7] summarized the properties of PNRSV and identified many of its symptoms. However, this did not eliminate some of the confusion that still exists over the use of the name. It is now recognized that viruses with properties similar to those of PNRSV occur naturally in several non-Prunus, perennial plants.

The early history of apple variegation diseases (mosaics) was reviewed by Bradford and Joley in 1933[11] and later by McCrum et al.[39] Rose mosaic diseases were examined in the 1930s by Thomas and Massey[14] and McWhorter.[40] In 1950, Cochran[41] reported reciprocal graft transmission of peach ringspot (= PNRSV) isolates to and from apple seedlings where they caused mosaic symptoms. Later, Fulton[42] transmitted a virus from mosaic-diseased rose to herbaceous plants which was purified and found to be distantly related to PNRSV.[43] A mosaic virus isolated from an apple tree in England was also found to be related serologically to PNRSV even though the symptoms on herbaceous hosts differed from those of PNRSV.[44] When viruses causing cherry necrotic ringspot, plum line pattern, rose mosaic, and apple mosaic were compared, the four viruses were found to be serologically related but separable into two serotypes.[9] ApMV and rose mosaic virus were closely related if not identical serologically, as were PNRSV and a plum line pattern virus referred to as Danish plum line pattern.[9] While the names ApMV and PNRSV were retained by Fulton,[9] the viruses were shown to be serotypes.

In 1959, Fridlund[45] transmitted a virus that resembled PNRSV from hop to cucumber. Later, Schmidt[21] transmitted a virus from hop with ringspot symptoms to a wide range of herbaceous plants. Bock[46,47] described two PNRSV serotypes that infect hops: serotype C, which resembled PNRSV from cherry, and serotype A, which resembled ApMV from apple. Both ApMV and a serotype intermediate between PNRSV and ApMV have been reported to occur widely in hops.[48,49]

In recent years, viruses serologically similar to ApMV have been isolated from almond,[50,51] birch,[52] blackberry,[53] hazelnut,[54,55] horse chestnut,[56] plum,[57] and red and black raspberry.[53]

15-2 DISTRIBUTION AND ECONOMIC IMPORTANCE

PNRSV occurs worldwide wherever stone fruits are grown.[25] All cultivated species and cultivars of Prunus appear to be susceptible to one or more strains of the virus.[25] The virus can cause serious disease in nurseries, where it causes poor stands and reduced tree growth.[58-60] It also reduces growth and yield of orchard trees.[61-66] However, with some species the effects may depend on the virus strains involved.[25] In roses, PNRSV is widely distributed,[67-71] and while the virus does not affect growth appreciably,[71] a number of deformed flowers are produced on some cultivars.[72] In hop, the effects of PNRSV depend on cultivar.[47]

ApMV is distributed worldwide in apples,[39] roses,[8] and probably occurs wherever hops are grown.[47] Many different strains of ApMV occur in apple, and varieties differ markedly in their reaction to them.[39,73] ApMV appears to be of economic importance in apples in the U.K.[73] but of little significance in roses there.[72,74]

ApMV infects a broader range of plant genera than PNRSV (Table 15-2).

15-3 SYMPTOMS

Because many perennial crop plants are naturally infected by either PNRSV or ApMV, or in the case of rose and hop by both, and because many biological variants of each virus occur in some of these crops, it is impossible to describe here in detail all of the diseases induced by Prunus necrotic ringspot viruses. General symptom types induced in some of their natural hosts are summarized in Table 15-3. While this information may provide a general guide for visual diagnosis, it is highly oversimplified. Only a few combinations of host cultivar and virus isolate may produce diagnostic symptoms.[25] In most virus-host combinations discrete or diagnostic symptoms are either very transient or do not occur. Chronic symptoms are often difficult to differentiate from symptoms caused by other viruses or by nonvirus disorders. In most crops, a definitive diagnosis for PNRSV or ApMV usually requires supporting information from bioassays or serological tests, or both.

15-4 CAUSAL AGENT

TABLE 15-2:   NATURAL HOSTS REPORTED FOR APPLE MOSAIC VIRUS (ApMV) AND PRUNUS NECROTIC RINGSPOT VIRUS (PNRSV).

Virus	Host		Reference
ApMV	Almond	*Prunus dulcis* (Mill.) D. A. Webb	50
	Apple	*Malus sylvestris* Mill.	8
	Birch (yellow)	*Batula papyrifera* Marsh.	52
	Birch (white)	*B. alleghaniensis* Britton	52
	Blackberry	*Rubus ursinus*	53
	Hop	*Humulus lupulus* L.	47
	Hazelnut	*Corylus avellana* L.	54
	Horse chestnut	*Aesculus* spp.	56
	Plum	*P. domestica* L.	57
	Raspberry (black)	*R. accidentalis*	59
	Raspberry (red)	*R. edaeus*	53
	Rose	*Rosa* spp.	14
PNRSV	All Prunus species tested		25
	Hop	*H. lupulus* L.	47
	Rose	*Rosa* spp.	7

TABLE 15-3: SYMPTOMS INDUCED BY PRUNUS NECROTIC RINGSPOT VIRUS (PNRSV) AND APPLE MOSAIC VIRUS (ApMV) IN NATURAL HOSTS.

Virus	Natural Host	Symptom types	Reference
PNRSV	Hop	May occur without symptoms	46
		Split leaf blotch in Fuggles type	47
		Ringspots	21
	Prunus	May occur without symptoms	25,75,76
		Necrotic shock symptoms	15,25,27
		Chronic mottle, necrosis, leaf deformity	3,23,25
	Rose	Line patterns, ringspots	72
ApMV	Almond	Mosaic	50,51
	Apple	Mosaic	8,11,39
	Birch	Line pattern, ringspots, recovery	52
	Hop	Line pattern, ringspots	47
		Split leaf blotch in Fuggles type	47
	Hazelnut	Mosaic, ring patterns	54,55
	Horse chestnut	Mosaic	56
	Plum	Line patterns, mosaic	23,50,57
	Rose	Chlorosis, puckering, leaf distortion	72
	Rubus	Yellow speckle, line pattern, or symptomless	53

15-4-1 Properties in Tissue Extract

Both PNRSV and ApMV can be transmitted from woody to herbaceous hosts by mechanical inoculation,[2-8] but like most ilarviruses they are unstable in tissue extracts. Early attempts to determine the general properties of PNRSV in extracts gave results that varied greatly with the isolates and methods used.[5,31,72-79] Essentially all infectivity was lost from extracts after 9 to 18 hours.[29] This loss of infectivity occurred more rapidly in undiluted than in diluted extracts.[5] However, addition of 0.1 M sodium diethyldithiocarbamate (NaDIECA), a polyphenol oxidase inhibitor, dramatically reduced this rapid loss of infectivity.[80] Using chemically stabilized extracts, Waterworth and Fulton[81] reported thermal inactivation points (10 minutes) for 23 PNRSV isolates to be between 55 and 62°C. Infectivity of ApMV can be stabilized by addition of 0.02 M 2-mercapto-ethanol.[9] In stabilized extracts, the thermal inactivation point for ApMV is reported to be 54°C.[8]

15-4-2 Purification

Both PNRSV and ApMV can be purified readily from infected cucumber cotyledons triturated in cold buffer consisting of 0.02 M phosphate, pH 8.0, containing 0.02 M 2-mercapto-ethanol and NaDIECA.[7,8] After low speed centrifugation at 10,000× g, the supernatant liquid can be clarified by mixing with 0.9 volumes of hydrated calcium phosphate[7,8] or by adjusting the pH to 4.5 for 30 minutes.[82] Following a second low-speed centrifugation, the virus can be sedimented by one or two cycles of differential centrifugation for 3 hours at $78,000 \times$ g. Further purification is achieved by rate zonal sucrose density gradient centrifugation.

15-4-3 Properties of Purified Virus

Particle shape. In early reports, particles of PNRSV and ApMV were described as isometric with diameters of 23 nm[7] or 26 nm,[8] respectively. However, recent electron micrographs indicate that only a portion of the particles associated with either virus is isometric.[3] Some particles appear quasi-isometric to short bacilliform (axial ratios ranging between 1.01 and 1.5).[83,84] With some PNRSV isolates, as many as 15% of the particles appear distinctly bacilliform (axial ratios greater than 2.2). Bacilliform particles up to 70 nm long have been observed.[84] The occurrence and proportion of bacilliform particles depend on the isolate tested.[84]

Sedimentation and electrophoretic properties. Three nucleoprotein components of PNRSV and ApMV have been resolved by rate zonal ultracentrifugation[83–86] and by agarose gel electrophoresis.[83,84] Top and middle centrifugal components were not infectious when tested individually,[84,85] whereas the bottom component was slightly infectious.

Top, middle and bottom centrifugal components were found to be equivalent to the fast, middle and slow components in agarose electrophoresis gels.[84] The electrophoretic mobility patterns on 2.5% agarose gels were characteristic of the isolates used.[84]

Viral genome. Like most ilarviruses, both PNRSV and ApMV have four molecules of positive-sense, single stranded RNA (RNAs 1 to 4) of M_r between 0.3 and 1.3 $\times$ 10^6 (Table 15-4). By analogy with other ilarviruses, the three largest RNAs are presumed to carry all the genetic information of the virus, and the nucleotide sequence contained on RNA 3 which specifies the coat protein is also carried on RNA 4, the so-called subgenomic RNA.[2]

While RNAs 1, 2, and 3 carry all the genetic information of the virus, these RNAs require the presence of RNA 4 or a small amount of coat protein for infectivity to occur. Coat proteins from other ilarviruses will activate the RNAs 1 to 3 of both PNRSV and ApMV.[86] PNRSV-RNA 4 efficiently activated the rose mosaic isolate of ApMV-RNA 1 + 2 + 3, but ApMV-RNA 4 had little ability to activate PNRSV-RNA 1 + 2 + 3.[86] RNA 4 preparations from alfalfa mosaic virus and citrus leaf rugose virus also activated ApMV-RNA 1 + 2 + 3.[86]

TABLE 15-4: SIZE OF RNAS AND COAT PROTEIN FOR PRUNUS NECROTIC RINGSPOT AND APPLE MOSAIC VIRUSES.

Virus	RNA				Ref.	Coat	Ref.
	1	2	3	4		Protein	
		($\times 10^6$)				($\times 10^3$)	
PNRSV	1.3	0.9	0.7	0.3	85	25	87
ApMV	1.2	1.0	0.7	0.3	2	25	86

15-5 DISEASE CYCLE

While numerous serotypes, strains, and biotypes have been reported for both PNRSV and ApMV,[3,7–9,73,75,84] the greatest diversity among isolates occurs within PNRSV. Despite this diversity, however, no PNRSV isolates have been reported to occur naturally in Malus or Pyrus species, and the ApMV isolates that have been found naturally in Prunus species occur, so far, only in almond[50] and plum[57] (Table 15-3).

PNRSV isolates distinct enough to be considered strains are those that incite necrotic ringspot disease of both sour and sweet cherry, recurrent ringspot disease of sour cherry, rugose mosaic disease of sweet cherry, calico disease of almond, and Stecklenberger disease of sour cherry.[25] Within each strain are many biological variants that induce symptoms which range from barely detectable to very severe.

In most Prunus species, initial symptoms of PNRSV appear early in the spring following infection. They begin as a shock reaction which consists of chlorotic or necrotic (or both) rings.[3,15,16,20,22,25,82] Centers of the necrotic rings later fall out, giving a shot-hole effect. With some isolates this effect is severe, leaving only skeletonized leaves or lace leaves. In plums, some strains cause chlorotic bands or oak-leaf patterns, while others cause bright yellow mosaic or calico.[25] Leaf development may be delayed in newly infected trees. In peach, color streaking may occur in flower petals.[88] Some PNRSV isolates induce various abnormalities in pollen.[89–96]

Shock symptoms may occur on only a few twigs or branches or may occur throughout a tree during the initial year of infection. During this period, virus can be transmitted readily from symptomatic tissues to herbaceous indicator plants by mechanical inoculation and can be detected by various serological procedures.

Typically, shock symptoms fade as the season progresses. While shock symptoms may reoccur in some hosts for years after initial infection,[25] in general they seldom appear on the same twig or limbs in subsequent years. Fruit yields and quality may be severely affected in the shock year.

Chronic symptoms in subsequent years are highly dependent on the host, strain, and environmental factors.[25] Symptoms range from leaf chlorosis and deformity to stunting, limb and trunk cankers, gumming, and terminal dieback. However, trees seldom, if ever, die as a direct result of PNRSV.[25] With some strains, chronically infected trees may exhibit few, if any, diagnostic symptoms. In sweet cherry orchards in Washington state, many PNRSV-infected trees may exhibit no symptoms on leaves or fruit.[75]

In many Prunus species, PNRSV is carried on pollen[25,36,97–106] and can infect seeds,[107–116] even seeds produced on healthy plants.[111] Seed transmission and propagation of infected budwood account for the worldwide distribution of PNRSV.[25] The role of pollen in disease epidemiology will be discussed later.

In apple, ApMV induces mosaic diseases of various intensity, depending on the strain.[73] While fruit yields can be reduced from slightly to as much as 55%[39] the disease is of little economic concern in most apple-growing areas because severely diseased trees are frequently removed and mildly diseased trees appear unaffected.

In rose and hop, the diseases caused by PNRSV and ApMV cannot be satisfactorily resolved without supplemental information from biological or serological assays.[46,47,72,74]

15-6 INDEXING FOR VIRUS

Confirmation that a plant is infected by either PNRSV or ApMV frequently requires more than just visual observation. Peach seedlings[15] and Montmorency sour cherry[24] were the first woody indicator plants used to detect PNRSV in symptomless trees.[25] However, the discovery in 1945 that Shirofugen flowering cherry (*P. serrulata* Lindl.) would develop necrotic reactions around PNRSV or PDV-infected buds that were inserted into the limbs of rapidly growing trees provided researchers with an important indexing tool.[117–119] Briefly, one or more buds or bark pieces from candidate plants are budded onto a vigorously growing branch at intervals of approximately 5 cm. When large Shirofugen trees are used, numerous candidate trees can be tested simultaneously on the same tree anytime for early spring until late fall. Bark tissues from PNRSV or PDV-infected trees induce a band of necrosis around the inserted bark patch which is easily visible within 30 to 40 days.[118] The necrosis continues to spread and can ultimately girdle the limb.[118] The test is rapid, accurate, and can be used to test large numbers of trees throughout most of the growing season. However, the test does not distinguish between PNRSV and PDV.[25]

Seedlings of *P. tomentosa* Thunb. have been used effectively as indicators for PNRSV,[120,121] especially under greenhouse growing conditions where some isolates of PNRSV and PDV can be distinguished by symptomatology.[122] PNRSV isolates induce necrotic spots, rings, and arcs on leaves shortly after budbreak followed by collapse and death of some growing points.[122] By contrast, PDV induced chlorotic mottle, rings, and arcs with only traces of necrotic spots. This latter virus is frequently encountered when indexing cherry and peach trees and seedlings.

Mechanical inoculation to differential herbaceous hosts can be used to detect most isolates of PNRSV and ApMV under suitable growing conditions.[3,25,123] Transfer of virus from leaves or petals is most successful in early spring but becomes less reliable as the season progresses.[25] While direct inoculations from woody hosts are usually made first to cucumber,[3,25,36,72,123,124] some PNRSV isolates from sweet cherry infect only *Chenopodium quinoa* on initial transfer.[125]

The most rapid method for detection and identification of Prunus necrotic ringspot viruses is with virus specific antisera.[3] However, the serological method one might choose is often determined by the objectives of the test. Antigens of PNRSV or ApMV strains in crude or clarified sap of herbaceous plants give visible precipitin lines in agar gel double diffusion tests with rabbit polyclonal antisera[3,9,38,43,46,47,49,52,81,126–128] but not with some monoclonal antibodies.[129] Fusion of reaction lines between two isolates or the formation of spur reactions of varying intensity can provide information on serological identity or degrees of relatedness to an extent not possible by other methods. The initial serological differences used to distinguish between PNRSV and ApMV were determined by this technique.[9] Later, Casper[128] reported one isolate that reacted strongly with antisera to both viruses and postulated the existence in rose of a continuous spectrum of virus isolates extending from PNRSV to ApMV. So-called intermediate serotypes have also been reported in hops.[48] Recently, Mink et al.[130] isolated three distinct serotypes of PNRSV from three different sweet cherry trees, each exhibiting symptoms typical of cherry rugose mosaic disease. These results illustrate how complex the serological relationships among Prunus necrotic ringspot viruses are.

While agar gel double diffusion tests can provide useful information on strain relationships and serotypes, the test has many limitations, especially for use in routine detection of virus. The test has been used to detect virus in Prunus tissues,[3,127] but results are often inconsistent.[3] In recent years, enzyme-linked immunosorbent assay (ELISA) has become the most common method for routine detection of Prunus necrotic ringspot viruses.[48,75,82,114,115,131–145] Although there are some modifications of the basic method in use,[142] the procedures are essentially those described by Clark and Adams.[146] Large-scale surveys have been done using ELISA.[75,82,135,137,140]

For special purposes, PNRSV can be detected by immunosorbent electron microscopy.[147,148]

15-7 EPIDEMIOLOGY

15-7-1 Transmission

Both PNRSV and ApMV are transmitted by tissue grafting,[3,25] and both are perpetuated by vegetative propagation of infected tissues.[3,25,39,47,72,149] Production of plants from virus-contaminated nursery stock was the principal method by which both viruses were distributed prior to the development of virus certification programs.[3,25,47,72,74,149] This continues to be the principal method for distributing these viruses in rose.[150]

PNRSV is readily transmitted through seed of some Prunus species[3,25,151] and, in some cases, is found even in seed produced on healthy trees.[110,111,152] The percentage of infected seed varies with the species but may be as high as 70%.[3,115] Seedlings arising from infected seed may not express symptoms and therefore are not likely to be rogued from nursery stock.[3] ApMV is apparently not transmitted through seed of apple.[3]

While PNRSV-infected seed can present problems for those commercial nurseries that intend to produce virus-certified plants, virus-infected seed produced on wild-growing woody or herbaceous plants appear to play no significant role in the epidemiology of the virus in commercial crops.[151,153–155]

Plant-to-plant spread by means other than root grafting has been reported only for PNRSV and then only in orchards of certain stone fruits.[3,25] Despite an intensive effort,[156] no conventional insect vector has been found to be associated with natural spread. Although one gall mite species[157] and one species of nematode[158] have been reported as vectors of PNRSV, their role in natural spread of the virus has not been confirmed.

15-7-2 Role of Pollen

Spread of PNRSV from tree to tree in Prunus orchards has been shown to be closely associated with pollen.[97,100,101,159–166] Early evidence for the role of pollen in virus spread included the fact that sour cherry trees which were deflowered before bloom each year remained virus free for several years,[101,161] as did trees kept under cage during bloom.[97,100,101] More convincing was the fact that at least 4 of 14 sour cherry trees that were emasculated and subsequently hand pollinated with pollen from virus-infected trees ultimately became infected.[97]

15-7-3 Source of Inoculum

External sources of virus for sour cherry orchards in the midwestern and eastern U.S. appear to be mainly pollen from PNRSV-infected trees located in adjacent orchards.[159,160,167-169] Apparent spread of PNRSV over a distance of 800 yards has been reported.[159] Intraorchard inoculum sources include infected trees obtained form noncertified sources,[75] an occasional certified tree that may have been inadvertently propagated on a virus-infected seedling,[160] and the initial trees infected from pollen originating outside the orchard.[160]

Although pollinating insects appear to play a major role in spreading PNRSV among and within orchards, little is known about the mechanism by which trees become infected. In the western U.S., many commercial beekeepers located in Washington and Oregon move hives of honeybees to California for the winter.[170] In early spring, the bees are used to pollinate various stone fruits before returning to the Pacific Northwest. Circumstantial evidence suggests that these bees provide a means for transporting PNRSV and PDV over long distances to initiate infections in Washington sweet cherry orchards.[170]

15-7-4 Orchard Spread

The rate of spread within Prunus orchards depends on several factors: the species or even cultivar involved,[101] tree age,[159] and the initial level of inoculum.[159,160] In Oregon, the rate of PNRSV spread in Montmorency sour cherry orchards was faster than in orchards of Italian prune but slower than in orchards of Brooks prune.[101]

In both sweet and sour cherry orchards, PNRSV spreads slowly when the trees are young. Spread accelerates as the trees mature and bloom more profusely.[159,160,162,168,171]

Most workers report no apparent pattern of spread within orchards,[101,159,160,162,165,167,168] except most new infections occur in trees adjacent to an infected tree.[159,160,162,165,171]

15-8 MANAGEMENT

15-8-1 General

Diseases of perennial plants that are caused by Prunus necrotic ringspot viruses can be controlled, at least in part, by efforts applied in two different ways. First, virus certification programs can provide growers with virus-free selections of important cultivars.[172] Such trees not only eliminate the detrimental effects of viruses for many years but they also greatly reduce potential inoculum sources for subsequent spread. Second, orchard management practices can prevent, eradicate, or contain spread within orchards.

15-8-2 Certification Programs

Use of virus-free planting stock has long been considered the most important control measure for fruit tree diseases caused by PNRSV.[25] The principle is simple. Select virus-free clones of important cultivars, propagate them, and plant them in orchards isolated from virus-infected trees.[173] However, for many of the most important cultivars of stone fruits and roses, it has been difficult, if not impossible, to find virus-free individuals.[150,174] Consequently, considerable time, effort, and money has been expended over the past 25 years to produce virus-free individuals of the various stone fruits.[173,174] Methods used include heat therapy and use of antiviral compounds.

Heat therapy.   The most successful procedure developed thus far to rid plants of the Prunus necrotic ringspot viruses involves selection and propagation of tissues after long exposures to high air temperatures.[150,173–181] Use of heat treatment alone or in combination with meristem culture techniques has provided virus-free individuals of most stone fruit, hop, and rose cultivars. Exposures to hot water treatments have not been successful.[182]

Antiviral compounds.   Relatively few chemicals have any demonstrable effect on viruses systemically distributed in perennial plants.[171] However, in recent years a few compounds have shown some promise against PNRSV and ApMV in limited tests.[183–186] At the moment, the most promising compound appears to be 2,4 dioxohexahydro-1,3,5-triazine (DHT) which, when injected under the cortex of sweet cherry trees, produced buds that appeared to be free of PNRSV and PDV when indexed on Shirofugen indicator plants. However, it remains to be established whether buds selected from DHT-treated trees will in turn produce trees that remain free of the virus.

15-8-3 Orchard Management Practices

Preventative measures.   There are several steps that growers can take to establish and maintain virus-free orchards.

1. Plant only virus-free trees.[173]
2. Locate virus-free plantings as far as possible from virus-infected trees—preferably one-half mile or more.[25]
3. Avoid interplanting with virus-infected trees.[173]
4. Use virus-free propagation wood for any topworking.[173]
5. If commercially available honeybees are used to aid pollination, avoid hives that were placed in other stone fruit orchards during the preceeding two weeks.[170]

Eradication.   The prompt removal of diseased plants can reduce and, in some cases, eliminate secondary spread of some viruses. For that reason, roguing of diseased trees is recommended to control spread of PNRSV in stone fruit orchards.[25] However, in both sweet and sour cherry orchards the results of eradication efforts have been erratic.[82] As mentioned earlier, tree-to-tree spread in cherry orchards has been associated with virus carried on pollen. Infections that occur on single branches, often high in the tree, are seldom noticed in the year that they occur. In subsequent years, as the virus becomes systemically distributed, flowers that bear virus-contaminated pollen are produced several days or weeks before disease symptoms are obvious. Consequently, contaminated pollen is available for at least one season before diseased trees can be marked for removal. The result is the continued appearance of new diseased trees in spite of rigorous eradication efforts.[171]

Procedures such as ELISA that could identify PNRSV-diseased cherry trees before they flower should improve the chances for controlling spread by tree removal. This was attempted for rugose mosaic disease of sweet cherry trees in Washington.[75,82] While PNRSV could be reliably detected in dormant cherry tissues during the winter months in adequate time for tree removal before bloom, the discovery that symptomless biotypes of the virus were widespread in commercial orchards made the serological tests ineffective for most orchards.[75]

Containment by cross-protection.   Early workers reported that infection by some mild variants (or strains) of PNRSV and ApMV could protect plants against the effects of more severe variants.[73,187–190] However, cross-protection as a method for controlling either virus received very little attention until recently. Howell and Mink[76] reported that symptomless variants of PNRSV could restrict field spread of cherry rugose mosaic disease in sweet cherry and therefore offered the potential to control a disease that tree removal had failed to control. Although tests currently in progress demonstrate that not all symptomless variants will protect against all disease-causing variants, one symptomless variant has protected graft-inoculated trees for five years.[191] Under orchard conditions, naturally occurring symptomless strains spread rapidly to healthy trees, but because they cause no economic effects the infected trees are not removed by growers. In orchards where cherry rugose mosaic disease occurs, disease spread ceases when surrounding healthy trees are infected with symptomless, protective isolates.[171] At this time, containment by use of protective strains appears to offer a practical method to control cherry rugose mosaic disease.

15-9 CONCLUSION

Prunus necrotic ringspot virus includes a diverse and very complex group of interrelated virus isolates. In spite of advances in technology since these viruses were first recognized some 50 years ago, much confusion still exists as to what constitutes a virus, a strain, or a biotype. This confusion is due in part to the great variability that exists among isolates in such intrinsic properties as biological activity, antigenicity, chemical composition, and even particle structure. The confusion that exists has not been helped by the fact that many isolates in this group exist in low concentrations in plant tissues and are difficult to handle *in vitro*. A large body of literature has accumulated over the years on the diseases induced by viruses in this group. However, in comparison to other virus groups, relatively little work has been done on either the comparative structure and composition of the different Prunus necrotic ringspot viruses or on their gene expression.

Collectively, Prunus necrotic ringspot viruses cause economically important diseases in a wide range of perennial plants; mostly woody, rosaceous species. Experimentally, these viruses can be transmitted to a wide range of herbaceous plants. However, perhaps because of a lack of conventional vectors involved in field spread, viruses in this group have not been reported to cause economic problems in herbaceous plants other than hops. An association between stone fruit pollen and spread of PNRSV has been demonstrated in many different ways, yet almost nothing is known about the mechanism by which the virus moves from pollen to flower and ultimately into and throughout the tree. The fact that male hop plants become infected naturally suggests that means other than pollen fertilization of an ovule may be involved in transmission.[2]

Serological techniques such as ELISA can be used on a clinical scale to detect infected plants rapidly and accurately. However, in crops such as sweet cherry, virus detection alone is not always sufficient to aid control efforts. The widespread occurrence of symptomless isolates of PNRSV that are of no economic concern to growers and that may actually be beneficial by protecting trees against disease-causing strains creates a challenging situation for virologists. To be useful for controlling PNRSV-induced diseases, new rapid detection techniques are needed that will not only detect viruses accurately but also recognize their biological variability. Thus far, serological techniques have not provided reliable information on the disease-causing potential of isolates. Most biological techniques currently available are either too slow or too cumbersome for routine clinical use.

As yet, no techniques exist to cure field-grown plants infected with any of the Prunus necrotic ringspot viruses. Consequently, primary control efforts involve virus-certified plants combined with various management practices to reduce primary and secondary spread. However, in the case of sweet cherry orchards, cherry rugose mosaic disease induced by PNRSV has been found to spread more rapidly in virus-free orchards than in orchards established before certified trees were available.[75] This observation and other recent information provides evidence that cross-protection by symptomless strains may be useful for containing disease spread under certain conditions. However, because of the biological variability among PNRSV isolates, deliberate distribution of any PNRSV isolate into orchard plantings for cross-protection has its inherent dangers. Isolates that induce mild or no symptoms in cherry trees may cause significant diseases if introduced into nearby hop fields.

So far as I am aware, no effort has yet been made to transform important cherry varieties with the coat protein gene of PNRSV. However, recent success in expressing alfalfa mosaic virus coat protein gene in tobacco[192] suggests that the concept is viable. Once the technical problems associated with the use of woody plants have been solved, transgenic plants that express one or more virus coat protein genes and exhibit protection against a spectrum of disease-causing isolates will have a definite place in cherry orchards of the future. Similar situations may well exist in other perennial crops that could make development of transgenic plants valuable not only academically but also commercially.

15-10 REFERENCES

1. Mink, G. I. and Uyeda, I., Ilarviruses: Suggested revision of subgroups, *Proc. Int. Conf. on Comp. Virol., Banff, Canada,* 262,1982.

2. Francki, R. I. B., Milne, R. G. and Hatta, T., *Atlas of plant viruses*, Vol. II, pp. 81–93, CRC Press, Boca Raton, Florida, 1985.

3. Fulton, R. W., Ilarviruses. In *Handbook of plant virus infections and comparative diagnosis* (E. Kurstak, ed.), pp. 387–413, Elsevier/North Holland Biomedical Press, 1981.

4. Fulton, R. W., Comparative host ranges of certain mechanically transmitted viruses of Prunus, *Phytopathology*, 47, 215, 1957.

5. Fulton, R. W., Properties of certain mechanically transmitted viruses of *Prunus, Phytopathology*, 47,683, 1957.

6. Fulton, R. W., Mechanical transmission of Prunus viruses to cherry, *Phytopathology*, 47, 12, 1957.

7. Fulton, R. W., Prunus necrotic ringspot virus, *CMI/ABB Descriptions of Plant Viruses*, No. 5, 1970.

8. Fulton, R. W., Apple mosaic virus, *CMI/ABB Descriptions of Plant Viruses*, No. 83, 1972.

9. Fulton, R. W., Serology of viruses causing cherry necrotic ringspot, plum line pattern, rose mosaic and apple mosaic, *Phytopathology*, 58, 635, 1968.

10. Valleau, W. D., A virus disease of plum and peach, *Ky. Agr. Exp. Sta. Bull.*, 327, 89, 1932.

11. Bradford, F. C. and Joley L., Infectious variegation in the apple, *J. Agr. Res.*, 46,901, 1933.

12. Thomas, H. E., Apple mosaic, *Hilgardia*, 10, 581, 1937.

13. Thomas, H. E. and Rawlins, T. E., Some mosaic diseases of Prunus species, *Hilgardia*, 12, 623, 1939.

14. Thomas, H. E. and Massey, L. M., Mosaic diseases of the rose in California, *Hilgardia,* 12, 647, 1939.

15. Cochran, L. C. and Hutchins, L. M., A severe ring-spot virosis on peach, *Phytopathology*, 31, 860, 1941.

16. Cation, D., Necrotic spot. A peach disease transmissible by budding, *Phytopathology*, 32, 2, 1942.

17. Zeller, S. M., Virus diseases of stone fruits, *Oregon State Hort. Soc. Annu. Rep.*, 34, 85, 1942.

18. Moore, J. D. and Keitt, G. W., Host range studies of necrotic ringspot and yellows of sour cherry, *Phytopathology*, 34, 1009, 1944.

19. Willison, R. S. and Berkeley, J. H., Tatter leaf of sweet cherry, *Phytopathology*, 36, 73, 1946.

20. Baumann, Gisela, Die Stecklenberger Krankheit, eine bisher nicht Beobahtete Viruskrankheit der Sauerkirsch, *Tijdschr. Plantenziekten*, 62, 51, 1946.

21. Schmidt, H. E., Die Isolierung eines mechanisch übertragbaren virus von ringfleckenkranken Hofen (*Humulus lupulus* L.), *Phytopathol. Z.*, 47, 192, 1963.

22. Allen, W. R., Comparison of strains of Prunus necrotic ringspot and prune dwarf viruses with viruses of Pfeffinger (rasp leaf) type, *Phytopathology*, 53, 1436, 1963.

23. Seneviratne, S. D. de S. and Posnette, A. F., Identification of viruses isolated from plum trees affected by decline, line pattern and ringspot diseases, *Ann. Appl. Biol.*, 65, 115, 1970.

24. Berkeley, G. H., Cherry yellows and necrotic ring spot of sour cherry in Ontario, I. The value of *Prunus persica* and *Prunus domestica* var. Italian prune as index hosts, *Phytopathology*, 37, 2, 1947.

25. Nyland, G., R. M. Gilmer and Moore, J. D., "Prunus" ringspot group. In *Virus diseases and noninfectious disorders of stone fruits in North America*, U.S. Dept. of Agr., Agr. Handbook 437, pp. 104–132, U.S. Government Printing Office, Washington, DC, 1976.

26. Fulton, R. W., Purification of sour cherry necrotic ringspot and prune dwarf viruses, *Virology*, 9, 522, 1959.

27. Willison, R. S., Berkeley, G. H. and Hilderbrand, E. M., Tatter leaf. In *Virus diseases and other disorders with viruslike symptoms of stone fruits in North America*, USDA Handb. 10, pp. 141–146, U.S. Government Printing Office, Washington, DC, 1951.

28. Gilmer, R. M., Nyland, G. and Moore, J. D., Prune dwarf. In *Virus diseases and noninfectious disorders of stone fruit in North America*, U.S. Dept. Agr., Agr. Handbook 437, pp. 179–190, U.S. Government Printing Office, Washington DC, 1976.

29. Fulton, R. W., Relationships among ringspot viruses of Prunus, *Dtsch. Acad. Landweilschaflwiss Berl. Tagungsber*, 97, 123, 1968.

30. Moore, J. D., Boyle, J. S. and Keitt, G. W., Mechanical transmissions of a virus disease to cucumber from sour cherry, *Science*, 108, 623, 1948.

31. Willison, R. S. and Weintraub, M., Studies on stone-fruit viruses in cucurbit hosts, I. A method of evaluating the infectivity of infectious juice, *Phytopathology*, 43, 175, 1953.

32. Boyle, J. S., Moore, J. D. and Keitt, G. W., Cucumber as a plant host in stone fruit virus research, *Phytopathology*, 44, 303, 1954.

33. Varney, E. H. and Moore, J. D., Tobacco and zinnia—two new herbaceous hosts for Prunus virus, *Phytopathology*, 44, 509, 1954.

34. Heinis, J. L., Correlation of ring spot virus reactions on stone fruits and cucumber, *Phytopathology*, 46, 163, 1956.

35. Milbrath, J. A., Squash as differential host for strains of stone fruit ring-spot viruses, *Phytopathology*, 46, 638, 1956.

36. Ehlers, L. G. and Moore, J. D., Mechanical transmission of certain stone fruit viruses from Prunus pollen, *Phytopathology*, 47, 519, 1957.

37. Fulton, R. W., Identity of and relationships among certain sour cherry viruses mechanically transmitted to Prunus species, *Virology*, 6, 499, 1958.

38. Fulton, R. W. and Hamilton, R. I., Serological differentiation of prune dwarf and sour cherry necrotic ringspot viruses, *Phytopathology*, 50, 635, 1960.

39. McCrum, R. C., Barrat, J. G., Hilborn, M. R. and Rich, A. E., Apple virus diseases, *Maine Agr. Exp. Sta. Bull.*, 595, 7, 1960.

40. McWhorter, F. P., Further report on rose mosaic in Oregon, *Plant Dis. Rep.*, 54, 1, 1931.

41. Cochran, L. C., Infection of apple and rose with the ring-spot virus, *Phytopathology*, 40, 964, 1950.

42. Fulton, R. W., Mechanical transmission and properties of rose mosaic virus, *Phytopathology*, 42, 413, 1952.

43. Fulton, R. W., Purification and serology of rose mosaic virus, *Phytopathology*, 57, 1197, 1967.

44. De Sequeira, A. O., Purification and serology of an apple mosaic virus, *Virology*, 31, 314, 1966.

45. Fridlund, P. R., A latent virus of hop detected by cucumber inoculation, *Plant Dis. Rep.*, 43, 594, 1959.

46. Bock, K. R., Arabis mosaic virus and Prunus necrotic ringspot viruses in hop (*Humulus lupulus* L.), *Ann. Appl. Biol.*, 57, 131, 1966.

47. Bock, K. R., Strains of Prunus necrotic ringspot virus in hop (*Humulus lupulus* L.), *Ann. Appl. Biol.*, 59, 437, 1967.

48. Barbara, D. J., Clark, M. F., Thresh, J. M. and Casper, R., Rapid detection and serotyping of Prunus necrotic ringspot virus in perennial crops by enzyme-linked immunosorbent assay, *Ann. Appl. Biol.*, 90, 395, 1978.

49. Smith, D. R. and Skotland, C. B., Host range and serology of Prunus necrotic ringspot virus serotypes isolated from hops (*Humulus lupulus*) in Washington, *Plant Dis.*, 70, 1019, 1986.

50. Lansac, M., Detienne, G., Bernhard, R. and Dunez, J., Analysis of nine isolates of almond mosaic, *Acta Phytopathologica Acad. Sci. Hungar.*, 15, 359, 1980.

51. Tirro, A. and Catara, A., Risultati di saggi immunoenzimatici (ELISA) per le virosi di una collezione di cultivar di mandorio, *Attii Giornate Fitopath.* Suppl., 77, 1982.

52. Gotlieb, A. R. and Berbee, J. G., Line pattern of birch caused by apple mosaic virus, *Phytopathology*, 63, 1470, 1973.

53. Baumann, G., Casper, R. and Converse, R. H., The occurrence of apple mosaic virus in red and black raspberry and in blackberry cultivars, *Acta Horticulturae*, 129, 13, 1982.

54. Barba, M. and Quacquarelli, A., Due ilar virus associati al mosaico del nocciolo, *Ann. 1st Sper. Patol. Veg. Roma* 9, 97, 1984.

55. Postman, J. D. and H. R. Cameron, Apple mosaic virus in U.S. filbert germplasm, *Phytopathology*, 71, 944, 1987.

56. Sweet, J. B. and Barbara, D. J., A yellow mosaic disease of horse chestnut (*Aesculus* spp) caused by apple mosaic virus, *Ann. Appl. Biol.*, 92, 235, 1979.

57. Gilmer, R. M., Probable coidentity of Shiro line pattern virus and apple mosaic virus, *Phytopathology*, 46, 127, 1956.

58. Milbrath, J. A., Latent ringspot virus of cherries reduces growth of nursery trees, *Plant Dis. Rep.*, 34, 374, 1950.

59. Pine, T. S., Growth reduction of peach trees graft-inoculated with different components of the stone fruit ringspot virus complex, *Phytopathology*, 52, 747, 1962.

60. Millikan, D. F., The influence of infection by ringspot virus on growth of one-year-old Montmorency nursery trees, *Phytopathology*, 45, 565, 1955.

61. Lewis, F. H., The effect of ring spot and yellows on yield of Montmorency cherry, *Phytopathology*, 41, 24 (Abst.), 1951.

62. Posnette, A. F. and Cropley, R., Virus diseases of cherry trees in England, II. Growth suppression caused by some viruses, *J. Hort. Sci.*, 31, 298, 1956.

63. Milbrath, J. A., Effect of some sour cherry viruses on growth of young orchard trees, *Phytopathology*, 47, 655, 1957.

64. Parker, K. G., Brase, K. D., Schmid, G., Barksdale, T. H. and Allen, W. R., Influence of ringspot virus on growth and yield of sour cherry, *Plant Dis. Rep.*, 43, 380, 1962.

65. Pine, T. S., Influence of necrotic ringspot virus on growth and yield of peach trees, *Phytopathology*, 54, 604, 1964.

66. Posnette, A. F., Cropley, R. and Swait, A. A., The incidence of virus disease in English sweet cherry orchards and their effect on yield, *Ann. Appl. Biol.*, 61, 351, 1968.

67. Kirkpatrick, H. C., Lindner, R. C., Cheney, P. W. and Graham, S. O., Rose as a source of Prunus ringspot virus, *Plant Dis. Rep.*, 46, 722, 1962.

68. Schmelzer, K., Zur Atiologie des Rosengelbmosaiks. Zentralblatt fur Bacteriologie, Parasilenkunde, Infectionskrankheiten und Hygiene, *Naturwisseuschaft*, 120, 233, 1966.

69. Traylor, J. A., Williams, H. E. and Nyland, G., Symptoms caused by strains of Prunus necrotic ringspot virus in rose resemble typical rose mosaic, *Phytopathology*, 56, 152, 1966.

70. Basit, A. A. and Francki, R. I. B., Some properties of rose mosaic virus from South Australia, *Aust. J. Biol. Sci.*, 23, 1197, 1970.

71. Bos, L., Symptom expression and variation of rose mosaic, *Netherlands J. Plant Pathol.*, 82, 239, 1976.

72. Thomas B. J., Studies on rose mosaic disease in field-grown roses produced in the United Kingdom, *Ann. Appl. Biol.*, 98, 419, 1981.

73. Posnette, A. F. and Cropley, R., Apple mosaic virus. Host reactions and strain interference, *J. Hort. Sci.*, 31, 119, 1956.

74. Thomas, B. J., The effect of Prunus necrotic ringspot virus on field grown roses, *Ann. Appl. Biol.*, 100, 129, 1982.

75. Mink, G. I. and Aichele, M. D., Use of enzyme-linked immunosorbent assay results in efforts to control orchard spread of cherry rugose mosaic disease in Washington, *Plant Dis.*, 68, 207, 1984.

76. Howell, W. E. and Mink, G. I., Control of natural spread of cherry rugose mosaic disease by a symptomless strain of Prunus necrotic ringspot virus, *Phytopathology*, 74, 1139, 1984 (Abst).

77. Willison, R. S. and Weintraub, M., Studies on stone-fruit viruses in cucurbit hosts, II. Some factors affecting the aging of inoculum in vitro, *Phytopathology*, 43, 324, 1953.

78. Willison, R. S. and Weintraub, M., Studies on stone-fruit viruses in cucurbit hosts, IV. Some effects of hydrogen ion concentration and dilution on infectivity, *Phytopathology* 44, 533, 1954.

79. Weintraub, M. and Willison, R. S., Studies on stone-fruit viruses in cucurbit hosts, V. Thermal inactivation of some isolates, *Phytopathology*, 44, 538, 1954.

80. Hampton, R. E. and Fulton, R. W., The relation of polyphenol oxidase to instability in vitro of prune dwarf and sour cherry necrotic ring-spot viruses, *Virology*, 13, 44, 1961.

81. Waterworth, H. E. and Fulton, R. W., Variation among isolates of necrotic ringspot and prune dwarf viruses isolated from sour cherry, *Phytopathology*, 54, 1155, 1964.

82. Mink, G. I., Identification of rugose mosaic diseased cherry trees by enzyme-linked immunosorbent assay, *Plant Dis.*, 64, 691, 1980.

83. Ong, C-A. and Mink, G. I., Biophysical characterization of two *Prunus* necrotic ringspot virus isolates from sweet cherry trees, *Phytopathology*, 75, 1352, 1985.

84. Ong, C-A., *Separation and characterization of nucleoprotein components of Prunus necrotic ringspot virus isolates,* Ph.D. thesis, Washington State Univ., Pullman, WA, 118 pp., 1987.

85. Loesch, L. S. and Fulton, R. W., *Prunus* necrotic ringspot virus as a multicomponent system, *Virology*, 68, 71, 1975.

86. Gonsalves, D. and Fulton, R. W., Activation of Prunus necrotic ringspot virus and rose mosaic virus by RNA 4 components of some ilarviruses, *Virology*, 81, 398, 1977.

87. Barnett, D. W. and Fulton, R. W., Some chemical properties of Prunus necrotic ringspot and Tulare apple mosaic viruses, *Virology*, 39, 556, 1969.

88. Cochran, L. C. and McClain, R. F., Breaking patterns produced on peach flowers by ringspot virus, *Phytopathology*, 41, 936, 1951.

89. Way, R. D. and Gilmer, R. M., Reductions in fruit set on cherry trees pollinated with pollen from trees with sour cherry yellows, *Phytopathology*, 53, 399, 1963.

90. Marenaud, C. and Desvignes, J. C., Effets de divers virus sur le pollen de prunus persica GF305, *C. R. Acad. Agric.*, 51, 782, 1965.

91. Marenaud, C. and Desvignes, J. C., Incidence de quelques varoses d'arbres fruitiers a noyau sur la qualite du pollen des pehers inocules, *Zastit bilga br.*, 16, 87, 1965.

92. Basak, W., Influence of Prunus necrotic ringspot virus on pollen of sour cherry, *Polska Akad. Nauk. Bull. Ser. Sci. Biol.*, 14, 797, 1966.

93. Nyeki, J. and Vertesy, J., Effects of different ringspot viruses on the physiological and morphological properties of Montmorency sour cherry pollen, *Acta Phytopathology*, 9, 23, 1974.

94. Marenaud, C. and Saunier, R., Action des virus du type ilar sur la pollen de l'espece Prunus persica, *Ann. Amelior. Plantes*, 24, 169, 1974.

95. Nicolaescu, M., Acutinea unor virusuri asupra capacitatii germinative a polenului de mar, *An. Inst. Cer. Pentru. Prot. Plant* 10, 31, 1974.

96. Schimanski, H. and Schade, C., Nachweis des nekrotischen ringflecken-virus der kirsche im pollen der vogel-kirsche (Prunus avium L.) und des chlorotischen ringflecken-virus der kirche im pollen der steinweichsel (Prunus mahaleb L.), *Arch. Phytopathology Pflanzenschutz*, 10, 3, 1974.

97. George, J. A. and Davidson, T. R., Pollen transmission of necrotic ring spot and sour cherry yellows viruses from tree to tree, *Can. J. Plant Sci.*, 43, 276, 1963.

98. Way, R. D. and Gilmer, R. M., Pollen transmission of necrotic ring spot virus in cherry, *Plant Dis. Rep.*, 42, 1222, 1958.

99. Gilmer, R. M. and Way, R. D., Pollen transmission of necrotic ring spot and prune dwarf viruses in cherry, *Tidsskr. Planteavl.*, 65, 111, 1961.

100. George, J. A. and Davidson, T. R., Further evidence of pollen transmission of necrotic ringspot and sour cherry yellows viruses in sour cherry, *Can. J. Plant Sci.*, 44, 384, 1964.

101. Cameron, R. R., Milbrath, J. A. and Tate, L. A., Pollen transmission of Prunus necrotic ringspot in prune and sour cherry orchards, *Plant Dis. Rep.*, 57, 241, 1973.

102. Kishi, K., Takanashi, K. and Abiko, K., Studies on virus diseases of stone fruits, III. Pollen transmission of peach necrotic ringspot virus and prune dwarf virus in peach trees and results of several related experiments, *Bull. Hort. Res. Sta. Hiratsuka*, 12, 185, 1973.

103. Cole, Anabel, Mink, G. I. and Regev, S., Location of Prunus necrotic ringspot virus on pollen grains from infected almond and cherry trees, *Phytopathology*, 72, 1542, 1982.

104. Hamilton, R. I., Nichols, C. and Valentine, B., Survey for Prunus necrotic ringspot and other viruses contaminating the exine of pollen collected by bees, *Can. J. Plant Pathol.*, 6, 196, 1984.

105. Meyer, S., Casper, R. and Bünemann, G., Localization of ringspot viruses in plum pollen in immunofluorescence, *Gartenbauwissenschaft*, 51, 125, 1986.

106. Kelly, R. D. and Cameron, H. R., Location of prune dwarf virus and Prunus necrotic ringspot virus in sweet cherry pollen and fruit, *Phytopathology*, 76, 791, 1986.

107. Cochran, L. D., Passage of the ringspot virus through peach seeds, *Phytopathology*, 40, 964, 1950.

108. Nyland, G., Separation of sour cherry yellows and ringspot viruses by passage through the seeds of Stockton Morello, *Phytopathology*, 42, 517, 1952.

109. Gilmer, R. M., Imported mahaleb seeds as carriers of necrotic ring spot virus, *Plant Dis. Rep.*, 39, 727, 1955.

110. Taylor, J. A., Williams, H. E., Weinberger, J. H. and Wagnon, H. K., Studies on the passage of Prunus ringspot virus complex through plum seed, *Phytopathology*, 53, 1143, 1963.

111. Fleisher, Z., Blodgett, E. C. and Aichele, M. D., Presence of virus (necrotic ringspot group) in mazzard and mahaleb cherry seedlings grown in the Pacific Northwest from various seed sources, *Plant Dis. Rep.*, 48, 280, 1964.

112. Fridlund, P. R., Transmission of latent viruses in commercial peach seed, *Plant Dis. Rep.*, 50, 740, 1966.

113. Williams, H. E., Jones, R. W., Traylor, J. A. and Wagnon, H. K., Passage of necrotic ringspot virus through almond seed, *Plant Dis. Rep.*, 54, 822, 1970.

114. Casper, R., Testing of Prunus avium seed for prune dwarf virus with the ELISA procedure, *Phytopathol. Z.*, 90, 91, 1978.

115. Mink, G. I. and Aichele, M. D., Detection of Prunus necrotic ringspot virus and prune dwarf virus in Prunus seed and seedlings by enzyme-linked immunosorbent assay, *Plant Dis.*, 68, 378, 1984.

116. Schimanski, H. H., Fuchs, E. and Kegler, H., Seed transmission of Prunus necrotic ringspot virus in blackthorn, *Zentrablatt fur microbiologie*, 139, 213, 1984.

117. Milbrath, J. A. and Zeller, S. M., Latent viruses in stone fruits, *Science*, 101, 114, 1945.

118. Milbrath, J. A., Selecting stone fruit trees free from virus diseases, *Oregon Agr. Exp. Sta. Bull.*, 522, 27, 1952.

119. Helton, A. W., Relative merits of shirofugen and peach trees as indicators for Prunus ringspot virus in prune trees, *Phytopathology*, 52, 846, 1962.

120. Fridlund, P. R., The reaction of Prunus tomentosa seedlings to latent Prunus virus cultures, *Phytopathology*, 44, 488, 1954.

121. Fink, H. C., Prunus tomentosa as an index plant for sour cherry yellows viruses, *Phytopathology*, 45, 320, 1955.

122. Fridlund, P. R., Symptoms of necrotic ringspot and prune dwarf in several *Prunus* species *Plant Dis. Rep.*, 49, 288, 1965.

123. Barba, M., Detection of apple mosaic and Prunus necrotic ringspot viruses in almond seeds by ELISA, *Arch. Phytopathology Pflanzenschutz*, 22, 279, 1986.

124. Cropley, R., Gilmer, R. M. and Posnette, H. F., Necrotic ringspot and prune dwarf viruses in *Prunus* and in herbaceous indicators, *Ann. Appl. Biol.*, 53, 325, 1964.

125. Mink, G. I., unpublished data, 1987.

126. Tremaine, J. H. and Willison, R. S., Estimation of the size of a stone-fruit virus antigen by the gel precipitation technique, *Can. J. Bot.*, 39, 1843, 1961.

127. Van Regenmortel, M. H. V. and Engelbrecht, D. J., The rapid diagnosis of necrotic ringspot virus infection of stone fruits by serological means, *South African J. Agr. Sci.*, 5, 607, 1962.

128. Casper, R., Serological properties of Prunus necrotic ringspot and apple mosaic virus isolates from rose, *Phytopathology*, 63, 238, 1973.

129. Halk, E. L., Tsu, H. T., Aebig, J. and Franke, J., Production of monoclonal antibodies against three ilarviruses and alfalfa mosaic virus and their use in serotyping, *Phytopathology*, 74, 367, 1984.

130. Mink, G. I., Howell, W. E., Cole, A. and Regev, S., Three serotypes of Prunus necrotic ringspot virus isolated from rugose mosaic-diseased sweet cherry trees in Washington, *Plant Dis.*, 71, 91, 1987.

131. Clark, M. F., Adams, A. N., Thresh, J. M. and Casper, R., The detection of plum pox and other viruses in woody plants by enzyme-linked immunosorbent assay (ELISA), *Acta Horticulturae*, 67, 51, 1976.

132. Thresh, J. M., Adams, A. N., Barbara, D. J. and Clark, M. F., The detection of three viruses of hop (*Humulus lupulus*) by enzyme-linked immunosorbent assay, *Ann. Appl. Biol.*, 87, 57, 1977.

133. Casper, R., Testung von Prunus avium-samen auf prune dwarf virus mit den ELISA-verfahren, *Phytopathol. Z.*, 90, 91, 1977.

134. Barbara, D. J., Detecting Prunus necrotic ringspot virus in rosaceous host by enzyme-linked immunosorbent assay, *Acta Phytopathology Acad. Sci. Hung.*, 15, 329, 1980.

135. Hardcastle, T. and Gotleib, A. R., An enzyme-linked immunosorbent assay for the detection of apple mosaic virus in yellow birch, Betula alleghaniensis, *Can. J. Forest Res.*, 10, 278, 1980.

136. Kaniewski, W. and Skotland, C. B., Detection and identification of hop viruses by enzyme-

linked immunosorbent assay (ELISA), *Probl. Pflanzenvirol.*, 19, 447, 1980.

137. Thomas, B. J., The detection of serological methods of viruses infecting the rose, *Ann. Appl. Biol.*, 94, 91, 1980.

138. Torrance, L., Use of forced buds to extend the period of serological testing in surveys for fruit tree viruses, *Plant Pathol.*, 30, 213, 1981.

139. McMorran and Cameron, H. R., Detection of 41 isolates of necrotic ringspot, apple mosaic and prune dwarf viruses in *Prunus* and *Malus* by enzyme-linked immunosorbent assay, *Plant Dis.*, 67, 536, 1983.

140. Torrance, L. and Dolby, C. A., Sampling conditions for reliable routine detection by enzyme-linked immunosorbent assay of three ilarviruses in fruit trees, *Ann. Appl. Biol.*, 104, 267, 1984.

141. Jordan, R. L., Aebig, J. and Hsu, H. T., Epitope specificity of seven monoclonal antibodies to apple mosaic virus and Prunus necrotic ringspot virus, *Phytopathology*, 74, 808 (Abstr.), 1984.

142. Edwards, M. L. and Cooper, J. I., Plant virus detection using a new form of indirect ELISA, *J. Virology Methods*, 11, 309, 1985.

143. Jordan, R. and Aebig, J., Evaluating the specific immunoreactivities of monoclonal antibodies to two plant viruses, *Acta Horticulturae*, 164, 185, 1985.

144. Meyer, S., Casper, R. and Buenemann, G., Detection of viruses in stone fruit trees by the indicator *Prunus avium* F-12-1 and ELISA, *Gartenbauwissenschaft*, 51, 179, 1986.

145. Stein, A., Levy, S. and Loebenstein, G., Detection of Prunus necrotic ringspot virus in several rosaceous hosts by enzyme-linked immunosorbent assay, *Plant Pathol.*, 36, 1, 1987.

146. Clark, M. F. and Adams, A. N., Characteristics of the microplate method of enzyme-linked immunosorbent assay for the detection of plant viruses, *J. Gen. Virology,* 34, 475, 1977.

147. Begtrup, J., Jorgensen, K. and Thomsen, A., Prunus necrotic ringspot virus identified with immunosorbent electron microscopy directly from the trees, *Tidsskr. Planteavl*, 89, 81, 1985.

148. Kalaschjan, J. A. and Werderewskaja, T. D., Elektronenmikroskopische untersuchungen von virender nekrotischen ringfleckenkrankheit in infirzerten gewebe der kirsche, *Taf.-Ber. Akad. Landwirtsch,-Wiss. DDR*, Berlin, 184, 61, 1980.

149. Thomas, B. J., Epidemiology of three viruses infecting the rose in the United Kingdom, *Ann. Appl. Biol.*, 105, 213, 1984.

150. Manners, M. M., The rose mosaic heat therapy program at Florida Southern College, *Proc. Fla. State Hort. Soc.*, 98, 344, 1985.

151. Schimanski, H. H., Schmelzer, K., Kegler, H. and Albercht, H. J., Wildwachsende Prunus-arten der untergattungen prunophora nad padus als naturliche wirstspflanzen for kirschenringflecken-viren, *Zb. Bakt.*, 130, 109, 1975.

152. George, J. A., A technique for detecting virus-infected Montmorency cherry seeds, *Can. J. Plant Sci.*, 42, 198, 1962.

153. Schimanski, H. H., Kegler, H. and Fuchs, E., Wild-growing woody rosaceous plants as natural reservoirs of pome and stone fruit viruses, *Zentralbl. Mikrobiol.*, 142, 239, 1987.

154. Demski, J. W. and Boyle, J. S., Absence of necrotic ringspot virus of sour cherry in weed hosts, *Phytopathology*, 58, 125, 1968.

155. Sweet, J. B., Hedgerow hawthorn (*Crataegus* spp.) and blackthorn (*Prunus spinosa*) as host of fruit tree viruses in Britain, *Ann. Appl. Biol.*, 94, 83, 1980.

156. Swenson, K. G. and Milbrath, J. A., Insect and mite transmission tests with Prunus necrotic ringspot virus, *Phytopathology*, 54, 399, 1964.

157. Proeseler, G., Ubertragungsversuche mit dem latenten prunus-virus und der gallmilbe vasateo fockeui nal, *Phytopathol. Z.*, 63, 1, 1968.

158. Fritzsche, R. and Kegler, H., Nematoden als vektoren von viruskrankheiten der obstgewachse, *Dtsch. Akad. Landwirtsch. Wiss. Berlin, Tagunksber*, 97, 289, 1968.

159. Davidson, T. R. and George, J. A., Spread of necrotic ringspot and sour cherry yellows viruses in Niagara peninsula orchards, *Can. J. Plant Sci.*, 44, 471, 1964.

160. Demski, J. W. and Boyle, J. S., Spread of necrotic ringspot virus in a sour cherry orchard, *Plant Dis. Rep.*, 52, 972, 1968.

161. Davidson, T. R., Field spread of Prunus necrotic ringspot in sour cherry orchards in Ontario, *Plant Dis. Rep.*, 60, 1080, 1976.

162. Marenaud, C. and Llager, G., Etude de la diffusion de virus de type ILAR (taches annulaires nécrotiques) dans un verger de cerisier (*Prunus avium*), *Ann. Amelior. Plantes*, 26, 357, 1976.

163. Gergenova, T., Spreading of Prunus necrotic ringspot virus and prune dwarf virus in a cherry orchard, *Gradinar Lozar Nauka Hortic. Vitic. Sci.*, 16, 60, 1979.

164. Gella, R. F., Research note on the diffusion of ILAR virus in a collection of varieties of *Prunus domestica* L., *Acta Phytopathologica Acad. Sci. Hungar.*, 15, 351, 1980.

165. Gerginova, T., Incidence of Prunus necrotic ringspot virus (PNRV) and prune dwarf virus in cherry orchards, *Acta Phytopathologica Acad. Sci. Hungar.*, 15, 223, 1980.

166. Pocsai, E., Szalay, K. and Mers, F., Study of Prunus necrotic and Prunus chlorotic ringspot virus by using serological methods in fruit-bearing stone orchards, *Novenyvedelem*, 16, 123, 1980.

167. Klos, E. J. and Parker, K. G., Yields of sour cherry affected by ring spot and yellows viruses, *Phytopathology*, 50, 412, 1960.

168. Willison, R. S., Berkeley, G. and Chambelain, G., Yellows and necrotic ringspot of sour cherries in Ontario, distribution and spread, *Phytopathology*, 38, 776, 1948.

169. Gilmer, R. M., The frequency of necrotic ringspot, sour cherry yellows and green ring mottle viruses in naturally infected sweet and sour cherry orchard trees, *Plant Dis. Rep.*, 45, 612.

170. Mink, G. I., The possible role of honeybees in long distance spread of *Prunus* ringspot virus from California into Washington sweet cherry orchards. In *Plant virus epidemiology*, (R. T. Plumb and J. M. Thresh, eds.), pp. 85–91, Blackwell Scientific Publications, Oxford, 1983.

171. Howell, W. E. and Mink, G. I. Natural spread of cherry rugose mosaic disease and two Prunus necrotic ringspot virus biotypes in a central Washington sweet cherry orchard, *Plant Dis.*, 72, 636, 1988.

172. Mink, G. I., Control of plant diseases using disease-free stocks. In *Handbook of pest management in agriculture*, Vol. 1 (D. Pimentel, ed.), pp. 327–346, CRC Press, 1981.

173. Welsh, M. F., Control of stone fruit virus disease. In *Virus diseases and non-infectious disorders of stone fruits in North America*, Agric. Handbook, No. 437, pp. 10–15, U.S. Dept. of Agr., 1976.

174. Fridlund, P. R., IR-2, the interregional deciduous tree fruit repository. In *Virus diseases and non-infectious disorders of stone fruits in North America*, Agric. Handbook, No. 437, pp. 16–22, U.S. Dept. of Agr., 1976.

175. Nyland, G., Heat inactivation of stone fruit ringspot virus, *Phytopathology*, 50, 380, 1960.

176. Adams, A. N., Elimination of viruses from hop (*Humulus lupulus*) by heat therapy and meristem culture, *J. Hort. Sci.*, 50, 151, 1975.

177. Bjarnason, E. N., Hanger, B.C., Moran, J. R. and Cooper, J. A., Production of Prunus

necrotic ringspot virus-free roses by heat treatment and tissue culture, *New Zealand J. Agr. Res.*, 28, 151, 1985.

178. Patrakosal, P., Detection and heat therapy of *Prunus* necrotic ringspot virus and prune dwarf virus in peach trees, *Memoirs of the Tokyo Univ. of Agr.*, 27, 36, 1985.

179. Lenz, F., Baumann, G. and Kornkamhaeng, P., High temperature treatment of *Prunus avium* L. 'F12/1' for virus elimination, *Phytopath. Z.*, 106, 373, 1983.

180. Kornkamhaeng, P., *Eliminierung des viruskomplexes prunus necrotic ringspot virus-prune dwarf in Prunus avium L. mit hilfe von hitzebehandlung und meristemkulture*, Diss., Bonn, 1983.

181. Mosella, Ch. L., Signoret, P. A. and Jonard, R., Sur la mise au point de techniques de micrograffage d'apex lu vue de l'elimination de deux types de particules virales chez le pecher (*Prunus persica* Batsch), *Pomptes Rendes Hebdomadaires des seances de l'academie des sciences*, 290, 287, 1980.

182. Fridlund, P. R., Failure of hot water treatment to eliminate Prunus viruses, *Plant Dis. Rep.*, 55, 738, 1971.

183. Cheplick, S. M. and Agris, G. N., Effect of injected antiviral compounds on apple mosaic, scar skin and dapple apple diseases of apple trees, *Plant Dis.*, 67, 1130, 1983.

184. Hansen, A. J., Effect of ribavirin on green ring mottle causal agent and necrotic ringspot virus in Prunus species, *Plant Dis.*, 68, 216.

185. Schuster, G. and Arenhovel, C., On the mode of action of the antiphytoviral compound 2,4-dioxohexahydro-1,3,5-triazine (-5 azadihydroouracil), *Intervirol.*, 21, 134, 1984.

186. Bogusch, L. J., Verderevskaja, T. D. and Schuster, Freeing sweet and sour cherry trees from Prunus necrotic ringspot virus by injection of 2, 4 dioxahexahydro-1,3,5-triazine (DHT), *Phytopath. Z.*, 114, 189, 1985.

187. Cochran, L. C., Interference between forms of the ring-spot virus in peach trees, *Phytopathology*, 42, 512, 1952.

188. Willison, R. S. and Berkeley, G. H., An experiment in cross protection with some stone fruit viruses, *Proc. Can. Phytopath. Soc.*, 19, 20, 1952.

189. Willison, R. S., Cross-protection and interference with some stone fruit viruses, *Can. J. Bot.*, 40, 1041, 1962.

190. Moore, J. D., and Slack, D. A., Interaction of strains of necrotic ring spot virus, *Phytopathology*, 42, 470, 1952.

191. Howell, W. E. and Mink, G. I., unpublished results.

192. Van Dun, C. M. P., Bal, J. F. and Van Vloten-Doting, L., Expression of alfalfa mosaic virus and tobacco rattle coat protein genes in transgenic plants, *Virology*, 159, 299, 1987.

16

MANGO MALFORMATION

J. KUMAR

Hill Campus of G. B. Pant University of Agriculture & Technology
RANICHAURI-249 199, Distt. Tehri, U.P., India

S. P. S. BENIWAL

International Center for Agricultural Research in the Dry Areas,
Douyet Research Station, P.B. 2335, FES, Morocco

16-1 INTRODUCTION

A number of stresses, both pathological and physiological, have been recognized in mango in different agroclimatic zones which have marred its productivity to various levels. Mango malformation, an age-old disease, is one of the most threatening diseases of recent times primarily because of its destructive and widespread nature and because of persistent lacuna in complete understanding of its nature.

16-2 DISTRIBUTION

Although the disease was recognized as early as 1891 by Maries (Watt, 1891)[1] in India in Bihar, some attention was devoted to it only in mid-1950s when it started assuming serious proportions in other parts of the country also. Burns,[2] one of the early workers, made a note of it in 1910, and stated "vegetative activity at the time of flowering overpowers the reproduction and as a result these inflorescences bear leaf buds instead of normal flowers." Again in their book on mango, Burns and Paryag[3] mentioned the disease, but no work was taken up, as it was sporadic and was not then regarded as serious and damaging. In subsequent years, the disease was recorded in other parts of India with as many causes assigned to it. This was primarily because symptoms of the disease have often been misleading and different causes have been assigned to it by workers of different disciplines and fields. Nevertheless, based on the symptoms reported from India and other countries where the disease is known to occur, it is safe to assume a common etiologic agent(s) which has been exploited differently by workers of different interests.

Presently, the disease is known to occur in most of the mango-growing areas of the world (e.g., Pakistan,[4] the Middle East,[5] South Africa,[6] Brazil,[7] Central America, Israel, Mexico, and the U.S.A.[8]). The Indian mango industry is badly hit by the malady, as more than 50% of trees are affected, causing heavy losses in yield, particularly in northern parts of India.[9]

16-3 SYMPTOMS

Mango malformation comprises two stages (1) vegetative malformation (Fig. 16-1), and (2) floral malformation[10] (Fig. 16-2).

16-3-1 Vegetative Malformation

Vegetative malformation is more pronounced on young seedlings but may also appear on mature trees. Young seedlings show symptoms when about five months old. In the initial stages of symptom development, vegetative buds in the axils of leaves or at the apex of seedlings swell and produce small shootlets bearing small, scaly leaves. In such seedlings, apical dominance is lost and as a result numerous vegetative buds sprout, producing hypertrophoid growth which constitutes vegetative malformation. More than one vegetative bud at one point gets activated, and as a result small shootlets arise bearing very small rudimentary leaves at short internodes. These leaves get crowded so that shootlets and their branches are not distinguishable, and the whole mass of rudimentary leaves gives a bunch-like appearance. Such symptoms, when present at the apex of the seedling, are referred to as bunchy-top (BT) stage. In some cases, thick shootlets arise from the swollen axillary buds which ultimately have secondary branches that elongate further and bear small rudimentary leaves at the internodes. Collectively, the whole structure looks like a witch's broom. In other cases, the vegetative buds get activated throughout the length of internodes and produce small, scaly leaves. The BT stage is not necessarily formed in such cases (Fig. 16-3).

The seedlings infected at the apex at an early stage do not recover but remain stunted and ultimately dry out. However, those infected at a later stage continue to grow normally and resist stunting and drying. In cases when the infection is localized as a bunch in the axil of the leaf, the seedling grows normally after the bunch is detached. In other cases, when infection is systemic, the seedlings continue to produce symptoms at different sites even after removal of the diseased tissues.[11]

In trees also, axillary buds get swollen and produce small shootlets bearing numerous scaly leaves giving a bunchy appearance. Such a bunch may be present at the apex of the growing shoot crowded with normal leaves, on shoot tips bearing malformed panicles, along with malformed floral buds, or all along the length of the internode. Such compact bunches dry up after some time and remain attached to the shoots as dry masses.

In certain cases, these dry masses again produce malformed growth during the next growing season. In trees when most of the branches are vegetatively malformed, no

Figure 16-1   A young mango seedling showing symptoms of vegetative malformation.

inflorescence appears at the terminal buds of the shoots, or flowering and fruiting is severely reduced.[11]

16-3-2 Floral Malformation

The disease makes its appearance with the emergence of inflorescences on the shoots. Normally, the healthy panicles emerge from terminal buds and bear hermaphrodite and male flowers which vary from 6 to 8 mm in diameter. Contrarily, the individual flowers in the diseased panicles are greatly enlarged, which constitutes a malformed inflorescence. At maturity, such panicles appear hypertrophoid, though diameter of the main and sec-

Figure 16-2   A 10-year-old severely infected mango tree showing floral malformation symptoms.

ondary axis does not always increase as compared with a healthy inflorescence. However, branching in the inflorescence is more crowded, flowering appears in cymose fashion, as usual, but individual flowers get hypertrophoid, giving the panicle a malformed appearance. Such panicles are heavier and generally much greener as compared to healthy ones (Fig. 16-4).

Malformed panicles showing severe infection produce far greater numbers of flowers as compared to healthy ones, although most flowers remain unopened. The ovaries in such flowers are either nonfunctional or, the fruits if borne, are shed off before attaining the pea stage.

There are several reports of decrease in sex-ratio of flowers in the malformed panicles,[12-19] however detailed observations on the unopened buds were probably not made in the past. The unopened malformed buds are not necessarily sterile, as the percentage of hermaphrodite buds in some cases was found to be equal or even greater than that of the healthy buds. Whether the ovaries in such flowers are functional or not remains to be determined. Even in cases where the sex-ratio is quite high and all the flowers buds in severely malformed panicles get opened during the season, very few fruits are set which drop off at mustard stage only.[11]

Depending on the cultivar, degree of hypertrophy and severity of symptoms, great variation in symptoms is observed in a malformed panicle (Fig. 16-5). They range from the small and compact type to the loose type that assumes the shape of a witch's broom. It is usual to observe both healthy and malformed inflorescences on the same tree. Sometimes a shoot tip may bear two panicles, one of which remains healthy and the other becomes malformed. Less frequently, a healthy panicle may have one or more branches

Figure 16-3 Vegetative malformation: (a) a young mango seedling showing bunchy-top symptoms; (b) a vegetatively malformed seedling showing sprouting of vegetative buds at various sites of internodes; (c) bunchy-top systems on the branch of a tree; (d) transformation of malformed panicles to vegetatively malformed apices.

or sometimes few flowers as malformed. Likewise, healthy branches may also be present in a malformed panicle. Such partially infected panicles bear fruits which may even attain maturity. In cases when only few flowers are malformed, the affected panicle is similar in length and diameter to the healthy panicle. At times it becomes difficult to differentiate between the two.

Some of the panicles that are initially malformed may recover as the inflorescences reach maturity and bear fruits, few of which even mature. Malformed inflorescences continue to bear flowers even after fruit setting in the normal panicles. Such inflorescences remain green for quite some time and hang up as dry masses until the next growing season.

Figure 16-4 Floral malformation: (a) A 10-year-old mango tree showing healthy (H) and malformed (D) panicles; (b) malformed (left) and healthy (right) panicles.

16-3-3 Relationship between Vegetative and Floral Malformation

Both vegetative and floral malformation have been suspected to be symptoms of same disease, since hypertrophy of tissues is involved in both cases, and vegetative malformation appears at times on tree bearing malformed inflorescences.[14,15] However, this had not been experimentally proved, because the specific cause of the disease was unknown and, thus, inoculation tests were not applicable. Therefore, an experimental proof[10] was achieved through veneer-grafting in mango seedlings by grafting of the diseased scions from trees onto healthy rootstocks (Table 16-1). After establishment of the union, the diseased scion that would have produced a malformed inflorescence in on-year (flowering year) produced symptoms typical of vegetative malformation (Fig. 16-6), and plants derived from such grafts (diseased scion on healthy rootstock) produced malformed panicles. This provided the first experimental proof that both vegetative and floral malformation are two different expressions of the same disease, the former commonly appearing on young seedlings and the later on flowering trees. Further, the results indicate one reason for the rapid spread of disease into new areas.

Figure 16-5 Floral malformation: (a) a malformed panicle showing the formation of numerous small heads due to clustering of hypertrophoid flowers; (b) a partially infected panicle showing healthy (H) and malformed (D) flowers. Note the size of malformed flowers.

16-4 ETIOLOGY

All along, etiology of the disease has been controversial. Different workers engaged in establishing its cause have opined freely and differently. The result was that a number of different causes have been assigned to it.

Earlier, the disease was considered (i) physiological in nature, because of symptomatology and physiological changes occurring inside the diseased tissues,[2,16,17] (ii) viral because of the resemblance of symptoms with the leaf hopper-transmitted viral diseases, association of some insects with it, and also because of the failure to isolate any pathogenic agent from diseased tissues,[13,18–20] and (iii), toxemia resulting from the feeding of eriophid mites.[21–23] But contradictions, though meaningful, are no less that refuted all the

TABLE 16-1: RESULTS OF VENEER-GRAFTING EXPERIMENT ON MANGO SEEDLINGS

Stock	Scion	No. seedlings grafted	No. successful grafts	No. grafted scions showing BT symptoms[a]	Infection rate (%)
Healthy	Healthy[b]	50	39	0	0
Healthy	Diseased[c]	50	39	30	76.8

[a]BT=bunchy top symptom
[b]Shoot tips that had healthy inflorescence in the previous year
[c]Shoot tips that had malformed inflorescence in the previous year
(From Kumar and Beniwal, 1987)[10].

Figure 16-6 Development of symptoms of vegetative malformation on diseased scions when grafted onto healthy rootstock; (a) diseased scion producing bunchy-top growth; (b) normal shoot arising out of bunchy-top growth.

aforementioned possibilities in one way or the other. Lately, there has been much emphasis on the role of *Fusarium* spp., that is, *F. moniliforme* Sheld. and *F. oxysporum* Schlecht., in the causation of the disease,[24–26] though validity of one or the other has been doubted because of the unreproducibility of the results at different places in producing both the types of symptoms by one or the other *Fusarium* sp.[11,27,28] Nevertheless, considerable reports have now been generated on the association of *F. moniliforme* f. sp. *subglutinans* with both the stages of the disease, though it is yet to receive acceptability at many scientific forums.

It would, therefore, be worthwhile to analyze the available information and arrive at some conclusion in light of it.

16-4-1 Causal Agent: Infectious or Noninfectious

Spread.   The annual recurrence of the disease in new seedlings and a rather slow increase in infection from year to year has been reported by several workers.[13,29–31] In one observation, it was recorded that growing seedlings for one to two years in an area where this abnormality was widespread did not result in a higher average incidence than about 10%.[30] A similar gradual increase from tree to tree within a variety and between varieties has also been recorded.[13,32,33]

In one study,[11] the incidence of vegetative malformation in the same mango nursery revealed that the plots showing maximum infection during first year also showed maximum disease incidence in the second year (Fig. 16-7). The increase in infection was gradual and not sharp. The causal factor, therefore, appears to be localized and spreads slowly. Similarly, observations on the incidence of floral malformation in young Dashehri trees (8 to 10-years-old) revealed that affected trees were localized in a particular zone (Zone I; Fig. 16-8), which showed maximum disease index with no healthy trees in this zone during three years of observation. Zone II, which was demarcated as a zone of low

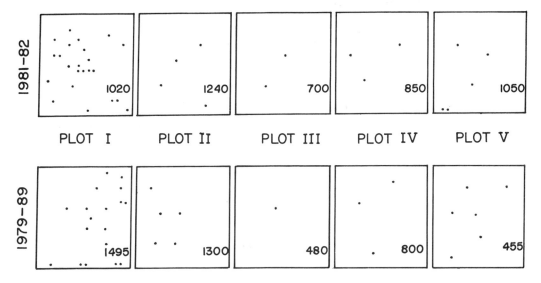

Figure 16-7 Incidence of vegetative malformation at the same mango nursery during a 2-year period (figure in plots indicates approximate number of seedlings).

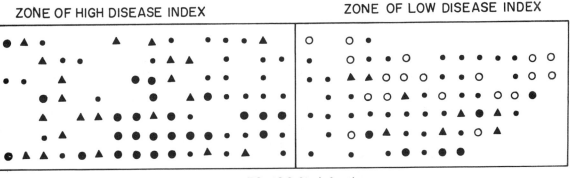

Figure 16-8 Incidence of floral malformation on the same mango orchard during 1980–1983. Diseased trees are clustered rather than randomly distributed.

disease index during all three years, comprised 30% of healthy trees. Spread of disease was quite slow, as only one new infection was recorded during three years, but disease severity (% malformed panicles) on individual trees varied in subsequent years. Also, no directional trend in the incidence or spread of malformation was found. Thus, the causal factor appears to be localized (soil borne?) and spreads slowly, as no geometric increase of disease spread from tree to tree was recorded with the increase of inoculum potential.

Graft transmission

Grafting diseased scions over healthy rootstocks.   From the grafting experiment discussed earlier (Table 16-1), two very important conclusions could be inferred from the results besides the evidence of relationship between BT and malformed-inflorescence (MI) symptoms.

1. The causal agent is localized at the tip of the scion (shoot tip) that ramifies in the new growth produced by the scion once vascular connection gets established with the rootstock.
2. The cells of the apical bud of the shoot tip are conditioned to produce malformed growth, and this condition continues along with the ramification of the causal agent in the new growth. Therefore, there is no dilution effect which would have been there had the causal agent been nonpathogenic.

Grafting healthy scions over diseased rootstocks.   When healthy scions (those obtained from healthy shoots) are grafted onto diseased rootstocks (those that produced BT growth), symptoms of vegetative malformation do not appear on the scion shoots.[11] However, a notable feature about such a union is the suppression of vegetative buds on the scions in those grafts where BT growth on the rootstock seedling was not excised after the establishment of the union. Even if sprouted, the new shoot remains dormant even after six months of union. When BT growth was excised, the young shoots (scions) started flourishing. This study reveals that

1. The disease involves a malformation-inducing principle (MIP) which does not get transferred into scions from rootstocks or vice versa through the vascular union.
2. There is another principle, i.e., toxic principle (TP) produced by the malformed tissue that could translocate from the rootstocks into the scions and cause suppression of growth of young tissues in scions. However, it is present only in the malformed tissue and has no existence as a sole factor responsible for the symptom productions.
3. Malformation-inducing principle (MIP) might be infectious.

Application of diseased tissue extract to healthy tissues.   When methanol, dichloromethane, or aqueous extract of malformed tissues was applied at the apex of the healthy shoots bearing floral buds, drying of three to four leaves down the application point on the shoots could be observed. In some cases, the new inflorescences that emerged remained stunted for some time and later recovered. This indicates that MIP could not be transmitted through diseased tissue extract, whereas the TP could be. Also, TIP could translocate downward to cause toxicity in the leaves below the inoculation point. The recovery of the stunted inflorescences indicates dilution of TP, as the source of toxin (TP) was not there. Since, the MIP was not transmitted, no symptoms were produced.

Symptomatology.   During early stages of symptom development, BT growth or vegetative malformation appears as luxuriant growth of young hypertrophoid tissues, which keeps on growing along with the seedling. In severe cases, when the seedling cannot withstand infection, it gradually stops growing and ultimately dries out.

Therefore, the MIP could either be some infectious (pathogenic) agent or its metabolic product. It acts at the time of vegetative or floral bud induction and conditions the cells to produce malformed growth. Second, the TP is produced inside the infected tissues as a result of infection and is able to translocate downward. It exhibits toxicity symptoms as retardation of growth of the seedling or inflorescence and, in severe cases, drying of the seedlings or destroying the floral ability of the trees. A higher amount of TP is present in vegetatively malformed as compared to floral malformed tissues.

16-4-2 Histopathology

BT-affected seedlings.   Sections of the stem of BT-affected seedlings showed cellular arrangements similar to those in sections of healthy seedlings.[11] However, differences were observed in bunchy top or malformed growth, in which cuticular thickening was more pronounced and underlying epidermal cells were radially elongated. At later stages, hypertrophy of different cells, as well as deformation of pith cells, was also observed. But no fungal hyphae or any other microorganism could be observed in the transverse or longitudinal sections of healthy or diseased tissues. Prasad et al.[20] also did not find any microorganisms in the healthy or diseased sections of vegetative shoots. Axillary buds on the diseased seedlings were enlarged and their cells much hypertrophoid. In some sections, few strands of fungal hyphae were observed at the zone of attachment of the axillary buds to the main stem. The hyphae were found to be inter- and intracellular.

Malformed inflorescence.   Increase in cell size, as well as their number, appeared to be a common phenomenon in the diseased tissues. The number of cortical and pith cells increased tremendously in the main and to a lesser extent in the secondary axis of MI. At later stages of infection, the pith cells appeared to lose their integrity and formed certain cavities. Fungal mycelium could not be located in any section of main and secondary axes, but it was present at the juncture of shoot tip and panicles where the hyphae were seen traversing the tissues. The hyphae were differentially stained from host tissues, and at some places microconidia of the fungus were also observed attached to the main hyphae. The hyphae were inter- and intracellular, and this has been observed in other studies also.[27]

Thin sections of flower buds, flowers, and mustard stage fruits from MI revealed that development of floral parts was similar to those of normal flowers except that cells of sepals, petals, disc, and ovary of malformed flowers were enlarged. The anther lobes of the fertile stamen contained pollen grains, and carpel was not degenerated in the malformed hermaphrodite flowers. Functioning of the ovary can be questioned because very few fruits are produced on malformed inflorescence, and even these do not reach beyond pea stage, as they drop prematurely. The abscission zones in the pedicel of the fruits from malformed and healthy panicles were similar, but some differences were observed in the embryo. In diseased fruits, the cells of the embryo were degenerating or were undeveloped, a situation which was not observed in healthy fruits.

Fungal hyphae were found in the axes of petals and sepals of undeveloped buds, though not frequently. Also, these were observed in the meristematic zone of flower buds. Presence of the fungus was also revealed by ultrathin sectioning and electron microscopy of BT and MI tissues. Presence of the fungal hyphae was frequently observed in sections from the malformed flower buds.

16-4-3 Cause

Bacteria. Repeated isolations from samples obtained from different areas did not yield any bacterial growth on artificial medium.[11] Also, application of antibiotics to diseased seedlings and malformed inflorescences that might have assisted in the definition of the pathogen proved ineffective.[11,34] No answer to this question could be obtained from antibiotic treatment of grafts because of the length of time needed for producing disease symptoms or remission of symptoms which may be more than persistence of compounds. Therefore, in such cases, the failure of antibiotic treatments does not always rule out the possibility of bacteria or bacteria-like organisms. However, no such organisms were noticed in extensive electron microscopy observation of ultrathin sections of BT and MI tissues.[11] In literature also, no report exists regarding any possible involvement of bacteria with the malformed tissues.

Virus and mycoplasma-like organisms (MLOs). Veneer-grafting experiments[11] using the combination diseased scions on healthy rootstocks and healthy scions on diseased rootstocks did not result in symptom production on the healthy part of either combination. Thus, failure of the agent to move from the diseased portion of the union to the healthy portion even after four years suggests the non-graft-transmissible nature of the causal agent. Similar observations have also been made by Kishtah et al.[34] Some of the early workers, however, have reported transmission of the causal agent by grafting or budding,[32,35-38] while others[13,20] did not find any encouraging results in efforts to transmit the disease from branches to seedlings by inarching, cleft-grafting, and bark patch budding. Similarly, efforts to transmit the causal agent mechanically or through dodder[11] and seed[11,39] proved futile. Seedlings grown in sterilized or unsterilized soil from the rhizosphere of diseased seedlings and trees did not show the development of any symptom.[11,39] Furthermore, heat treatment at 50 and 60°C for varying periods did not free the budwood from malformation agent.[39]

Ultrathin sectioning and electron microscopy of malformed tissues did not reveal any virus or mycoplasma-like organism.[11,39] Nonetheless, treatment of malformed trees and seedlings with oxytetracycline HCl did not prevent or inhibit malformation of vegetative or floral growth. Culturing an extract of malformed tissues did not change the red color of PPLO medium. Results were negative when such extracts were tested by ELISA with conjugated antisera of *Spiroplasma citri*, corn stunt spiroplasma, and the causal agents of peach yellows, leaf roll, and aster yellows.[34] Thus, none of the published work indicates the possibility of any virus or mycoplasma-like organism being involved in the causation of disease.

Fungus

Isolations.   Repeated isolations made over the years have yielded more than one species of *Fusarium*,[11,27,28] although *F. moniliforme* var. *subglutinans* has been found to be most frequently associated with the diseased tissues.[11,24,25,40,41] Association of *F. oxysporum* Schlecht. has also been recorded.[11,26] Other less frequently isolated *Fusarium* spp. are *F. fusaroides, F. solani,* and *F. equisiti.*[11] Based on the available information, it could be safely concluded that

1. The fungus *Fusarium* is invariably associated with the diseased tissues in a much higher proportion than in the healthy ones.
2. The presence of more than one species of *Fusarium* in the malformed tissues is not uncommon.

The presence of *Fusarium* sp. appears to be of common occurrence even in healthy trees situated in the vicinity of diseased ones, though to a much lesser extent. It suggests that accumulation of fungus occurs in the diseased tissues, and its certain level might be necessary to produce malformation symptoms. This is emphasized by the finding that healthy panicles from disease-free trees contained much less population of *Fusarium* sp. as compared to healthy panicles on diseased trees, which was in turn less than found in malformed panicles of the same trees. Also, shoots bearing malformed panicles yielded much higher amounts of *Fusarium* sp. In partially infected panicles, healthy tissues yielded the same *Fusarium* sp. as was found in the diseased portion, though in a lesser amount. Fungus, therefore, appears to be systemically distributed inside the tree, which has also been observed earlier.[25,26] Isolation from BT tissues also revealed two *Fusarium* spp. (i.e., *F. oxysporum* and *F. moniliforme* var. *subglutinans*).[11] Systematic isolation from three types of seedlings—namely, Type I (BT-affected seedlings), Type II (healthy seedlings with no apparent malformation symptom and growing next to BT-affected seedlings in the same row), and Type III (healthy seedlings with no apparent malformed growth and growing at a distance from the BT-affected seedlings)—exhibited higher Fusarial population in the BT-affected tissue, collar portion, and shoots of Type I seedlings (Fusarium Growth Index, or FGI, 245, 300, and 300, respectively), while the connecting stem revealed lower Fusarial population. This indicates that there should be a continuous uptake of *Fusarium* propagules from soil into the seedlings which multiply and/or accumulate in BT-affected seedlings. Type II seedlings which grew in close vicinity of diseased (Type I) should have a similar microbial environment to that of diseased, yet the *Fusarium* population accumulated at the top (apex 2.5 cm) was less (FGI 66.66) and was still less (FGI 16.66) in Type III seedlings. This indicates that there is a minimum level of pathogen population which causes diseased condition through the production of MIP.

It was further observed[11] using radio-labeled *Fusarium* propagules that these propagules are taken up by the damaged roots and translocated acropetally to the stem of seedlings irrespective of their diseased or healthy condition. However, no *Fusarium* propagule could enter leaves through petioles, which was confirmed by isolation from leaves of infected seedlings which did not reveal the presence of any fungus. This information yields the following generalizations:

1. *Fusarium* propagules are drawn up by the seedlings through their damaged root system when present in the zone of high *Fusarium* population.

2. Higher population of *Fusarium* is associated with infected tissues.

3. Population of *Fusarium* in the close vicinity of infected plant part is, probably, the critical factor for symptom development.

4. *Fusarium* propagules do not traverse through petioles into the leaves.

Based on the above, it may now be emphasised that the MIP, as mentioned earlier, may constitute the pathogenic cause, and the fungus comprises it. The TP is also associated with the fungus and it may be responsible for the symptoms like, retardation of tissue growth. This hypothesis explains production of symptoms like partially infected panicles, where few secondary axes or main axis at the tip remain healthy in an otherwise diseased panicle. Similarly, it also explains the situation where malformed and healthy panicles arise from the same shoot tip.

Pathogenicity tests.   The invariable association of *Fusarium* spp. with the diseased tissues reflects the fungal nature of the disease. Summanwar et al.[24] were the first to associate the fungus *F. moniliforme* Sheld. with malformed inflorescence and indicated its involvement in the causation of disease, as they could induce vegetative malformation after inoculating the aerial parts of 8 to 10-month-old seedlings. Later, their work was supported by other workers.[8,40] Varma et al.[25] identified the fungus as *F. moniliforme* var. *subglutinans* and reportedly reproduced vegetative and floral malformation by inoculating aerial parts of the plants by slit method during January and February.

In one study,[11] inoculation of the aerial parts of the seedlings in February separately by six different isolates (UDP-1, *F. oxysporum* Schlecht. isolated from malformed panicles obtained from Udaipur: UDP-2, *F. moniliforme* var. *subglutinans* isolated from malformed panicles obtained from Udaipur; PNT, *F. oxysporum* Schlecht. isolated from malformed panicles obtaining from Pantnagar; HSR, *F. moniliforme* var. *subglutinans* isolated from malformed panicles obtained from Hissar.; DLI, *F. moniliforme* var. *subglutinans* isolated from malformed tissues obtained from Delhi; and FOC, *F. oxysporum* f. sp. *ciceri* the chickpea isolate obtained form Pantnagar) did not result in symptom production even after 1 year of inoculation. Inoculations were made at three places: (1) the apex, (2) the axis of the uppermost leaf, and (3) 15 cm below the apex. The only symptoms that could mark the presence of the inoculated fungus were shortening of internodes (in some cases), burning of leaf margins, and shriveling of leaves. In seedlings where the apex was damaged due to inoculation of the fungus, three or four branches developed due to disturbance of the apical dominance. In some of these seedlings, the internodes remained shortened, and thus clustering of leaves occurred. However, these symptoms did not resemble with the typical BT symptoms. Inoculation with the chickpea isolate similarly caused leaf necrosis, which therefore appears to be a common symptom in mango seedlings as a result of inoculation of any *Fusarium* sp.

Inoculation through soil proved successful to some extent in producing some of the symptoms of vegetative malformation, if not all. At the highest dose of inoculum (i.e., 300 g per pot, 9" dia), all the six isolates proved highly toxic, as all the seedlings dried within a fortnight. At the lowest dose (i.e., 100 g per pot) of inoculum, the surviving seedlings showed burning of leaves, which marked the toxicity symptoms. In such cases, leaves were shriveled and curled. The other noteworthy symptom in such seedlings was clustering of leaves and production of more than one branch at the apex. Shortening of internodes was also observed in some cases. Nevertheless, the two seedlings each inoculated by the lowest dose of UDP-1 and UDP-2 isolates showed some of the symptoms after more than one year that at least resembled incipient BT symptoms under normal conditions. Growth of such seedlings remained arrested; instead, a few branches appeared from the apex. Such branches produced scale leaves at the apex at short internodes. Small hypertrophoid growth of vegetative tissues developed at the apex of one of such branches of a seedling inoculated with the UDP-1 isolate. Such symptoms could be considered those of vegetative malformation, although luxuriant malformed growth consisting of hypertrophoid shootlets could not be observed in any of the cases as observed under natural conditions. The probable reasons could have been arrested growth of the seedling due to toxicity of *Fusarium* sp., which could not thrive best under such a stressed condition. On reisolation, the respective *Fusarium* sp. was isolated from both the seedlings. Although the number of seedlings maintained per isolate was less (15), this does not seem to affect the validity of the experiment in any way.

Growth malfunctions as a result of pathogen interference are numerous. Plants become dwarf or giant, plant parts assume new and exotic shapes, buds that ordinarily are held in check are suddenly released, individual cells escape from organism control and grow as a cancerous mass.[42] In such diseases, excess or imbalance in plant growth substances have often been found, and the pathogen involved has been shown to be capable of producing growth-regulating substances *in vitro*. Thus, it is tempting to speculate that a direct contribution of pathogen to growth substance imbalance in the host cell is very important in the redirection of the development. The aforementioned symptoms seem to befit those of vegetative malformation. Hormonal imbalance has been found to occur in BT and MI tissues. Levels of auxins and gibberellins were found to be decreased in the diseased tissues as a result of infection, while those of inhibitors and IAA oxidase were much higher. The latter was assumed to be of fungal origin.[43] Therefore, it is very likely that the causal fungus interferes with the hormonal balance needed for symptom development.

Zinc deficiency in plants is also known to inhibit auxin synthesis and produce symptoms like small, scaly leaves on small shootlets arising from the leaf axil or at the apex of seedlings in young mango seedlings (Fig. 16-9). Dieback of branches constitutes the typical symptom found in nearly all plants. But as zinc interferes with other metabolic processes, luxuriant malformed growth do not arise from such small shootlets. Therefore, it becomes necessary to differentiate such symptoms from those produced by seedlings under artificial inoculated conditions. Moreover, it is always worthwhile to negate this possibility while experimenting in pots because fusarial toxicity might add to the Zn deficiency, since fusaric acid chelates metal ions and symptoms produced might be the result of both. Therefore, any attempt to control malformation through the application of Zn chelates can only be speculative and misleading.

Figure 16-9   Zinc deficient (right) and normal (left) mango seedling. Note resemblence of these symptoms with those of the vegetative malformation.

The method of soil inoculation with each of the six different isolates failed under field conditions.[11] Such seedlings did not show even the toxicity symptoms, probably the toxic effect was diluted in soil. Also, attempts to produce floral symptoms on the flowering trees using all the six isolates failed. Earlier, some of the workers who found the association of *F. moniliforme* with BT and MI tissues could not produce the symptoms of floral malformation in flowering trees.[23,27,28,40] However, Varma et al.[25] could produce symptoms of floral malformation by inoculating *F. moniliforme* var. *subglutinans* in some of the grafted plants.

Successful reproduction, though at a very low frequency, of vegetative malformation was, however, observed when seedlings were raised in a nursery plot already inoculated with a sufficient quantity of malformed inflorescences.[11] The symptoms were exactly similar to those observed under natural conditions. Also, the appearance of symptoms synchronized to symptoms appearing under natural conditions.

Cultural filtrates of all the six *Fusarium* isolates were similar in the respect that none produced characteristic malformation symptoms in seedlings or trees irrespective of the method of inoculation. However, the symptoms such as wilting of seedlings, shortening of internodes in some cases, and burning of leaf margins were the obvious symptoms in seedlings, while burning of leaves and stunting of emerging panicles, which recovered later, were obvious in trees. Similar symptoms were also observed when diseased tissue extract was inoculated in young seedlings and flowering trees.

In the witch's broom type of diseases, which show loss of apical dominance, numerous buds on the shoots are released. These form new shoots with new buds, and the new buds are again released. In such type of disease in cacao caused by the fungus *Marasmius perniciosus*, the fungus extracts were found particularly effective in destroying auxins and an excess production of cytokinins.[44] Symptoms production in vegetative malformation may have such a probability. Culture filtrates of *Fusarium* sp. associated with diseased tissues could induce the production and/or release of IAA oxidase which might be involved in lowering auxin concentration in the plant.[43]

At this juncture, it would be worthwhile to review the scientific progress made toward understanding the physiology of malformed plants as it is likely to influence any conclusion drawn on the etiology of disease.

16-4-4 Physiology of the Malformed Plant

Considerable physiological changes have been observed in malformed tissues as a result of infection. Leaves of BT-affected plants were found to contain more ash, silica, calcium, and water but lesser potassium.[45] Analysis of the free and bound amino acid content of healthy and malformed tissues revealed that free and bound amino acids were always present in appreciably higher amounts in stem than in leaf, irrespective of the diseased or healthy condition of plant parts.[46] The leaves from malformed shoots contained a greater number of free amino acids than those from healthy shoots, with a concomitant fall and rise in the quantity of bound amino acids in healthy and malformed leaves, respectively.

El-Ghandour et al.[47] analyzed extracts from normal and malformed shoots of susceptible and healthy cultivars and observed higher levels of total phenols and reducing sugars in healthy shoots of susceptible cultivars than in those of resistant ones. But malformed shoots of susceptible cultivars contained higher total phenols, O-dihydroxy phenols, and reducing sugars than healthy shoots of susceptible cultivars. Similarly, Pandey et al.[48] also observed an accumulation of excessive amounts of acid hydrolyzable polysaccharides. ^{14}C-sucrose translocation studies[49] have revealed that the normal and the elongated malformed panicles are larger sinks than the compact malformed panicles. Bound label recovery from the rachis of compact panicles was more, and distribution behavior of radioactivity was similar in normal and elongated malformed panicles. The water potential of malformed rachis was higher than that of the normal ones, showing high water resistance in malformed rachis. The respiratory rate of malformed panicles was found lower than that of normal panicles. Malformed tissues were found to contain decreased amounts of chlorophyll and total polysaccharides but an increased level of proteins, RNA, and DNA as compared to healthy tissues.[50-53] An increased phosphorus mobilization at the malformed region of mango was observed which correlated with increased RNA activity.[54] It was also observed[52] that the cellulose and lignin content of malformed inflorescences were 60 and 30% of the dry tissues, respectively. The fungus (*F. moniliforme* var. *subglutinans*) caused considerable differences in degradation of various components of cellulose.

An investigation on the etiology of malformation by Ghosal et al.[55] revealed that mangiferin, a normal metabolite of *Mangifera indica* L., that functions as a vegetative growth-promoting compound[56] and also in the transportation of micronutrients from bark to leaves, plays a significant role, when immobilized, in the development of malformation. Monthly measurements showed that the twigs malformed by *F. moniliforme* var. *subglutinans* contained less mangiferin than twigs of healthy plants. Ghosal et al. suggested that *F. moniliforme* var. *subglutinans* caused considerable imbalance in mangiferin concentration in mango plants by causing accumulation of mangiferin at the site of its synthesis.[57] The fungus has been found to deplete micronutrients (Zn, Cu, Fe) required for the translocation of mangiferin in the form of mangiferin-metal ion complexes.[58] When the micronutrients (e.g., Zn) are made available to the diseased organs in the form of mangiferin-Zn complex, much of the disease symptom disappeared.[58]

Significant changes in the nature and content of the phenolic and steroidal constituents of healthy and malformed florets and those artificially inoculated with the fungus *F. moniliforme* var. *subglutinans* have been reported.[59] Malformed tissues contained a mixture of zoosterols (namely, pregnenolone, progestrone, and a lanostance derivative), the quantities of which were correlated with the severity of the disorder. However, phenolic and steroidal compounds may not be considered as a primary incitant of malformation. They may induce toxicity symptoms but not typical malformation, as they may be the consequences of malformed growth.

Mangiferin, ellagic acid, gallic acid, and a galloyl derivative of glucose were isolated from malformed and normal panicles.[49] Two flavones, tetrahydroxy (Kaempferol) and pentahydroxy (quercetin), were detected in flowers of panicles. There was more turnover of phenols in flowers as compared to rachis in both types of panicles. Bound phenols were more in rachis at various stages. The level of mangiferin was lower in panicles than in the vegetative shoots and leaves, and the level in malformed panicles was not supraoptimal to cause abnormalities. Sprays of mangiferin-metal chelates were found to be ineffective in controlling malformation. The accumulation of phenolic substances was considered not as phytoalexins; rather, it was the consequence of altered metabolism.

The hormonal status of healthy and malformed tissues has been compared at various developmental stages of panicles. It has generally been observed that high levels of auxins occur in healthy as compared to malformed inflorescences, whereas levels of inhibitors show the reverse trend.[60–62] Even the shoots bearing malformed panicles maintained lower levels of IAA when compared with shoots bearing healthy panicles at all the corresponding stages of panicle development.[62] Contrary to this, higher levels of auxins in extracts of malformed tissues as compared to healthy ones have also been observed.[63] According to Rajan,[49] auxin (acidic) activity was higher in malformed buds at balloon stage, but it remained lower than the normal ones at rapid growth phase of the panicles and increased after 32 days. Nonacidic auxin, 3-indole acetontrile (IAN) was 50 times more in malformed panicles at 48 days after bud burst. Differences in auxin levels in healthy and BT-affected seedlings has also been observed.[43,64] It was found that levels of IAA and IAN were decreased in BT-affected tissues by 98.4 and 92.6%, respectively. Further, no auxins could be detected in culture filtrates of *F. oxysporum* isolated from BT tissues.[43]

Levels of gibberellins have also been compared in healthy and malformed tissues. Very low levels of gibberellins but considerable growth-inhibiting activity in malformed tissues of susceptible cultivars was observed.[47] Gibberellin activity was found to decrease at the rapid growth stage of the panicle.[49] Similarly, in BT-affected tissues either none or only low level gibberellin-like substances were detected, whereas these were easily detected in healthy tissues.[64,65] Contrarily, there are also reports on the presence of higher levels of gibberellins in the malformed panicles.[63,66,67]

An increased production of peroxidase and polyphenoloxidases in malformed mango inflorescences were observed.[68] The activity of IAA oxidase, peroxidase, and polyphenoloxidase was 350, 118, and 32% higher, respectively, in BT as compared to healthy tissues. IAA oxidase activity in BT-affected tissues could be of fungal origin.[43] The decreased level of auxins in BT-affected tissues was correlated to their oxidation by IAA oxidase, peroxidase, and polyphenol oxidases. Malformation-like symptoms were induced on mango panicles within 15 days of spray with morphactins.[41] It was found that morphactins increased IAA oxidase activity, which in turn had a negative correlation with catalase and growth parameters. It was suggested that malformation occurred due to decreased levels of auxins and hormonal imbalance.[69]

Enzymic studies[49] evidenced for the protein synthesis involved in physiological functions showed that the sequential alterations in the activity of enzymes did not appear to be typical as observed in the host during infection and progression of disease. The alteration in oxidative enzymes was due to phenolic substances that exhibited a positive correlation with enhanced enzyme activity in malformed panicles rather than host-parasite interaction. The investigation suggested nonexistence of defensive mechanisms associated with polyphenol oxidase and peroxidase activities. IAA oxidase from malformed and normal panicles showed different mobility in electrophoresis although more than one form of the enzyme within a tissue was not detected.

Qualitative and quantitative changes in cytokinins of healthy and malformed panicles at different growth periods have been assessed,[49,67] and the presence of additional cytokinins different from zeatin and iso-pentenyl adenosine (IPA) were recorded. The cytokinin levels in both healthy and malformed panicles followed a similar pattern, but cytokinin concentrations were always higher in malformed panicles than healthy ones.[67] The cytokinin complement from malformed flowers was also different from that isolated from healthy flowers; the most obvious difference being the absence of trans-zeatin (tZ), dihydrozeatin (DHZ), and ribosyldihydrozeatin (DHZR) in malformed flowers. Iso-pentenyladenine (2iP), which was not detected in healthy flowers, was present in malformed ones.[70] The studies with F. moniliforme var. subglutinans cultures showed that the fungus had the capacity to synthesize cytokinins, notably 2iP derivatives. Cultures of the fungus metabolized both tZ and 2iP rapidly but did not accumulate the applied compound or the metabolites formed in the hyphae. The authors considered the major contribution of the fungus to cytokinin production as its ability to rapid production of iso-pentenyl derivatives and/or convert tZ to such derivatives. This may reduce production of DHZ compounds necessary for normal flower development and fruit production.[71,72] In their[72] opinion, the reason for malformation could be (1) the production of 2iP by the fungus; (2) the increased production of ribosyl zeatin (ZR) and glycosyl-O-zeatin (ZOG) in malformed flowers; or (3) blocking of the production of DHZ-like compounds in malformed flowers.

Levels of ethylene are higher in malformed panicles at various developmental stages, in shoots bearing malformed panicles at corresponding developmental stages of panicles, and in leaves borne on shoots bearing malformed panicles as compared with corresponding healthy tissues.[73] Ethylene might play some role in causing malformation in suppressing apical dominance, more isodiametric growth of rachis, and the shortening and thickening of secondary branches of malformed panicles. The higher endogenous levels of ethylene were attributed to (1) malformin production in the malformed panicles, and (2) abscisic acid induced ethylene production.[73] Noticeably, high levels of abscisic acid in BT and MI tissues as compared to corresponding healthy tissues have also been observed.[74]

Presence of malformin-like substances in the malformed tissues have been recorded at various developmental stages of panicles.[75,76] These substances have also been detected in BT-affected but not in healthy mango seedlings.[77] Although the chemical nature of these substances is yet to be ascertained, one of these has been identified as malformin A, which occurs at 42.0 $\mu g/g$ in stem and 41.0 $\mu g/g$ in roots of malformed seedlings.[77]

Malformin, a cyclic pentapeptide (cyclo-D-cysteinyl-D-cysteinyl-L-valyl-D-leucyl-L-isoleucyl[78]) has been reported to be produced by *Aspergillus niger*, which causes malformation of maize roots[79] and bean plants.[80] The presence of a similar compound was detected in the culture filtrates of *F. moniliforme* var. *subglutinans*, the fungus most frequently isolated from malformed mango tissues.[81] Culture extracts of the fungus caused typical curling and stunting of maize and pea roots and formation of flap-like growth at the tip of wheat roots, but failed to induce malformation in bean plants, as observed by Curtis[80] with the culture filtrates of *A. niger*. Hence, it is yet to be confirmed that the compound from *F. moniliforme* var. *subglutinans* that caused root malformation in maize and pea was the same as the one isolated from *A. niger* causing curvature and root growth retardation of maize and malformation of stem and petioles of bean plants.[79,80,82]

The root malformation factor as observed in culture filtrates of *F. moniliforme* var. *subglutinans* was different from the growth substances, GA_3, NAA, kinetin, and reduced glutathion, as none of those produced a similar response. Similarly, no such response could be produced in the culture filtrates of other fungi (e.g., *F. oxysporum* Schlecht. *F. moniliforme* var. *intermedium* Neish & Leggett, and *A. flavus* Link. Fr.) less frequently isolated from mango malformed tissues.[11] The root malformation factor reported from malformed inflorescences and BT-affected mango seedlings may constitute malformation-inducing principle (MIP), which causes growth substance imbalance and conditioning of host cells to produce malformed growth. However, its role could not be established in mango malformation, as it failed to induce the typical malformation symptoms.[81]

Malformins are highly active plant growth regulators, and when applied to apical buds of various plants, induce malformation on stems and petioles.[80] From histological observations, malformation resulted from an apparent partial loss of polarity of the cells in rapidly growing regions of young internodes and petioles.[82] It was concluded that malformin differs from other growth substances since it affected cellular orientation. Severely malformed plants resume normal growth above malformed areas and produce flowers which do not differ in appearance from those on normal plants. Malformin inhibits cell wall synthesis and acts through an ethylene mediated system,[83] and the enhanced ethylene production might inhibit auxin transport through affected plant tissues.[84,85] In *Phaseolus vulgaris*, malformin stimulated the efflux of IAA-2-^{14}C or its metabolites, altered slightly the metabolism of IAA, but had little effect on polar transport.[86] Similar studies employing mango malformed tissues should be helpful in characterizing metabolic changes induced by malformin. While this avenue would not directly lead to the identification of the etiologic agent of the disease, it would certainly reveal the biological mechanism of action of the malformation-inducing principle.

Nevertheless, based on the available information where fungus mediated growth disturbances in vegetative and floral parts are seemingly occurring, the role of *Fusarium* sp. can be broadly ascribed through the functioning of two principles in the mango mal-formation process:

1. Malformation-inducing principles (MIP), which work through growth substance imbalance and conditioning of cells; and
2. Toxic principle (TP), which causes growth retardation and toxicity symptoms.

Conidia of the fungus are taken up by the damaged mango roots from the soil and are apoplastically transported to the growing points (shoot tips, axillary buds, etc.) where they multiply saprophytically/parasitically and gradually release the MIP for an extended period of time. MIP may condition the host cells to produce malformed growth. Once exponential growth phase of the fungus is over due to depletion of nutrients, it produces toxic secondary metabolites (TP) which are translocated in the plant and result in toxicity symptoms such as reduced growth and necrosis of malformed tissues, seedling necrosis and loss of flowering.

16-4-5 Conclusion

From the available information one may be tempted to conclude that *Fusarium* sp. partic-ularly *F. moniliforme* var. *subglutinans* is the most likely cause of mango malformation as:

1. It is most often isolated from malformed tissues where its population is much higher than the healthy tissues.
2. Its conidia are taken up by the damaged roots and are translocated to different plant parts except leaves.
3. Physiology of the fungus and/or its metabolic products may explain several physiological alterations observed in the infected tissues some of which individ-ually or collectively may constitute malformation-inducing principle (MIP). These are (i) fungus produces malformin-like compound which is also present in infected tissues and may act as MIP. Inability of the compound to produce typi-cal malformation symptoms on exogenous application does not totally rule out its involvement as a constituent of MIP. Because, effect of a growth regulating compound like malformin, when released gradually by the fungus inside the host tissue near the growing point for an extended period of time, may be quite different than the one obtained by its exogenous application. In a woody plant like mango, penetration and translocation of any organic compound is likely to be very poor particularly when applied on stem and shoot tips because of its adsorption by the chemical constituents of the plant like, lignin, etc., (ii) fungus produces *iso*-pentenyladenine which is present only in infected tissues. It can also metabolize *trans*-zeatine resulting in reduced production of dihydrozeatine (DHZ)-like compounds which are necessary for the normal flower and fruit development. *Trans*-zeatine and DHZ-like compounds are absent in malformed tissues, and (iii) fungus produces IAA oxidase which may be responsible for the reduced IAA in infected tissues.

4. Culture filtrates of the fungus, when inoculated on seedlings or flower buds, produce similar toxicity symptoms as induced by the extracts of malformed tissues. Therefore, it may also serve as a source of toxicity principle (TP). These toxicities may be the part of malformation syndrome.

5. There is no strong evidence to refute the involvement of fungus, more specifically *F. moniliforme* var. *subglutinans*, as the causal agent of mango malformation.

Unfortunately, none of the aforementioned characteristics are associated exclusively with *F. moniliforme* var. *subglutinans* except for its invariably higher population in infected as compared to healthy tissues. However, this could simply be a consequence of malformed growth. Conidia of the fungus are already present in the vascular system of plant as they are taken up from the soil through the damaged roots. Malformed tissues may simply support its saprophytic multiplication by providing congenial growth conditions including nutritional factors, resulting in its increased population. The positive pathogenicity tests with *F. moniliforme* var. *subglutinans* are outnumbered by the failures under similar inoculation conditions. In light of the fact that fungus induces some of the toxicity symptoms found to be associated with malformation (which may simply be the consequences of saprophytic multiplication of fungus in malformed tissues rather than part of the actual malformation syndrome) as a result of direct toxicity of its secondary metabolites or by inducing Zn deficiency which may be often confused with typical malformation symptoms, none of the positive pathogenicity tests with *F. moniliforme* var. *subglutinans* could be taken seriously unless they are reproducible. In one of the opinion surveys involving 20 mango pathologists from all over India unanimous conclusion was that, 'mango malformation should still be considered a disease of unknown etiology.'

16-5 DISEASE CYCLE

Not many earlier experiments were successful in producing infection either of vegetative or floral malformation following inoculations. Few, if successful, could not be reproduced by other able investigators. Therefore, sufficient information on the causal organism, its biology and infection court, mode of entry, mechanism of dissemination, most vulnerable links in the chain of events, etc. have not been established. Nevertheless, the recorded developments toward establishing the cause have put forth growing evidences on the involvement of fungus-directed principles in the etiology of the malady. Our understanding of the way fungus induces disease in mango plants has not progressed as quickly as the reports regarding the control have accumulated. It indicates that a set of complex factors govern the development of a syndrome that characterizes mango malformation. Ostensibly, it cannot be so simple as would spraying mango plants with spore suspension of *F. moniliforme* (*Gibbrella fujikuroi*) var. *subglutinans* leading to the production of vegetative and floral malformation symptoms, as has been claimed.[116] Apparently, more information is required on pathogenicity tests and artificial inoculation techniques (that are reproducible) which could establish knowledge on relevant aspects of disease cycle. Until then, a cycle is proposed that gives at least the chain of events in the rapid development of disease in newer areas (Fig. 16-10).

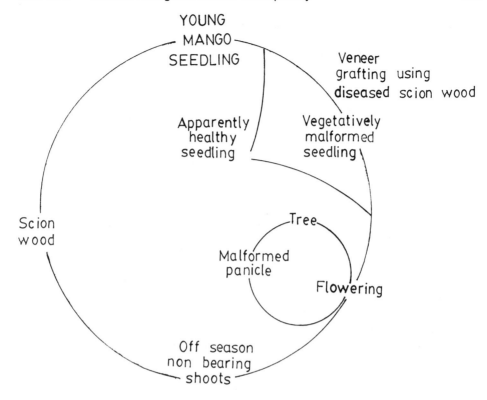

Figure 16-10   Chain of events in the development of mango malformation disease.

16-6 DISEASE RATING AND CULTIVAR SUSCEPTIBILITY

The existing cultivars of mango show a great variation in susceptibility to the disease.[15,19,87,88] Only one monoembryonic variety, Bhaduran, was observed to be completely free from the disease out of more than 100 mono and polyembryonic varieties that were examined.[4,14,19,33,89] It was observed that most of the early and mid-season mango cultivars had moderate to high incidence of disease, whereas late cultivars showed low incidence.[90] In most of the investigations, the percentage of disease incidence has been taken as the criterion for the expression of varietal susceptibility.[13,32,91,92] However, this is not an appropriate method for the quantitative estimation of cultivar susceptibility, as no consideration seems to have been given to the uniqueness of both the crop and the disease.

A characteristic feature about malformation disease of mango is that a tree once infected never recovers, and the disease severity gradually increases with time (Table 16-2)[93]. Because of the perennial nature of mango trees, the situation is quite different from annual field crops. Another characteristic problem with mango trees is their habit of alternate bearing; that is, twigs that bear an inflorescence in one year (on-year) fail to bear in the next year (off-year). As a result, a major part of the tree bears flowers in the on-year and poor or no flowering in the off-year. Thus, it seems imperative that rating for disease estimates should be done only in the on-year. But whatever may be the relative distribution, a tree may take two consecutive years to flower completely. So accuracy of

TABLE 16-2: ALTERATION IN DISEASE SEVERITY (DS) OF 86 MANGO PLANTS (CV. DASHEHRI) FROM SAME ORCHARD, INFECTED WITH FLORAL MALFORMATION, DURING 1983 OVER 1981*.

Alteration in disease severity (%)	Number of plants
Increase	
<1	2
1–5	36
5.1–10	18
>10	19
Decrease	
<1	7
1–5	3
5.1–10	1

*Increase in DS was significant (P=.05) during 1983 over 1981 when compared by paired *t* test after subjecting the percentage severity values to Arc sine   percentage transformation.
(From Kumar and Beniwal).[93]

disease estimates based only on on-year might be affected by uneven distribution of mal-formed inflorescence in an individual tree between the on- and off-years or its restricted distribution to a few twigs, which may not flower in the on-year when the disease ratings are done. Therefore, the pooled disease index of two consecutive years proves useful. The percent disease severity (DS-I and DS-II) can thus be calculated as

$$\text{DS-I} = \frac{D1 + D2}{T1 + T2} \times 100$$

$$\text{DS-II} = \frac{D2 + D3}{T2 + T3} \times 100$$

where D1, D2, and D3 represent the number of infected inflorescences per plant during first, second, and third years, respectively; and T1, T2, and T3 represent the correspond-ing number of total inflorescences per tree.

The disease incidence, an approximate measure of variability in inherent suscepti-bility of individual trees of a particular cultivar/germplasm, can be calculated as

$$\text{Disease incidence} = \frac{N1}{N2}$$

where N1 and N2 represent the number of infected plants and the total number of plants observed for each mango variety/cultivar, respectively. Maximum disease index (DI) is taken as a product of disease incidence and severity. Only the higher of the two disease severities (DS-I or DS-II) is considered. The formula used for calculation is given by

$$\text{DI} = \text{DS-I or DS-II} \times \frac{N1}{N2}.$$

A 1 to 9 rating scale is adopted where 1 = free from disease (resistant); 3 = 0.1 to 1% disease index (DI) (moderately resistant); 5 = 1.1 to 10% DI (tolerant); 7 = 10.1 to 20% DI (moderately susceptible); and 9 = >20% DI (susceptible).

Yield losses in a disease such as mango malformation are not just a linear function of DS, because it includes both direct (complete loss of fruiting on malformed inflorescence) as well as indirect losses (decreased flowering and shedding of fruits before maturity in healthy inflorescence of trees with a higher DS) probably due to the production of systemic toxic compounds in malformed tissues.[11]

Any tree with vegetative malformation even on 5% twigs was graded under the susceptible category considering the adverse effects on flowering and fruiting (Kumar, 1983).[11] The method was used to determine the susceptibility and resistance of 29 germplasms and cultivars of mango.[93]

Observations on some of the entries (Table 16-3) were available for just three trees, which may not provide a correct estimate, particularly if a germplasm is grouped as resistant, moderately resistant, or tolerant. In all such cases, results should be confirmed on a higher number of trees. However, when a germplasm is grouped as susceptible or moderately susceptible, a smaller number of trees may not affect the overall rating to any significant extent, as the method followed gives more weightage to the highest DS than to the disease incidence. Using this rating procedure, three cultivars (Asaugia Devban, Langra Rampur, and Malda Handle) were considered in the category "resistant," and six were moderately resistant ($<1\%$ DI). Popular mango cultivars such as Dashehri, Chausa, Mallihabadi, and Langra Gorakhpur were graded as susceptible.

Because the specific cause of the disease is not clearly understood artificial inoculation is not applicable. Therefore, field evaluation of susceptibility as described is at present the only way to screen donors for resistance. Since breeding for resistance in mango may take years, only resistant (DI = 0) germplasms should be used as donors. In the absence of any practical control measures, use of moderately resistant and agronomically acceptable varieties could be the next best solution.

16-7 EPIDEMIOLOGY

Very little is known about the epidemiology of the disease because of the lack of uniformity in its occurrence and variations in the severity of diseases from season to season. But a tree once infected does not seem to recover in subsequent years. Nevertheless, attempts were made to correlate the seasonal disease variation with ambient temperature at the time of flowering.[94,95] Less incidence of disease was found when temperature was raised around the trees artificially during the flowering period. It was also observed that flower buds that emerged earlier were heavily affected with the disease, whereas those emerging later escaped disease.[90] Malformation was maximum in the floral buds emerging during the first 10 days and was reduced in buds emerging later. Such a reduction was attributed to a relatively high temperature prevailing during panicle development.[96] However, no evidence exists in support of such an observation except that incidence of disease is quite scarce in the southern part of India, where the temperatures remain high during most part of the year.[11] Such a trend may also be visualized in the world distribution of the malady, since its incidence appears to be recorded most where mean temperature during winter remains around 16°C. However, more experimental evidence is needed in support of such a generalization.

TABLE 16-3: GRADING OF DIFFERENT GERMPLASMS/CULTIVARS OF MANGO FOR THEIR REACTION TO MANGO MALFORMATION.

Rating	Maximum disease index (%)	Gerplasm/Cultivar*
1 (Resistant)	0	Asaugia Davban,** Langra Rampur, Malda Handle
3 (Moderately resistant)	0.1–1	Hammif Pasand (0.ᴗ),** Tamancha (0.5), Amin (0.2), Karutha Colomban (0.3),** Kalahapus (0.5), Sannakulu (0.5)**
5 (Tolerant)	1.1–10	Guruvan (8),** Mithua Malda (5),** Hazoor Pasand (1.3), Eurovadi Banglora (1.7),** Kazalio (1.5)
7 (Moderately susceptible)	10.1–20	Neelam (11.0), Goa Mankur (10.2), Mallika (20)
9 (Susceptible)	>20	Keshar (50),** Hathizhul (45),** Zafranigola (60), N×Chausa (70),** Langra Gorakhpur (60), Mypilian (80),** Mallihabadi (56), N×Himayuddin (60), Ashadio (56), Braniko (90), Chausa (75)§, Dashehri (74)§

*Figure in parenthesis is the maximum disease index (%) of individual germplasm/cultivar.
**Observations recorded on three trees only.
§Susceptible checks
(From Kumar and Beniwal).[93]

16-8 CONTROL

As there is no scarcity of opinions regarding the cause of the disease, there are numerous reports on its control. Different workers have assigned different causes to this malady, tried to control the disease accordingly, and proved or refuted the cause. Attempts have been made by several workers to cure the diseased plant mostly be pruning, spraying nutrients, and using growth regulators and pesticides. The real difficulty has been the vagaries of practical results that have so far been obtained and reported. This is because, irrespective of the nature of problem and its cause or causes in synergism, the intensity and rapidity of application of the aforementioned methods on a country-wide basis has been practiced. Unless utility of any of the methods is principally accepted in light of the etiology of disease, such indiscriminate selection of control procedures is inadvertent and must be largely conjectural.

16-8-1 Vegetative Malformation in Young Seedlings

Because of the constant association of *Fusarium* sp. with the malformed tissues, control of disease by the use of two systemic fungicides, viz., benomyl and carbendazim was attempted,[11] based on their *in vitro* effectiveness against *Fusarium* sp.[97,98] However, none of the seedlings treated at the highest dose (15 g per plant) of either of the two fungicides showed any signs of recovery. However, *Fusarium* population could be reduced by 99% in soil and 56% in seedlings receiving the highest dose of carbendazim. This revealed the possibility of poor translocation of the fungicide in the seedlings, which was confirmed by using ^{14}C carbendazim.[99] Second, the *Fusarium* population surviving in the seedling after fungicide application probably again multiplies to the threshold level, which probably continues the diseased condition; or conditioning of the tissues, once started, does not necessitate the presence of fungi. Failure of antibacterial antibiotics to control the disease underlines the fact that the disease is possibly not bacterial or mycoplasmal in nature. This possibility was better explained when 17.5- to 20-cm-long scions were dipped in the solution of antibiotics before they were veneer-grafted on healthy stocks, with no remission of symptoms in any of the successful grafts. However, dipping of scions in antibiotic solution does not ensure their translocation up to the tip where MIP works. At the same time, failure to get any control when scions were dipped in carbendazim solution suggests two possibilities[11]: (1) poor translocation of fungicide which does not reach to the target, (2) non-stay of fungicide (because of its rapid translocation to leaves) in vascular stream particularly the xylem vessles, where the fungus is present, for duration enough to eliminate fungus, and (3) MIP is insensitive to fungicides.

16-8-2 Floral Malformation

Control of floral malformation has been variously attempted using different fungicides, nutrients, and hormones and through cultural practices like pruning. But the results are varied. Validity of all these and some of the new treatments was tested in a comprehensive trial.[100] The treatments included root pruning; application of FYM; NPK and FeSO_4; and NAA spray and fungicide (carbendazim) application in 14 different combinations. However, all proved ineffective in controlling floral malformation to any appreciable extent.[11]

Cultural practices and nutrients.   Root pruning is a practice suggested to enhance flowering as the operation provides a shock treatment to the tree.[101] This practice alone or in combination with other treatments proved ineffective.[11] Application of FYM, NPK, and FeSO_4 was done simply to promote nutritional status of the trees, as some of the workers have reportedly reduced floral malformation and increased fruit yield by similar treatments.[102–105] Those who believed nutritional imbalance was the cause of the disease tried to correct the same by injecting or spraying solutions of boron, calcium, copper, iron, magnesium, manganese, potassium, sodium, and zinc, without any obvious success.[13,31,45,89,106] The intensity of the disease was reportedly decreased by increasing levels of nitrogen,[20,107–109] while it increased in plants receiving PK or K alone. However, an increase in disease with rising N rates and also no reduction in the severity of the disease by N, P, and K treatments and soil application of Bayfolan, (containing N, P, K, Fe, Cu, Mn, B, Zn, Cu, and Mo) were also observed.[110–114]

Those who believed accumulation of mangiferin in response to infection by *F. moniliforme* var. *subglutinans* as the cause of the disease claimed a partial disease control by sprays of mangiferin Zn^{2+} and mangiferin-Cu^{2+} chelates.[55,116] According to them, these treatments revived the normal balance of mangiferin and micronutrients in the diseased part, caused a decline in the fungal population, and enabled the emergence of healthy shoots. The validity of such experiments depends on the sometime doubtful and misleading assumption that mere spray of spore suspension of *F. moniliforme* var. *subglutinans* (*Gibberella fujikuroi*) on mango plants could produce bunchy top, dieback, abnormal inflorescence and blossom blight symptoms. Second, there is apparently the need to differentiate such symptoms from those produced by Zn deficiency and typical malformation. Contrarily, Rajan[49] observed that sprays of mangiferin-metal chelates were ineffective in controlling the disease.

Pruning of shoots bearing malformed panicles (two-year growth) has been found as quite an effective treatment in suppressing the disease satisfactorily.[11] Pruning of shoots probably removes MIP, which remains localized at the shoot tip. But such trees have also been found to produce malformed inflorescences again after two years. Possibly, MIP is generated again by the associated fungi, and reaches to the threshold level during this period. Complete pruning (heading back) of the branches and/or Bavistin drenching (100 g/tree) cannot be adjudged superior to pruning of two-year growth until the cut trees reach full bloom and effect is longer lasting. Even at this stage, application of fungicies to the pruned trees appeared unnecessary because of poor translocation of fungicide inside the trees.[99] Pruning treatment was found effective by some early workers also;[110,132,133,134] although, the operation was not found useful by others.[19,31,97] The success probably depends on the stage and extent of infection in the treated trees.

Fungicides.   The use of pesticidal chemicals in the field vis-à-vis the laboratory depends on their *in vitro* efficacy at minimal and economically acceptable dosage and their efficient and rapid transport to the infection site. Invariable association of *Fusarium* spp. with the malformed tissues necessitated the use of systemic fungicides for controlling the disease.[97] Carbendazim was highly effective (ED^{50} $< 1\mu g$ ml^{-1}), in *in vitro* tests against *Fusarium* sp. isolated from malformed mango tissues.[11,98] Application of carbendazim as spray has also been reported to be effective in reducing severity of floral malformation,[98,117,118] though there are conflicting reports also.[119] Application of carbendazim (Bavistin 50WP) as soil drench (100 g/tree) or trunk injection (35 g/tree) alone or in combination with other treatments failed to reduce disease severity to any appreciable extent. Poor translocation of fungicide or noninvolvement of *Fusarium* sp. could be the explanations for such a situation. Although carbendazim, methyl 2-benzimidazole carbamate (MBC), was readily absorbed when applied through roots, stem, or leaves of mango seedlings, the major part of the applied fungicide remained with the treated plant part, or had a very poor translocation within the plant.[99] It was found that MBC was rapidly (80% within 14 minutes) and readily (1075 μg g^{-1} liginin from 8.0 μg ml^{-1} MBC solution) adsorbed by mango lignin. Adsorption was reversible, as about 73% of MBC adsorbed on lignin could be desorbed in seven washings with 0.05 N HCl. Nevertheless, about 27% of MBC could not be desorbed even after seven washings, indicating a comparatively higher strength of the bonding of this residual fraction. Thus, comparatively strong binding of MBC with lignin molecule, the major constituent of xylem tracheary elements and accounting for more than 15% of stem fresh weight, might be responsible for the poor translocation of the fungicide within the plant, in spite of its good uptake.

Plant growth regulators.   The malformation-inducing principle (MIP), which probably involves hormonal imbalance, could be theoretically corrected to normalcy through the extraneous application of hormones. However, unless the source of MIP is controlled it would be generated again making the effect of exogenous hormones as temporary phenomenon. Ever since mango malformation was considered as a physiological disorder[120] due to imbalance of auxins and antiauxins caused by vectors, diseases, and nutritional deficiencies, instant correction of imbalance has been sought by exogeneous application of auxin formulations (NAA 200 ppm) at flower bud differentiation and/or deblossoming (manual removal) at budburst stage.[17] Deblossoming helped to induce emergence of auxillary buds at a later date when temperatures were higher.[41,121,122] This was hailed as the most promising practice to control mango malformation.[41,90,123-128] However, there are controversial reports on the usefulness of this treatment.[11,111] The severity of disease in the tree is an important factor that limits the usefulness of the practice. NAA (200 ppm), when sprayed alone or in combination with fungicides, did not show any encouraging results.[11] The only promising treatment was deblossoming alone or in combination with NAA spray, though the former showed better effects. Panicles appearing after deblossoming as they arise from latent secondary buds were shorter, and more than one panicle emerged from one shoot tip which bore fruits, some of which were retained till maturity. Even in these treatments, some of the panicles appearing after deblossoming remained malformed.[11] Moreover, such a treatment does not seem practical under field conditions, particularly in large trees.

It has been observed that the incidence of floral malformation is higher in the panicles that emerge early and declined subsequently.[96] GA reportedly delays the emergence of panicles by nearly two weeks when applied at the concentration of 10^{-1} to 10^{-2} M before flower bud differentiation.[129] Accordingly, Shawky et al.[130] tried to delay the time of flower bud opening by spraying GA_3 at various concentrations on malformed panicles in mid-November and found 50 ppm as the best concentration, which resulted in an increased number of perfect flowers, increased pollen viability, and resulted in about four times greater yields. On the contrary, El-Beltagy et al.[111] found GA_3 at 100 or 250 ppm completely ineffective in reducing the percentage of malformed panicles which, on the other hand, increased significantly. Similarly, they tried urea 0.25 to 2.0% to delay the time of flowering and found improved pollen viability, fruit set, and yield. The best results were obtained with urea (1 to 2%), which resulted in an eight-fold increase in yield as compared with the untreated control. This substantiated an earlier report[131] that a foliar spray of urea at monthly intervals during winter could reduce the disease intensity. Further, application of CCC (Chloromequat) at 20 or 50 ppm to the Bayfolan-treated trees (100 ml/tree) was found effective in reducing the percentage of malformed panicles and increasing the healthy ones.[111] Application of ethephon (400 ppm) at bud inception stage has also been reported to reduce malformation and improve yield and fruit quality.[73]

Malformin. The activity of malformin, an antimetabolite, is inhibited by reducing agents, thiol compounds, or SH reagents.[135] Recent investigations on growth substance imbalance with particular reference to malformins has prompted the use of antimalformins to counteract malformin response generated in the malformed panicles.[11,49,117,136,137] Although the treatment was found ineffective in reversing normal growth out of the floral buds that were destined to bear malformed panicles,[11,49] others staked very astounding claims about achieving control.[117,136,137] The latter claim that spraying twice with potassium metabisulfite (560 ppm) silver nitrate (2100 ppm), ascorbic acid (1055 ppm), and reduced glutathione (GSH, 560 ppm) increased the percentage of fruiting panicles and the number of fruits per panicles, and that the extent of malformed panicles converting into healthy ones after the aforementioned treatments were 93.3, 80.0, 93.3, and 86.6%, respectively.[137] The results need further testing before they are advocated.

The biological activity of malformin reflects dissimilar mode of action in roots and stems, as has been demonstrated in *Phaseolus vulgaris* and *Zea mays*,[86] and malformin acts at more than one locus and has more than one mode of action. Therefore, additional research on this compound is necessary using mango malformed tissues. In view of the inhibiting action of sulphydryl compounds on the biological activity of malformin and the reaction of malformin with sulphydryl compounds, it was observed that reduced glutathione effectively antagonized malformin-induced corn root curvature.[135] IAA has been shown to markedly and rapidly increase the concentration of GSH in plant tissues,[138] and this phenomenon may explain why some activities of malformin are successfully antagonized by IAA.[86] This suggestion does not preclude the involvement of thiol groups in activities of malformins which are not antagonized by IAA.

With the available information on the mode of action of malformin and its inhibition, it would be tempting to speculate the working of a similar mechanism in mango malformed tissues when IAA (NAA) and antimalformin compounds are applied to seek control of the malady. Such a proposition may satisfy those who have recorded significant control in floral malformation using these chemicals. Others who have been unable to get satisfactory results should realize that *Fusarium* induced MIP could be a combination of many factors (growth regulators), and malformin could be just one. At the same time it may be difficult for any organic/inorganic compound to reach to the site of production/action of MIP in concentration enough to neutralize its effects unless the source of MIP has been hit. It must, however, be taken into consideration by both that any control measure that is suggested for the floral malformation should also work against the vegetative malformation, since both are symptoms of the same disease, and the latter contributes significantly to inhibiting floral development.

From the foregoing account, it seems that restoration of normal growth from a panicle once it has been malformed seems improbable. Once the tissue has been conditioned to produce malformed growth, it cannot resume normalcy unless some alterations are brought before the flower bud initiation. This may be the reason why different control measures fail; the origin of MIP is not hit, but the consequences of MIP are being neutralized, which are natural to develop every season. In countless experiments, plant growth regulators have been applied to diseased plant in pharmacological approach. Needless to say, the fruits of thousands of hours of research efforts do not provide us with any conclusive results. Perhaps the overriding impediment is the paucity of information about the mechanism(s) that leads to the malformed growth. Until the nature of MIP is clearly established, it is unlikely that any consensus could be achieved about the control method. Such a vantage point could provide the insights for necessary breakthroughs.

16-8-3 Conclusion

None of the control measures experimented till date is non controversial and practical enough for recommendation. However, rapid spread of disease in new plantations could be effectively checked by using disease free plants generated from scions from perfectly healthy trees. Substantial evidence exists to show that veneer-grafting, the most common method of mango propagation plays a decisive role in the spread of the disease, as it is impossible to distinguish diseased scions on a tree in an off-year unless they are marked in the on-year. Indiscriminate use of scions helps spread the disease and, therefore, careful selection of potential scions from disease-free shoots of adult tree will go a long way toward preventing establishment of the disease in mango orchards. Till such time when practical methods of curing infected plants are available, the approach holds promise and should be practised.

ACKNOWLEDGMENT

We thank Y. L. Nene, D. N. Srivastava and U. S. Singh for reviewing the manuscript, and Girish Chandra for typing.

16-9 REFERENCES

1. Watt, G., *A dictionary of economic products of India,* Govt. Printing Press, Calcutta, p. 149, 1891.

2. Burns, W., A common malformation of mango inflorescence, *Poona Agric. Coll. Mag.,* 2, 38, 1910.

3. Burns, W. and Paryag, S. H., *The book of the mango,* Dept. Agric. Bombay, p. 103, 1920.

4. Khan, M. D. and Khan, A. H., Studies on malformation of mango inflorescences in West Pakistan, *Punjab Fruit J.,* 23, 247, 1960.

5. Hassan, A. S., Notes on *Eriophyes mangiferae* S. N. (Acarina), *Bull. Soc. Fouad, Ent.,* 28, 179, 1944.

6. Schwartz, A., A new mango pest, *Farming in S. Africa,* 9, 7, 1968.

7. Flechtmann, C. H. W., Kimati, H., Medcale, J. C. and Ferre, J., Preliminary observation on mango inflorescence malformation and the fungus, insect and mites associated with it, *Anais da Fscola Superior de Agricultura 'Luiz de Qveiroz,'* 27, 281, 1970.

8. Malo, S. E. and McMillan, R. T. J., A disease of *Mangifera indica* L. in Florida similar to mango malformation, *Florida State Hort. Soc.,* 85, 264, 1972.

9. Anon., *Mango cultivation,* Indian Institute of Horticultural Research, Extension Bull., 9, 1983.

10. Kumar, J. and Beniwal, S. P. S., Vegetative and floral malformation: Two symptoms of the same disease on mango, *FAO Plant Prot. Bull.,* 35, 21, 1987.

11. Kumar, J., *Studies on symptomatology, etiology and control of mango malformation,* Ph.D. thesis, G. B. Pant University of Ag. & Tech., Pantnagar, 1983.

12. Majumdar, P. K. and Sinha, G. C., Studies on the effect of malformation on growth, sex-ratio, fruit set and yield of mango, *Acta Horticulturae,* 24, 230, 1972.

13. Singh, L. B., Singh, S. M. and Nirvan, R. S., Studies on mango malformation. Review, symptoms, extent, intensity and cause, *Hort. Adv.* 5, 197, 1961.

14. Tripathi, R. D., Bunchy top and malformation disease of mango, *Indian J. Hort.,* 11, 122, 1954.

15. Schlösser, E., Mango malformation: Symptoms, occurrence and varietal susceptibility, *FAO Plant Prot. Bull.* 19, 12, 1971.

16. Khan, M. D. and Khan, A. H., Some chemical studies on malformation of mango inflorescence in West Pakistan, *Punjab Hort. J.,* 3, 229, 1963.

17. Majumdar, P. K., Sinha, G. C. and Singh, R. N., Effect of exogenous application of NAA on mango malformation, *Indian J. Hort.* 27, 130, 1970.

18. Sattar, A., Diseases of mango in the Punjab, *Punjab Fruit J.*, 10, 56, 1946.

19. Singh, K. K. and Jawanda, J. S., Malformation in mangoes, *Punjab Hort. J.*, 1, 18, 1961.

20. Prasad, A., Singh, H. and Shukla, T. N., Present status of mango malformation disease, *Indian J. Hort.*, 22, 254, 1965.

21. Narasimhan, M. J., Malformation of panicles in mango incited by a species of *Eriophyes*, *Curr. Sci.*, 23, 297, 1954.

22. Nariani, T. K. and Seth, M. L., Role of eriophyid mites in causing malformation disease of mango, *Indian Phytopath.*, 15, 231, 1962.

23. Puttarudriah, M. and Channa Basavana, G. P., Mango bunchy-top and the eriophyid mite, *Curr. Sci.*, 30, 114, 1961.

24. Summanwar, A. S., Raychaudhuri, S. P. and Pathak, S. C., Association of the fungus *Fusarium moniliforme* Sheld. with the malformation in mango (*Mangifera indica* L.), *Indian Phytopath.*, 19, 227, 1966.

25. Varma, A., Lele, V. C., Raychaudhuri, S. P. and Ram, A., Mango malformation: A fungal disease, *Phytopath. Z.*, 79, 254, 1974.

26. Bhatnagar, S. S. and Beniwal S. P. S., Involvement of *Fusarium oxysporum* in causation of mango malformation, *Plant Dis. Reptr.*, 61, 894, 1977.

27. Ibrahim, A. N., Satawi, M. M., El-Tobsh, S. M. and Abdel Sattar, M. A., Pathological and histological note on mango malformation in Egypt, *Curr. Sci.*, 44, 443, 1975.

28. Salma, E. A., Khader, A. S. and Hussain, S. A., Effect of artificial inoculation of mango floral buds with 3 sp. of *Fusarium* and with a virus on the development of malformation, *Egyptian J. Hort.*, 6, 23, 1979.

29. Nirvan, R. S., Bunchy top of young mango seedlings, *Sci. & Cul.*, 18, 335, 1953.

30. Schlosser, E., Mango malformation: Incidence of "bunchy top" on mango seedlings in West Pakistan, *FAO Plant Prot. Bull.*, 19, 41, 1971.

31. Saeed, A. and Schlösser, E., Effect of some cultural practices on the incidence of mango malformation, *Zertsscherft fur Pflanzentran Kherten and Pflanzenschutz*, 79, 349, 1972.

32. Mallik, P. C., Mango malformation—symptoms, causes and cure, *Punjab Hort, J.*, 3, 292, 1963.

33. Jagirdar, S. A. P. and Shaik, M. R., Control of malformation of mango inflorescences, Souvenir mango and summer fruit show, Mirpur Khas, West Pakistan, 1968.

34. Kishtah, A. A., Nyland, G., Nasr El-Din, T. M., Tolba, M. A., Lowe, S. K., Khalil, E. M., Tadros, M. R. and El-Amrely, A., Mango malformation disease in Egypt. I. Electron microscopy, effect of antibiotics, cultural and serological studies, *Egyptian J. Phytopath.*, 17, 151, 1985.

35. Ahmed, G. D. and Sattar, A., Some studies on malformation of mango inflorescences in Punjab. In *Pak. Assoc. Adv. Sci.*, Second Conf. Agric. Sec., p. 9, 1950.

36. Vasudeva, R. S., Virus diseases of fruits and vegetables, Third Hort. Res. Workers Conf., Simla, 1957.

37. Kausar, A. C., Malformation of inflorescence in mango, *Punjab Fruit J.*, 22, 19, 1959.

38. Bindra, O. S. and Bakhetia, D. R. C., Studies on the population dynamics of mango bud mite, *Aceria mangiferae* Sayeed, in relation to incidence of malformation, *J. Res. Punjab. Agric. Univ.*, 6, 200, 1969.

39. Kishtah, A. A., Eid, S. A. and Nasr El-Din, T. M., Mango malformation disease in Egypt. 2. Studies on transmission and thermotherapy, *Egyptian J. Phytopath.*, 17, 159, 1985.

40. Prasad, A., Nirvan, R. S. and Singh, S., Mango malformation—a review of work done at the Horticultural Research Institute, Saharanpur, India, *Acta Horticulturae*, 24, 227, 1972.

41. Chadha, K. L., Pal, R. N., Om Prakash, Tandon, P. L., Singh, H., Singh, N. P., Rao, M. R. K. and Lal, B., Studies on mango malformation, its cause and control, *Indian J. Hort.*, 36, 359, 1979.

42. Wood, R. K. S., *Physiological plant pathology*, Blackwell, Oxford, p. 228, 1967.

43. Kumar J., Beniwal, S. P. S. and Ram, S., Depletion of auxins in mango seedlings affected with bunchy top stage of mango malformation, *Indian J. Exp. Biol.*, 18, 286, 1980.

44. Karupasagar, V. and Sequeira, L., Auxin destruction by *Marasmius perniciousus*, *Am. J. Bot.*, 56, 390, 1969.

45. Tripathi, R. D., Malformation disease of the mango as related to deficiency of mineral nutrients, *Indian J. Hort.*, 12, 173, 1955.

46. Sadhu, M. K., Changes in the free and bound amino acids and amides in healthy and malformed mango shoots, *Sci & Cult.*, 41, 111, 1975.

47. El-Ghandour, M. A., Hamawi, H. A. and Fadl, M. S., Biological studies for exploring interaction between some native growth regulators and abnormal growth of mango shoots, *Egyptian J. Hort.*, 3, 69, 1976.

48. Pandey, R. M., Rao, M. M. and Pathak, R. A., Biochemical changes associated with floral malformation in mango, *Scientia Hortic.*, 6, 37, 1977.

49. Rajan, S., *Biochemical basis of mango (Mangifera indica L.) malformation*, Ph.D. thesis. Indian Agricultural Research Institute, Delhi, 1986.

50. Pandey, R. M., Singh, R. N. and Rao, M. M., Nucleic acid and protein level in healthy and malformed panicles of mango cultivars, *Sci & Cult.*, 41, 386, 1976.

51. Chattopadhyay, N. C. and Nandi, B., Nucleic acid contents in malformed mango inflorescence caused by *Fusarium moniliforme* var. *subglutinans*, *Phytopath. Z.*, 89, 256, 1977.

52. Chattopadhyay, N. C. and Nandi, B., Degradation of cellulose and lignin in malformed mango inflorescences by *Fusarium moniliforme* var. *subglutinans*, *Acta Phytopathologica Academiae Scientiarum Hungaricae*, 12, 283, 1977.

53. Chattopadhyay, N. C. and Nandi, B., Changes in total contents of saccharides, proteins and chlorophyll in malformed mango inflorescences induced by *Fusarium moniliforme* var. *subglutinans*, *Biologia Plantarum*, 20, 468, 1978.

54. Chattopadhyay, N. C. and Nandi, B., Mobilization at the malformed region of mango caused by *Fusarium moniliforme* var. *subglutinans* wr. et. Rg., *Indian J. Microbiol.*, 17, 43, 1977.

55. Ghosal, S., Chakrabarti, D. K., Biswas, K. and Kumar, Y., Toxic substances produced by *Fusarium*. X. Concerning the malformation disease of mango, *Experientia*, 35, 1633, 1979.

56. Ghosal, S., Biswas, K., Chakrabarti, D. K. and Basuchaudhary, K. C., Control of *Fusarium* wilt of safflower by mangiferin, *Phytopathology*, 67, 548, 1977.

57. Chakrabari, D. K. and Ghosal, S., Effect of *Fusarium moniliforme* var. *subglutinans* infection on mangiferin production in the twigs of *Mangifera indica*. *J. Phytopath.*, 113, 47, 1985.

58. Chakrabarti, D. K. and Ghosal, S., Control of malformation disease of *Mangifera indica* incited by *Fusarium moniliforme* var. *subglutinans* with mangiferin metal chelates, Abstract, Indian Science Congress, Mysore, p. 70, 1982.

59. Ghosal, S. and Chakrabarti, D., Differences in phenolic and steroidal constituents between healthy and infected florets of *Mangifera indica, Phytochem.*, 27, 1339, 1988.

60. Pandey, R. M., Rathore, D. S. and Singh, R. N., Hormonal regulation of mango malformation, *Curr. Sci.*, 43, 694, 1974.

61. Pandey, G., *Role of malformins and auxins in the floral malformation of mango (Mangifera indica L.)*, Ph.D. thesis. G. B. Pant University of Agriculture & Technology, Pantnagar, 1988.

62. Dhillon, B. S. and Singh, Z., Depletion of indole-3-acetic acid in malformed tissues of mango and its alleviation, *Acta Horticulturae*, 239, 371, 1989.

63. Abou-Hussein, M. R., Fadl, M. S. and Ghandour, M. A., Some morphological and physiological features of floral malformation in mangoes, *Egyptian J. Hort.*, 2, 199, 1975.

64. Singh, Z. and Dhillon, B. S., Hormonal changes associated with vegetative malformation of mango (*Mangifera indica*), *J. Phytopath.*, 125, 193, 1989.

65. Kumar, J. and Beniwal, S. P. S., Note on gibbeellin-like substances and inhibitors in mango seedlings affected with bunchy top stage of mango malformation, *Pantnagar J. Res.*, 4, 1979.

66. Mishra, K. A. and Dhillon, B. S., Levels of endogenous gibberellins in the healthy and malformed panicles of mango (*mangifera indica* L.), *Indian J. Hort.*, 37, 33, 1980.

67. Bist, L. D. and Ram, S., Effect of malformation on changes in endogenous gibberellins and cytokinins during floral development of mango, *Scientia Hortic.*, 28, 235, 1986.

68. Chattopadhyay, N. C. and Nandi, B., Peroxidase and polyphenol oxidase activity in malformed mango inflorescences caused by *Fusarium moniliforme* var. *subglutinans.*, *Biologia Plantarum*, 18, 321, 1976.

69. Pal, R. N., Kalra, S. K., Tandon, D. K. and Chadha, K. L., Activity of IAA oxidase, catalase and amylase in morphactin induced malformation of mango inflorescences, *Scientia Hortic.*, 19, 271, 1983.

70. Nicholson, R. I. D. and Staden, V. J., Cytokinins and mango flower malformation. I. Tentative identification of the complement in healthy and malformed inflorescences, *J. Pl. Physiol.*, 132, 720, 1988.

71. Van Staden, J. and Nicholson, R. I. D., Cytokinins and mango flower malformation. II. The cytokinin complement produced by *Fusarium moniliforme* and ability of the fungus to incorporate (8^{-14}C) adenine into cytokinins, *Physiol. Mol. Plant Path.*, 35, 423, 1989.

72. Van Staden, J., Bayley, A. D. and Macrae, S., Cytokinins and mango flower malformation. III. The metabolism of (^{3}H) *iso*-pentenyladenine and (8^{-14}C) zeatin by *Fusarium moniliforme*, *Physiol. Mol. Plant Path.*, 35, 433, 1989.

73. Singh, Z. and Dhillon, B. S., Floral malformation, yield and fruit quality of *Mangifera indica* L. in relation to ethylene, *J. Hort. Sci.*, 65, 215, 1990.

74. Singh, Z., *Biochemical aspects and control of mango malformation*, Ph.D. thesis, Punjab Agricultural University, Ludhiana, 1986.

75. Ram, S. and Bist, L. D., Occurrence of malformin-like substances in the malformed panicles and control of floral malformation in mango, *Scientia Hortic.*, 23, 331, 1984.

76. Singh, Z. and Dhillon, B. S., Presence of malformin-like substances in malformed floral tissues, *J. Phytopath.*, 125, 117, 1989.

77. Singh, Z. and Dhillon, B. S. Occurrence of malformin-like substances in seedlings of mango (*Mangifera indica* L.), *J. Phytopath.*, 120, 245, 1987.

78. Bodanszky, M. and Stall, G. L., The structure and synthesis of malformin A, *Proc. Natl. Acad. Sci.*, USA, 71, 2791, 1974.

79. Curtis, R. W., Root curvature induced by culture filtrate of *Aspergillus niger, Science,* 128, 661, 1958.

80. Curtis, R. W., Curvature and malformations in bean plants caused by culture filtrate of *Apergillus niger, Plant Physiol.* 33, 17, 1958.

81. Kumar, J., Singh, U. S. and Beniwal, S. P. S., Presence of a root malformation factor in culture filtrates of *Fusarium moniliforme* var. *subglutinans, J. Phytopath.,* 119, 7, 1987.

82. Postlethwait, S. N. and Curtis, R. W., Histology of malformations produced on bean plants by culture filtrates of *Aspergillus niger, Am. J. Bot.,* 46, 31, 1959.

83. Curtis, R. W., Effect of malformin on the major constituents of *Phaseolus vulgaris, Plant Cell Physiol.,* 10, 203, 1969.

84. Byer, E. and Morgan, P., Abscission: The role of ethylene modification of auxin transport, *Plant Physiol.,* 48, 208, 1971.

85. Ernest, L. C. and Valdovinos, J. G., Regulation of auxin levels in *Coleus blumei* by ethylene, *Plant Physiol.,* 48, 402, 1971.

86. Curtis, R. W. and Fellenberg, G., Effect of malformin on adventitious root formation and metabolism of indole-acetic acid-2-^{14}C by *Phaseolus vulgaris, Plant Cell Physiol.,* 13, 715, 1972.

87. Azzous, S., Hamdy, Z. M. and Dahshan, I. M., Studies on malformed inflorescences of mango—the degree of susceptibility among different varieties, *Agric. Res. Rev.,* 56, 17, 1978.

88. Nath, R., Kamalwanshi, R. S. and Sachan, I. P., Studies on mango malformation, *Indian J. Mycol. Plant Path.* 17, 29, 1987.

89. Sharma, B. D., Studies on the diseases of *Mangifera indica, Proc. Indian Sci. Cong.* (Abs.), 70, 1953.

90. Singh, U. R., Gupta, J. H. and Dhar, L., Performance of mango cultivars against malformation in Uttar Pradesh, *Prog. Hort.* 8, 65, 1977.

91. Puttoo, B. L., Gupta, B. K. and Tegh, H. S., Extent of mango malformation in Jammu, *Indian J. Mycol. Plant Path.,* 5, 181, 1975.

92. Chib, H. S., Andotra, P. S. and Gupta, B. R., Survey report on incidence and extent of mango malformation in mango growing areas of Jammu and Kashmir State, *Indian J. Mycol. Plant Path.,* 14, 86, 1984.

93. Kumar, J. and Beniwal, S. P. S., A method of estimating cultivar susceptibility against mango malformation, *Trop. Pest Management,* 33, 208, 1987.

94. Varma, A., Lele, V. C., Majumdar, P. K., Ram, A., Sachchidananda, J., Shukla, U.S., Sinha, G. C., Yadav, T. D. and Raychaudhuri, S. P., Report of the work done on mango malformation (horticultural aspects), presented at ICAR workshop on Fruit Research, Ludhiana, April 28 to 30, 1969.

95. Majumdar, P. K. and Sinha, G. C., Seasonal variation in the incidence of malformation in *Mangifera indica* L., *Acta Horticulturae,* 24, 221, 1972.

96. Singh, U. R., Dhar, L. and Gupta, J. H., Note on the effect of time of bud burst on the incidence of floral malformation in mango, *Prog. Hort.,* 11, 41, 1979.

97. Varma, A., Raychaudhuri, S. P., Lele, V. C. and Ram, A., Preliminary investigation on epidemiology and control of mango malformation, *Proc. Indian Nat. Sci. Acad.,* 37(B), 291, 1971.

98. Sharma, O. P. and Tiwari, A., Studies on mango malformation, *Pesticides,* 9, 44, 1975.

99. Kumar, J., Singh, U. S., Beniwal, S. P. S. and Srivastava, P. C., Binding of carbendazim by lignin, *Indian J. Mycol. Plant Pathol.,* 17, 24, 1987.

100. Kumar, J. and Beniwal, S. P. S., Control of the malformation disease of mango. In *Abst. Second Int. Symp. Mango*, May 20 to 24, p. 96, 1985.

101. Singh, L. B., *The mango: Botany cultivation and utilization,* Leonard Hill (Books) Limited, London, 1960.

102. Minessy, F. A., Biely, M. P. and El-Fahl, A., Effect of iron chelates in correcting malformation of terminal bud growth in mango, *Sudan Agric. J.*, 6, 71, 1971.

103. Minessy, F. A., Biely, M. P. and El-Fahl, A., Effect of iron chelates in correcting malformation of terminal bud growth in mango in Sudan, *Egyptian J. Hort.*, 1, 212, 1974.

104. Abou-El Daheb, M. K., Correcting malformation symptoms of mango trees in Egypt by soil application of iron chelates, *Egyptian J. Phytopath.* 7, 97, 1977.

105. Peswani, K. M., Bhutani, D. K., Attri, B. S. and Bose, B. N., Preliminary studies on the role of potassium on inhibition of mango malformation, *Pesticides*, 13, 48, 1979.

106. Abdel-Motaleb, M., Ibrahim, A. E. M., Bayoumi, N. A., Preliminary study on the relation between soil properties and mango malformation, *Mihufiya, J. Agric. Res.*, 7, 395, 1983.

107. Jagirdar, S. A. P. and Shaik, M. R., Role of NPK in minimizing the intensity of malformation of mango inflorescences, *Agric. Pakist.* 20, 175, 1969.

108. Mallik, M. M. and Raza, M., Effect of different doses of NPK and NAA and time of deblossoming on the intensity or malformation of mango inflorescences, *J. Agric. Res. Pakist.*, 23, 97, 1985.

109. Kanwar, J. S. and Kahlon, G. S., Effect of nitrogen, phosphorus, and potassium fertilization on panicle malformation in mango (*Mangifera indica* L.) cv. Dashehri, *Punjab Hort. J.*, 27, 12, 1987.

110. Bindra, O. S. and Bakhetia, D. R. C., Investigation on the etiology and control of mango malformation, *Indian J. Hort.*, 28, 80, 1971.

111. El-Beltagy, M.S., El-Ghandour, M. A. and El-Hanawi, Effect of Bayfolan and some growth regulators on modifying flowering and the incidence of flowering malformation of mango (*Mangifera indica* L.), *Egyptian J. Hort.*, 6, 125, 1980.

112. Shawky, L., Zindan, Z., El-Tomi, A. and Dahshan, A. D. J., Effect of urea sprays on time of blooming, flowering malformation and productivity of Taimour mango trees, *Egyptian J. Hort.*, 5, 132, 1978.

113. Shawky, I., Dahshan, D. I. and El-Shiekh, A. F., Effect of Alar and urea sprays on flowering behaviours, panicle malformation and productivity of 'Hindy be Sinnara' mango, *Ann. Agric. Sci.*, Moshtohr, 27, 225, 1982.

114. Cheema, S. S. and Malhi, C. S., Effect of N, P and K fertilizers on the incidence of mango malformation, anthracnose and powdery mildew, *Punjab Hort. J.,* 26, 6, 1986.

115. Deleted in proof.

116. Chakrabarti, D. K. and Ghosal, S., The disease cycle of mango malformation induced by *Fusarium moniliforme* var. *subglutinans* and the curative effects of metal chelates, *J. Phytopathol.*, 125, 238, 1989.

117. Mehta, N., Sandooja, J. K., Madaan, R. L. and Yamdagni, R., Role of different chemicals in mango malformation and related physiological factors, *Pesticides,* 20, 17, 1986.

118. Siddiqui, S., Sandooja, J. K., Mehta, N. and Yamadagni, R., Biochemical changes during malformation in mango cultivars as influenced by various chemicals, *Pesticides,* 21, 17, 1987.

119. Diekman, F., Manicom, B. Q. and Coetze, K., An attempt to control blossom malformation of mango with chemical sprays, *Information Bulletin*, Citrus and Subtropical Fruit Institute, 117, 15, 1982.

120. Jagirdar, S. A. P. and Jafri, N. R., Malformation of mango inflorescences, *Agric. Pakist.*, 17, 351, 1966.

121. Singh, R. N., Majumdar, P. K., Sharma, D. K., Sinha, G. C. and Bose, P. C., Effect of deblossoming on productivity of mango, *Scientia Hortic.*, 21, 393, 1974.

122. Majumdar, P. K., Sharma, D. K., Singh, M. P. and Singh, R. N., Improve productivity of malformed mango trees, *Indian Hort.*, 20, 7, 1976.

123. Shant, P. S., Preliminary studies on the effect of Planofix on mango malformation, *Bangladesh Hort.*, 3, 49, 1975.

124. Bajpai, D. N. and Shukla, H. S., Combating mango malformation through exogenous application of NAA, *Plant Science*, 10, 135, 1978.

125. Dang, J. K. and Daulta, B. S., A note on screening of mango cultivars against floral malformation, *Haryana J. Hort. Sci.*, 11, 215, 1982.

126. Singh, Z. and Dhillon, B. S., Effect of naphthalene acetic acid, ethrel, dikegulac and hand deblossoming on floral malformation, flowering, yield and fruit quality of mango (*Mangifera indica* L.), *Acta Horticulturae*, 175, 307, 1986.

127. Singh, Z. and Dhillon, B. S., Effect of plant growth regulators on floral malformation, flowering, productivity and fruit quality of mango (*Mangifera indica* L.), *Acta Horticulturae*, 135, 315, 1986.

128. Mehta, N., Madaan, R. L., Sandooja, J. K. and Daulta, B. S., A note on effect of NAA on different cvs. of mango against malformation, *Haryana J. Hort. Sci.*, 15, 62, 1986.

129. Kachru, R. B., Singh, R. N. and Chacko, B. K., Inhibition of flowering in mango by gibberellic acid, *Hort. Sci.*, 6, 140, 1971.

130. Shawky, I., Zidan, Z., El-Tomi, A. and Dahshan, A. D. J., Effect of GA$_3$ sprays on time of blooming and flowering malformation in Taimour mango, *Egyptian J. Hort.*, 5, 123, 1978.

131. Khan, M. D. and Khan, A. H., Some studies on malformation of mango inflorescence, *Pak. Assoc. Adv. Sci.*, 10th Conference, Agriculture Section, p. 33, 1958.

132. Narasimhan, M. J., Control of mango malformation disease, *Curr. Sci.*, 28, 254, 1959.

133. Mallik, D. C., Studies on the malformation of the mango inflorescence. In *Proc. Bihar A. C. of Agri. Sci.*, p. 8, 1959.

134. Desal, S. L., Kaul, C. K. and Mathur, B. P., Note on control of mango malformation, *Indian J. Agri. Sci.*, 46, 545, 1976.

135. Suda, S. and Curtis, R. W., Studies on the mode of action of malformin: inhibition of malformin, *Plant Physiol.*, 39, 904, 1964.

136. Bist, L. D. and Ram, S., Effect of some antimalformins and growth regulators on the control of floral malformation in mango, *Indian J. Hort.*, 42, 161, 1985.

137. Bist, L. D. and Ram, S., Chemical treatment to control floral malformation in mango (*Mangifera indica* L.), *Hort. Sci.*, 21, 664, 1986.

138. Marre, E. and Arrigoni, O., Metabolic reactions to auxin. I. The effects of auxin on glutathione and the effects of glutathione on growth of isolated parts, *Physiol. Plant.*, 10, 289, 1957.

17

BACTERIAL CANKER
OF STONE FRUITS

M. J. HATTINGH

Department of Plant Pathology, University of Stellenbosch
Stellenbosch 7600, South Africa

ISABEL M. M. ROOS

Fruit and Fruit Technology Research Institute, Private Bag X5013
Stellenbosch 7600, South Africa

17-1 INTRODUCTION

Bacterial canker (gummosis) of stone fruit crops is caused by three pathovars of *Pseudomonas syringae:*[1] *P.s.* pv. *syringae* van Hall incites the disease on all stone fruit species grown commercially; *P.s.* pv. *morsprunorum* (Wormald) Young, Dye, and Wilkie is virtually restricted to cherry and plum; and *P.s.* pv. *persicae* (Prunier, Luisetti, and Gardan) Young, Dye, and Wilkie causes leaf spots, cankers, and gummosis of fruit on peach trees in France. *P. syringae* is a common epiphyte that occurs on many plant species,[2] including deciduous fruit trees.[3]

Bacterial canker of stone fruit develops when the host is stressed or dormant. The subtle nature of the predisposing factors and the intimate association of *P. syringae* with symptomless trees[4] complicate disease management. This contribution focuses on the general nature of bacterial canker of stone fruit trees. Bacterial canker per se, as well as more specialized aspects of the relationship of *P. syringae* with host and nonhost plants, has also been covered in other review papers.[2-7]

17-2 DISTRIBUTION AND ECONOMIC IMPORTANCE

Many strains of *P. syringae* that cause bacterial canker of stone fruit might have been introduced to new commercial nurseries and orchards in the temperate and Mediterranean regions of the world through latently infected propagating material. The pathogens probably occur in all countries where stone fruit crops are grown commercially. However, if predisposing factors do not favor disease development, symptoms may be inconspicuous, absent, or difficult to distinguish from other minor irregularities commonly seen on deciduous fruit trees. Therefore, unless trees or branches are killed outright, the damage caused by pathovars of *P. syringae* on stone fruit trees can rarely be expressed accurately in monetary terms. Apart from favoring disease development, adverse soil and environmental conditions reduce tree productivity in many other ways which cannot readily be distinguished from pathological effects.

Disease damage varies from subtle, almost undetectable effects to death of trees in some nurseries and orchards. In South Africa, bacterial canker caused by *P.s.* pv. *syringae* is one of the most important diseases of stone fruit crops, and annual damage probably exceeds $10 million (U.S.).[4] In southeast England, bacterial canker has been regarded as the most important single factor limiting the successful cultivation of stone fruit crops.[3] In the southeastern United States, bacterial canker is associated with the peach tree short life (PTSL) syndrome.[8] The peach tree population in Georgia alone dropped from 16 million in 1930 to 3 million in 1965.[9] *P.s.* pv. *persicae* has killed more than a million peach trees in France.[7] This pathovar also causes problems on nectarine, peach, and Japanese plum in New Zealand.[10] However, globally *P.s.* pv. *syringae* is regarded as the most damaging of the three pathovars.

17-3 SYMPTOMS

The type of symptom associated with bacterial canker of stone fruit caused by *P.s.* pv. *syringae* or *P.s.* pv. *morsprunorum* depends on the host cultivar, rootstock, age of the tree, origin of inoculum and plant part invaded, strain of pathogen, horticultural practices, and nature of the predisposing factors.[3-5,7] However, the essential features of the disease caused by either of the two pathovars are the same.[3] Cankers usually develop at the bud union, in pruning wounds, and at the base of infected spurs. New cankers are typically formed in late winter or early spring. Gum is often exuded from cankers, especially early in the growing season (Fig.17-1). Terminal shoots or twigs of cankered trees often die back (Fig.17-2). A diseased lateral or trunk girdled by a canker could die within weeks. Root systems of diseased trees usually remain functional, thus allowing development of suckers in the crown region.

The pathogen may be present in dormant leaf and flower buds.[11,12] Infected, dormant buds are often killed, but some invaded buds open normally in spring, only to collapse in early summer. Leaves from these buds wilt and fruit tends to dry out, yet leaves and flowers arising from other infected buds frequently remain symptomless. Blossom infection is followed by development of cankers on twigs and spurs. Dead flowers remain attached to trees.

Leaf infections, especially on cherry, appear as water-soaked spots which later become brown and dry.[5] Shot holes might be seen later. However, symptoms on leaves are apparently limited to some stone fruit cultivars growing in geographic regions with wet summers and high humidity.

Figure 17-1   Gum exuded from canker on plum branch.

17-4 CAUSAL ORGANISMS

P.s. pv. *syringae* and *P.s.* pv. *morsprunorum* fall in RNA group I of the genus *Pseudo-monas*.[13] They are polar-flagellated, Gram-negative straight rods, 0.7–1.2 × 1.5 μm. They do not accumulate poly-β-hydroxybutyrate as a carbon reserve material and are negative for arginine dihydrolase and oxidase. Most strains produce fluorescent pigments.

   Procedures for the isolation and identification of plant pathogenic fluorescent pseudomonads have been documented in detail.[14-16] Simple diagnostic tests, such as the GATTa scheme (for gelatin liquefaction, aesculin hydrolysis, tyrosinase activity, and tartrate utilization) can be used to distinguish between *P.s.* pv. *syringae* (G+A+T-Ta-) and *P.s.* pv. *morsprunorum* (G-A-T+Ta+).[17] However, intermediate forms are often en-

Figure 17-2   Cankered plum tree with symptoms of dieback.

countered. This is not surprising if the marked phenotypic heterogeneity of the two pathovars, particularly of *P.s.* pv. *syringae*, is considered.[18]

    P.s. pv. *persicae* is more closely related to *P.s.* pv. *morsprunorum* and to *P.s.* pv. *morsprunorum*. Unlike *P.s.* pv. *morsprunorum*, *P.s.* pv. *persicae* is nonfluorescent on King's medium B, but fluorescence has been recorded on casamino acid-sucrose agar.[19]

17-5 INOCULATION AND DISEASE RATING

Pathogenicity of *P. syringae* can conveniently be determined by inoculating immature fruit. Gross et al.[20] found that stone and pome fruit strains of *P.s.* pv. pv. *syringae* were pathogenic to immature cherry as well as pear fruit. In contrast, the strains of *P.s.* pv. *morsprunorum* tested were pathogenic to cherry fruit only. Pathogenic strains injected into fruit caused brown and sunken lesions that spread from the areas of application. Lesions produced by highly aggressive strains engulfed nearly the whole fruit after three days' incubation at 25 °C.

Care should be taken when interpreting results obtained with inoculated fruit. Endert and Ritchie[21] reported that lesions induced on cherry fruit correlated poorly with lesions on potted peach trees in the greenhouse, but cherry fruit were capable of detecting pathogenicity in some moderately to highly aggressive strains. However, pathogenicity as well as aggressiveness could be determined with confidence by inoculating etiolated peach seedlings.

Apart from aggressiveness and pathogenicity, the ability of *P. syringae* to invade deciduous fruit trees systemically is also important. Determination of invasiveness in symptomless trees is cumbersome and requires quantitative and qualitative detection of the pathogen in host tissue.[4] Pathogens in symptomless, invaded trees can cause disease under stressful conditions or can be transmitted to new trees through infected budwood material.

17-6 EPIDEMIOLOGY AND DISEASE CYCLE

The epidemiology of bacterial canker of stone fruit features briefly in the disease cycle shown in Fig.17-3. The complexity of the cycle is due to the excellent ability of *P. syringae* to adapt to an intricate perennial, deciduous host. Previous views[3] that canker is simply a cyclic disease in which a winter stage in the bark of the stems and branches alternates with a summer phase in the leaves and other tissues needs to be extended to accommodate more recent findings.

P. syringae overwinters in cankers, buds, and systemically inside other symptomless host tissue.[4] In addition, *P.s.* pv. *syringae* survives epiphytically on weeds in orchards.[22,23] The pathovar is now also known to spread from colonized peach blossoms that are not killed, to developing seed.[4] In turn, infected seed can give rise to infected seedlings. Symptomless, infected seedlings may serve as effective, yet unsuspected, carriers of the disease to new plantings.

Epiphytic populations of pathogenic *P. syringae* establish on the surface of symptomless leaves.[3] The bacteria are present in the buds and colonize new leaves as they emerge in spring. Population levels can fluctuate sharply from day to day and might even change within hours.[2] Early in the season, periods of frequent rainfall and high humidity, cool temperatures, and wind favor infection and dispersal of the pathogen. Secondary dispersal during the growing season ensures that inoculum is available throughout the orchard. Bacterial activity subsequently declines during the dry, hot summer months, and then increases in autumn.

Epiphytic populations of *P. syringae* are not restricted to the external surface of leaves. Substomatal chambers serve as protected sites[24] which enable the pathogen to survive adverse atmospheric conditions during warm, dry spells. After gaining entry through stomata, *P.s.* pv. *syringae* colonizes the intercellular spaces of the spongy parenchyma.[25] When uninvaded substomatal chambers are reached, bacteria can multiply profusely, and masses of these cells are extruded through stomata. It is possible that epiphytic populations are constantly replenished in this way. The pathogen probably moves through the parenchyma of the bundle sheath into the vascular system of veins. Aggressive strains migrate to axillary buds and to the twig supporting the leaf, thereby promoting long-term survival of the pathogen.

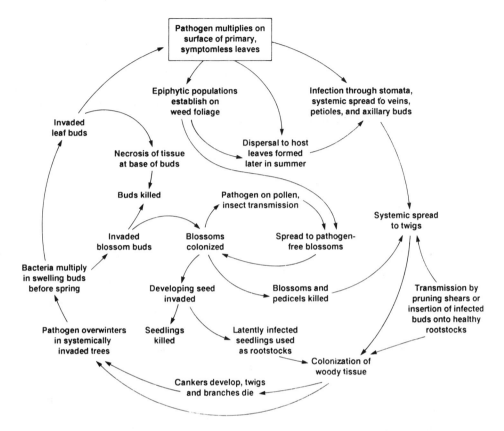

Figure 17-3 Disease cycle of bacterial canker of stone fruit caused by *Pseudomonas syringae*.[4] Reprinted with permission from *Plant Disease*, 73, 784; copyright 1989 by The American Phytopathological Society.

The trunk and laterals of trees are usually infected in autumn and winter.[3] Pruning cuts and other injuries provide points of entry. In England, most cankers caused by *P.s.* pv. *morsprunorum* on cherry trees originate through the leaf scars on fruiting spurs and extension shoots during autumn.[3] However, *P.s.* pv. *morsprunorum* or *P.s.* pv. *syringae* may also reach axillary buds by systemic spread well before leaf fall occurs. Cankers subsequently appear at the base of invaded buds. Canker development slows during winter but recommences at the beginning of the growing season. Further development is arrested later in spring when callus tissue forms. Populations of the pathogens in cankers decline during summer. Cankers can become completely inactivated or, if the pathogen has not been contained effectively, they can be perennial.

An aggressive strain of *P.s.* pv. *syringae* inoculated into plum petioles in spring also spreads to the xylem and other elements of leaf veins.[26] The pathogen multiplied and was exuded from stomata of these invaded leaves. In orchards, internal migration of the pathogen from stems and shoots to leaves might compensate for epiphytic populations lost during unfavorable conditions. Pockets of bacteria can readily be seen in invaded stems.

Differential interactions occur between strains of *P.s.* pv. *syringae* and plum cultivars, and probably also other stone fruit cultivars.[3] This can be shown by recording the response in expanding shoots inoculated with low population levels of the pathogen. Four extremes can be mentioned, but many intermediate reactions occur as well: (1) virtually no growth of the pathogen and no symptoms; (2) systemic spread but no symptoms; (3) pathogen confined to the lesion (canker); (4) extensive spread of pathogen beyond the canker. The nature of the response depends on both the pathogen strain and host cultivar. In general, populations of *P. syringae* decline in cankers on deciduous fruit trees during summer.[3] The rate of decline and level of the population inside the tree at the end of summer seem to depend on the particular host-pathogen combination.

17-7 PREDISPOSITION

Some of the stress factors that favor bacterial canker in the United States and elsewhere include freeze injury, wounds, nematode damage, and dual infections of *P.s.* pv. *syringae* and plant-pathogenic fungi such as *Cytospora* and *Nectria*.[7] Most if not all, factors that reduce tree vigor probably promote disease development.

Ice-nucleation-active strains of *P. syringae* occur commonly on stone fruit trees.[6,7,20] Lindow[6] proposed that *P. syringae* has evolved the capacity to predispose plant tissue to ice damage and subsequent bacterial penetration and disease development. Canker development has been related to low temperatures,[27-29] but it should be noted that the disease also occurs in some regions where freezing temperatures do not occur. Moreover, a recent study[30] has questioned the involvement of ice-nucleation-active bacteria in frost injury to flower buds. The intrinsic wood-associated ice nucleus was implicated as the initial and primary source of ice-nucleation during injurious frosts. However, freezing-thawing cycles in winter create water soaking in the bark of peach and apricot shoots.[28] This allows the rapid uptake and spread of bacteria in host tissue.

17-8 MANAGEMENT

As mentioned previously, trees with latent infections probably occur in all of the important fruit-producing regions of the world. Nevertheless, canker is unlikely to be a serious problem, provided crucial predisposing factors are avoided or counteracted. In this regard, the comprehensive 10-point plan[8] developed for controlling PTSL should also be consulted.

17-8-1 Avoidance of Marginal Conditions

New orchards established under marginal conditions are at risk. Trees are particularly susceptible in some sandy soils, in waterlogged soils that drain poorly, and during prolonged periods of drought. In the southeastern United States, PTSL has been greatly reduced by maintaining soil pH of 6.5 or above in orchards.[9] Also, the dormancy requirements of some cultivars are probably not properly met in regions that experience mild winters.

17-8-2 Horticultural Practices

Precautions should be taken during pruning to limit spread of the pathogens. If pruning is done in summer rather than autumn and winter, trees are likely to become infected.

Major outbreaks of bacterial canker in young orchards are often attributed to poor horticultural practices. The situation is undoubtedly aggravated if the pathogen has been introduced uniformly into young nursery trees through infected buds or rootstocks. These trees might not become diseased in nurseries where optimal growth conditions can be maintained with little difficulty. However, the trees are severely stressed during transplanting and therefore are more prone to disease development until they have become established in their new environment.

17-8-3 Selection of Rootstock

Selection of proper rootstocks for a particular geographical region is important.[31] In California, for example, plums on Lovell peach root, and French prune on both Lovell and Nemaguard, suffer less damage from bacterial canker than do those on Myrobalan or Marianna rootstocks.[32]

17-8-4 Chemical Sprays

Chemical control of bacterial canker in Europe and North America is based primarily on protective sprays with fixed copper or Bordeaux mixture in autumn and in spring before blossoming.[33,34] However, the wide distribution of copper-resistant strains of *P.s.* pv. *syringae* recovered from cherry blossoms in Michigan[35] indicates that fixed copper cannot be used indefinitely to reduce epiphytic populations. In general, only limited success in controlling the disease has been achieved by use of chemical sprays.[32] This might be understood if one considers the heterogeneous nature and versatility of the pathogens,[18] especially their outstanding ability to populate symptomless leaves and to invade host trees systemically.

17-9 CONCLUSIONS

Although bacterial canker of stone fruit can be highly destructive, *P. syringae* is regarded as being a weak pathogen. By growing trees under marginal or stressful conditions, humans have fostered the disease. In addition, there seems little doubt that movement of material required for propagating new trees has allowed the pathogen to be dispersed to all regions of the world where stone fruit is grown commercially. In this regard, the occurrence of bacterial canker in several Southern Hemisphere countries such as Australia, Chile, New Zealand, and South Africa is noteworthy. Stone fruit crops were introduced to these countries and United States by early settlers from Europe.

The development of more reliable methods to manage bacterial canker requires two broad approaches. First, more accurate quantification of stress factors is needed. We also need to know how these factors operate singly and in concert. Second, breeding for resistance to bacterial canker should be based on new basic knowledge of how the pathogen caused disease. For example, the toxin syringomycin produced by *P.s.* pv. *syringae* has been implicated in canker development. Two genes involved in syringomycin production by this pathovar have recently been identified and analyzed.[36] Finally, one of the major challenges facing plant breeders is to produce trees that are resistant to bacterial canker without compromising fruit quality. Application of modern methods to produce transgenic plants might overcome some of the obstacles to be expected in such an endeavor.

17-10 REFERENCES

1. Bradbury, J. F., *Guide to plant pathogenic bacteria*, CAB International Mycological Institute, Slough, United Kingdom, pp. 167, 169, 175, 1986.

2. Hirano, S. S. and Upper, C. D., Ecology and epidemiology of foliar bacterial plant pathogens, *Annu. Rev. Phytopathol.*, 21, 243, 1983.

3. Crosse, J. E., Epidemiological relations of the pseudomonad pathogens of deciduous fruit trees, *Annu. Rev. Phytopathol.*, 4, 291, 1966.

4. Hattingh, M. J., Roos, I. M. M. and Mansvelt, E. L., Infection and systemic invasion of deciduous fruit trees by *Pseudomanas syringae* in South Africa, *Plant Dis.*, 73, 784, 1989.

5. Cameron, H. R., Diseases of deciduous fruit trees incited by *Pseudomonas syringae* van Hall, *Oreg. Agric. Exp. St. Tech. Bull.*, 66, 1962.

6. Lindow, S. E., The role of bacterial ice nucleation in frost injury to plants, *Annu. Rev. Phytopathol.*, 21, 363, 1983.

7. Moore, L. W., *Pseudomonas syringae:* Disease and ice nucleation activity, *Ornamentals Northwest*, 12(2), 3, 1988.

8. Ritchie, D. F. and Clayton, C. N., Peach tree short life: A complex of interacting factors, *Plant Dis.*, 65, 462, 1981.

9. McGlohon, N. E., Management practices that are controlling peach diseases, *Plant Dis.*, 66, 7, 1982.

10. Young, J. M., *Pseudomonas syringae*, pv. *persicae* from nectarine, peach, and Japanese plum in New Zealand, *Bull. OEPP*, 18, 141, 1988.

11. Roos, I. M. M. and Hattingh, M. J., Pathogenic *Pseudomonas* spp. in stone fruit buds, *Phytophylactica*, 18, 7, 1987.

12. Sundin, G. W., Jones, A. L. and Olson, B. D., Overwintering and population dynamics of *Pseudomonas syringae* and *P.s.* pv. *morsprunorum* on sweet and sour cherry trees, *Can. J. Plant Pathol.*, 10, 281, 1988.

13. Palleroni, N. J., Genus I. *Pseudomonas* Migula. In *Bergey's manual of systematic bacteriology*, Vol. 1 (N. R. Krieg, and J. G. Holt, eds.), p. 141, Williams and Wilkins, Baltimore, 1984.

14. Fahy, P. C. and Lloyd, A. B., *Pseudomonas*: The fluorescent pseudomonads. In *Plant bacterial diseases—a diagnostic guide* (P. C. Fahy, and G. J. Persley, eds.), p. 141, Academic Press, Sydney, Australia, 1983.

15. Hildebrand, D. C., Schroth, M. N. and Sands, D. C., *Pseudomonas*. In *Laboratory guide for identification of plant pathogenic bacteria*, 2nd ed. (N. W. Schaad, ed), p. 60, APS Press, St. Paul, 1988.

16. Lelliott, R. A. and Stead, D. E., *Methods for diagnosis of bacterial diseases of plants*, Blackwell, Oxford, ch. 3, 1987.

17. Latorre, B. A. and Jones, A. L., *Pseudomonas morsprunorum,* the cause of bacterial canker of sour cherry in Michigan, and its association with *P. syringae, Phytopathology,* 69, 335, 1979.

18. Roos, I. M. M. and Hattingh, M. J., Pathogenicity and numerical analysis of phenotypic features of *Pseudomonas syringae* strains isolated from deciduous fruit trees, *Phytopathology,* 77, 900, 1987.

19. Luisetti, J., Prunier, J.-P. and Gardan, L., Un milieu pour la mise an évidence de la production d'un pigment fluorescent par *Pseudomonas mors-prunorum* f. sp. *persicae, Ann. Phytopathol.,* 4, 295, 1972.

20. Gross, D. C., Cody, Y. S., Proebsting, E. L., Jr., Radamaker, G. K. and Spotts, R. A., Ecotypes and pathogenicity of ice-nucleation-active *Pseudomonas syringae* isolated from deciduous fruit tree orchards, *Phytopathology,* 74, 241, 1984.

21. Endert, E. and Ritchie, D. F., Detection of pathogenicity, measurement of virulence, and determination of strain variation in *Pseudomonas syringae* pv. *syringae, Plant Dis.,* 68, 677, 1984.

22. Latorre, B. A. and Jones, A. L., Evaluation of weeds and plant refuse as potential source of inoculum of *Pseudomonas syringae* in bacterial canker of cherry, *Phytopathology,* 69, 1122, 1979.

23. Roos, I. M. M. and Hattingh, M. J., Weeds in orchards as potential source of inoculum for bacterial canker of stone fruit, *Phytophylactica,* 18, 5, 1986.

24. Roos, I. M. M. and Hattingh, M. J., Scanning electron microscopy of *Pseudomonas syringae* pv. *morsprunorum* on sweet cherry leaves, *Phytopathol. Z.,* 108, 18, 1983.

25. Roos, I. M. M. and Hattingh, M. J., Systemic invasion of cherry leaves and petioles by *Pseudomonas syringae* pv. *morsprunorum, Phytopathology,* 77, 1246, 1987.

26. Roos, I. M. M. and Hattingh, M. J., Systemic invasion of plum leaves and shoots by *Pseudomonas syringae* pv. *syringae* introduced into petioles, *Phytopathology,* 77, 1253, 1987.

27. Weaver, D. J., Interaction of *Pseudomonas syringae* and freezing in bacterial canker on excised peach twigs, *Phytopathology,* 68, 1460, 1978.

28. Vigouroux, A., Ingress and spread of *Pseudomonas* in stems of peach and apricot promoted by frost-related water-soaking of tissues, *Plant Dis.,* 73, 854, 1989.

29. Klement, Z., Rozsnyay, D. S., Báló, E., Pánczél, M. and Prileszky, Gy., The effect of cold on development of bacterial canker in apricot trees infected by *Pseudomonas syringae* pv. *syringae, Physiol. Plant Pathol.,* 24, 237, 1984.

30. Proebsting, E. L., Jr. and Gross, D. C., Field evaluations of frost injury to deciduous fruit trees as influenced by ice nucleation-active *Pseudomonas syringae, J. Am. Hortic. Sci.,* 113, 498, 1988.

31. Garrett, C. M. E., Influence of rootstock on the susceptibility of sweet cherry scions to bacterial canker, caused by *Pseudomonas syringae* pvs *morsprunorum* and *syringae, Plant Pathol.,* 35, 114, 1986.

32. English, H., Devay, J. E., Ogawa, J. M. and Lownsbery, B. F., Bacterial canker and blast of deciduous fruits, *Div. Agric. Sci. Univ. Calif.,* Leaflet 2155, 1980.

33. Agrios, G. N., *Plant pathology,* 3rd ed., Academic Press, San Diego, p. 565, 1988.

34. Olson, B. D. and Jones, A. L., Reduction of *Pseudomonas syringae* pv. *morsprunorum* on Montmorency sour cherry with copper and dynamics of the copper residues, *Phytopathology,* 73, 1520, 1983.

35. Sundin, G. W., Jones, A. L. and Fulbright, D. W., Copper resistance in *Pseudomonas syringae* pv. *syringae* from cherry orchards and its associated transfer *in vitro* and *in planta* with a plasmid, *Phytopathology*, 79, 861, 1989.

36. Xu, G.-W. and Gross, D. C., Physical and functional analyses of the *syrA* and *syrB* genes involved in syringomycin production by *Pseudomonas syringae* pv. *syringae*, *J. Bacteriol.*, 170, 5680, 1988.

18

POSTHARVEST DISEASES
OF TREE FRUIT

JOSEPH M. OGAWA, RONALD M. SONODA, AND HARLEY ENGLISH

First and third authors are from Department of Plant Pathology, University of California, Davis, California, and the second author is from the University of Florida, IFAS, AREC, Ft. Pierce, Florida.

18-1 INTRODUCTION

Fresh fruits serve as key sources of nutrition for many of the over 4 billion human inhabitants of our world. In some areas, fruit crops are the main dietary staple. In many regions, fresh fruits, with their high water content, serve as a clean source of that essential ingredient and, in addition, supply food energy, protein, fat, carbohydrate, ash, calcium, phosphorous, iron, sodium, potassium, vitamin A, thiamine, riboflavin, niacin, and ascorbic acid.[1] In modern technological societies, fruits with their relatively high fiber content and low fat content have become an essential ingredient for healthier nutrition.[1]

In less developed economies, fruits, which are the most perishable of commodities, often are only seasonally available. They are picked when nearly ripe and sold mostly to consumers at the farm or at local markets. In these situations, losses to postharvest disease can be heavy, but accurate estimates of fruit losses are difficult to obtain. In more technically advanced societies, a diversity of fresh fruit is available on a year-round basis. The availability of these fruit is the result of relatively recent advances in harvesting, handling, storage, transportation, and disease-control technologies, as well as advances in marketing techniques. These advances in technology have changed the eating habits of a majority of the people in the more advanced economies. In fact, the availability of wholesome, tasty, fresh fruit is now expected at all times of the year by these consumers. However, even in economies with advanced technologies, postharvest diseases frequently are a significant problem to the ultimate consumer and also interfere with the economic well-being of the various people along the chain from the farm to the consumer.

Postharvest losses for crops of various types are estimated to be between 10 and 30%[2] and for perishable tree fruit estimates can reach higher percentages. Prevention of decay has been a major challenge in the marketing of fresh fruit.[2] In this discussion, we highlight the factors in postharvest fruit decay which relate to the economics of marketing decay-free high-quality fruit.

18-2 DISTRIBUTION AND ECONOMIC IMPORTANCE

The main tree fruit crops grown in temperate zones include stone fruits (apricots, nectarines, peaches, plums, and sweet cherries) and pome fruits (apples and pears). Semitropical fruits include avocados, citrus, feijoas, figs, and kiwis. Important tropical tree fruits include bananas and papayas. The marketable period for these fresh fruits ranges from a few days to a few months, and longer when they are held under cold storage or in a modified atmosphere environment. The year-round availability of some of these crops benefits from the harvest period coupled with the use of long-distance shipping, especially between the northern and southern hemispheres.

Postharvest losses, including those due to parasitic diseases, increase with the increase in time between harvest and consumption. Delay in consumption may be due to the distance from farm to market, or it may be due to lack of demand or a temporary oversupply, with the consequent extension of storage in the hope for a better market. Multiple handling of fruit also increases postharvest losses caused by diseases and other postharvest problems. When fruit losses occur, it is often difficult to determine whether the primary causal factor was biotic or abiotic. Some postharvest pathogens of fruit are natural microflora of healthy produce and become a problem only when one or more physical, physiological, or biological factors come into play. Wounding or bruising of fruit by improper or excessive handling often provides avenues of infection for fruit rot pathogens. Bruising and wounding also may be the more important physical steps in the development of some nonpathogen-induced postharvest problems.

With some types of fruit, a few decay lesions may result in loss of profit to the grower, packer, shipper, broker, grocer, or consumer. The point at which the direct loss occurs may have a different impact on the total economic loss suffered.[3] The major pathogens of important fruit crops are presently known, but the interactions when more than one pathogen is involved are not clear. Identification of the causes of losses is often difficult because interaction of both biotic and abiotic factors is involved in most postharvest loss situations.[4] At shipping and receiving points, current diagnosis of the fungal or bacterial agent responsible for decay is largely based on the ability of the diagnostician to identify disease signs and symptoms. Since this type of disease assessment is not wholly accurate, more sophisticated techniques for positive identification of decay pathogens are required and, fortunately, some are currently being developed.

Decreases in the storage period and increases in transportation speed have permitted wider distribution of fresh fruit, both domestically and internationally. Yet new methods that would increase the storage life of fruit would make international marketing even more desirable without the increased costs associated with rapid transportation.

18-3 CAUSAL ORGANISMS

In this discussion, we have not included postharvest disorders of crops caused by abiotic agents such as nutrition, respiration, temperature, or chemicals. Abiotic diseases are well covered in texts by Snowdon,[5] Jones and Aldwinckle[6] and Ogawa and English.[7]

The most prevalent postharvest decay-causing organisms of fruit throughout the world are fungi (Fig.18-1).[5,7] The *Erwinia carotovora-chrysanthemi* complex of bacteria and *E. amylovora* are occasionally associated with fruit decay, and a few other bacteria also have been reported. Viruses and other biotic agents are rarely implicated in fruit decay situations. The genera and species of fungi that cause fruit decay are numerous.[4,8,9] The fungi that cause the most important postharvest diseases of temperate-zone, subtropical, and tropical crops will be used in our discussion of the factors that should be considered in studies on the nature and management of these diseases.

The decay-causing fungi of temperate-zone fruit that will be considered are *Monilinia fructicola* (Wint.) Honey, *Rhizopus stolonifer* (Ehrenb. : Fr.) Vuill., *Mucor piriformis* E. Fisch., *Gilbertella persicaria* (E. D. Eddy) Hesseltine, *Botrytis cinerea* Pers. : Fr., *Penicillium expansum* Link, *Alternaria alternata* (Fr. : Fr.) Keissl., and *Glomerella cingulata* (Stonem.) Spauld. and Schrenk (anamorph *Colletotrichum gloeosporioides* (Penz.) Penz. & Sacc. in Penz.).

The important postharvest pathogens of the tropical and subtropical fruits considered are *G. cingulata, Collectotrichum musae* (Berk. and Curt.) Arx, *Physalospora rhodina* Berk. and Curt. apud Cooke (anamorph *Botryodiplodia theobromae* Pat.), *Cerato-*

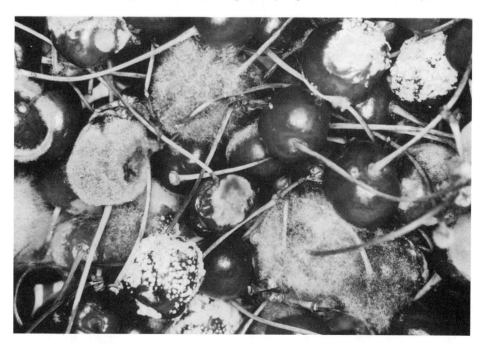

Figure 18-1   Sweet cherry fruits with decay from a complex of fungal decay organisims (*Botrytis, Rhizopus, Monilinia, Penicillium, Alternaria*, etc.).

cystis paradoxa (Date) Moreau, *Botryosphaeria ribis* Grossenb. and Duggar, *Penicillium digitatum* Sacc., *P. italicum* Wehmer, and *Diaporthe citri* Wolf (anamorph *Phomopsis citri* Fawc.). There are excellent sources of lists and descriptions of the many other fungi that can cause significant postharvest losses under certain circumstances.[10-14]

18-4 SYMPTOMS

Monilinia fructicola and the related species *M. laxa* (Aderh. and Ruhl.) Honey and *M. fructigena* (Aderh. and Ruhl.) Honey cause brown rot of stone and pome fruit and are scattered throughout the temperate areas of the world where their hosts are grown. *M. fructicola* causes major postharvest decay problems on stone fruit in the United States, Australia, Brazil, and Japan. However, *M. fructicola* has not been reported from Europe. In addition to causing pre- and postharvest decay, the pathogen infects blossoms, young twigs, and green fruit. In the orchard, symptoms on green fruit are minute necrotic lesions which may become quiescent. In the orchard, developing lesions are initially small, superficial, circular, brown spots that continue to increase in size when the environment is suitable. Conidiophores and conidia are soon produced. The infected area becomes brownish-black (pseudosclerotia), and when a fruit is completely rotted it becomes shriveled and mummified. In packed boxes, infections are most common on the stem end of the fruit, and the lesions, in general, are similar to those that occur in the orchard. Sporulation (sporodochia), however, varies under different storage environments and may not develop when fruit are not exposed to light. The rotted tissue is firm, and parts of the fruit flesh not yet involved may have an off flavor.[7]

Rhizopus stolonifer causes postharvest diseases of many fruits and vegetables. Also, it occasionally attacks growing plants.[5] Although generally reported from temperate areas, it is reportedly destructive on papayas in tropical regions.[9,10] *R. stolonifer* is a good colonizer of plant debris, and thus its spores are almost ubiquitous and also are easily carried by air currents. Because of the rapidity with which it spreads from a decayed fruit to an adjacent fruit at room temperature, the pathogen is one of the most feared of postharvest decay fungi. The fungus produces a watery, soft rot but is not active at temperatures below 4.4°C. The whisker-like mycelium plus extending stolons and sporangiophores of the fungus soon cover the fruit. *M. piriformis* and *G. persicaria* behave like *R. stolonifer,* in general, but *M. piriformis* can decay fruit at 0°C and *G. persicaria* at 36°C; also the latter species do not form nest rot as quickly as *R. stolonifer* because stolons are not produced. They, like *R. stolonifer*, cause postharvest decay of several temperate-zone fruit crops. The decay symptoms caused by *R. stolonifer, M. piriformis,* and *G. persicaria* are similar, but the signs (external fungus structures) are characteristic for each pathogen (Fig.18-2).

Gray-mold rot caused by *Botrytis cinerea* has symptoms similar to those induced by *M. fructicola.*[7] The surface of infected areas of the fruit usually has a soft-grayish appearance due to sporulation structures of the fungus. *B. cinerea* is found on fruit crops throughout the temperate zone, where, in addition to postharvest decay, it sometimes causes blossom blight, green fruit rot, and, at times, shoot blight. It is a common pathogen of many other crops and is especially severe in a cool, wet environment. It is an important postharvest decay pathogen in temperate regions because of its extensive host

Figure 18-2 Peach fruit decaying with *Rhizopus stolonifer* (area with black sporangia) and *Gilbertella persicaria* (stem-end area with collapsed sporangia).

range and its ability to cause nest rot even under cold-storage conditions.

Members of the genus *Penicillium* are important postharvest pathogens of fruit.[11,13,14] Blue-green mold, *P. expansum*, is an important postharvest pathogen of temperate-zone stone and pome fruit, while green mold, *P. digitatum*, and blue mold, *P. italicum*, are important pathogens of the subtropical citrus crops. On stone fruit, blue-green mold is an important postharvest pathogen of sweet cherries and apricots but is usually of minimal concern on other species. Field infections by *Penicillium* are rare, but the pathogen is ubiquitous and a few infected and sporulating fruits on the orchard floor can produce massive numbers of wind-dispersed spores. Thus, harvested fruit are contaminated with spores of *Penicillium*, and injured areas resulting from harvesting operations and transportation to the packing shed serve as infection sites. On lemons, the cold-storage holding rooms serve as a locale for *Penicillium* infections. The fungus can decay fruit in cold storage and produce massive numbers of spores which contaminate other fruit, the storage area, and the packing shed (Fig.18-3).

Alternaria alternata causes black mold of many fruits[5,7] and vegetables. Decay problems with *Alternaria* on fruit are most serious on those that are ripe or overripe. On sweet cherries, *Alternaria* is most serious on fruit doubles and spurs and on rain-damaged

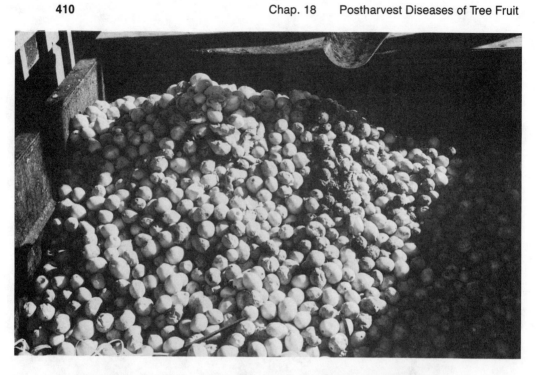

Figure 18-3 Lemon fruit with *Penicillium* decay being culled after long-term storage.

Figure 18-4 Nontreated peach fruits with *Rhizopus* and *Monilinia* decay (left) compared to fungicide treated fruit (right).

Figure 18-5   Pineapple fruit with basal shoot attached to keep stem alive and prevent fungal decay.

fruit as the fungus enters through wounds and weakened tissue. At times *Alternaria* decay can be confused with a similar disease caused by *Cladosporium* sp., but the spore characteristics of the two fungi provide easy identification.

Fungi identified as *Glomerella cingulata* or its anamorph, *Colletotrichum gloeosporioides*, are extremely common and important postharvest pathogens of many fruit crops. This pathogen is most common in warm climates with high rainfall. It is a major postharvest decay fungus on tropical and subtropical fruit. The pathogen has not been a problem on fruit crops in California (detected on peaches in 1990 by Dr. A. Feliciano), although it is serious on stone fruit in the eastern United States and in Brazil. It is the most important postharvest pathogen on mangoes, avocados, papayas, guavas, and some other tropical and subtropical fruit. It can also be a problem on mandarin and on mature-green citrus fruit treated with ethylene during ripening.[11] A closely related fungus, *C. musae*, is an important postharvest decay agent on banana.[12] Most of the *C. gloeosporioides* strains that attack growing plant parts have relatively defined host ranges. However, the relationship among the *C. gloeosporioides* strains that cause postharvest decay has not been extensively studied. Postharvest diseases caused by this group of fungi can with infection at almost any stage of fruit development, which start is usually followed by a quiescent period. As fruit ripen after harvest, dark spots develop and increase in size. The fruiting structures (acervuli) of these pathogens, with initially salmon-pink spore masses which eventually turn brownish, can soon be seen on the dark lesions. Most strains of this fungus produce setae which are visible with a low-power lens. The lesions on the various kinds of fruit may be either superficial or deeply penetrating.

Important postharvest diseases of many types of fruit in tropical and subtropical areas with heavy rainfall, especially near the time of fruit harvest, are the stem-end rots. The various stem-end rots are caused by a complex of fungi. Probably the most frequently encountered and most important is a fungus currently referred to as *Physalospora rhodina* Berk. and Curt. apud Cooke or as *Botryosphaeria rhodina* (Cooke) Arx in its teleomorph and as *Botryodiplodia theobromae* Pat. or *Diplodia natalensis* in its anamorph.[7] The fungus penetrates harvest wounds and also causes infection and a discoloration around the stem end in most fruits. In citrus, the fungus spreads down the axis of the fruit to the stylar-end.[11,13,14] The rind eventually turns dark brown. Other important fungi in the stem-end complex are *B. ribis* and *Diaporthe citri* Wolf on citrus and *D. citri* on mango. *P. rhodina* causes finger rot and several other diseases of banana. The most important pathogen of stem-end rot of papaya in Hawaii is *Mycosphaerella* sp.[10]

The important postharvest diseases of banana are those caused by several relatively weak pathogens that enter the cut surface of the main stalk and spread into the crown and pedicel of fruit transported on main stems or through the cut surface of the crown tissue when shipped as hands.[10] These wound pathogens include *Ceratocystis paradoxa, C. musae, P. rhodina*, and more recently the benomyl resistant *Fusarium oxysporum* pv. *roseum* and *Acremonium* sp.[9] Symptoms of infection of the stalk or crown are a black soft rot that begins at or near the cut surface. Mycelial growth may be visible on the surface of decaying tissue.

18-5 DISEASE CYCLE AND EPIDEMIOLOGY

Disease cycles for postharvest diseases can usually be divided into three general categories: (1) Inoculum is present in the orchard; infection occurs before the fruit is harvested and sometimes is followed by a quiescent period in which little if any fungus activity occurs; then a period of aggressive pathogen growth occurs in mature and ripe fruit tissue, followed by production of inoculum (e. g., *Monilinia* decay of stone fruits, *Colletotrichum* rot of tropical fruits); (2) most inoculation, infection, and aggressive growth of the fungus occurs after the fruit is harvested; infection can occur immediately after the fruit is harvested or when it is being transported, and it also can take place in the packing house or the storage room (e. g., *Rhizopus* rot on fruit crops, *Penicillium* decay of citrus fruit), or (3) fruit are infected in the orchard and the fungus grows aggressively inside the core of the fruit, but symptoms sometimes may not become visible for many months (e. g., *Alternaria* or *Mucor* decay on apples). Most of the pathogens involved in postharvest diseases follow only one of these cycles, but some may be involved in all three. For most postharvest pathogens, one of the three cycles is predominantly followed. In addition to the foregoing, secondary spread from fruit to fruit (nest rot) can occur in storage when fruit are loose packed or when special devices to separate them are not used.

Inoculum for pathogens that infect fruit before harvest can be derived from within the crop or from other infected plant tissue, including plant debris on the orchard floor or in an adjacent field. The inoculum of a few of these pathogens occurs only sporadically and is interrelated with the host life cycle, while that of others is present throughout the year. The environment under which fruit have grown, cultural practices, age of plant, ripeness of fruit, and application of pesticides are some of the factors that interact to influence the susceptibility of fruit to postharvest diseases and the rate and extent of their development. Many of the postharvest pathogens that attack fruit after they are harvested are part of the normal microflora of healthy produce. These fungi may also be present in debris or decayed fruit in the orchard, packing house, or storage unit. They generally become a problem once a trigger situation (appropriate temperature and free moisture) or a series of physical, physiological, or biological steps occurs.

For stone fruit, the major postharvest decay pathogen is *M. fructicola,*[7] although other *Monilinia* species can also be serious.[15] In comparison with most other postharvest pathogens, *Monilinia* has a relatively restricted host range which includes only stone and pome fruit. Its disease cycle is closely related to the host cycle of development. *M. fructicola* overwinters in rotted dried fruit (mummies), infected fruit peduncles, blighted blossoms, and twig cankers, all of which can serve as sources of conidial inoculum in the spring. Mummified fruit which fall to the ground may produce ascospores (from apothecia), which also cause spring infection. During periods of wetness and appropriate temperatures, all blossom parts and, at times, young leaves usually can become infected. With peach blossoms, however, only the anthers are ordinarily infected. Rain during the bloom period results in sporodochial formation and conidial inoculum buildup on blighted blossoms. Occasionally, young fruit are infected and the pathogen survives as quiescent infections until fruit are mature, but in other instances the lesions develop and the fruit become mummified with heavy sporulation.[16,17] Blighted blossoms and mummified green fruit serve as sources of inoculum for contamination and infection of mature fruit. Conidial inoculum may be windborne or rain splashed. Insects, such as nitidulid beetles, may serve as vectors of conidia from decaying fruit on the orchard floor to injured fruit on the tree. Free moisture is required for germination of *M. fructicola* conidia, and although infection can take place through the uninjured fruit surface it more commonly occurs through injured tissues. The temperature optimum for infection and disease development is about 24°C, but decay can develop in storage at temperatures of 0 to 5°C. Rot develops rapidly at commercial fruit-ripening temperatures (20°C), and the fungus can move by contact from decaying fruit to adjacent healthy fruit. Once infection is established, decay symptoms can be expressed within 24 hours at room temperature. Spores often form in abundance on decaying fruit in storage and serve as inoculum for infection of other fruit. The harvesting, packaging, and retailing processes can open infection sites (injuries) on the fruit, and free moisture from fruit injuries or condensation (changes in ambient temperature) is sufficient to establish new infections.

Inoculum of *C. gloeosporioides* can survive in previously infected plant parts that have become debris on the orchard floor or in infected tissue still on the tree. Although both sexual and asexual spore production occur with most biotypes of the fungus, most infection of fruit is thought to be from conidia, as these represent the principal inoculum present during infection periods. Invasion of most tropical fruits can occur at any time during the growing season, as rainfall usually is common and temperatures are generally favorable. Infections on most immature tropical fruits (mangoes, avocados, papayas, etc.) are quiescent.[9] Visible lesions develop when fruit become mature. Fruit at harvest are often contaminated with conidia that can cause infection during and after harvest. The fungus grows at a wide range of temperatures, including those used for storage of tropical and subtropical crops. These crops are stored at relatively warm temperatures, as they are prone to chilling injury.

Botrytis cinerea is a ubiquitous fungus that survives well in plant debris. The pathogen infects many different crops.[5,8] It attacks host tissues that are either very young, senescing, or injured. In temperate areas, in the spring, the fungus attacks young living tissue but prefers senescing tissue, especially that of blossom parts. In subtropical areas, injury of tissue during cool, wet weather may initiate an epidemic, as conidial inoculum builds up rapidly on damaged tissue, such as blighted blossoms, decaying fruits, or senescing vegetative parts. Infection by the fungus can be quiescent but usually is not. The fungus produces massive numbers of conidia soon after lesions begin to develop. Although the pathogen is favored by cool, moist weather, inoculum can form on infected fruit under most climatic conditions. This inoculum can provide both preharvest and postharvest fruit infection.

Rhizopus stolonifer, M. piriformis, and *G. persicaria* are mucoraceous fungi which survive in soil and debris in the orchard or in and around packing houses.[7] Released spores can survive under long dry periods and relatively high temperatures. However, they usually do not cause direct infection of healthy tissue. In California peach orchards, they usually colonize either injured fruit or overripe fruit that fall to the orchard floor. Open wounds on these fruit serve as infection courts. Preharvest infection by these pathogens commonly occurs in fruit injured by insects or affected by physiological disorders such as split pit. Insects are vectors of spores for infection of injured fruit both before and after harvest. These fungi sporulate profusely on diseased fruit. Their spores are airborne and present in both orchard and packaging house air; thus, they are able to contaminate fruit both preharvest and postharvest. Free moisture is required for spore germination and infection. Rot caused by *R. stolonifer* develops rapidly from infected to adjacent healthy fruit by stolons, making it one of the most feared postharvest pathogens of temperate-zone crops.

The disease cycle of the *Penicillium* species is similar to that of the mucoraceous fungi. *P. expansum* on apple, and *P. digitatum* and *P. italicum* on citrus are pathogens that produce tremendous numbers of conidia that are easily spread in air currents, especially when the substrate they arise from is physically moved. *Penicillium* spp. survive on organic residues in the soil and infect fruit that drop to the orchard floor. The abundant spores produced on the diseased tissue are windborne and contaminate the fruit on the tree. *P. expansum* usually does not attack apples on the tree; however, *P. digitatum* and *P. italicum* are ubiquitous in citrus orchards and easily infect wounds on citrus fruit still on the tree. Conidia in the packing house atmosphere can infect fruit wounded during the packaging operation. *P. digitatum* and *P. italicum* not only decay fruit, but the masses of conidia they produce taint fruit surfaces and make such fruit unmarketable because of the extreme difficulty of removing them. *P. italicum* can spread by contact from an infected fruit to healthy unwounded fruit and cause extensive losses in both storage and transit. Ethylene production during fruit decay by *P. digitatum* can accelerate the ripening of healthy fruit and make them more prone to infection and decay.

Physalospora rhodina and *Diplodia citri* are involved in many postharvest problems of tropical and subtropical fruits. These fungi are relatively weak pathogens. They are saprophytes that inhabit dead branches or bark of various fruit trees and are present in plant debris on the orchard floor. Both are common in causing postharvest diseases in high rainfall areas. Inoculum for infection is produced during humid weather, and spore dispersal occurs during rain. *D. citri* can cause quiescent infection by direct penetration, but *P. rhodina* infection commonly occurs in the button, and the mycelium then invades the axis of the fruit and grows toward the stylar end; whereas *D. citri* spores are lodged underneath the button, and infection and decay begin to develop after harvest. Both fungi may be present in the same fruit. *D. citri* is slower than *P. rhodina* as a fruit rotter. Disease caused by these fungi may occur occasionally on the tree, but it usually develops 10 to 20 days after harvest. In banana, *P. rhodina* causes a finger rot. The fungus apparently can infect banana through decaying flowers or through wounds at the time of harvest.

18-6 MANAGEMENT OF POSTHARVEST DECAY

18-6-1 Preharvest Treatments

There are many opportunities during fruit production to minimize postharvest losses due to fruit-decay organisms. This begins with selection of both genetic material to be planted and the planting site. There are differences in susceptibility of cultivars to both production diseases and postharvest diseases.[18] For *Monilinia*, recent studies have shown differences in resistance between the peach cultivar Bolinha from Brazil and commercial peach cultivars grown in California.[19] Differences in resistance may be due to differences in epidermal morphology.[20] Studies on cherries show a correlation between cuticular thickness and resistance to infection following surface inoculation.[21] However, in general, among the major commercial fruit crops, differences in cultivar resistance to pathogen-induced fruit decay are, with few exceptions, minimal.[18]

Selection of planting sites where the inoculum potential of both production and postharvest diseases is minimal is important. The interaction of planting site and cultivar also is an important consideration. Some cultivars tolerate diseases well in one area but not in another. Cultivars may be selected on the basis of how much postharvest disease can be tolerated by the intended market. Cultivar selection should also be based on what sorts of postharvest treatments the fruit will have to undergo to be marketed. There also are cultivar differences in tolerance to various handling practices and to refrigeration temperature. For instance, some tropical fruit cultivars are more sensitive than others to low temperature, and physiological changes in these cultivars during storage may lead to greater losses from disease. The planting site, in terms of distance from the packing house or storage facility, condition of roadways, and types of hauling equipment available are important considerations. There is a need to minimize the time required to move harvested fruit from the orchard to the storage facility and to minimize the damage that can result from transporting fruit on rough roads.

The planting material used may affect the amount of postharvest disease that develops; this is especially true when transplants are used. *Phytophthora* spp. that cause brown rot of citrus fruit are often brought into new planting areas on transplants. There is evidence that other pathogens, such as *Colletotrichum* and *Monilinia*, may be introduced in a similar manner. This may be a common mechanism of moving disease susceptible crops as well as postharvest tree-fruit pathogens to new areas.

Research efforts to determine if reducing field inoculum would be an effective means of controlling decay have been minimal. Management techniques to reduce inoculum throughout the growing season should reduce problems with postharvest pathogens that infect fruit during the growing period and survive in quiescent infections until the fruit mature. Inoculum reduction by sanitation should be useful, especially in cases where fallen fruit and other plant debris become infected and serve as sources of inoculum. Reduction of the source (fruit mummies) of primary inoculum at the end of the cropping period and before the next crop is set has been attempted with brown rot of peaches caused by *Monilinia fructicola*, but the results in terms of disease reduction, although showing some promise, have been difficult to evaluate (unpublished data by authors). Regardless of these results, this principle is a valid one to explore and possibly put into practice. It is an example of an approach that can be made with a postharvest fruit disease that affects other parts of the host. Inoculum reduction by sanitation should be useful in cases where plant debris, including fallen fruit, becomes infected and serves as a source of inoculum. The effect of pruning dead and dying branches which serve as sources of inoculum for pathogens such as *P. rhodina* and *D. citri* also needs study. Many sanitation procedures, such as the pruning out of diseased material, have been suggested, but little research has been done on the effectiveness of these procedures in reducing either inoculum or disease. Studies have shown that there is a correlation between the population of spores of *Penicillium* in the air in packing sheds and the incidence of postharvest decay. Further studies of this sort are needed to determine if greater effort should be placed on the reduction of inoculum in postharvest situations, especially at the site of fruit packaging.

The application of preharvest sprays should be carefully considered and used only in cases where there is a historical problem with pathogens that are amenable to control with fungicides and where research has indicated that a preharvest spray is necessary. Thorough studies should be made so fungicides are applied only when there is a benefit. Monitoring and forecasting systems being developed should aid in fulfilling these requirements in the future.

The maturity of fruit at harvest is a critical factor in the postharvest life of fruit. Fruit maturity at harvest affects not only its quality but also its susceptibility to different sorts of physical and physiological damage and its susceptibility to different pathogens during the postharvest period. Different cultivars of the same fruit species can have different optimum times for picking. Moisture and temperature conditions at harvest can have differential effects on the postharvest disease by increasing pathogen inoculum and spore germination as well as by preconditioning fruit for infection. Also critical in the decision to harvest is the ready availability of storage space and the rapidity with which the fruit can be processed and marketed.

Many observers believe that harvesting is the most critical process in terms of postharvest shelf life and postharvest susceptibility to disease. The bruising and wounding of fruit during harvest and transportation to the packing shed provide sites of infection for postharvest pathogens. There are three important factors that affect the level of wounding: the actual harvesting mechanism—whether by hand or machine; the training and ability of the people involved in the harvest operation; and the container into which the harvested fruit are placed. For hand picking, personnel must be trained to pick fruit as damage-free as possible. At present, mechanical harvesting is much more prone than hand picking to cause damage that can serve as ingress sites for postharvest pathogens. Fruit for fresh market generally are not mechanically harvested.

Selection of the type of container in which fruit are initially placed should be based on whether this is to be the container for the fruit when it is sold to the consumer or only for its temporary holding. Minimal handling of fruit obviously should be the course of choice. Regardless of how often the fruit are handled, containers should be designed and filled in such a way that bruising and wounding are minimized. They should be clean and when reused should be free of debris (washed) and sanitized to counteract the possibility that they serve as a source of inoculum. Containers can harbor common postharvest pathogens[22] for extended periods.

Containers with fruit should be protected from heat and rain and moved as quickly as possible from the orchard to the packing house or storage facility. Roadways should be selected or prepared so movement of fruit within containers is minimized. Transporting machinery must be rid of residues or rubbish that could harbor postharvest pathogens. Fruit that have been harvested at high temperature should be cooled as quickly as possible to an acceptable temperature.

18-6-2 Postharvest Treatments

Fruit requiring postharvest fungicide treatment[23-26] should be treated as soon as possible after harvest. This will minimize establishment of infections during or immediately following the harvesting process. Different fruit species as well as, in some cases, different cultivars of the same species may require different types of treatments based on preharvest conditions or their susceptibility to postharvest pathogens.[27] Sanitization of the packing house is one of the key requirements for reducing postharvest decay. Belts used for transporting the fruit must be kept clean, dump tank water sanitized, and debris removed. If chlorine is used in the dump tank water, or in the hydrocooler, its concentration as well as the acidity of the solution must be monitored. Effectiveness of the chlorine treatment is determined by the concentration of available hypochlorous acid. The equilibrium of relative concentration of HOCl depends on the pH of the solution. Hypochlorous acid becomes more unstable as pH decreases and chlorine is lost as a gas from the solution. Also, as soil and debris accumulate in the dump tank, chlorine is gradually inactivated because of its tendency to combine with this extraneous material. The efficiency of chlorine depends not only on HOCl availability but also on the length of contact time between the target organisms and the active ingredient. In the dump tank and in hydrocooling systems, contact between HOCl and organisms on fruit surfaces may range up to 20 minutes. When a chlorine solution is sprayed on fruit, the contact time is relatively short; therefore, concentrations need to be higher. Free spores of pathogens in the dump tank are more easily killed than spores on fruit surfaces, and spores in injuries are even more difficult to kill.

The efficacy of the postharvest fungicide depends on several factors, such as its chemical structure, formulation, penetrating quality, residual half-life, and mixing properties with waxes and oils which are commonly used to reduce moisture loss of fruit. In addition, when biocontrol agents become available, their mixture with fungicides must not reduce their viability. Yet the spectrum of activity of these treatments must be sufficient to kill, prevent, or suppress the complex of major decay pathogens under postharvest handling procedures and storage conditions.

Before the fruit are packaged, they often are sprayed with water containing a detergent, rinsed in fresh potable water, and treated with a fungicide-wax formulation. The fungicides used have different spectra of activity against postharvest pathogens (Table 18-1). Of the various fungicidal treatments, HOCl (chlorine) is the most widely applied

TABLE 18-1: SPECTRUM OF ACTIVITY OF POSTHARVEST FUNGICIDE TREATMENTS ON DECAY PATHOGENS[a]

Pathogens	HOC1	DCNA	Benzimidazoles	Iprodione	Triforine
Alternaria	+	−	−	++	−
Botrytis[b]	+	+	++	++	−
Cladosporium	+	−	++		−
Colletotrichum	+		++		
Gilbertella	+	−	−	−	−
Monilinia[bc]	+	+	++	++	++
Mucor	+	−	−	−	−
Penicillium[b]	+	−	++	++	−
Rhizopus stolonifer	+	++	−	+	−
Other *Rhizopus* spp.	+	−	−		−

[a] + = partial control; ++ = effective control; − = not effective.
[b] Some strains resistant to benzimidazoles.
[c] Some strains resistant to dicarboximides (iprodione and vinclozolin).

treatment. It is used as a general disinfestant (dip, drench, or spray) and does not act as a disinfectant, nor is it capable of killing all spores located in injuries. Also under study as postharvest disinfestants of wash water and the fruit surface are ozone and chlorine dioxide, but at present the feasibility of their commercial use has not been established. The greatest benefit of chlorine is its ability to kill decay pathogens provided its concentration is adequately maintained (fruit crops, 50 to 100 ppm with pH at approximately 7.0). DCNA (Botran, The Upjohn Company, Kalamazoo, Michigan) is highly effective in the control of *Rhizopus stolonifer* but not other *Rhizopus* species such as *R. arrhizus* and *R. oryzae* or other mucoraceous pathogens. Iprodione (Rovral, Rhone-Poulenc in France) is less active on *R. stolonifer*, but its activity against other *Rhizopus* species has not been tested. Iprodione, however, is active against a wide range of other important decay pathogens, including *Monilinia, Botrytis,* and *Alternaria*, and has been registered for postharvest use in spray or dip treatments on sweet cherries, peaches, plums, and nectarines in the United States. Iprodione shows localized penetration similar to that of thiophanate methyl. The combination of benomyl or thiophanate methyl with DCNA was considered the most effective postharvest treatment for stone fruit until the development of benzimidazole-resistant strains of *Monilinia* that resulted from repeated orchard applications of benzimidazole compounds. Triforine is an effective alternate for control of benzimidazole-resistant strains of *Monilinia* but is not active against *Botrytis. Botrytis* decay is serious on pears and sweet cherries and is becoming more important on other stone fruits.

18-6-3 Management in a Sustainable Agricultural System

The most important principle to consider in the management of postharvest fungal decay without the use of pesticides is to harvest and market mature, nonblemished fruit free of surface contamination and infection and to transport and store them in an environment not conducive to decay or physiological stress. We emphasize the need for critical examination of fruit in the packaging line and the removal of diseased and defective fruit to minimize subsequent decay and provide top-quality fruit for the market.

The following are examples of the effect of certain postharvest handling practices on the development of fungal decay of fruit: When stone fruit such as peaches and nectarines are harvested firm-mature, they must be held for two to three days at room temperature before they are ripe; under these conditions decay losses can sometimes reach as high as 50%. Bananas are harvested three quarters ripe and exposed to ethylene gas to trigger the ripening process; the ripened fruit are much more susceptible to decay than were the immature fruit. Control of finger rot of bananas, caused by *Botryodiplodia theobromae*, is important because during the decay process ethylene is produced which causes ripening of other bananas in storage. Pome fruit are harvested mature and can be held in cold or modified-atmosphere storage for periods up to a year. Pathogens such as *M. piriformis* and *B. cinerea* can cause decay of this fruit under these favorable storage conditions. Sweet cherries are harvested ripe, but because of their high susceptibility to fungal decay they can be held in storage for only a limited period. Special harvest and storage conditions exist for each of the tree-fruit species to prevent excessive decay in fruit to be stored for long periods or shipped to distant markets; the proper observance of these techniques is critical. Thus, for some international markets, fresh fruit which require postharvest ripening cannot be shipped without a postharvest fungicide treatments. For home consumption fruit can be harvested tree ripe, and since they usually are consumed rather quickly there is less need for quality control or special packaging. Yet, even for home use, fruit that are not consumed immediately should be carefully selected for quality and diseased ones discarded before packaging; they also should be examined during the storage period for removal of those with decay. One option which has not been critically examined is to accept crops with some decay and lower quality provided they have been shown not to be detrimental to our health.

Some of the disease-control methods currently under study are heat treatments (hot water and hot air), irradiation, modified atmospheres,[28] and low barometric pressure, none of which leave a chemical residue on the fruit. Heat treatments are used to some extent on papayas to reduce anthracnose and on peach fruit to eradicate *Monilinia* infections. Modifications of the environment generally have a greater effect on the crop and its susceptibility to infection than on the pathogen.

As another alternative to use of fungicides, much effort is being made to introduce biocontrol agents,[29] but at present their efficacy in decay control does not match that of currently registered fungicide treatments. Some of the microbial organisms under test are *Bacillus subtilis, Pseudomonas cepacia, P. syringae, Enterobacter aerogenes, Trichoderma* sp., and *Debaryomyces hansenii*.[29]

18-7 CONCLUSIONS

There has been a general lack of interest in supporting research in postharvest pathology. Eckert and Ogawa[23,24] and Kelman[2] pointed out that only recently has concern been focused on postharvest losses of the various commodities. It is too early to tell if this concern will translate into significant support for postharvest disease research. In recent years, assessments of small parts of the total loss sustained at wholesale markets have been made in major cities of the United States.[8,30] If more accurate estimates of losses were available on a world-wide basis, they probably would confirm that 200 million or more people could be fed if crops were not lost after harvest.

Fresh tree fruit are more prone to postharvest losses than are vegetables because of their higher water content and the fact that they are organs whose biological function depends on their decomposition. As previously pointed out, losses to postharvest diseases can be large even in technologically advanced societies. Greater protection of fruit against postharvest diseases is one way to increase the source of nutrients for the world's population and to decrease the economic cost to both the consumer and the fruit industry. Efforts to save harvested fruit would cost less than efforts to produce more fruit when one considers the constant decrease in availability and constant increase in price of cultivable land.

Research emphasis for tomorrow should be directed toward obtaining more basic information on the etiology,[4] biology,[31] epidemiology,[32] and mycotoxin production of the pathogens involved in the decay complex.[33,34] Furthermore, studies on the mechanism of tissue degradation[35] by decay pathogens would enhance our understanding of host susceptibility and resistance. Included in such research should be the effects of weather, plant nutrition, cultural practices, chemical treatments, and host genotype on the incidence and severity of postharvest diseases. Information derived from these studies should aid in the forecasting of disease incidence and in developing more effective management strategies for coping with these disorders. Disease-control chemicals should be assessed for their safety, for an effective method of application, and for their potential to induce the selection of fungicide-resistant strains.[36] Those determined to be hazardous to the fauna or flora of the world should be eliminated from registration in both developed and developing countries. Programs in alternative agriculture[37] demand the prevention of undesirable environmental consequences from farming operations, especially those that result in pesticide or antibiotic residues in food. This is a high-priority issue that remains to be solved. Equally important is to put more emphasis on postharvest pathology in our educational system so new scientists can be trained and the public better educated on the benefits and risks of providing quality fruit and vegetables to feed the world's increasing population.

18-8 REFERENCES

1. Oakley. H., ed., Blue Goose, Inc., *The Buying Guide* for fresh fruits, vegetables, herbs, and nuts, Educational Department, Blue Goose, Inc., P.O. Box 1118, Hagerstown, MD 21740, p.136, 1955.

2. Kelman, A., The importance of research and postharvest losses in perishable crops. In *Postharvest pathology of fruits and vegetables: Postharvest losses in perishable crops* (H. E. Moline ed.), pp.1–3, University of California, Bull. 1914, 1984.

3. Baritelle, J. I. and Gardner, P. D., Economic losses in the food and fiber system: From the perspective of an economist. In *Postharvest pathology of fruits and vegetables: Postharvest losses in perishable crops* (H. E. Moline, ed.), pp.4–10, University of California, Bull. 1914, 1984.

4. Moline, H. E., Diagnosis of postharvest diseases and disorders. In *Postharvest pathology of fruits and vegetables: Postharvest losses in perishable crops* (H. E. Moline, ed.), pp.17–23, University of California, Bull. 1914, 1984.

5. Snowden, A. L., Post-harvest diseases and disorders of fruits and vegetables, Vol. I: *Introduction to fruits,* CRC Press Inc., Boca Raton, FL p.440, 1990.

6. Jones, A. L. and Aldwinckle, H. S., *Compendium of apple and pear diseases,* APS Press, The American Phytopathological Soc., St. Paul, MN, p.100, 1990.

7. Ogawa, J. M. and English, H., *Diseases of temperate-zone fruit and nut crops,* University of California, Agric. Div. and Natur. Resources, Publ. 3345, p. 464, 1991.

8. Cappellini, R. A., Ceponis, M. J. and Lightner, G. W., Disorders in avocado, mango, and pineapple shipments to the New York market, 1972–1985, *Plant Disease,* 72, 270, 1988.

9. Sun, Shou-Kung, Postharvest diseases of tropical and subtropical fruits and their control. In *Postharvest handling of tropical and subtropical fruit crops* (J. Bay-Peterson, ed.), pp.113–122, Foods and Fertilizer Technology Center for the Asian and Pacific Regions, 1988.

10. Alvarez, A. M. and Nishijima, W. T., Postharvest diseases of papaya, *Plant Disease* 71. 681, 1987.

11. Eckert, J. W. and Eaks, I. L., Postharvest disorders and diseases of citrus fruits. In *The citrus industry,* Vol. 5 (W. Reuther, E. C. Calavan and G. E. Carman, eds.), pp.179–260, Div. Agr. and Nat. Res. University of Calif., 1989.

12. Slabaugh, W. R. and Grove, M. D., Postharvest diseases of bananas and their control, *Plant Disease,* 66, 746, 1982.

13. Whiteside, J. O., Garnsey, S. M. and Timmer, L. W., *Compendium of citrus diseases,* APS Press St. Paul, MN, p.80, 1988.

14. Smoot, J. J., Houck, L. G. and Johnson, H. B., *Market diseases of citrus and other subtropical fruits,* Agr. Handbook No. 398, USDA ARS, p. 115, 1971.

15. Ogawa, J. M., Manji, B. T. and Sonoda, R. M., *Management of the brown rot disease on stone fruits and almonds in California,* New York Agricultural Exp. Sta., Special Report No. 55, pp. 8–15, 1985.

16. Manji, B. T. and Ogawa, J. M., Quiescent infections and disease control in the shipping container, *Proc. National Peach Council,* pp. 23, 24, 1985.

17. Manji, B. T., Ogawa, J. M., Buchner, R. P. and Sawamura, K., Quiescent infections of *Monilinia fructicola* on immature fruit of French prune (Abstr.), *Phytopathology,* 80, 890, 1990.

18. Manji, B. T., Ogawa, J. M., Adaskaveg, J. E. and Osorio, J. M., Variability of disease development in nectarine cultivars for export (Abstr.), *Phytopathology,* 79, 1171, 1989.

19. Feliciano, A., Feliciano, A. J. and Ogawa, J. M., Monilinia fructicola resistance in peach cultivar Bolinha, *Phytopathology,* 77, 776, 1987.

20. Adaskaveg, J. E., Feliciano, A. J. and Ogawa, J. M., Comparative studies of resistance in peach genotypes of *Monilinia fructicola* (Abstr.), *Phytopathology,* 79, 1183, 1989.

21. Northover, J. and Biggs, A. R., Susceptibility of immature and mature sweet and sour cherries to *Monilinia fructicola, Plant Disease,* 74, 280, 1990.

22. Sonoda, R. M., Hayslip, N. C. and Stoffella, P. J., Tomato fruit rot infection cycle in a fresh market packing operation, *Proc. Fla. Sta. Hort. Soc.,* 94, 281, 1981.

23. Eckert, J. W. and Ogawa, J. M., The chemical control of postharvest diseases: Deciduous fruits, berries, vegetables and root/tuber crops, *Ann. Rev. Phytopathology,* 26, 433, 1988.

24. Eckert, J. W. and Ogawa, J. M., The chemical control of postharvest diseases: Subtropical fruits, *Ann. Rev. Phytopathology,* 23, 421, 1985.

25. Ogawa, J. M., Canez, V. M., Jr. and Walls, K. M., Fungicides and bactericides used for management of plant diseases. In *Fate of pesticides in the environment*, pp.25–43, University of California, Agr. Exp. Sta. Publ. 3320; and Appendix B, Representative groups of fungicides and bactericides, pp.149–155, 1987.

26. Ogawa, J. M. and Manji, B. T., Control of postharvest diseases by chemical and physical means. In *Postharvest pathology of fruits and vegetables: Postharvest losses in perishable crops,* pp. 55–66, University of California, Bull. 1914, 1984.

27. Berger, R. D., Epidemiology of postharvest disease. In *Postharvest pathology of fruits and vegetables: Postharvest losses in perishable crops* (H. E. Moline, ed.), pp.31–35, University of California, Bull. 1914, 1984.

28. Spotts, R. A., Environmental modification for control of postharvest decay. *Postharvest pathology of fruits and vegetables: Postharvest losses in perishable crops* (H. E. Moline, ed.), pp.67–72, University of California, Bull. 1914, 1984.

29. Wilson, C. L. and Wisniewski, M. E., Biological control of postharvest diseases of fruits and vegetables: An emerging Technology, *Ann. Rev. Phytopathol.*, 27, 425, 1989.

30. Cappellini, R. A. and Ceponis M. J., Postharvest losses in fresh fruits and vegetables. In *Postharvest pathology of fruits and vegetables: Postharvest losses in perishable crops* (H. E. Moline, ed.), pp.24–30, University of California, Bull. 1914, 1984.

31. Conway, W. S., Preharvest factors affecting postharvest losses from disease. In *Postharvest pathology of fruits and vegetables: Postharvest losses in perishable crops* (H. E. Moline, ed.), pp.11–16, University of California, Bull. 1914, 1984.

32. Leach, S. S., Integrated pest management systems in postharvest technology. In *Postharvest pathology of fruits and vegetables: Postharvest losses in perishable crops* (H. E. Moline, ed.), pp.73–75, University of California, Bull. 1914, 1984.

33. Phillips, D. J., Mycotoxins as a postharvest problem. In *Postharvest pathology of fruits and vegetables: Postharvest losses in perishable crops* (H. E. Moline, ed.), pp.50–54, University of California, Bull. 1914, 1984.

34. Kelman, A., Opportunities for future research in postharvest pathology. In *Postharvest pathology of fruits and vegetables: Postharvest losses in perishable crops* (H. E. Moline, ed.), pp.76–80, University of California, Bull. 1914, 1984.

35. Stelzig, D. A., Physiology and pathology of fruits and vegetables. In *Postharvest pathology of fruits and vegetables: Postharvest losses in perishable crops* (H. E. Moline, ed.), pp.36–41, University of California, Bull. 1914, 1984.

36. Ogawa, J. M., Manji, B. T., Adaskaveg, J. E. and Michailides, T. J., Population dynamics of benzimidazole-resistant *Monilinia* species on stone fruit trees in California. In *Fungicide resistance in North America*, pp.36–39, APS Press, St. Paul, MN, 1988.

37. Committee on role of alternative farming methods in modern production agriculture, in *Alternative agriculture,* National Academy Press, Washington, DC, p.448, 1989.

19

PHYTOPHTHORA ROOT ROT OF AVOCADO

MICHAEL D. COFFEY

Department of Plant Pathology
University of California
Riverside, California

19-1 INTRODUCTION

Phytophthora root rot of avocado (*Persea americana* Mill.) is caused by *Phytophthora cinnamomi* Rands. The disease is widespread and occurs in the majority of countries where avocados are cultivated. It is generally a major limiting factor in production of this fruit crop. As an example of this, in California it is estimated to affect between 60 and 75% of orchards, causing approximately $30 million in annual losses in 1986.[1] Avocados are produced in many subtropical and tropical regions, notably in Mexico, Guatemala, California, Florida, Hawaii, Puerto Rico, Dominican Republic, Brazil, Chile, Indonesia, Israel, Spain, Morocco, Algeria, South Africa, Australia, and New Zealand. The origins of the avocado, an evergreen subtropical and tropical fruit tree, are in Mexico and Central America, particularly Guatemala. Three horticultural varieties of avocado are recognized. These are the Mexican (var. *drymifolia*), Guatemalan, and West Indian. The Mexican race, which as the name suggests originates from Mexico, is still widely grown in that country, especially in the state of Puebla. Around Atlixco are still found many different types of the native criollo avocado with its characteristic smooth, thin-skinned fruit with their pointed seed. The leaves of the Mexican avocado have a distinctive anise scent when crushed. The Guatemalan avocado is found primarily in Guatemala, southern Mexico, El Salvador, and Honduras. Its fruit has a rough, thick skin, and the seed is oblate in shape; the leaves lack any obvious anise scent. Fruit of the West Indian race are generally larger than the other types, their skin is thin, and they have a large, pointed seed. Its origins are more uncertain, though the tropical lowlands of Central America are a likely place; it has been long established in the Caribbean islands. In addition to the three races, many culti-vars are hybrids (Mexican × Guatemalan, Guatemalan × West Indian, etc.). There are also near relatives of avocado such as *P. schiedeana* Nees, with large edible fruit, that grow as large trees through much of tropical Central America and Mexico. The Mexican cultivars are most frost tolerant, followed by the Guatemalan types, both being widely grown in the subtropics, whereas the West Indian cultivars are distinctly tropical in their behavior. However, some Guatemalan trees appear extremely coldhardy, having been found growing in the Guatemalan Highlands at elevations of 8000 feet or more, in cli-mates where subzero temperatures can occur in winter. Some of these trees are over 50 years old.

The avocado is a recent introduction into the horticultural world, outside of its native home in Mexico and Central America, In California, the first trees were intro-duced from Mexico into the Santa Barbara area in 1871 by Dr. Franceschi, an early pioneer in the introduction of exotic subtropical fruits to the area.[2] Following the discov-ery in 1911 of the cultivar Fuerte, a Mexican × Guatemalan hybrid, which was well adapted to even the coldest winters in Southern California, a small horticultural industry soon became established. A problem known as avocado decline became evident in the 1920s and 1930s. By the late 1930s, this decline or die back problem had become quite serious, especially on trees ten years or older.[3] Trees were particularly prone to this malady when grown on heavy soils with poor drainage, though it also occurred on some sandy soils. However, careful examination of the profile of such soils usually revealed an impervious subsoil.[3] In the early 1940s, a South African plant pathologist, Vincent Wager, working at the University of California Citrus Experiment Station at Riverside, was the first to record *P. cinnamomi* on avocados in the U.S.A.[3] He conducted inoculation experi-ments with pot-grown avocado trees and determined that when such plant material was waterlogged for only two to three days, they died within a few weeks. In contrast, trees without *Phytophthora* but similarly waterlogged were not adversely affected. Early ex-periments on the relationships between irrigation practices and root rot severity on avoca-dos further emphasized the role of waterlogging in accelerating disease development.[4] Five-year-old budded avocado trees irrigated once a week by sprinklers succumbed much more rapidly than those only watered every two weeks. Despite the severe problem with root rot, the avocado industry in California continued to expand, spreading out into new areas where the pathogen had not yet reached. During the 1970s, hectarage expanded from 9000 ha to 33,000 ha by the end of that decade.[1] This was made possible by the introduction of drip (trickle) irrigation technologies from Israel, permitting a major ex-pansion of speculative plantings on the steep hillsides previously covered with native chaparral. Many of the soils of these hillsides are shallow, however, raising doubts as to their ultimate suitability for sustained avocado production. Relatively low water costs and high fruit prices for green gold in the 1960s and 1970s made avocado a potentially attrac-tive investment for the adventurous investor. In addition, the coastal mediterranean cli-mates where avocados are grown in southern California represent some of the most at-tractive examples of rural real estate in the area.

By the 1980s, avocados had occupied most of the suitable cultivatable land, land prices had begun to escalate, water prices had skyrocketed, and root rot was reasserting its presence. Finally, most of the plantings were still on susceptible seedling rootstocks. California losses in 1989 due to root rot were estimated at $44 million.

Elsewhere in the world *P. cinnamomi* has wreaked a similar havoc. In Mexico, the largest producer of avocados in the world, the disease has been known for a long time. In the state of Queretaro, the avocado was virtually eliminated by the 1970s.[5] In Australia, *P. cinnamomi* is widespread, not only causing havoc on avocados but also causing major ecological damage to different indigenous plant communities.[6] This pathogen has had it most devastating effects in the Jarrah eucalyptus forests of Western Australia, destroying the trees and the majority of the understorey plants.

One country, Israel, has so far escaped the excessive ravages of *P. cinnamomi.* Avocados are grown primarily on alkaline soils, and root rot has not been a problem. Recently, an outbreak of root rot was recorded in one orchard; however, it was apparently contained.[7] In addition, some isolates of *Phytophthora* spp. recovered from avocados were described as only mildly pathogenic on that host.[8] Currently, there is some uncertainty as to whether these isolates are *P. cinnamomi.*[9]

In general, *P. cinnamomi* has proven to be one of the most troublesome soilborne pathogens not only on avocado but also on many other predominantly woody hosts, including azalea, camellia, *Chamaecyparis lawsoniana,* chestnut, cinchona, heather (*Erica* sp.), *Eucalyptus* sp., fir (*Abies fraseri*), pines, pineapples, rhododendron, and walnut. In fact, it has been recorded on nearly 1000 hosts in over 70 countries (Fig. 19-1).

19-2 SYMPTOMS

Phytophthora root rot can attack trees of all ages including young nursery trees on tolerant rootstocks. It is most severe in heavy soils and lighter sandy soils with impeded drainage caused by a clay or rock subsoil. Both excessive soil moisture and severe drought stress can accelerate the symptoms. The most obvious symptoms of the disease on avocado consist of a gradual decline or dieback of the foliage.[1,3] The leaves are smaller, instead of being dark green are pale or yellow green, and are often wilted in appearance. Progressive defoliation occurs, and the tree displays a sparse appearance (Fig. 19-2). A lack of new growth flushes is typical of trees at an advanced stage of the disease. Soon after decline symptoms appear, trees may set an abnormally high crop of very small fruit, though ultimately fruit production on diseased trees is very poor. Eventually, entire branches dieback. Early symptoms of root rot are often detectable at the top of the tree, where the first signs of branch dieback may be visible to the educated eye. Eventually, the tree will become totally defoliated and die. Below ground, the tell-tale symptoms consist of a distinct absence of healthy feeder roots. At the onset of the disease, many of the small feeder roots are black and brittle. They break easily in the hand. Others have brownish lesions from which the pathogen *P. cinnamomi* can be readily isolated.

Occasionally, *P. cinnamomi* will also attack larger roots.[3] Rarely, it may also form trunk cankers originating at, or slightly below, the soil line. These can be quite extensive in some situations, growing up the trunk for approximately 2 meters. However, the more common cause of trunk cankers in California is *P. citricola.*[10]

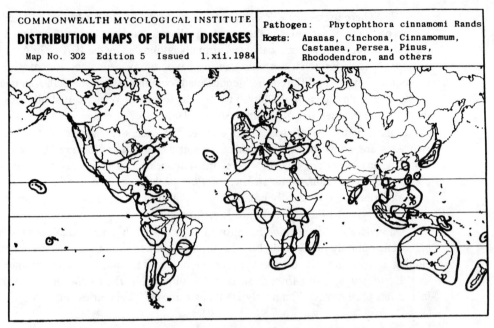

COMMONWEALTH MYCOLOGICAL INSTITUTE	Pathogen: Phytophthora cinnamomi Rands
DISTRIBUTION MAPS OF PLANT DISEASES	Hosts: Ananas, Cinchona, Cinnamomum, Castanea, Persea, Pinus,
Map No. 302 Edition 5 Issued 1.xii.1984	Rhododendron, and others

Figure 19-1 Commonwealth Mycological Institute Map (revised 1984) showing the distribution of *P. cinnamomi* (courtesy of CMI). The fungus, especially the A2 type, is widespread in temperate, subtropical, and tropical climates.

19-3 DIAGNOSIS AND ASSESSSMENT

Critical diagnosis of Phytophthora root rot requires isolation of the pathogen in pure culture. Short segments of infected roots, approximately 1 cm in length, dipped for a few seconds in 70% ethanol and blotted dry, are directly plated on a solid medium such as cornmeal agar. Ideally, this medium should be amended with a Phytophthora-selective cocktail of specific fungicides and antibiotics. Various selective media are available for this purposes.[11] In our own laboratory, we have routinely used a modified PARPH cornmeal agar.[12,13] The active ingredients of PARPH are (P) pimaricin (10 mg/L), (A) ampicillin trihydrate (125 mg/L), (R) rifampicin (10 mg/L), (P) pentachloronitrobenzene (100 mg/L), and (H) hymexazol (20 mg/L). The addition of hymexazol is especially useful for isolation of *P. cinnamomi* from soils since it suppresses many *Pythium* spp. The PARPH cocktail can be made up in 70% ethanol and added directly to the molten, cooled (50°C) agar medium immediately prior to pouring the plates. This procedure can be developed into a quantitive assay for determining percent root infection by plating out between 20 to 40 randomly selected 1 to 2 cm root pieces per avocado root system.[13]

In soil, qualitative analysis of the presence of *P. cinnamomi* is best done using a trap method.[14,15] Susceptible *Persea indica* L. seedlings,[14] or blue lupin (*Lupinus angustifolius* L.) seedlings or cotyledons, can be utilized.[16,17] In the *P. indica* trap method, a soil sample taken in three or four locations from a tree's root system is placed in a small container.[14,15] The soil sample, approximately 2 to 3 cm deep, is then covered with water to the top of the container. A young bare-rooted seedling of *P. indica* is placed in the container. If *P. cinnamomi* is present, its roots will generally become infected and brittle-black within

Figure 19-2 A Hass avocado tree, approximately 10 years old, showing severe symptoms of Phytophthora root rot.

seven days. A stem canker may also develop. *P. citricola* can also cause a stem canker, though it seldom causes a severe root rot. Infected root and stem pieces can then be plated out on PARPH cornmeal or other selective agar medium. Another method for trapping from soil involves the use of mature, thin-skinned Mexican or Mexican hybrid fruit such as Fuerte.[14,15] The soil surface is flooded with water in a container, and the fruit is placed on the distal (nonstem) end about one quarter deep into the mud. If *P. cinnamomi* is present, brown lesions will appear on the fruit at the water line within four to six days. *P. citricola* may also cause a similar fruit rot.

For quantification of *P. cinnamomi* in soils, a modification of the soil-dilution plating technique of Kannwischer and Mitchell can be used.[12,13] A 50-g (wet weight) soil sample is mixed very thoroughly with 100 ml of 0.1% to 0.25% Difco water agar.[13,18] The dry weight of a 1-ml sample from the soil agar slurry is determined. One millimeter of the slurry if spread over 5 to 10 PARPH cornmeal agar plates (9 cm width). After two to three days in the dark at 24°C, the plates are rinsed free of soil. The colonies of *P. cinnamomi*, recognized by production of hyphae with distinctive hyphal swellings, are then counted. Population densities of *P. cinnamomi* in soil are generally expressed as colony-forming

units (cfu/g) or propagules (ppg) on a gram dry-weight soil basis. Field soils infected with
P. cinnamomi may be expected to have population of 0.3 to 6 cfu/g.[13,18]

Visual assessment of disease severity, using a numerical index of either 0–5 or 0–
10, based on foliar symptoms, is a routine method used for evaluating tree performances
in fungicide and rootstock trials.[6,19,22] The index is based on a visual rating where 0-
healthy and either 5 or 10-dead. The scale is subjective and considers the vigor of the tree
(e. g., evidence of growth flushes), defoliation and branch dieback, wilting, and leaf
color (yellowish green, pale green, or dark green). However, it can be made much less
subjective by reference to a standard set of color plates.

Fruit-production data (weight or numbers) has only been used to a limited extent in
field assessments.[19,20] This is because avocado is markedly alternate bearing, and fruit
production per individual tree, even in the same planting, is highly variable, frequently
making such data relatively meaningless unless large numbers of trees (75 to 100) are
used per treatment.

19-4 CAUSAL AGENT

Phytophthora cinnamomi was first described by Rands in 1922 as causing a stripe canker
on cinnamon in Sumatra.[23] It is most likely of Asian origin.[15] Since that time, it has been
found causing root rot and/or canker problems on approximately 1000 different plants in
about 70 countries. It is most destructive on avocado and some other fruit or nut trees
(peach, pear, walnut), chestnut, woody ornamental plants (azalea, camellia, rhododendron), and susceptible eucalyptus.[1,3,5,6,14,15]

19-4-1 Morphology and Identification

The fungus produces nonpapillate sporangia and is heterothallic with two compatibility
types designated A1 and A2. It is classified in Group VI of the Waterhouse key to *Phytophthora* species.[24] Detailed taxonomic descriptions of the fungus have been given in
three articles.[15,25,26] Although *Phytophthora* is often difficult to characterize to the species
level, *P. cinnamomi* possesses several morphological and cultural features which make it
fairly easy to identify. Key diagnostic features include the following:

- Young hyphae which are distinctly coralloid, (i.e., possess frequent rounded nod-
  ules or vesicles);

- Mature hyphae which are broad (8 μm or more) and tough;

- Clusters of terminal chlamydospores in grape like bunches with thin walls, which
  are readily produced on a rich medium such as V8® juice agar;

- Large nonpapillate, noncaducous (persistent) sporangia, ellipsoid to ovoid in
  shape, which are *not* normally produced on solid agar media, but do form in
  nonsterile aqueous solutions such as 1% soil extract;

- Colony morphology on potato dextrose agar (PDA) is frequently characteristic,
  with a distinctive camellioid or rosette pattern (Fig. 19-3).

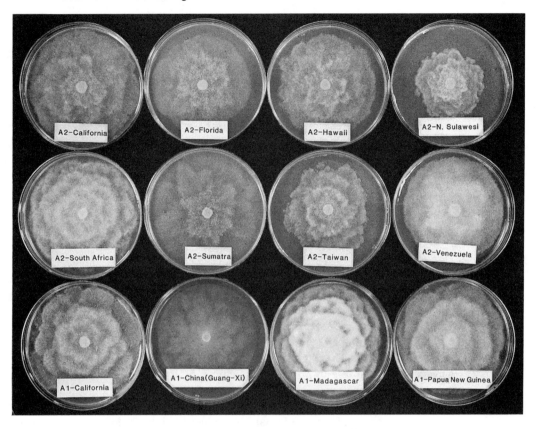

Figure 19-3 Colony morphology on PDA of representative isolates of A1 and A2 compatibility types from worldwide sources. All isolates are from avocado.

Oospores are not formed in single culture, since the fungus is heterothallic, but are produced in intraspecific and interspecific pairings with a strain of opposite compatibility type.[15] A2 strains, but not A1 types, have also been induced to form homothallic oospores by addition of an avocado root extract,[27] or oleic acid,[22] to the medium. In paired A1 × A2 cultures, the amphigynous oospores of *P. cinnamomi* range in diameter from 23 to 49 μm.[15]

19-4-2 Isozyme Analysis and Compatibility Type

In a study of 183 isolates of *P. cinnamomi* from many hosts in Australia and Papua New Guinea, low levels of isozyme variability were detected in both A1 and A2 compatibility types.[29] Only two genotypes were detected among the A2 isolates. Somewhat greater genetic variability was found in the A1 type isolates from Papua New Guinea, but not Australia. A follow-up study with 280 isolates from native vegetation in southeastern Australia confirmed the earlier findings.[30] In one situation, both A1 and A2 isolates were recovered from a 50-g soil sample.[30] A1 and A2 isolates could be distinguished isozymically, indicating that sexual hybridization does not occur between compatibility types in the field.[29,30] In a recent isozyme study of isolates of *P. cinnamomi* from worldwide sources and a broad host range, eight distinct genotypes were identified.[31] The A2 isolates belonged to two genotypes; the A1 isolates to six genotypes. The A2 isolates could be readily distinguished from the A1 isolates based on the migration pattern of isocitric dehydrogenase.[31]

Worldwide, avocado isolates are mainly A2 and possess the common CINN4 genotype based on isozyme analysis.[31] Only a few avocado isolates have been determined to be the A1 compatibility type,[32] and these belong to three different isozyme genotypes CINN1, CINN2, and CINN6.[31] These rare A1 isolates were recovered from avocados in California,[31,32] China (Guangxi Region),[31] Madagascar[31,32] and Papua New Guinea.[31]

19-5 DISEASE CYCLE

The pathogen survives primarily as *chlamydospores* in decaying feeder roots. Such spore structures are eventually released into the soil, where they persist for long periods.[15] They germinate by producing several germ tubes.[33] *Sporangia* production is stimulated by various soil microorganisms, particularly bacteria.[15] Production of both chlamydospores and sporangia occurs in the soil over a temperature range of about 12 to 30°C, optima being 21 to 24°C.[15] Sporangia germinate indirectly by releasing 10 to 30 motile, naked biflagellate *zoospores*. Upon contacting a solid surface such as a host root zoospores encyst; the resulting spherical *cyst* then germinates and infects the host feeder root. It rapidly causes a black, brittle root rot. The pathogen can survive in moist soils for many years, mainly as chlamydospores. The role of *oospores,* which can be found in avocado roots,[27] is less certain. Their dormancy and ability to survive under extreme conditions make them a potentially highly persistent spore type. *P. cinnamomi* on avocado in California is mainly recovered from the top 15 cm of soil, though occasionally it has been recovered at depths up to 1 meter.[34] The disease cycle of *P. cinnamomi* on avocado is presented diagramatically in Fig. 19-4.

Moisture level is a key factor in the disease cycle. High moisture contents in the soil provide the free water conditions favoring release of from sporangia. This facilitates movement of the motile zoospores to the root surface. Release of root exudates sets up chemotactic gradients that can then attract large numbers of zoospores to the zone of elongation.[15] Infection and rapid death of the root follows quickly, and sporangia and chlamydospores can be formed within a few days in the decaying host, perpetuating the disease cycle.

19-6 DISEASE MANAGEMENT

The four components of integrated control of Phytophthora root rot of avocado are sanitation and hygiene, cultural and biological control, tolerant rootstocks, and fungicides.

19-6-1 Sanitation and Hygiene

Due to its extensive host range, *P. cinnamomi* can be introduced onto a property through the planting of a variety of plants, including many woody ornamentals (e.g., azalea, camellia, heather), pines, and deciduous fruit or nut trees (e.g. peach, pear, walnut). In

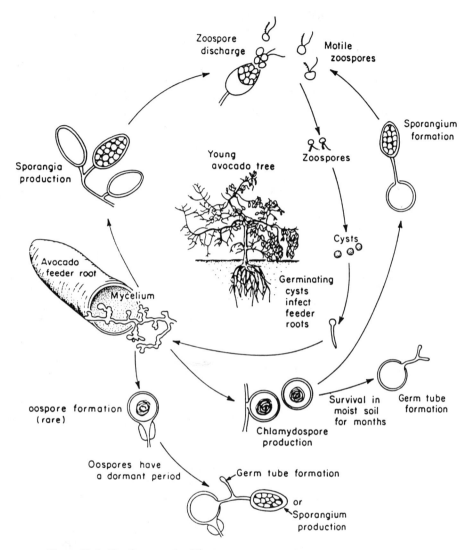

Figure 19-4 The disease cycle of *P. cinnamomi* on avocado roots.

addition, since many ornamental plants are produced together in the same nursery environment, even nonhosts of *P. cinnamomi* may inadvertently be carriers of the pathogen, due to use of contaminated potting mixes. Consequently, it is advisable to restrict planting of ornamentals and landscape trees to a minimum. If shrubs and trees are planted, it is recommended that they first be tested for freedom from *P. cinnamomi,* ideally using the type of sensitive baiting method outlined in Section 18-3 (e.g., *Persea indica* trap method).

Since *P. cinnamomi* is commonly present in most avocado-growing areas, it is also necessary to try to prevent movement of contaminated water and soil into a property. In California, growers generally surround their property with a strong wire mesh fence to restrict unauthorized access of humans and movement of animals onto the avocado planting. At the entrance to the orchard, to prevent movement of contaminated soil, tire baths containing fresh hypochlorite solution are provided for vehicles. Trays containing dry Bordeaux mixture are often available for treatment of footwear. Alternatively, the soles of boots and shoes can be treated with a 70% ethanol solution. Movement of contaminated water is even more difficult to control. Provision of drainage ditches which prevent flooding is important, both in high rainfall areas and in any situation where *P. cinnamomi* is already present, either within the property, or on adjacent properties.

To restrict further spread of the pathogen from contaminated areas, it is recommended that a ditch be dug around the zone of root rot infestation, ensuring that all infected trees are isolated. This strategy only works where initial infections are detected at any early stage, before extensive spread has occurred. It follows that regular testing of the roots and soil under avocado tress for *P. cinnamomi* must be practiced if such a containment procedure is envisaged as a component of disease management.

In the nursery, trees should be raised in well-drained potting mixes. This mix should be fumigated with methyl bromide, or alternately subjected to thermal treatment, either steam at 100°C for 30 minutes, or ideally aerated steam at 60°C for 30 minutes.[35,36] Avocado seed used directly in the production of seedling rootstocks, or indirectly as a nurse stock in clonal propagation of cuttings as rootstocks, should be dipped in a hot-water bath for 30 minutes at 50°C to kill *P. cinnamomi*.[15] Trees should be produced from carefully selected sunblotch-viroid-free, budwood and seed, if possible.[37,38] Since clonally propagated avocados may take 18 to 24 months to produce a strong root system, it is vital that the trees are produced on benches, well above the ground, in an area where flooding is unlikely to occur. Since water is also a medium that facilitates spread of the pathogen, it is essential that good sources of irrigation water free from contamination are available to a nursery. Alternatively, the water must be chemically treated with fresh hypochlorite solution or low levels of copper sulfate prior to use.

19-6-2 Cultural and Biological Control

A key factor in cultural control is the selection of a suitable planting site. Ideally, the soil should be deep, well-drained, high in organic matter, low in salinity, not excessively alkaline, and in a site not prone to flooding. In California, soils considered only slightly hazardous for root rot are well-drained sandy loams over 1.5 m deep.[14] However, more and more plantings are made on shallower soils with poor internal drainage, or those with impervious clay or rock subsoils. Most importantly, the soil in the immediate vicinity of the planting site should be free of *P. cinnamomi*. Unfortunately, in most avocado-growing areas, finding suitable soils for avocado cultivation free of the pathogen is no longer easy. Cook and Baker (1983) regard contamination of soils by *P. cinnamomi* as "a particularly dangerous type of environmental pollution,"[39] since once the pathogen is established it is virtually impossible to eradicate. In an attempt to reduce the inoculum level in such soils, growers in the past have resorted to the use of various fumigants. Experience in California with such procedures has been unsatisfactory, due in part to the difficulty of getting adequate penetration of the chemical into soils. In addition, many new avocado plantings are on steep rocky hillsides where such measures are often impractical.

Undoubtedly, one of the most beneficial cultural practices is the use of mounds

(Fig. 19-5) or ridges for planting new trees, especially in *Phytophthora*-contaminated areas where heavy soils with poor internal drainage are prevalent.[1] Such practices help to reduce excessive soil moisture and help to reduce soil compaction in the immediate vicinity of the tree's root zone. Mounding or ridging procedures also provide the opportunity for incorporation of organic amendments into the soil. Such an addition of organic matter, especially poultry manure, is an important component of the Ashburner system used by some Australian avocado growers.[40] Trees in Queensland are often planted on *Phytophthora*-suppressive soils cleared of tropical rainforest. To partially maintain their suppressiveness, the Ashburner system utilizes organic amendments, straw mulching, and cover crops to maintain high organic matter, and the addition of dolomitic limestone to maintain soil pH above 6.0.[40] The suppressive properties of these soils are attributed to the high microbiological activity of various spore-forming bacteria and actinomycetes. however, no individual antagonistic microorganisms have been isolated, and the suppressive effect is not transferable to a conducive soil.[41] It may be that this phenomenon is one of general suppressiveness, where the amount of microbiological activity determines its effectiveness and no one specific group of organisms is responsible.[40]

Figure 19-5 An avocado tree on a clonal rootstock planted on a mound in a *P. cinnamoni*-infected grove with a heavy soil with poor internal drainage in southern California.

Unfortunately, the majority of soils on which avocados are grown are conducive to *Phytophthora* root rot. However, addition of organic matter, particularly the use of grain straw (barley, wheat, sorghum, etc.) as a mulch, together with the natural leaf litter that forms, can create a feeder root zone which has not only improved microbiological activity, but also increased soil porosity and nutrient-holding capacity.[16] In semiarid zones such as California and Western Australia, mulching also provides an effective means of conserving soil moisture.

The addition to soils of specific microbial antagonists to control *P. cinnamomi* may become a possibility in the future. Certainly, the practice of fumigation or thermal inactivation of nursery potting mixes provides an opportunity for introduction of such antagonists. In addition, use of straw mulches, or organic amendments incorporated into the soil, may enhance the colonization of those microbial antagonists capable of decomposing such materials. However, it is uncertain whether a surface mulch would be effective. In greenhouse experiments using alfalfa meal as an amendment, it was determined that the material had to be incorporated into the soil, to suppress *P. cinnamomi* on avocado seedlings.[42] Incorporation of a specific antagonist growing on a suitable substrate such as straw, alfalfa meal, or cotton waste in a mound at the time of planting might provide an answer to this problem.

There are very few records of specific microorganisms active against *Phytophthora* root rot in soils. Certain ectomycorrhizae may play a role, especially with pines and other conifers in suppressing *P. cinnamomi,* but avocados do not develop such mycorrhizal associations. Generally, *Trichoderma* spp. have not been recorded as effective against *P. cinnamomi,*[42] and their reputed poor performance in wet soils might partially explain this. Kelley (1976) was unable to control *P. cinnamomi* on pine seedlings with a strain of *T. harzianum.*[43]

Recently, a rare example of a presumptive *Phytophthora*-suppressive soil was identified in an avocado grove in southern California in Santa Barbara county.[18] *P. cinnamomi* had originally been recovered from this site in 1942. The majority of the groves in the area, all planted on an extremely well drained, deep (1.5 meter) sandy loam soil, had eventually declined due to root rot over a 40-year period. However, in this grove the majority of the trees showed no signs of root rot, being rated 0 on a scale of 0 to 5. Examination of the feeder roots revealed little evidence of root rot. However, upon plating out these roots on cornmeal-PARPH selective medium, it was possible to recover *P. cinnamomi*. It was conceivable that the pathogen was being suppressed by components of the soil microflora. Among 36 fungi and 110 bacteria, only one microorganism demonstrated any consistent ability to suppress *P. cinnamomi* utilizing the highly susceptible avocado relative *Persea indica* in a seedling bioassay. The microbial antagonist was a strain of the fungus *Myrothecium roridum* Tode ex Fr.[18] In both peat-perlite mixture and UC mix no. 4, results were good; in the latter, *M. roridum* suppressed root rot by 50 to 94% compared to uninoculated controls (Fig. 19-6). In addition, in three different natural, conducive field soils, incorporation of *M. roridum* inoculum grown on a wheat-bran substrate resulted in 15 to 54% root rot, compared with 58 to 93% for *P. cinnamomi*-infected controls over a four week period.[18] Using a carbendazim-resistant mutant of this strain of *M. roridum*, it was determined that the root tips of *P. indica* seedlings were colonized by the fungus up to four weeks after initial inoculation.

These results clearly indicate that potential biocontrol agents for *P. cinnamomi* are

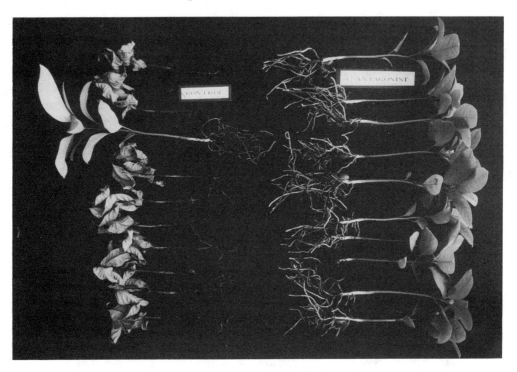

Figure 19-6 Biological control of *P. cinnamoni* on *Persea indica* seedlings using strain TW of *Myrothecium roridum*.

rare in soils, but do exist and may give reproducible results, even in natural soils where they must compete with many different microorganisms.

The ultimate aims of research in the biological control must be to develop suitable technologies for using such microbial antagonists in the field. This will require development of in-depth knowledge of the ecology and population biology of the antagonist in different field soils. The development of appropriate delivery technologies, such as formulations, that will be both economic and efficacious over extended periods of time will also be necessary. As previously mentioned, the avocado system offers some opportunities, particularly since the potting mix is fumigated or heat treated. Colonization of such a mix is rendered much easier. In addition, incorporation of a biocontrol agent, growing on a suitable organic substrate, into a planting mound may be feasible. In California, it is planned to test strain TW of *M. roridum* in the field as soon as good formulation procedure is worked out.

19-6-3 Tolerant Rootstocks

The search for resistant germ plasm which might be used as a rootstock for avocados has continued for 40 years at the University of California, Riverside. At least 100,000 seedlings of *Persea* spp. have been screened for resistance to *P, cinnamomi* in this time. High resistance has been detected in collections of species in the subgenus *Eriodaphne*: *P. alba, P. borbonia, P. caerulea (P. skutchii), P. cinerascens, P. donnell-smithii,* and *P. pachy-*

poda.[44] However, these species, which produce extremely tiny fruit, are only distantly related to the subgenus *Persea* in which belongs *P. americana,* the avocado. In fact, there is no graft compatibility between species in the two subgenera;[45] hybridization between species in the two groups has also proved ineffective.[46]

The first evidence of some moderate resistance to *P. cinnamomi* in the avocado came in the early 1950s with the discovery that some seedlings of the Duke cultivar survived tests with the pathogen.[47] The Duke cultivar is a Mexican variety introduced into California in the early part of this century. Following extensive testing, two seedling selections, Duke 6 and 7, emerged as showing some potential for use as rootstocks. Unfortunately, the majority of their seedling progeny are susceptible to *P. cinnamomi,* and it is therefore necessary to produce the rootstock as cuttings. For most fruit trees, this is not a difficult task. However, with avocado the direct rooting of cuttings proved virtually impossible.

Consequently, further impetus in the research on rootstocks awaited development of a good clonal propagation method. Frolich at UCLA finally produced a suitable procedures in the late 1960s, based on use of a nurse seedling to which a resistant budwood cutting is grafted; juvenile, etiolated shoots are then produced, and one of these is selected, girdled at its base, and rooting can then be induced.[48] This clonal propagation method is now used commercially in California, Israel, Australia, New Zealand, South Africa, Mexico, and Spain to produce clonal rootstocks.

In commercial practice in California, the Duke rootstocks have been widely planted since the mid 1970s. In 1986, it was estimated that over a half million trees on clonal rootstocks had been planted, mainly the Duke 7 selection.[1] Much of the land suitable for avocados in California is already contaminated with *P. cinnamomi,* and many of the soils are rated as a "moderate to severe root rot hazard"[14] if the pathogen is present. Consequently, long-term performance of this rootstock has proved mixed, with many instances of failure.[1] As Zentmyer indicated in 1980,[15] the level of moderate "horizontal" resistance or tolerance in rootstocks like Duke 7 may not be sufficient over prolonged periods of time, and especially on soils with the most severe root rot hazard. One complicating factor is that avocado, in the absence of root rot, is an extremely long-lived tree, with a potentially useful production span of 30 to 40 years. Even on deep, well-drained soils with only a slight hazard regarding root rot, conditions may become very severe for short periods of time. Excessive irrigation in semiarid regions, or a deluge of rainfall in tropical regions, can temporarily make even an ideal planting site extremely hazardous with regard to root rot development. Insufficient irrigation or even relatively short drought periods can predispose roots that otherwise might possess tolerance to *P. cinnamomi* to increased, even severe, root rot.

The emphasis in rootstock research in the last 20 years has been on the discovery of selections with superior tolerance to *P. cinnamomi,* using Duke 7 as a minimum standard. Increased collecting activity in Central America, especially Guatemala, led to the identification of the G6. Despite its origin, G6 is typical of the wild avocados found growing in Mexico,[1,15] and belongs to the horticultural variety *drymifolia.* In non-root-rot soils, it can have a high growth rate[20] and it is a prodigious seed producer compared to most rootstock selections. Unfortunately, it has not been as consistent in its performance as Duke 7, and in fact a recent field test indicated that it might have little tolerance to *P. cinnamomi* in some severe root situations.[20]

Two selections in particular have emerged so far with a superior tolerance than Duke 7: Martin Grande and Thomas.[1,20]

In September 1975, ten seeds of a close relative of avocado, *P. schiedeana* Nees, were collected at the market in Coban, Guatemala, by Eugenio Schieber. At Riverside, eight of them were tested for resistance to *P. cinnamomi,* and the strongest seedling, G755c, was selected for further testing. Two seedlings were grown up without testing, and these were later labeled G755a and G755b. It was believed that they came from San Juan Chamelco, a village northeast of Coban. The vegetative growth of all three seedlings was different from previous selections of *P. schiedeana,* in possessing stem and leaf characters intermediate between that species and *P. americana.* Eventually, isozyme analysis confirmed that all three G755 selections were natural hybrids between the two closely related species.[49] Since the predominant type of avocado in the Coban region is Guatemalan, this was presumably the avocado parent of G755. Thousands of avocado trees on G755 rootstocks have been tested in commercial plants in California since 1984.[50] In 1987, it was proposed that the G755 rootstock be renamed Martin Grande in honor of one of Eugenio Schieber's Mayan helpers.[51] Martin Grande rootstocks have generally proved to be more tolerant of root rot than Duke 7 and have been established successfully in sites where the latter had previously failed.[50,51] Failures have occurred, as with Duke 7, and recently some concern has been expressed over its apparent susceptibility to chlorosis in some high pH soils.[50]

Thomas is a Mexican selection recovered from an apparently healthy tree in a grove in Escondido, California, where root rot had caused severe damage.[1,52] In a field test where rootstocks were deliberately inoculated with *P. cinnamomi* and evaluated without fungicide treatment, the Thomas rootstock performed as well as any other (Fig. 19-7), including Martin Grande, and was superior to Duke 7.[20]

Limited field testing since 1984 under commercial situations, where fungicides such as metalaxyl (see next section) are used routinely, also indicates that Thomas is a potentially useful rootstock.[53]

Two additional rootstocks have also shown some preliminary promise under root rot conditions: Barr Duke and D9.[20] Barr Duke is a seedling of the Duke 6 rootstock selection,[54] and appears to have superior resistance to Duke 7.[20] D9 is a semidwarfing rootstock, propagated from Duke budwood subjected to a gamma irradiation; it also appears to have superior tolerance to root rot.[20]

The type of resistance to *P. cinnamomi* in avocado rootstocks has been described as both moderate resistance and tolerance. The mechanisms underlying such resistance have been investigated in some detail in recent years under laboratory or greenhouse conditions.[13,55–57] Duke 7 and particularly Thomas possess an exceptional ability to regenerate new feeder roots following root pruning and this appears to be key component of their tolerance to *P. cinnamomi.*[13,57]

No evidence could be obtained to suggest that either rootstock is capable of restricting development of *P. cinnamomi;* in fact, if anything they supported higher populations of the pathogen that a susceptible rootstock.[13,57] In contrast, the Martin Grande (G755) rootstocks demonstrated a moderate ability to restrict development of *P. cinnamomi* within their tissues.[55,56] This property of restricted development of *P. cinnamomi* has also been detected in three selections of *P. schiedeana* from Guatemala,[55] perhaps indicating

Figure 19-7 A comparison of Hass avocados growing either on a *Phytophthora*-tolerant rootstock Thomas (top) or a susceptible rootstock (bottom) under severe root rot conditions without fungicides.

that in the Martin Grande (G755) hybrid this resistance character is derived from the latter *Persea* species, rather than *P. americana*.

Martin Grande, specifically G755c, also exhibited an enhanced ability to regenerate feeder roots as compared to Duke 7.[57]

Breeding for rootstock resistance is still in its infancy because of the long generation times involved in field evaluation of their performance, and most pertinently because potential parental lines are only just beginning to be identified.

19-6-4 Fungicides

Twenty years ago, the concept of using *Phytophthora*-selective fungicides with good soil mobility properties and, even more remote, the idea of producing a fungicide that was truly systemic or ambimobile, was entirely theoretical. In recent years, however, two exciting classes of systemic fungicides have been discovered with good activity against *P. cinnamomi* and other *Phytophthora* spp.[1,6,12,19,21,22,58,59] These are the phenylamides, notable metalaxyl, and the phosphonates, especially potassium phosphonate (phosphonic acid buffered with KOH) and fosetyl-aluminum.[58,59] Metalaxyl, an acylalanine,[58] is highly soluble in aqueous solution (7.1 mg/ml) and possesses a low soil absorption coefficient, giving it the potential for high mobility in many soils at concentrations sufficient to inhibit sporulation and growth of *Phytophthora*.[58] A study of the effective concentration needed to cause 50% inhibition (EC_{50}) revealed values of 0.007 to 0.29 μg/ml for mycelial growth of 100 isolates of *P. cinnamomi*.[50] Both chlamydospore and sporangium production were completely inhibited by 0.25 μg/ml.[50] Metalaxyl has proved highly efficacious in establishing clonal Duke 7 rootstocks in replant situations.[1,19] However, with mature, bearing trees on susceptible seedling rootstocks, it proved uneconomical[1,19] and prone to enhanced biodegradation in some soils.[61,62] Accelerated biodegradation of metalaxyl has been detected in avocado soils in semiarid California and the more typical regions of Queensland (Australia) and South Africa.[6,61] It appears to develop within a few years of initial application of the fungicide.[61] A range of different microorganisms, both bacteria and fungi,[62] appear capable of cometabolizing metalaxyl, primarily to its acid metabolite.[63] A sensitive bioassay for metalaxyl using an isolate of *P. boehmeriae* has proven very useful in surveying soils for evidence of accelerated biodegradation.[64]

Phosphonate fungicide, both fosetyl-aluminum.[66] and more particularly its active metabolite phosphonate,[59] also possess good activity against *P. cinnamomi*.[65–67] On a low phosphate medium, the EC_{50} for inhibition of mycelial growth of *P. cinnamomi* by phosphonate was 4 to 12 μg/ml.[65–67] Sporangium production was even more sensitive to phosphonate, with an EC_{50} of 1.8 μg/ml.[67] For chlamydospore production in three isolates of *P. cinnamomi*, EC_{50} values were in the range of 15 to 44 μg/ml.[67] Fosetyl-aluminum and potassium phosphonate, applied at equivalent phosphonate levels either as a foliar spray or soil drench, gave equivalent control of root rot on the avocado relative, *P indica*.[65] Further experiments on avocado seedlings demonstrated that fosetyl-aluminum was metabolized to phosphonate in both soil and plant foliage following soil drench and foliar application, respectively.[68] Fosetyl-aluminum was rapidly broken down to phosphonate in avocado foliage, following foliar application of the fungicide.[68] Over an eight-week-period, the lowest levels of phosphonate detected in feeder roots (8 to 18 μg per gram fresh weight tissue) were after foliar application of either fosetyl-aluminum or potassium phosphonate.[68] Such levels can readily account for the direct inhibition of *P. cinnamomi* based on known EC_{50} values for phosphonate against mycelial growth, sporangium production, or zoospore release.[65-67] With soil drenches, much higher levels of phosphonate (208 to 1399 μg per gram fresh weight tissue) were detected in roots over the same eight-week-period.[68] One additional feature of phosphonate is its persistence in plant tissues.[58,59] In a field experiment, application of fosetyl-aluminum (1 liter per tree, 3 g ai) at planting provided up to 40 weeks of control of *P. cinnamomi* on Duke 7 rootstocks.[1] Applied as an initial soil drench at planting and then as three or four foliar sprays, fosetyl-aluminum has given excellent control of root rot in many commercial situations in California where clonal tolerant rootstocks are being replanted. On larger, bearing avocados, foliar sprays of fosetyl-aluminum have given superior control of *P. cinnamomi* when compared to metalaxyl.[19] In part, this is due to the tendency of metalaxyl efficacy to drop off after several years usage due to its accelerated biodegradation.[6,61]

A major breakthrough in fungicide application technology came in 1984, when results following the trunk injection of fosetyl-aluminum were described for the first time.[22] Excellent control of *P. cinnamomi* was obtained on Fuerte avocado trees in the second year of treatment using a trunk injection of 7 to 10% fosetyl-aluminum, applied twice a year.[22] Similar results have been obtained in Australia using a 20% potassium phosphonate solution again on Fuerte avocados.[21] On Hass avocados in California, potassium phosphonate and fosetyl-aluminum have given equivalent results with good control of root rot, in some instances but not all, after about three years of injection.[1] The technique of trunk injection is elegant in its simplicity and involves the use of a modified plastic hypodermic syringe to inject the fungicide under hydrostatic pressure into the tree trunk. The method is cost-effective since only small amounts of fungicide, 2 to 20 g per application, dependent on the size of tree being treated, are required. The only drawback is that with prolonged usage, moderate to severe injury to sections of the tree trunk may occur.

19-7 CONCLUSIONS

There does not appear to be any single solution to Phytophthora root rot on avocado. Use of phosphonate fungicides as trunk injections comes closest to a short-term answer. However, prolonged and continuous usage of this method is not be recommended since it can lead to serious damage of the tree trunk. Furthermore, the potential for development of metalaxyl and phosphonate-resistant strains of *Phytophthora* spp. has been demonstrated in laboratory experiments using mutagenic agents.[69] Some mutant strains were similar to parental wild types in growth and sporulation and could compete *in vivo* with the latter.[70] More information is required to evaluate the possible implications of these findings for field application, but obviously sustained continuous usage of fungicide such as metalaxyl or phosphonate is unwise. In addition, trunk injection with phosphonate fungicides is not always successful. On poorer soils with a severe root rot hazard, where the disease rating is already high, and in climates where excessive stress can occur due variously to flooding, drought, or high salinity, results may be marginal at best. Tolerance to *P. cinnamomi* exists in a few rootstocks: Duke 7, D9, Barr Duke, Martin Grande, and Thomas. On their own, such rootstocks offer moderate resistance to the pathogen, particularly on low-hazard, well-drained soils. Again, results in more severe root rot situations may be unacceptable. Combined with good sanitation and hygiene practices, cultural and biological control, and carefully timed applications of fungicides, the use of tolerant rootstocks offers the best opportunity for effective and sustained control of root rot on avocados. Such an integrated approach is not simple, however, since it does require that the grower stays on top of the situation, managing the planting on a day-to-day basis.

19-8 REFERENCES

1. Coffey, M. D., Phytophthora root of avocado: An integrated approach to control in California, *Plant Disease,* 71, 1046, 1987.

2. Condit, I. J., History of the avocado and its varieties in California with a checklist of all named varieties, *Annual Report 1916 California Avocado Association,* 105, 1916.

3. Wager, V. A. *Phytophthora Cinnamomi* and wet soil in relation to the dying-back of avocado trees, *Hilgardia,* 519, 1942.

4. Zentmyer, G. A. and Richards, S. J., Pathogenicity of *Phytophthora cinnamomi* to avocado trees, and the effect of irrigation on disease development, *Phytopathology,* 42, 35, 2952.

5. Zentmyer, G. A., Origin of *Phytophthora cinnamomi:* Evidence that it is not an indigenous fungus in the Americas, *Phytopathology,* 67, 1373, 1977.

6. Pegg, K. G. and While, A. W., *Phytophthora* control in Australia, *South African Avocado Grower's Association Yearbook,* 10, 94, 1987.

7. Pinkas, Y., The arrest of avocado root rot in Israel, Conference on Phytophthora Disease of Citrus and other Crops in the Mediterranean Area, EPPO, Palermo, Italy, 1989.

8. Kariv, A., Zilberstein, M. and Pinkas, Y. Variability in virulence among isolates of *Phytophthora cinnamomi* from avocado trees in Israel, The 11th Congress of Phytopathological Society of Israel, *Phytoparasitica,* 16, 1, 1990.

9. Mao, G., Coffey, M. D. and Pinkas, U., unpublished data, 1990.

10. Coffey, M., Oudemans, P. and Ouimette, D., *Phytophthora citricola:* Another cause of avocado decline, *California Avocado Society Yearbook,* 72, 127, 1988.

11. Tsao, P. H., Factors affecting isolation and quantitation of *Phytophthora* from soil, In *Phytophthora: Its biology, taxonomy, ecology, and pathology,* (D. C. Erwin, S. Bartnicki-Garcia and P. H. Tsao, eds.), p 219, American Phytopathological Society, St. Paul, MN, 1983.

12. Kannwischer, M. E. and Mitchell, D. J., The influence of a fungicide on the epidemiology of black shank of tobacco, *Phytopathology,* 68, 1760, 1978.

13. Kellam, M. K. and Coffey, M. D., Quantitative comparison of the resistance to Phytophthora root rot in three avocado rootstocks, *Phytopathology,* 75, 230, 1985.

14. Zentmyer, G. A. and Ohr, H. D., Avocado root rot, University of California Agriculture Experiment Station Cooperative Extension Leaflet, 2440, 1978.

15. Zentmyer, G. A., *Phytophthora cinnamomi* and the disease it causes, *Phytopathological Monograph No. 10,* American Phytophthora Society, St. Paul, MN 96 pp., 1980.

16. Chee, K. H. and Newhook, F. J., Improved methods for use in studies of *Phytophthora cinnamomi* Rands and other *Phytophthora* species, *New Zealand Journal of Agricultural Research,* 8, 88, 1965.

17. Kliejunas, J. T. and Ko, W. H., Dispersal of *Phytophthora cinnamomi* on the island of Hawaii, *Phytopathology,* 66, 457, 1976.

18. Gees, R. and Coffey, M. D., Evaluation of a strain of *Myrothecium roridum* as a potential biocontrol agent against *Phytophthora cinnamomi, Phytopathology,* 79, 1079, 1989.

19. Coffey, M. D., Ohr, H. D., Campbell, S. D. and Guillemet, F. B., Chemical control of *Phytophthora cinnamomi* on avocado rootstocks, *Plant Disease.* 68, 956, 1984.

20. Gabor, B. K., Guillemet, F. B., and Coffey, M. D., Comparisons of field resistance to *Phytophthora cinnamomi* in twelve avocado rootstocks, *HortScience,* 25, 1655, 1990.

21. Pegg, K.G., Whiley, A. W., Saranah, J. B. and Glass, R. J., Control of Phytophthora root rot of avocado with phosphorus acid, *Australasian Plant Pathology,* 14, 25, 1985.

22. Darvas, J. M., Toerien, J. C. and Milne, D. L., Control of avocado root rot by trunk injection with phosethyl-Al, *Plant Disease,* 68, 691, 1984.

23. Rands, R. D., Streepkamker van kaneel, veroorzaakt door *Phytophthora cinnamomi* n. sp., *Mededelingen van het Instuut Plantenziekten ten,* 54, 41, 1922.

24. Waterhouse, G. M., Key to the species of *Phytophthora* de Bary, *Commonwealth Mycological Institute Mycological Papers, No.* 92, 1963.

25. Waterhouse, G. M. and Waterston, J. M., *Phytophthora cinnamomi, Commonwealth Mycological Institute Descriptions of Pathogenic Fungi and Bacteria,* No. 113, 1966.

26. Tucker, C. M., Taxonomy of the genus *Phytophthora* de Bary, *Missouri Agriculture Experiment Station Research Bulletin,* 153, 208 pp., 1931.

27. Zentmyer, G. A., Stimulation of sexual reproduction in the A mating type of *Phytophthora cinnamomi* by a substance in avocado roots, *Phytopathology,* 55, 69, 1129, 1979.

28. Zaki, A. I., Zentmyer, G. A., Sims, J. J. and Keen, N. T., Stimulation of sexual reproduction in the A2 mating type of *Phytophthora cinnamomi* by oleic acid and lipids from avocado roots, *Phytopathology,* 73, 199, 1983.

29. Old, K. M., Moran. G. F., and Bell, J. C., Isozyme variability among isolates of *Phytophthora cinnamomi* from Australia and Papua New Guinea, *Canadian Journal of Botany,* 62, 2016, 1984.

30. Old, K. M., Dudzinski, M. J. and Bell, J. C., Isozyme variability in field populations of *Phytophthora cinnamomi* in Australia, *Australian Journal of Botany,* 36, 355, 1988.

31. Oudemans, P. and Coffey, M. D., Isozyme comparison within and among worldwide sources of three morphologically distinct species of *Phytophthora, Mycological Research,* 95, 19, 1991.

32. Zentmyer, G. A., Distribution of the A1 mating type of *Phytophthora cinnamomi, Phytopathology,* 66, 701, 1976.

33. Hemmes, D. E. and Wong, L. D., Ultrastructure of chlamydospores of *Phytophthora cinnamomi* during development and germination, *Canadian Journal of Botany,* 53, 2945, 2975.

34. Brodrick, H. T., Zentmyer, G. A. and Wood, R., Comparison of various methods for the isolation of *Phytophthora cinnamomi* from avocado soils, *California Avocado Society Yearbook,* 59, 87, 1976.

35. Baker, K. F., ed., The U. C. System for producing healthy container-grown nursery plants, *California Agriculture Experiment Station Manual,* 23, 1957.

36. Baker, K. F., Principles of heat treatment of soil and planting material, *Journal of the Australian Institute of Agricultural Science,* 28, 118, 1962.

37. Anon., Avocado sunblotch viroid, Commonwealth of Australia Department of Health, Plant Quarantine, Leaflet, No. 10, 1981.

38. Semancik, J. S. and Desjardins, P. R., Multiple small RNA species and the viroid hypothesis for sunblotch disease of avocado, *Virology,* 104, 117, 1980.

39. Cook, R. J. and Baker, K. F., Why biological control? In *The nature and practice of biological control of plant pathogens,* Ch. 1, The American Phytopathological Society, St. Paul, MN, 1983.

40. Cook, R. J. and Baker, K. F., The soil ecosystem. In *The nature and practice of biological control of plant pathogens,* ch. 7, The American Phytopathological Society, St. Paul, MN, 1983.

41. Broadbent, P. and Baker, K. F., Behaviour of *Phytophthora cinnamomi* in soil suppressive and conducive to root rot, *Australian Journal of Agricultural Research,* 25, 121, 174.

42. Kelly, W. D. and Rodriguez-Kabana, R. Competition between *Phytophthora cinnamomi* and *Trichoderma* spp. in autoclaved soil, *Canadian Journal of Microbiology,* 22, 1120, 1976.

43. Kelly, W. D., Evaluation of *Trichoderma harzianum*-impregnated clay granules as a biocontrol for *Phytophthora* causing damping-off of pine seedlings, *Phytopathology,* 66, 1023, 1976.

44. Kopp, L. E., A taxonomic revision of the genus *Persea* in the Western Hemisphere (Persea-Lauraceae), *Memoirs of the New York Botanical Garden,* 14, 1, 1966.

45. Frolich, E. F., Schroeder, Ca. A. and Zentmyer, G. A., Graft compatibility in the genus *Persea, California Avocado Society Yearbook,* 42, 102, 1958.

46. Bergh, B. O., Zentmyer, G. A., Whitsell, R. H., Boswell, S. B. and Storey, W. B., Avocado rootstock breeding, especially in relation to *Phytophthora, Acta Horticultura,* 57, 237, 1976.

47. Zentmyer, G. A. and Thorn, W. A., Resistance of the Duke variety of avocado to Phytophthora root rot, *California Avocado Society Yearbook,* 40, 169, 1956.

48. Frolich, E. F. and Platt, R. G., Use of the etiolation technique in rooting avocado cuttings, *California Avocado Society Yearbook,* 55, 97, 1971.

49. Ellstrand, N. C., Lee J. E., Bergh, B. O., Coffey, M. D. and Zentmyer, G. A., Isozymes confirm hybrid parentage for G755 selections, *California Avocado Society Yearbook,* 70, 199, 1986.

50. Coffey, M. Guillemet, F., Schieber, G. and Zentmyer, G., *Persea schiedeana* and Martin Grande, *California Avocado Society Yearbook,* 72, 107, 1988.

51. Zentmyer, G. A., Schieber, E. and Guillemet, F. B., History of the Martin Grande rootstock, *California Avocado Society Yearbook,* 72, 121, 1988.

52. Coffey, M. D. and Guillemet, F. B., Avocado tree called Thomas, United States Patent, Plant 6628, 1989.

53. Guillemet, F., Gabor, B. and Coffey, M., Field evaluations of some new avocado rootstocks, *California Avocado Society Yearbook,* 72, 133, 1988.

54. Coffey, M. D. and Guillemet, F. B., Avocado tree called Barr Duke, United States Patent, Plant 6627, 1989.

55. Dolan, T. E. and Coffey, M. D., Laboratory screening technique for assessing resistance of four avocado rootstocks to *Phytophthora cinnamomi, Plant Disease,* 70, 115, 1986.

56. Gabor, B. K. and Coffey, M. D., Rapid methods for evaluating resistance to *Phytophthora cinnamomi* in avocado rootstocks, *Plant Disease,* 75, 118, 1991.

57. Gabor, B. K. and Coffey, M. D., Quantitative analysis of the resistance to *Phytophthora cinnamomi* in five avocado rootstocks under greenhouse conditions, *Plant Disease,* 74, 882, 1990.

58. Cohen, Y. and Coffey, M. D., Systemic fungicides and the control of Oomycetes, *Annual Review of Phytopathology,* 24, 311, 1986.

59. Coffey, M. D. and Ouimette, D. G., Phosphonates: Antifungal compounds against Oomycetes, In *Nitrogen, phosphorus and sulphur utilization by fungi*, (L. Boddy, R. Marchant and D. J. Read, eds.), p. 107, Cambridge University Press, Cambridge, 1988.

60. Coffey, M. D. , Klure, L. J. and Bower, L. A., Variability in sensitivity to metalaxyl of isolates of *Phytophthora cinnamomi* and *Phytophthora citricola*, *Phytopathology*, 74, 417, 1984.

61. Bailey, A. M. and Coffey, M. D., Biodegradation of metalaxyl in avocado soils, *Phytopathology*, 75, 135, 1985.

62. Bailey, A. M. and Coffey, M. D., Characterization of microorganisms involved in accelerated biodegradation of metalaxyl and metolachlor in soils, *Canadian Journal of Microbiology*, 32, 562, 1986.

63. Droby, S. and Coffey, M. D., unpublished data, 1987.

64. Bailey, A. M. and Coffey, M. D., A sensitive bioassay for quantification of metalaxyl in soils, *Phytopathology*, 74, 667, 1984.

65. Fenn, M. E. and Coffey, M. D., Studies on the in vitro and in vivo antifungal activity in fosetyl-Al and phosphorous acid, *Phytopathology*, 74, 606, 1984.

66. Coffey, M. D. and Bower, L. A., In vitro variability among isolates of eight *Phytophthora* species in response to phosphorous acid, *Phytopathology*, 74, 738, 1984.

67. Coffey, M. D. and Joseph, M. C., Effects of phosphorous acid and fosetyl-Al on the life cycle of *Phytophthora cinnamomi* and *P. citricola*, *Phytopathology*, 75, 1042,1985.

68. Ouimette, D. G. and Coffey, M. D., Phosphonate levels in avocado (*Persea americana*) seedlings and soil following treatment with fosetyl-Al or potassium phosphonate, *Plant Disease*, 73, 212, 1989.

69. Bower, L. A. and Coffey, M. D., Development of laboratory tolerance to phosphorous acid, fosetyl-Al, and metalaxyl in *Phytophthora capsici*, *Canadian Journal of Plant Pathology*, 7, 1, 1985.

70. Lucas, J. A., Bower, L. A. and Coffey, M. D., Fungicide resistance in soilborne *Phytophthora* species, *EPPO Bulletin*, 20, 199, 1990.

INDEX

A

N

O